Phase Transitions and Relaxation in Systems with Competing Energy Scales

NATO ASI Series

Advanced Science Institutes Series

A Series presenting the results of activities sponsored by the NATO Science Committee, which aims at the dissemination of advanced scientific and technological knowledge, with a view to strengthening links between scientific communities.

The Series is published by an international board of publishers in conjunction with the NATO Scientific Affairs Division

A	**Life Sciences**	Plenum Publishing Corporation
B	**Physics**	London and New York
C	**Mathematical and Physical Sciences**	Kluwer Academic Publishers Dordrecht, Boston and London
D	**Behavioural and Social Sciences**	
E	**Applied Sciences**	
F	**Computer and Systems Sciences**	Springer-Verlag
G	**Ecological Sciences**	Berlin, Heidelberg, New York, London,
H	**Cell Biology**	Paris and Tokyo
I	**Global Environmental Change**	

NATO-PCO-DATA BASE

The electronic index to the NATO ASI Series provides full bibliographical references (with keywords and/or abstracts) to more than 30000 contributions from international scientists published in all sections of the NATO ASI Series.
Access to the NATO-PCO-DATA BASE is possible in two ways:

– via online FILE 128 (NATO-PCO-DATA BASE) hosted by ESRIN,
Via Galileo Galilei, I-00044 Frascati, Italy.

– via CD-ROM "NATO-PCO-DATA BASE" with user-friendly retrieval software in English, French and German (© WTV GmbH and DATAWARE Technologies Inc. 1989).

The CD-ROM can be ordered through any member of the Board of Publishers or through NATO-PCO, Overijse, Belgium.

Series C: Mathematical and Physical Sciences - Vol. 415

Phase Transitions and Relaxation in Systems with Competing Energy Scales

edited by

T. Riste
Institutt for energiteknikk,
Kjeller, Norway

and

D. Sherrington
Department of Physics,
University of Oxford, U.K.

Kluwer Academic Publishers

Dordrecht / Boston / London

Published in cooperation with NATO Scientific Affairs Division

Proceedings of the NATO Advanced Study Institute on
Phase Transitions and Relaxation in Systems with Competing Energy Scales
Geilo, Norway
13-23 April 1993

Library of Congress Cataloging-in-Publication Data

```
NATO Advanced Study Institute on Phase Transitions and Relaxation in
  Systems with Competing Energy Scales (1993 : Geilo, Norway)
    Phase transitions and relaxation in systems with competing energy
  scales : proceedings of the NATO Advanced Study Institute on Phase
  Transitions and Relaxation in Systems with Competing Energy Scales,
  Geilo, Norway, 13-23 April 1993 / edited by T. Riste and D.
  Sherrington.
        p.   cm. -- (NATO ASI series. Series C, Mathematical and
  physical sciences ; no. 415)
    Includes bibliographical references and index.
    ISBN 0-7923-2504-4 (acid-free paper)
    1. Superconductors--Congresses.  2. High temperature
  superconductors--Congresses.  3. Solids--Congresses.  4. Glasses-
  -Congresses.  5. Phase transformations (Statistical physics)-
  -Congresses.   I. Riste, Tormod, 1925-    . II. Sherrington, D. C.
  III. Title.  IV. Series.
  QC611.9.N38 1993
  530.4'16--dc20                                              93-31136
```

ISBN 0-7923-2504-4

Published by Kluwer Academic Publishers,
P.O. Box 17, 3300 AA Dordrecht, The Netherlands.

Kluwer Academic Publishers incorporates the publishing programmes of
D. Reidel, Martinus Nijhoff, Dr W. Junk and MTP Press.

Sold and distributed in the U.S.A. and Canada
by Kluwer Academic Publishers,
101 Philip Drive, Norwell, MA 02061, U.S.A.

In all other countries, sold and distributed
by Kluwer Academic Publishers Group,
P.O. Box 322, 3300 AH Dordrecht, The Netherlands.

Printed on acid-free paper

All Rights Reserved
© 1993 Kluwer Academic Publishers
No part of the material protected by this copyright notice may be reproduced or utilized in any form or by any means, electronic or mechanical, including photocopying, recording or by any information storage and retrieval system, without written permission from the copyright owner.

Printed in the Netherlands

CONTENTS

Preface   vii

Organizing committee and participants   ix

Low temperature phases, ordering and dynamics
in random media   1
    D.S. Fisher (invited)

Statics and dynamics of the flux-line lattice
in high-T_c superconductors   27
    E.H. Brandt (invited)

Magnetic flux line lattices in the cuprate
superconductors
    P.L. Gammel (invited) and D.J. Bishop   55

Theory and experiment of flux pinning and flux
creep in high T_c superconductors   71
    P.H. Kes (invited)

Vortex line fluctuations in superconductors from
elementary quantum mechanics   95
    D.R. Nelson (invited)

Anisotropy and strong pinning in $YBa_2Cu_3O_7$ with
Y_2BaCuO_5 inclusions   119
    M.G. Karkut, L.K. Heill, M. Slaski, L.T. Sagdahl
    and K. Fossheim

Interstitial and vacancy proliferation in flux line lattices   123
    E. Frey, D.R. Nelson and D.S. Fisher

Simulations of relaxation, pinning, and melting in
flux lattices   129
    H.J. Jensen (invited)

I-V characteristics of high temperature superconductors
with correlated defects   187
    M. Wallin and S.M. Girvin

Relaxation near glass transition singularities   191
    W. Götze (invited)

Neutron scattering at the glass transition   233
    U. Buchenau (invited)

Susceptibilty studies of supercooled liquids
and glasses   259
    S.R. Nagel (invited)

Hierarchical melting of one-dimensional
incommensurate structures   285
    R. Schilling (invited)

Charge density waves, phase slips, and instabilities   317
    S.N. Coppersmith (invited)

Universality in commensurate-incommensurate
phase transitions   335
    B. Hu, J. Shi and B. Lin

Glassy behaviour of the charge/spin density
wave ground state   339
    K. Biljaković (invited)

The critical behavior of 1-d charge density waves   359
    Z. Olami

Electron solidification in two dimensions   367
    R.L. Willett (invited)

Two-dimensional Wigner solid versus incompressible
quantum liquid   401
    G. Meissner

Domain growth and coarsening   405
    A.J. Bray (invited)

Critical wrinkling of depinned interfaces,
strings and membranes   437
    K. Sneppen and M.H. Jensen

Index   445

PREFACE

This volume comprises the proceedings of a NATO Advanced Study Institute held in Geilo, Norway, between 13-23 April 1993. It was the twelfth in a series held biannually, usually in the area of cooperative behaviour and phase transitions. Its purpose was to bring together and explore the conceptual and technical interrelations between various systems with competing energy scales, which are both widespread and the subject of considerable current interest, challenging conventional thinking.

The Institute brought together many lecturers, students and active researchers in the field from a wide range of countries, both NATO and non-NATO, including several participants from the former Soviet Union. Financial support was principally from the NATO Scientific Affairs Division but also from Institutt for energiteknikk, The Norwegian Research Council for Science and the Humanities (NAVF) and The Nordic Institute for Theoretical Physics (NORDITA). The organizers would like to thank all these contributors for their help in promoting an exciting and rewarding meeting, and in doing so are confident that they echo also the appreciation of all the participants.

Competing energy scales arise from a variety of mechanisms; from different interactions, intrinsic and extrinsic, and from conflicts with external forces or geometry. The main examples considered in these proceedings are high-temperature superconductors, charge and spin density-wave solids, supercooled liquids and glasses, and Wigner electron solids. The common features discussed are slow dynamics, metastable states, phase ordering, instabilities and relaxation.

In superconductors the emphasis is on the flux line lattice and liquid phases, on the role of point and columnar defects, and on the vortex glass phase, combining considerations of studies of experimental manifestations with theoretical and computer simulational modelling.

For "ordinary" glasses and their liquid precursors a review is given of mode coupling theory. Experimental studies by neutron scattering, specific heat and dielectric susceptibility are presented, covering not only the neighbourhood of the glass transition but several decades of temperature and relaxation times.

Experimental evidence for glassy states in charge (CDW) and spin (SDW) density waves is also considered. Special attention is paid to CDW dynamics in the presence of external field, with serious concern raised over some conventional assumptions. Resultant differences between driven dynamics and statics overlap conceptually similar simulation observations in superconductor vortex dynamics.

Another lecture shows how discommensurations in incommensurate systems could lead to hierarchical ordering of energy levels, in turn leading to hierarchical melting of the structure.

Wigner (quantum) crystallization of electrons at low densities is another topic of considerable interest, particularly with the availability of good semiconductor heterostructures. Evidence for solidification from several experimental measurements is reviewed, as also that for the quasi-particle Fermi liquid and Hall quantization at even fractional fillings.

Yet another example of the interconnection between disordered structure and slow dynamics is provided by the problem of growth of order when a system is quenched rapidly from a disordered to an ordered phase.

The lectures at the Study Institute set the scene for an environment for learning and for discussion, about both the spesific systems and their interrelationsships. The invited lecturers were supplemented by contributed seminars and posters. Some of the most relevant seminars are also summarized in these proceedings.

We are most grateful J.D. Axe, J. Feder, H.J. Jensen, R. Pynn, A.T. Skjeltorp and H. Thomas who helped us plan the programme.

Finally, we would like to express our deep gratitude to Gerd Jarrett of Institutt for energiteknikk, Kjeller, Norway, who did all the practical organization before, during, and after the school, including the preparation of these proceedings, with incredible efficiency, as indeed she did for the previous eleven Institutes and, we hope, she might for a few yet to come.

Tormod Riste   David Sherrington

LIST OF PARTICIPANTS

ORGANIZING COMMITTEE:

Riste, Tormod, director
  Institutt for energiteknikk, POB 40, N-2007 Kjeller, Norway

Sherrington, David, co-director
  Inst. for Theoretical Physics, 1, Keble Road, Oxford OX1 3NP, UK

Skjeltorp, Arne
  Institutt for energiteknikk, POB 40, N-2007 Kjeller, Norway

Jarrett, Gerd, secretary
  Institutt for energiteknikk, POB 40, N-2007 Kjeller, Norway

PARTICIPANTS:

Alstrøm, Preben
  Physics Lab, Ørsted Institute, DK-2100 Copenhagen, Denmark

Baysal, Nihal
  Polymer Research Center, Bogazici University, Tr-80815 Istanbul, Turkey

Berezovsky, Arkady
  Dept. of Mechanics, Inst. of Cybernetics, Akadeemia te 21, EE-0180 Tallin, Estonia

Biljakovic, Katica
  CNRS-CRTBT, BP 155, F-38042 Grenoble, France

Birovljev, Aleksandar
  Dept. of Physics, University of Oslo, POB 1048, N-0316 Oslo, Norway

Brandt, Ernst H.
  Max-Planck Institut f. Metallforschung, D-7000 Stuttgart, Germany

Braun, Oleg
  Inst. of Physics, Ukrainian Academy of Science, 46 Science Ave., Kiev 252028, Ukraina

Bray, Alan
  Dept. of Theoretical Physics, The University, Manchester M13 9PL, UK

Buchenau, Uli
  Institut f. Festkørperforschung, KFA, POB 1513, D-1570 Julich, Germany

Christensen, Kim
  Dept. of Physics, University of Oslo, POB 1048, N-0316 Oslo, Norway

Coppersmith, Susan
  AT&T, Bell Laboratories, Murray Hill, N.J. 07974, USA

Dahmen, Karin
  117 Clark Hall, Cornell University, Ithaca, N.Y. 14851, USA

Dmitriev, Alexey
  Dept. of Low Temperature Physics, M.V. Lomonosov State University, Moscow 119899, Russia

Dominguez, Daniel
  Internat. Centre for Theoretical Physics, POB 586, I-34100 Trieste, Italy

Eisinger, Siegfried
  Statens Sikkerhetshøgskole, Skåregt. 103, N-5050 Haugesund, Norway

Feder, Jens G.
  Dept. of Physics, University of Oslo, POB 1048, N-0416 Oslo, Norway

Fernandez, Ariel
  Dept. of Biochemistry, University of Miami, POB 016129. Miami, Florida. 33101-6129, USA

Fisher, Daniel
  Dept. of Physcs, Harvard University, Cambridge, Mass. 02138, USA

Florian, Martin
  Sarisska 7, 82109 Bratislava, Slovakia

Fogedby, Hans
  Dept. of Physics, University of Aarhus, DK-8000 Aarhus C, Denmark

Fossheim, Kristian
  Dept. of Physics, Technical University, N-7034 Trondheim, Norway

Franosch, Thomas
  Physik-Dept., T30, Technical University Munchen, D-8046 Munchen, Germany

Frey, Erwin
  Physik-Dept., T30, Technische Universität Munchen, D-8046 Munchen, Germany

Frontera, Carlos
   Dept. d'Structura i Con. de la Materia, University of Barcelona,
   Diagonal 647, E-08028 Barcelona, Spain

Galperin, Yuri
   Dept. of Physics, University of Oslo, POB 1048, N-0316 Oslo, Norway

Gammel, Peter L.
   AT&T, Bell Laboratories, Murray Hill, N.J. 07974, USA

Gingras, Michel
   TRIUMF, Theory Group, 4004 Wesbrook Mall, Vancouver, BC V6T-2A3 Canada

Gjølmesli, Svein
   Dept. of Physics, Technical University, N-7034 Trondheim-NTH, Norway

Götze, Wolfgang
   Physik Dept., Technische Universität Munchen, D-8046 Garching, Germany

Hetland, Per Otto
   Dept. of Physics, University of Oslo, POB 1048, N-0316 Oslo, Norway

Hu, Bambi
   Dept. of Physics, University of Houston, Houston, TX 77004, USA

Ilhan, Canan
   Polymer Research Center, Bogazici University, Tr-80815 Istanbul, Turkey

Janik, Janina, Fac. of Chemistry, Jagellonian University,
   ul. Ingardena 3, Pl-30 060 Krakow

Janik, Jerzy
   Inst. of Nuclear Physics, ul. Radzikowskiego 152,
   Pl-31 242 Krakow, Poland

Jensen, Henrik Jeldtoft
   Dept. of Mathematics, Huxley Bld., 180 Queen's Gate, London SW7 2BZ, UK

Jensen, Mogens Høgh
   NORDITA, Blegdamsvej 17, DK-2100 Copenhagen, Denmark

Johansen, Tom Hennning
   Dept. of Physics, University of Oslo, POB 1048, N-0316 Oslo, Norway

Jonsson, Anna
   Dept. of Theoretical Physics, Umeå University, S-90187 Umeå, Sweden

Jug, Giancarlo
   S.I.S.S.A., Via Beirut 4, I-34014 Trieste, Italy

Karkut, Michael
  Dept. of Physics, Technical University, N-7034 Trondheim-NTH, Norway

Kartha, Sivan
  Dept. of Physics, Cornell University, Ithaca, N.Y. 14851, USA

Kes, Peter H.
  Kammerlingh Onnes Lab., Rijksuniversiteit te Leiden. POB 9506,
  NL-2300 RA Leiden, The Netherlands

Koukiou, Flora
  CPT Ecole Polytechnique, F-91128 Palaiseau, France

Kutnjak-Urbanc, Brigita
  J. Stefan Institute, Jamova 39, Ljubljana, Slovenia

Levinsen. Mogens
  Niels Bohr Institut, Ørsted Lab., Universitetsparken 5,
  DK-2100 Copenhagen, Denmark

Marchetti, M. Christina
  Physics Department, Syracuse University, Syracuse, N.Y. 13244, USA

McCauley, Joseph
  Dept. of Physics, University of Houston, Houston, TX 77004, USA

Meissner, Gunther
  Theor. Physik, Universität des Saarlandes, D-66 Saarbrucken, Germany

Melker, Alexander
  St. Petersbureg State Technical University, Polyrekhnicheskaya 29,
  195251 St. Petersburg, Russia

Mendes, Jose F. Ferreira
  Dept. de Fisica da Universidade do Porto, Praca Gomes Teixeira,
  P-4000 Porto, Portugal

Nagel, Sidney R.
  James Franck Institute, University of Chicago, Chicago, Ill 60638, USA

Nelson, David R.
  Dept. of Physics, Harvard University, Cambridge, Mass. 02138, USA

Olaussen, Kaare
  Dept. of Theor. Physics, Technical University, N-7034 Trondheim-NTH,
  Norway

Olami, Zeev
  Dept. of Chemical Physics, Weizmann Institute, Rehovot, Israel

Oxaal, Unni
  Dept. of Physics, University of Oslo, POB 1048, N-0316 Oslo, Norway

Pépy, Gérard
  Lab. Leon Brillouin, CEN Saclay, F-91191 Gif-sur-Yvette, France

Pócsik, István
  Research Institute for Solid State Physics, POB 49, H-1525 Budapest, Hungary

Pynn, Roger
  LANSCE, MS-H804, Los Alamos National Laboratory, Los Alamos, NM 87545, USA

Radzihovsky, Leo
  AEP 212 Clark Hall, Cornell University, Ithaca, N.Y. 14853-2501, USA

Riise, Anjali
  Dept. of Physics, University of Oslo, POB 1048, N-0316 Oslo, Norway

Roberts, Bruce
  AEP 212 Clark Hall, Cornell University, Ithaca, N.Y. 14853-2501, USA

Rosenblum, Victor
  State Technical University of St. Petersburg, Polytekhnicheskaya St. 29, 195256 St. Petersburg, Russia

Schilling, Rolf
  Johannes Gutenberg Universität, Inst. f. Physik, POB 3980, D-6500 Mainz 1, Germany

Sellers, Howard
  Dip. di Metodi e Modelli Matematici, Univ. Degli Studi de Padova, via Belzoni 7, I-35131 Padova, Italy

Sibani, Paolo
  Dept. of Physics, Campusvej 55, DK-5230 Odense, Denmark

Steinsvoll, Olav
  Institutt for energiteknikk, POB 40, N-2007 Kjeller, Norway

Spencer, Steven
  Dept. of Mathematics, Imperial College, London SW7 2BZ, UK

Stinchcombe, Robin
  Dept. of Theoretical Physics, 1, Keble Road, Oxford OX1 3NP, UK

Tabiryan, Nelson
  Inst. f. Angewandte Physik, Hochschulestr. 6, D-6100 Darmstadt, Germany

Täuber, Uwe
  Physik-Dept., T30, Technische Universität Munchen, D-8046 Garching, Germany

Thomas, Harry
  Dept. of Physics, University of Basel, Klingelbergerstr. 82,
  CH-4056 Basel, Switzerland

Thrane, Bård
  LEPES, CNRS, BP 166, F-38042 Grenoble, France

Udovik, Oleg
  71, Apt, 3 Dobrohotova St., Kiev 252142, Ukraina

Wagner, Geri
  Dept. of Physics, University of Oslo, POB 1048, N-0316 Oslo, Norway

Wallin, Mats
  Dept. of Theoretical Physics, KTH, S-100 44 Stockholm, Sweden

Weber, Hans
  Dept. of Physics, Luleå University of Technology,
  S-95187 Luleå, Sweden

Willett, Robert L.
  AT&T, Bell Laboratories, Murray Hill, N.J. 07974, USA

Zelenskaya, Irina
  Inst. of Physics, Ukrainian Academy of Sciences, 46 Science Ave.,
  Kiev 252028, Ukraina

LOW TEMPERATURE PHASES, ORDERING AND DYNAMICS IN RANDOM MEDIA

DANIEL S. FISHER
Physics Department
Harvard University
Cambridge, MA 02138
USA

ABSTRACT. A pedagogical review of the current theoretical understanding of various collective phenomena in materials with quenched disorder is given. The emphasis is on the nature of ordered phases, their equilibrium dynamics, and development of long-range order and other non-equilibrium dynamic phenomena. The important role played by domain walls in pure and random magnets is used as the primary example, but the concepts developed in this context are applied, briefly, to spin glasses and to the vortex glass phase of dirty Type II superconductors.

Introduction

Slow dynamic processes are both the cause and consequence of structurally disordered materials. The slow kinetics of diffusion, ordering, phase segregation, grain growth and motion are responsible for the structural disorder in glasses, random alloys, composites and granular materials. In wide ranges of temperature, the resulting structural randomness is often effectively static or "quenched." This quenched structural randomness can itself be responsible for slow dynamics at low temperatures of other degrees of freedom, such as electronic or magnetic degrees of freedom in solids, or fluids in porous media. The physics of these *effects* of quenched randomness will be the subject of this paper.

In pure, non-random, systems such as perfect solids or fluids, the understanding of collective dynamic phenomena usually proceeds in three stages: first, the equilibrium statistical mechanics needs to be understood, then the dynamics near to local equilibrium, and finally, perhaps, the non-equilibrium dynamics. Often, such as in high temperature disordered phases, there is one characteristic "equilibration" time scale τ which separates equilibrium from non-equilibrium phenomena. However, near to second order phase transitions, in the presence of important conservation laws, or if the equilibrium state is phase separated or has some other broken symmetry, there will be a whole range of time scales, corresponding, roughly, to processes on a range of length scales. In these situations, the time scale $\tau(L)$ associated with processes on length scale L usually grows as a power of L: $\tau(L) \sim L^z$ with z a dynamic exponent which is often non-trivial at critical points but takes on simple values in other cases.[1,2] Perhaps the simplest situation is the diffusion of concentration of a conserved species which results in $z = 2$. An important consequence of such a scaling law is that unless z is large or the basic microscopic time scale t_0 is anomalously long, (as it can be at low temperatures) macroscopic—or at least mesoscopic—regions can equilibrate in macroscopic times, which can easily be $10^{15} t_0$.

In the presence of quenched randomness, the clean separation between equilibrium and non-equilibrium phenomena often does not occur, even in the absence of a phase transition or conservation law! The fundamental reason for this is the randomness itself: in an inhomogeneous medium the dynamic processes can proceed at different rates in different regions of the system. If the distribution of the characteristic times for processes in different regions is very broad, then on the time scale of an experiment, the degree (and concommitant spatial scale) of equilibrium in different regions can vary widely. As we shall see, this can result in the apparently paradoxical situation in which many of the macroscopic properties of an inhomogeneous material are dominated by rare regions of the sample which collectively occupy a very small fraction of its volume.

In spite of the difficulties of separating equilibrium from non-equilibrium phenomena in random media—which result in strong history dependence and disagreements between different experiments—it is nevertheless instructive to proceed in the conventional manner and first analyze equilibrium properties before analyzing the dynamic phenomena. For ease and unity of presentation, we will focus primarily on Ising models of magnets, which are both the simplest and most familiar models of collective phenomena.[3] Nevertheless, we shall see that they exhibit a rich variety of phenomena, many of which are prototypical of more complicated systems.

We first focus on a fundamental property of low temperature phases: the "stiffness" of the system. This will lead us naturally into a discussion of the properties of domain walls (or interfaces), on whose static and dynamic properties we will spend a majority of the paper. The kinetics of phase separation and low temperature excitations are dominated by the domain walls and we will discuss their effects on both near- and far-from-equilibrium phenomena. At the end of the paper we will make connections with other systems—especially vortices in dirty Type II superconductors—and raise some of the key questions.

Models

We will focus on the behavior of Ising magnets consisting of spins $\sigma_i = \pm 1$ on a d-dimensional hypercubic lattice with nearest neighbor couplings $\{J_{ij}\}$ and (sometimes) a uniform magnetic field H. The hamiltonian is simply

$$\mathcal{H} = -\sum_{(ij)} J_{ij}\sigma_i\sigma_j - H\sum_i \sigma_i \ . \tag{1}$$

We distinguish three basic cases:
(i) *pure* (non-random) Ising model with all J_{ij} the same value, $J > 0$;
(ii) *random exchange* with J_{ij} independently random drawn from some distribution $\rho(J)dJ$ with all J_{ij} *positive* (ferromagnetic); and
(iii) *spin glass* with J_{ij} random in *sign* and magnitude drawn independently from a *symmetric* (i.e. even) distribution.

In the absence of a magnetic field, each of these has an (at least) doubly degenerate ground state corresponding to breaking the symmetry of all $\{\sigma_i\} \to \{-\sigma_i\}$.

Stiffness of Ferromagnets

Perhaps the most important property of ordered phases is their resistance to spatial variations of the local order. This is a form of generalized *"stiffness"* of the system which is *absent* macroscopically in high temperature disordered phases. Since we will need to consider phenomena on various length scales, it is natural to consider the length scale dependence of the stiffness. In pure or random exchange Ising models this can be probed in a simple Gedanken experiment: The ground states just consist of all spins up (+1) or all spins down (-1). [4] The stiffness is then the (free) energy cost of the spins being up in one region and down in another. This is probed by defining the stiffness Σ_L of a (hyper)cube of size L^d as the difference in free energies between a system with + boundary conditions on the top and − on the bottom and one with + on both top and bottom. At $T = 0$ in the ground state, the $\pm$ boundary condition simply forces in a domain wall between a + region and − region. For the pure case, the minimum energy wall will be a $d − 1$ dimensional horizontal (hyper)plane, resulting in

$$\Sigma_L = \sigma L^\theta \qquad (2)$$

with a *stiffness exponent*
$$\theta = d - 1 \qquad (3)$$

and *stiffness coefficient* $\sigma = 2J$ which is just the interfacial free energy per unit area, i.e. the interfacial tension of the domain wall. In the presence of random (positive) exchange, the position of the interface will adjust to go through the weaker bonds (smaller J_{ij}) even, as we shall see, at the cost of increasing its area. [5] A naive, but qualitatively correct, argument suggests that Σ_L will be the sum of roughly L^{d-1} random positive variables so that it will have a distribution with mean proportional to L^{d-1} and width of order $\sqrt{L^{d-1}}$. Thus, we expect a similar behavior to the pure case with again $\Sigma_L \approx \sigma L^{d-1}$ with high probability for large L. Indeed, for a distribution $\rho(J)$ with finite non-zero minimum, J_1, and maximum, J_2, lower and upper bounds on σ are trivially obtained: $2J_1 \leq \sigma \leq 2J_2$. As we shall see, however, the interface will not be flat but will be "rough" and the variations in Σ_L will actually turn out to be much less than $\sqrt{L^{d-1}}$ for $d \leq 3$. [5]

The scaling of the ground state stiffness with length scale determines the energy of excitations away from the ground state in large systems: in pure and random exchange Ising models these are *"droplet"* excitations consisting of a simply connected region of spins which are reversed from the ground state. [6] Because of the existence of a surface tension, the minimum energy droplets of diameter L will have a compact shape with surface area of order L^{d-1} and hence energy of order L^θ. From this, we immediately see that the *sign* of θ is crucial: if θ is *positive*, large excitations are costly and thus very unlikely to occur thermally and the system will have an ordered phase at low temperatures which is predominantly (say) up spins with a low density of small thermally fluctuating droplets of down spins. If on the other hand, θ is *negative*, larger droplets cost little energy so the entropy of the many possible droplet configurations and positions will overwhelm their energy resulting in a proliferation of droplets and droplets within droplets, thereby destroying the broken symmetry. [The marginal cases $\theta = 0$ which occurs, for example, in two-dimensional pure systems with a continuous broken symmetry in the ground state, needs more detailed analysis to determine what happens at low temperatures. [7]]

From a renormalization group point of view, $-\theta$ is the eigenvalue of temperature at the zero-temperature ordered fixed point: if $\theta > 0$, thermal fluctuations are irrelevant on long length scales and order persists at low temperatures, while if $\theta < 0$, thermal fluctuations are relevant and the order is destroyed. The lower critical dimension of ferromagnetic Ising models is thus given by $\theta = 0$, yielding the well-known result $d_\ell = 1$. In dimensions greater than one, an ordered phase exists at low temperatures in ferromagnetic Ising models, both in the pure and random exchange cases. This phase has macroscopic stiffness. As the temperature is increased from zero, the entropy of small droplet excitations decreases the interfacial tension $\sigma(T)$ until a critical point is reached at which the entropic and energetic contributions balance on all length scales, σ vanishes and droplets proliferate; this is the critical point $T = T_c$.

At the critical temperature, the balance between energy and entropy suggests that the stiffness Σ_L will be independent of length scale so that $\theta_c = 0$. (This is the source of hyperscaling-like laws which relate the scaling of the surface tension $\sigma(T \leq T_c)$ to the correlation length ξ via $\sigma \approx cT_c\xi^{1-d}$.) [8] In random exchange Ising magnets in three dimensions, the critical behavior is modified by the randomness [9] and Σ_L is a random variable with length scale independent distribution with width comparable to its mean. [10] This suggests that near this random critical point, the correlations at scale L—which are mediated by Σ_L—will vary considerably from one region of the sample to another; this indeed occurs, [11] although it is not well understood.

Above the transition, the system is disordered, the stiffness vanishes exponentially with $\Sigma_L \sim e^{-L/\xi}$ (corresponding to $\theta = -\infty$), and interfaces and droplets are no longer well-defined.

Stiffness of Spin Glasses

Defining and understanding the stiffness of the ground state of spin glasses is rather more subtle than for ferromagnets. [4,12] Although the global up-down symmetry is broken in the ground states, each ground state will have half the spins up and the other half down with the exact arrangement determined by the delicate competition between the $\{J_{ij}\}$ known as "frustration." [13] Indeed, even whether there are only two ground states or many very different ones (in a subtle and poorly understood but well-defined sense) [4] is still a subject of heated controversy. [4,12-14]

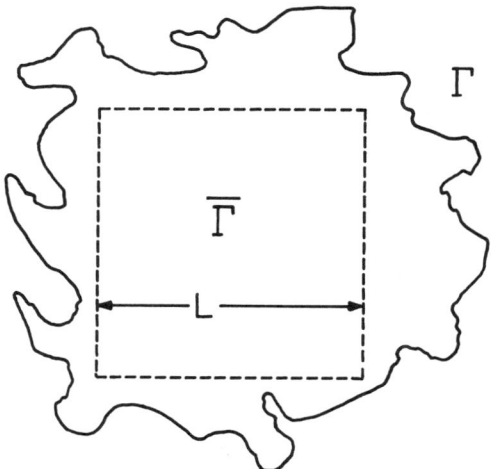

Figure 1. Droplet excitation in a spin glass. The spins outside of the domain wall are in the ground state configuration, Γ, while those inside the domain wall are reversed from the ground state, i.e. configuration $\bar{\Gamma}$ with $\sigma_i^{\bar{\Gamma}} = -\sigma_i^{\Gamma}$. The domain wall position adjusts to the lowest energy configuration with the constraint that all the spins in the box of size L^d are reversed, i.e. the configuration is $\bar{\Gamma}$. The surface of the droplet is fractal with surface area $A \sim L^{d_s}$ with $d - 1 < d_s < d$.

One natural way to probe the stiffness of a spin glass is to let it find its own minimum energy configuration, Γ and then to force the spins in some region of diameter L to reverse, allowing the boundary between the reversed and un-reversed regions to adjust to minimize the constrained energy, and then measuring the energy difference from the configuration Γ as shown schematically in Fig. 1. This defines a droplet excitation away from the ground state—enclosed by a *relative* domain wall—with excitation energy $\Delta_L > 0$ which will of course vary from one region to another. By analogy with Ising ferromagnets, we conjecture that typically $\Delta_L \sim L^\theta$ but anticipate that the distribution of Δ_L will be broad—in fact, the full *distribution* of Δ_L will scale as L^θ.

As for the ferromagnetic case, it is convenient to probe the stiffness of the system by how it responds to changing boundary conditions. Unfortunately, we are faced with a problem for spin glasses: unlike in the ferromagnetic case, we do not know what boundary conditions the system would "like" to have. One alternative is to adjust the boundary conditions to find the optimum match for the specific sample configuration of $\{J_{ij}\}$. This is rather cumbersome, however. An easier strategy is to make the finite system try to look as much as possible like a bulk system and then force a relative domain wall into the resulting configuration. We thus define the stiffness as the difference in free energy between periodic and antiperiodic boundary conditions in one of the directions (with periodic boundary conditions in the other $d-1$ directions best to avoid free surface effects). This is illustrated schematically in Fig. 2.

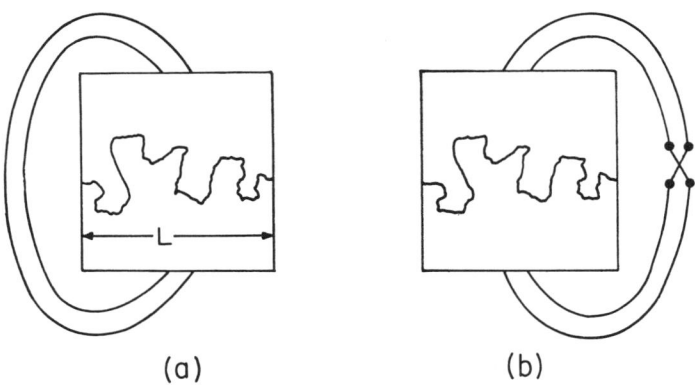

Figure 2. Configurations used to define the "stiffness" Σ_L of a spin glass on length scale L. In (a) the spins on the top and bottom layers are equal, i.e. periodic boundary conditions, while in (b) the exchange couplings $\{J_{ij}\}$ are identical to (a) but the spins on the top layer are reversed from those on the bottom layer, i.e. antiperiodic boundary conditions. The minimum energy spin configurations are the *same* in both cases *below* the relative domain wall that is shown, and *reversed* (i.e. all spins $\sigma_i^b = -\sigma_i^a$) *above* the domain wall. The energy difference between (a) and (b) defines Σ_L, which depends on the specific values of the $\{J_{ij}\}$.

For ferromagnets, the antiperiodic boundary condition forces in a domain wall and yields the same result as before, but for spin glasses, the antiperiodic boundary condition is just equivalent to changing the sign of all the vertical J_{ij}'s in one horizontal plane; this yields an equally likely realization of the same ensemble. Hence the mean $\bar{\Sigma}_L = 0$ and the distribution of Σ_L will be symmetric about zero. If we call the lower energy configuration the (approximate) ground state, and use this spin configuration as a reference, then the change in boundary conditions is seen to simply force in a *relative domain wall* spanning the system. Thus we expect that again $\Sigma_L \sim L^\theta$ with a broad symmetric distribution of Σ_L and the *same θ* as the droplet excitations. [12,15] Because of the frustration, θ is less than for ferromagnets, with a rigorous upper bound, $\theta \leq \frac{d-1}{2}$. [12] By analogy with the ferromagnetic case, we expect that there will be an ordered spin glass phase with broken spin reversal symmetry if and only if $\theta > 0$ (or perhaps =0).

The properties of the Ising spin glass phase are very rich, involving the effects of large rare anomalously-low-free-energy thermally active droplet excitations with $\Delta_L \sim T$, extreme sensitivity of the spin configuration to small temperature changes, and very slow dynamics due to distributions of barriers for formation and motion of domain walls. [12,16] For spin glasses, these phenomena are both difficult to analyze and controversial, [13] so we will focus instead on a much simpler system which exhibits many of the same features: a *single interface* in a random exchange Ising ferromagnet.

[Note that in spin glasses the domain walls on scale L will have *fractal* surface area $A_L \sim L^{d_s}$ with $d_s > d - 1$. Qualitative arguments [12] and numerical studies [17] suggest $d_s < d$. If the domain walls had been space filling, with $d_s = d$, this might plausibly have led to the existence of many ground states, although a consistent picture of such a scenario has not been put together.]

Interfaces in Random Media

If a domain wall is forced into a random exchange ferromagnetic Ising system by a change in boundary conditions or other means, the interfacial tension, σ, will try to make it as flat as possible, although this will be resisted both by thermal fluctuations and randomness in the exchange. Nevertheless, the interface will remain flat enough that the angles away from flat remain small on long length scales and we can parametrize its position by a "height" variable $h(\vec{x})$ which is the deviation away from a flat reference configuration normal to the z-direction, as a function of the $d-1$ dimensional-coordinate, $\vec{x}$, within the interface. For simplicity, we treat h as a continuous variable. The effects of the random exchange can be simply represented by a random potential $V(\vec{x}, z)$ with short-range correlations, [5] at the position $z = h(\vec{x})$ of the interface. We thus have an approximate hamiltonian for the interface [5]

$$\mathcal{H}_I = \int d^{d-1}x \left[\frac{K}{2}(\nabla h)^2 + V[\vec{x}, h(\vec{x})] \right] \qquad (4)$$

with K the interface stiffness (which is related to σ and its angular derivatives).

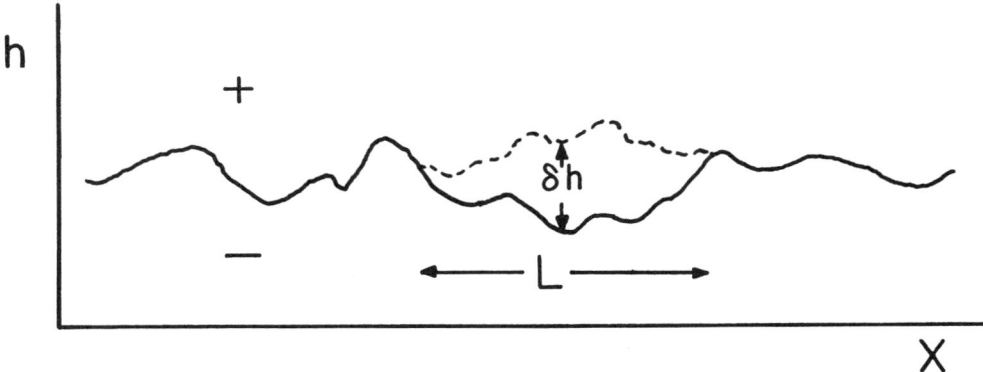

Figure 3. Schematic illustration of the interface in a two-dimensional random exchange Ising ferromagnet with the spontaneous magnetization positive above the interface and negative below. The lowest energy configuration with deviations from a flat interface $h(x)$ is indicated by the solid line. An excitation on scale L is shown: the interface is in a metastable configuration shown by the dashed line with relative displacement $\delta h \sim L^\alpha$ and optimal excitation energy for a given mean displacement, $\Delta_L \sim L^{\theta_I}$.

In the absence of a random potential, thermal fluctuations cause the interface to roughen with

$$\left\langle [h(x) - h(y)]^2 \right\rangle \sim \frac{T}{K}|x - y|^{2\zeta_T} \tag{5}$$

defining a thermal roughness exponent ζ_T relating the "width" $w(L)$ of the interface on scale L to its length. By balancing the temperature with the energy cost $F_E(L)$ of a distortion on scale L of size $h(L)$ with corresponding gradient $\nabla h(L) \sim \frac{h(L)}{L}$:

$$F_E(L) \sim K L^{d-1} \left(\frac{h}{L}\right)^2 \tag{6}$$

one obtains

$$\zeta_T = \frac{3-d}{2} \tag{7}$$

with logarithmic behavior in Eq(5) in three dimensions and $\zeta_T = \frac{1}{2}$ in 2D. Since $\zeta_T < 1$ for $d > 1$, the small ∇h approximation in Eq(4) is valid on long scales and the interface is well-defined and indeed looks flatter and flatter when viewed with less and less spatial resolution. [The fact that $\zeta_T \to 1$ as $d \to 1$ is another way of seeing that $d_\ell = 1$ is the lower critical dimension for Ising ferromagnets: if $\zeta_T = 1$, the interface fluctuates all over the system, the gradient approximation in Eq(4) breaks down and the system disorders.]

The scaling of the elastic deformation energy of a section of size L^{d-1} as in Eq(6) persists in the presence of the random potential due to a statistical symmetry of replacing $h(x)$ by $h(x) + \gamma \cdot x$ corresponding to a uniform tilt by angle γ. [5] However, this deformation energy cost must now compete with the gains in energy from finding more optimal regions of the random potential, i.e. weaker bonds. A naive balancing of the variations of the energy in a L^{d-1} section due to the randomness;

$$F_R(L) \sim \tilde{V} L^{\frac{d-1}{2}} \tag{8}$$

with the elastic energy Eq(6), yields a roughness exponent $\zeta = \zeta_R = \frac{5-d}{4}$ which turns out to be an upper bound. [18,19] Physically, this arises from the expectation that minimizing the energy of the interface over a large number of different positions will lead to variations in its total energy

$$\delta F(L) \sim L^{\theta_I} \tag{9}$$

which are *reduced* from the variations with a single *fixed* position. From the scaling of the elastic energy Eq(6), we expect that

$$\theta_I = 2\zeta + d - 3 \tag{10}$$

is the stiffness exponent of the interface in the random potential. The above argument suggests that

$$\theta_I \leq \frac{d-1}{2} . \tag{11}$$

Another bound can be found [18,19,20] by considering the variations of the potential averaged over a region of volume $L^{d-1} \times L^\zeta$. This yields

$$d - 1 - \theta_I \leq \frac{d-1+\zeta}{2} . \tag{12}$$

Combining these inequalities, we have

$$\frac{5-d}{5} \leq \zeta \leq \frac{5-d}{4} \tag{13}$$

and hence

$$\frac{3d-5}{5} \leq \theta_I \leq \frac{d-1}{2} . \tag{14}$$

[In more than five dimensions, the interface becomes flat, with only O(1) width. An ϵ-expansion has been carried out in $5-\epsilon$ dimensions, using functional renormalization group techniques, [18,21] yielding $\zeta \approx 0.2083\epsilon$ in disagreement with the formal field-theoretic result $\zeta = \frac{1}{2}\epsilon$. [21]] We thus see that the interfaces will roughen due to randomness by more than they will due to thermal fluctuations in any physical dimension, i.e. for $d \geq 2$, $\zeta > \zeta_T$. The concommitant positivity of θ_I implies that thermal fluctuations are irrelevant at long length scales and the interface is in a pinned, randomness dominated "phase" (which persists, in fact, at all temperatures). [21,22] Indeed, the mean square thermal *fluctuations* around the thermally-averaged configuration for a given sample, scale in the *same* way as in the absence of the random potential

$$\overline{\langle [h(x) - h(y) - \langle h(x) - h(y) \rangle]^2 \rangle} \sim \frac{T}{K}|x-y|^{2\zeta_T} \tag{15}$$

with the overbar denoting averaging over the random potential, i.e. different samples, and $\zeta_T = \frac{3-d}{2}$. Although this can be derived exactly and straightforwardly from the statistical tilt symmetry, [23] it arises from very interesting physics. [22,24]

Consider excitations away from the ground state configuration of an interface (fixed for definiteness on a $d-2$ dimensional boundary far away). A section of length L typically deviates away from flat by an amount $h_L \sim L^\zeta$. The dominant excitations from the ground state on scale L will differ from it also by $\delta h_L \sim L^\zeta$, as shown in Fig. 3. Typically, these cost a large excitation energy $\Delta_L \sim L^{\theta_I}$. However, the distribution of Δ_L will extend all the way down to zero with the probability that $\Delta_L \leq T$ being of order T/L^{θ_I}. These rare active low energy excitations occur because of the delicate balances which result in the lowest energy configuration for the section of interface under consideration: if the random potential in the region is changed gradually, the section will initially not change much, but then can suddenly find that a quite different configuration has now become lower in energy. If the potential in a given region is close to such a local threshold, then there will be an atypically low energy excitation which takes the region to the configuration which would become the new minimum if the potential were changed slightly. This can be explicitly seen for the one-dimensional interface in $2D$, by plotting the ground state configuration as a function of the position of one end of the interface line with the other held fixed, as shown in Fig. 4. [25]

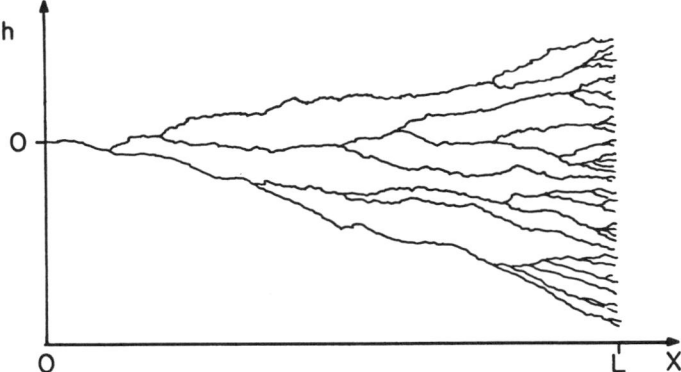

Figure 4. Minimum energy configurations of an interface in a random exchange ferromagnet with interfacial hamiltonian given by Eq(4), as a function of the position of the right end, h_L. [From Ref. 25.] Note that if h_L is increased by a small amount, the configuration will usually only change for $x \approx L$. But occasionally, changes will occur over much of the range from $0 \leq x \leq L$. Near to these values of h_L, there will be a large low energy excitation which takes the interface to what-will-become the minimum energy configuration when h_L is increased slightly.

What are the consequences of this picture? The excitations with energy $\Delta_L \gg T$ (most of them for large L) will not be thermally excited and hence not contribute to $\langle (h-<h>)^2 \rangle$. However, the rare active excitations with $\Delta_L \leq T$ will contribute to Eq(15) an amount $L^{2\zeta}$ if x and y are *both* in the *same* active excitation which must therefore have $L \geq (x-y)$. The probability of this happening is T/L^{θ_I} yielding dominance of the mean square thermal fluctuations by $L \sim |x-y|$ and hence

$$2\zeta_T = 2\zeta - \theta_I = 3 - d \tag{16}$$

as claimed!

This domination of correlations, and the associated susceptibilities (such as that to an applied field which pushes the interface in one direction) by rare regions of the system occurs as well in spin glasses and other randomness-dominated ordered phases such as the vortex glass phase in the mixed state of dirty type II superconductors which is discussed at the end of this paper. The anomalously low energy excitations which occur in the ordered phases of these random systems exist because of the broken statistical symmetry, in the case of an interface its statistical translational invariance. They are thus the analog of Goldstone modes in pure systems with broken continuous symmetries.

Concomitant with the low energy excitations is an extreme sensitivity to small changes of the temperature. Although thermal fluctuations are formally irrelevant, the system must in fact minimize its free energy rather than its energy; the former will be altered by the effects of small scale thermal excitations. The sensitivity of the optimal configuration to changes in the random potential gives rise to similar sensitivity to changes in the effective random contributions to the free energy due to changes in temperature. [22] With a small change δT, these change the configuration dramatically on length scales longer than

$$L_{\delta T} \sim \frac{1}{\delta T^{\nu_\delta}} \tag{17}$$

with

$$\nu_\delta = \frac{1}{\frac{d-1}{2} - \theta_I} \, . \tag{18}$$

Our understanding of the behavior discussed above has been enormously enhanced by analytical solution [5,22] and supporting numerical work [22,25] on a one-dimensional interface in a two-dimensional system. The exponents $\zeta = 2/3$ and $\theta_I = 2\zeta - 1 = 1/3$ have been obtained exactly, as well as information on correlations and response functions. [22,26]

Domain Wall Dynamics

With all of this background, we are now in a position to study the dynamics of the ordered phases of Ising systems. Both the dynamics near equilibrium [27] and the establishment of equilibrium long-range order [2] are dominated by the dynamics of domain walls. We first discuss the *pure case*. In zero magnetic field, a single flat domain wall has no net force acting on it. However, if a field is present or if the wall is curved, it will feel a force trying to move it to lower the total free energy. We are thus led to consider the effects of a normal force per unit area, f, on an interface. We expect that for small f, the interface will respond linearly, moving with mean velocity.

$$v \sim f \, . \tag{19}$$

The surface of a droplet excitation of radius R will feel a force due to the curvature [2]

$$f \sim \frac{\sigma}{R} \tag{20}$$

which acts to shrink the droplet, on average. The mean lifetime of such a droplet will, from $\frac{dR}{dt} \propto -f$, be

$$\tau_R \sim R^2 \, . \tag{21}$$

In three dimensions, the effect of thermal activation and subsequent decay of large droplets contributes to, but does not alter, the form of the autocorrelation function of a spin

$$C(t) \equiv \langle \sigma_i(0)\sigma_i(t)\rangle - <\sigma_i>^2 \; ; \qquad (22)$$

In the ordered phase, this simply decays exponentially as $C(t) \sim e^{-t/\tau}$. In two dimensions, however, the effects of large droplets are more dramatic; they result in the anomalous "stretched-exponential" decay [27,28]

$$C(t) \sim e^{-(t/\tau)^{\frac{1}{2}}} \; . \qquad (23)$$

Although this is a marked effect, the effects of interface dynamics on the formation and evolution of the long-range order are even more interesting.

If an Ising magnet is cooled rapidly to below T_c, initially there will be a very high density of domain walls separating the initially random up and down regions. The curvature of these walls causes them to move to reduce the total interfacial area. If the characteristic scale of the regions of a given sign magnetization is $R(t)$, then the curvatures will be of order $1/R$ and will evolve under the forces $f \sim \sigma/R$ eliminating the higher curvature regions and smaller droplets and causing a coarsening of the domain pattern with the scale of the remaining domain structure evolving as

$$\frac{dR(t)}{dt} \sim \frac{\sigma}{R} \qquad (24)$$

implying [2]

$$R(t) \sim t^{\frac{1}{2}} \qquad (25)$$

corresponding to a dynamic exponent of $z = 2$. In macroscopic times of minutes, the scale of the ordered regions in a pure Ising magnet can thus reach macroscopic sizes of order a centimeter.

Dynamics of Random Magnets

In the presence of randomness, the dynamics of interfaces is enormously slower. [29] The fundamental reason for this is closely related to the reason for the randomness induced roughening described above: the interface has regions in which it can have low energy and others with a similar shape of the interface with higher energy. To move from a region with low energy to one with even lower energy, the interface must move through regions in which it has higher energy, i.e. it must surmount energy (actually free energy) barriers. How big are these barriers? As for the other properties, the height of the barriers will depend on the size of the section of interface which is moving. The energy of an excitation of a section L^{d-1} is typically of order L^{θ_I}. However, the excitations are best thought of as the lowest energy configuration given a constraint that they differ from the optimal configuration in the section of size L^{d-1}. Their energy is thus a lower bound for the energy of the configurations which must be passed through to form the excitation. Thus we conjecture that the barriers to move a section of size L^{d-1} a distance of order L^ζ scale as [5,21]

$$B_L \sim L^{\psi_I} \qquad (26)$$

with

$$\theta_I \leq \psi_I \leq d-1 \; , \qquad (27)$$

the upper bound coming from the worst possible intermediate barrier configuration. [It is quite plausible that $\psi_I = \theta_I$, although the author is not aware of a good general argument for this.]

The time τ_L for an excitation of scale L to disappear is dominated by thermal activation over the barrier, i.e.

$$\tau_L \sim e^{B_L/T} \sim e^{cL^{\psi_I}} . \tag{28}$$

More precisely, since B_L is a random variable, τ_L will be extremely broadly distributed and the correct scaling is [30]

$$\ell n \tau_L \sim \frac{B_L}{T} \sim L^{\psi_I} . \tag{29}$$

This exponential dependence of time scales on length scales is a general property of low temperature phases in random systems. [12,30] It has been termed "activated dynamic scaling" to distinguish it from conventional dynamic scaling [1] with $\tau_L \sim L^z$. Equation (29) corresponds, crudely, to $z = \infty$. In equilibrium, the presence of large rare thermally active excitations of the interface with $\Delta_L \sim T$ (which occur with probability T/L^{θ_I}) and the large barriers for their relaxation, result in very slow dynamics of the autocorrelations:

$$\overline{\langle (h(x,t) - h(x,0))^2 \rangle} \sim L_t^{2\zeta - \theta_I} \sim (\ell n t)^{\frac{3-d}{\psi_I}} \tag{30}$$

where $L_t \sim (T \ell n t)^{\frac{1}{\psi_I}}$ is the typical size of excitations which can change in time of order t. Similar physics yields very slow decay of correlations in the spin glass ordered phase caused by large rare active droplet excitations, [12] resulting in

$$\overline{C(t)} = \overline{\langle \sigma_i(t)\sigma_i(0) \rangle} - \langle \sigma_i \rangle^2 \sim \frac{1}{(\ell n t)^{\theta/\psi}} \tag{31}$$

with ψ the barrier exponent for the spin glass ordered phase, which lies in the range $\theta \leq \psi \leq d - 1$. [12]

In random exchange ferromagnets, the formal irrelevance of the randomness at low temperatures results in the autocorrelations being dominated by extremely rare events in which a droplet with interfacial energy per unit area which is anomalously small is excited, yielding [27]

$$C(t) \sim e^{-\pi(T)\ell n t} \sim \frac{1}{t^{\pi(T)}} \tag{32}$$

with $\pi(T)$ a non-universal temperature dependent exponent which diverges as $T \to 0$.

Development of Order in Random Magnets.

Returning to the problem of a single domain wall in a random exchange ferromagnet, we consider the effects of a small uniform driving force f, which can be realized by an applied magnetic field which favors one phase over the other. If a section of the interface of size L moves in the direction of the force by an amount h_L, the gain in energy from the force will be $fh_L L^{d-1}$. However, this will typically cost the energy of formation of an excitation of the interface. For the optimal scaling of h_L, which is $h_L \sim L^\zeta$, the typical excitation energy cost will be of order L^{θ_I}. If L is small, this will usually dominate the energy gain, and the excitation will cost energy and thus decay away. If, however, L is sufficiently large, the total energy change will be negative so that if the section moves it will not move back. The characteristic scale L_f beyond which this occurs is obtained by balancing the two energies yielding [29]

$$L_f \sim f^{-\frac{1}{2-\zeta_I}} \qquad (33)$$

where we have used Eq(10) for θ_I. [The size L_f is analogous to the radius of a critical size droplet for nucleation near a first order phase transition.] The motion of one section of size L_f can lower the energy resulting in a pulling forward of neighboring sections. [29] The interface can thus creep along by motion of sections of size L_f. The barriers for this motion are of order $B_{L_f} \sim L_f^{\psi_I}$ resulting in an exponentially long time scale for the motion of a section and a mean velocity

$$v \sim \exp\left(-\frac{c}{Tf^\mu}\right) \qquad (34)$$

with

$$\mu = \frac{\psi_I}{2-\zeta_I} \; . \qquad (35)$$

It is interesting to note that this behavior is dominated by relatively typical excitations, in contrast to the behavior of the *ac linear* response which is simply related to the autocorrelation function in Eq(30).

We are now in a position to understand the evolution of long range order following a quench into the ordered phase of a random exchange magnet: since the curvature of domain walls of size $R(t)$ results in a typical force $f(t) \sim \frac{1}{R(t)}$, we see that the evolution of the domain size $R(t)$ from Eq(34) is very slow: [29,30]

$$R(t) \sim (\ell nt)^{\frac{1}{\mu}} \; . \qquad (36)$$

This logarithmic evolution of correlations is one of the most important features of ordered phases with quenched randomness. Its consequence is that, even on macroscopic time scales, long-range order on macroscopic length scales is difficult to establish. The only saving feature is the behavior near to the critical temperature when a random exchange ferromagnet is cooled slowly: Near T_c, the evolution is much more rapid, with $R(t) \sim t^{1/z_c}$ for times up to of order the critical correlation time; thus on slow cooling, large domains are established just below T_c, but they only grow very slightly at lower temperatures. In spin glasses, even the faster (although still slow) dynamics near T_c does not save the day: the extreme dependence of the long-distance correlations in the ordered state on temperature [12] means that the large "domains" which are established at temperatures just below T_c are the wrong phase at lower temperatures and the system has to start again essentially from scratch after the temperature is lowered further. [16] This has all sorts of interesting consequences for the non-equilibrium history dependent dynamics of spin glasses which are, unfortunately, beyond the scope of this paper. [16,31,32]

Driven Interface Motion

Can one speed up the motion of domain walls in random media? From the above discussion, the answer is clearly yes. If an external field H is applied which favors one of the phases in a random ferromagnet, then the interfaces will feel a driving force $f \propto H$. Even at zero temperature where the thermally activated creep discussed above cannot occur, a sufficiently large force, which is larger than all the pinning forces, will make the interface move. Let us consider, for simplicity, the driven behavior in the absence of thermal fluctuations. [33,34]

If a small force is applied, some small sections of the interface may jump forward only to be held back by their neighboring regions. As the force is increased further, a series of local instabilities will occur and other regions will move forward, sometimes pulling larger regions with them. The relaxation towards a new metastable configuration as the field is increased will thus consist of "avalanches" in which sections of the interface move forwards. For small f, large avalanches will be exponentially rare. However, as f is increased, the chance that a large region will have weak enough pinning that it is above its local threshold grows, thus the probability of large avalanches also grows. The resulting distribution of avalanche diameters falls off as a power law up to a characteristic correlation length, ξ:

$$\text{Prob (Avalanche diameter} > L) \sim \frac{1}{L^\kappa} \hat{P}(L/\xi) \tag{37}$$

with $\hat{P}(x) \to$ constant for small x and $\hat{P}(x) \sim e^{-x}$ for large x. At some critical value of f, f_T, the correlation length diverges and a small local instability can cause the whole system to move forward in a self-sustaining avalanche. This is the *threshold force* for motion of the interface.

The behavior near to the threshold force is a non-equilibrium dynamic critical phenomenon which has recently been analyzed by a combination of non-linear dynamics and renormalization group techniques. [33,34] It is found that, at threshold, the interface is rougher than in equilibrium with a roughness exponent [34]

$$\zeta_{th} = \frac{5-d}{3} \tag{38}$$

to all orders in $\epsilon = 5 - d$, and perhaps exact. The correlation length below threshold diverges as [33,34]

$$\xi \sim \frac{1}{(f_T - f)^{\nu_{th}}} \tag{39}$$

with

$$\nu_{th} = \frac{1}{2 - \zeta_{th}} \; . \tag{40}$$

Above threshold, the interface moves forward, initially in a very jerky non-uniform manner. The mean velocity is

$$v \sim (f - f_T)^\beta \tag{41}$$

with β calculated in an ϵ expansion in $d = 4 - \epsilon$ dimensions, yielding [34]

$$\beta = 1 - \frac{1}{9}\epsilon + O(\epsilon^2). \tag{42}$$

Thermal fluctuations will round out the sharp threshold and cause v to be non-zero for any $f \neq 0$. However, if the thermal fluctuations are weak, even just below the threshold the creep will be very slow and the threshold will appear to be quite sharp.

Non-linear Collective Transport

The threshold phenomena of driven interfaces is the best understood example of a wide class of non-linear collective transport phenomena in random media. This includes fluid invasion into porous media, [35] fluid flow down a rough inclined plane, [36] critical current phenomena in superconductors, [37,38] sliding charge density waves in solids, [39] and rough-solid-on-rough-solid friction. [40]

These phenomena fall very crudely into two broad categories: [40,41] driven motion of *elastic* media—such as the interfaces discussed above, [33] for which averaged over long times, the flow is roughly homogeneous in space; and fluid or plastic-like flow in which the system breaks up into flow channels and is thus intrinsically inhomogeneous. [36,41] For the especially interesting case of fluid invasion in which one fluid drives another fluid out of a porous medium under the influence of a pressure head, both types of behavior can occur. [41] If the invading fluid preferentially wets the surfaces, a well-defined interface is seen to be present on long length scales and the behavior is like that of a domain wall in a random ferromagnet. If, on the other hand, the invading fluid does not wet the surfaces, the interface will break up and the fluid will invade along weak links, forming, near threshold, critical-percolation-like regions with a non-trivial fractal dimension. [41] As Martys, Robbins and Cieplak [41] have found, at threshold there is a sharp transition, as a function of the surface-fluid contact angle, between these two types of behavior! This, and other non-linear collective transport phenomena, including models of earthquakes and sandpiles [43] and hysteresis in random magnets, [44] are currently the subject of intense study.

Metastability and Quasi-equilibrium Phenomena

We have seen that the low temperature phases of random magnetic systems can be characterized by several fundamental properties: some type of "stiffness" which resists spatial variations of the local order and increases as—in general—a non-trivial power of the length scale, excitations with a distribution of free energies, and scale dependent barriers for evolution of the correlations and development of the order which result in extremely broad distributions of time scales and logarithmic dependence of many properties on time.

The broad spectrum of time scales results in a mixing of quasi-equilibrium and non-equilibrium phenomena, causing history dependence and various kinds of "memory" effects. It does, however, cause a simplification: on any fixed macroscopic time scale, t, almost all processes in a random system are either almost completely in local equilibrium, i.e. with $\tau \ll t$ or almost totally out of equilibrium, i.e. with $\tau \gg t$, with only rare regions of the system, which collectively occupy a small fraction of the volume, active on time scales of order t. This separation of time scales can be used to make substantial progress in analyzing the interplay between equilibrium and non-equilibrium phenomena. [16]

We have focused, in this paper, primarily on the properties of low temperature phases in random systems. In pure systems, understanding the low temperature phases is often relatively simple, with the main difficulties associated with understanding the critical phenomena near to phase transitions. [1] In contrast, in random systems the main new physics and new challenges involve understanding the *ordered phases* with the critical behavior, although interesting and still difficult, often differing from pure systems only in more subtle ways. [11,45] An interesting question naturally arises from this observation: if, in random systems, much of the interesting physics occurs well below the phase transition, how much does it matter whether or not a phase transition actually occurs?

We thus are led to a brief discussion of the behavior of a random magnetic system *without* a phase transition. This occurs, for example, in two-dimensional Ising spin glasses for which $\theta < 0$. [46,47] Thermal fluctuations at low temperatures cause excitations to occur frequently if they are large enough that $L^\theta < T$. This yields a correlation length

$$\xi \sim \frac{1}{T^\nu} \qquad (43)$$

with $\nu = -1/\theta$. On scales smaller than ξ, the spins are correlated as they would be in the ground state, but regions of size of order ξ fluctuate between the two equivalent spin-reversed ground state configurations acting somewhat like independent "super paramagnetic" spin clusters. On scales up to ξ, there will be barriers for droplet excitations with $B_L \sim L^\psi$ and $\psi \geq 0$ (or perhaps $B_L \sim \log L$) leading to an overall equilibration rate τ given by [46]

$$\ln \tau \sim B_\xi / T \sim \xi^\psi / T \sim \frac{1}{T^{1+\psi/|\theta|}} \ . \qquad (44)$$

For $\psi > 0$, this yields faster-than-Arhennius temperature dependence of the equilibration time which, in the absence of measurements covering a very wide frequency range, may appear to diverge at finite T. On times much smaller than τ, the behavior will be qualitatively like that of a spin glass ordered phase with logarithmic decay of autocorrelations, hysteresis, etc. [46,47] Thus the presence of a true ordered phase in a random system should *not* be inferred from the existence of such phenomena at low temperatures, but rather by a careful analysis of whether or not the relaxation time on cooling (or an appropriate susceptibility, if measurable) diverges at a finite temperature.

Order in Spin Glasses

Before turning to superconductors, we briefly raise the key open question concerning the equilibrium behavior of random Ising magnets: the nature of the low temperature ordered phase in spin glasses.

David Huse and the author [4,12,16] have argued that a consistent phenomenological picture of the ordered phase of Ising spin glasses can be constructed in which, in the thermodynamic limit, there are exactly two ground states—in a sense which can be precisely defined. [4] These are simply related by global spin reversal. At finite temperatures below the transition, there are again just two symmetry related equilibrium states. However, the spin correlations at large distances will be extremely sensitive to temperature changes, being essentially independent (at long enough distances) at each temperature! [12] Any magnetic field is argued to destroy the ordered phase leading to a unique state. Qualitatively, and often quantitatively, virtually all the equilibrium and non-equilibrium phenomena which have been observed in spin glasses can be analyzed in terms of this framework. Nevertheless, it is highly controversial. [13] The origin of this controversy is the famous conjecture by Parisi [14] for the solution—via "replica symmetry breaking"—of the Sherrington-Kirkpatrick model [48] in which all spins are coupled to all others. In some sense which has never really been made precise, this model has *many* equilibrium states below the transition temperature; [14] these are not related by symmetry. In addition, similar behavior is predicted to persist up to a critical magnetic field. [14,49]

To the author's knowledge, no one has yet succeeded, for finite-dimensional systems with short-range interactions, in constructing a consistent picture of an ordered phase of spin glasses that has many states. Nevertheless, many workers in the field—probably a majority—steadfastly believe that this is what occurs. [13] What about experiments? In principle, if there were competing predictions from a picture with many states to compare with, experiments using local probes of magnetic correlations or fluctuations [50,51] could help to address the question of the existence of many equilibrium states. However, these are very difficult and would be tricky to analyze convincingly due to the very slow dynamics and existence of very metastable quasi-equilibrium "states." [16]

Another strategy appears to be rather more feasible: using the lesson mentioned above, that the *approach* to a putative ordered phase from above [52] is the best way to test its existence in a random medium, an experiment which tests one of the key disagreements is possible: a systematic search—using a wide frequency range—for a diverging relaxation time at a non-zero magnetic field. This could help answer the question of the existence of a phase transition in a field [49] which is likely to be closely linked to the presence or absence of many equilibrium states. [14] Such an experiment has many difficulties, but I am optimistic that they can be overcome. In the meantime, we are left with a theory based on the concepts of droplet excitations and scaling which has only two states but many experimental predictions [12,16] and another based on the infinite range model [48] which has infinitely many states [16] but, unfortunately, almost no direct predictions for the properties of the ordered phase that have not already been ruled out. [53]

It is interesting to note that recent theoretical developments on the properties of interfaces in random media discussed earlier suggest that when the dust settles there may no longer be a disagreement. [18,22,24] After suggestions to the contrary in earlier papers, the predictions based on "replica symmetry breaking" ideas for the properties of these systems [24] seem to have converged to a picture which is virtually identical to that of the scaling picture discussed above, with the "states" analogous to those of the Sherrington-Kirkpatrick model [48] becoming, here, just the large rare active excitations. [22] It appears, therefore, quite conceivable that such a resolution will also occur in spin glasses with the large rare active droplet excitations [12] being the finite dimensional equivalent of the "states" of Parisi et al. [14]

Superconductors and Vortex Glasses

In this last section, we briefly apply some of the ideas developed here to perhaps the most interesting problem so far studied in this field: the properties of the mixed state of dirty Type II superconductors. In Type II superconductors—which most useful superconductors, including the high-T_c cuprates, are—there are two distinct superconducting phases within the standard Ginzburg-Landau picture. [37] At low magnetic fields, $H < H_{c1}(T)$, magnetic flux is completely excluded from the sample so that $B = 0$ except in a surface layer of thickness the penetration length λ. Above H_{c1}, the magnetic field penetrates but, instead of doing so uniformly as in a normal metal, it forms an array of parallel flux tubes, each containing one quantum $\Phi_0 = \frac{hc}{2e}$ of magnetic flux. Associated with each flux tube is a vortex of circulating supercurrents and a core of radius the coherence length ξ in which the Cooper pair wave function—the complex superconducting order parameter, ψ,—vanishes, with its phase φ varying by 2π around any closed loop which circumscribes the vortex line. This intermediate field phase is called the "mixed state." In the absence of impurities and thermal fluctuations, the vortex lines form a perfect triangular lattice, as predicted by Abrikosov. As the field is increased, the vortices become closer and closer together, eventually ceasing to exist as discrete entities at the upper critical field $H_{c2}(T) = \frac{\Phi_0}{2\pi\xi^2}$ above which the field is uniform and the system a normal metal. The conventional phase diagram is shown in Fig. 5.

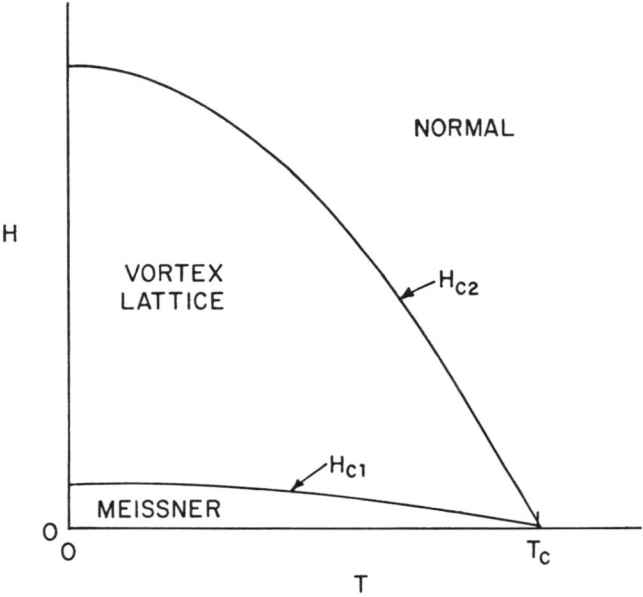

Figure 5. Schematic phase diagram of a conventional ideal (i.e. clean) Type II superconductor as a function of applied magnetic field H and temperature T. In the Meissner phase, the magnetic field, B, in the sample is zero, while in the vortex lattice phase, the field penetrates in a regular triangular array of quantized flux tubes—(or vortex lines).

Thermal fluctuations can melt the Abrikosov vortex lattice [38,54] at temperatures close to the phase boundary $H_{c2}(T)$, but in conventional superconductors this only occurs in an immeasurably small region of the phase diagram. Nevertheless, in the cuprate superconductors, due to a combination of factors, the fluctuations are orders of magnitude larger and the melted vortex fluid regime—which is qualitatively like the normal metal phase to which it is continuously connected—exists over a region more than 10K wide at intermediate fields in the Tesla range. [38,54]

What are the transport properties of the vortex lattice and vortex fluid phases? A transport current through the system with current density J causes a Lorentz force on the vortex lines which then move across the current causing electric fields proportional to the vortex velocity and hence resistance. [37] In a perfectly clean system, in both the vortex fluid and vortex lattice phases, the vortices respond linearly to the current, yielding *finite* linear conductivity inversely proportional to the vortex mobility. [38] Thus ideally, the mixed state is *not* superconducting. What then is the cause of the experimental observation that in all conventional superconductors the resistivity drops to an immeasurably small value as soon as the mixed state is entered on cooling? [37] As has been known since the 1950's, this is because impurities and other defects in the superconducting material act to pin the vortices and impede their motion. [55,56]

The behavior of a single vortex line in the random potential caused by impurities is very similar to the problem of an interface in a two-dimensional random exchange ferromagnet that we discussed earlier. [5,22,25] The primary difference is that the vortex line can distort in *two* transverse directions. This reduces the roughness exponent ζ from 2/3 to approximately 0.6 and $\theta = 2\zeta - 1$ from 1/3 to 0.2. [57] Nevertheless, the vortex line is always in a pinned state [22,25] with only a non-linear response to a force. This yields a vortex velocity v_v and hence, from the moving magnetic flux, an electric field

$$E \sim v_v \sim J e^{-J^{-\mu}/T} \; ; \qquad (45)$$

with, for the single vortex line, μ given by Eq(35). [38,58] Thus, the *linear resistivity vanishes*.

What about the behavior of interacting vortex lines in a random potential? Two decades ago, Larkin [56] pointed out that any amount of randomness destroys the long-range vortex lattice order above some scale length L_p. Although it was not studied in detail at the time (since this was before the era of spin glasses when randomness became "respectable"), it appears to have been implicitly assumed [55,56] that on scales larger than L_p the crystalline regions of size L_p^3 would move roughly independently, forming a sluggish fluid-like phase, recently dubbed a "vortex slush." [59] The barriers for motion of each region through the impurity potential would result in very small thermally-activated vortex motion and linear resistivity [55]

$$\rho \sim e^{-U_p/T} \; . \qquad (46)$$

In conventional superconductors, the barriers U_p are much much larger than T, [55] thus whether ρ has the form Eq(46) or is really zero is rather moot.

A novel and very different suggestion was put forward by Matthew Fisher [60] a few years ago and developed by him, David Huse and this author: [38] Instead of a vortex slush, a new phase—dubbed a "vortex glass," by analogy with spin glasses—was postulated. [60] In this phase, the competition between the vortex-vortex interactions and the random potential results in a phase in which the vortex lines fluctuate about particular, sample specific, positions, with the complex superconducting order parameter, ψ, having long-range order analogous to that of spins in a spin glass. [12,13]

In this magnetic analogy, the Meissner phase with $B = 0$, (which exists for $H < H_{c1}$), is analogous to a ferromagnet since $<\psi>$ points in the same direction in the complex plane at every point in space. The Abrikosov lattice phase, with $<\psi(r)>$ varying in a regular periodic way reflecting the periodic array of vortices, [61] is then analogous to an antiferromagnet, while the vortex glass phase, with $<\psi(r)>$ varying in space in a seemingly random way around the vortices, is analogous to a spin glass.

The proposed vortex glass phase can only exist at finite temperatures if it has a stiffness exponent $\theta > 0$ yielding a free energy stiffness $\Sigma_L \sim L^\theta$ for variations of the phase of ψ, and excitations that consist of rearrangements of vortex lines in regions of size L^d, with excitation free energies $\sim L^\theta$. [38] As for the individual vortex line [38,58] and interfaces in random magnets, [5,33] we expect a non-linear response to a driving force due to activation over a distribution of scale dependent barriers caused by the randomness. Although the nature of the dominant excitations and the low current creep behavior is not understood, arguments similar to those used earlier for interfaces yield mean vortex velocities and hence, electric fields of the form Eq(45)

$$E \sim J \exp - (J_T/J)^\mu \qquad (47)$$

with $\mu \leq 1$ and $J_T \sim T^{-1/\mu}$ a characteristic current scale. [38,60] Thus, the vortex glass phase is a *linear superconductor*, although there will be a finite resistance at any non-zero current. [38,60] [Similar non-linear dissipation due to nucleation of vortex loops occurs in the Meissner phase, yielding $\mu = 1$.]

What, then, is the apparent "critical current" observed in superconductors? [37] As the current is increased in a vortex glass phase, the forces will eventually become large enough to overcome the pinning forces and the vortices will start to flow without having to thermally activate over barriers. At zero temperature, this yields a sharp critical current J_c—analogous to the critical threshold force for interface motion discussed earlier. [36] The non-linear critical behavior near to the critical current is an interesting, but difficult, problem, [40] which has received surprisingly little attention. [62] If the thermal fluctuations are small, then $J_T \gg J_c$ and the creep rate of vortices just below the critical current will still be small and the critical current will appear sharp, [38] requiring measurement of very small resistances or very slow vortex creep to see the resistance below J_c. This is the case in conventional superconductors, [55] as illustrated in Fig. 6a, and the low current behavior of Eq(45) has never been observed. In the cuprate superconductors, in contrast, the fluctuations are large enough in substantial regions of the phase diagram [38] that dissipation can be measured well below the nominal critical current, thereby probing the low current creep regime; the current-voltage behavior in this regime is illustrated in Fig. 6b. Tests of Eq(47) are currently underway. [63]

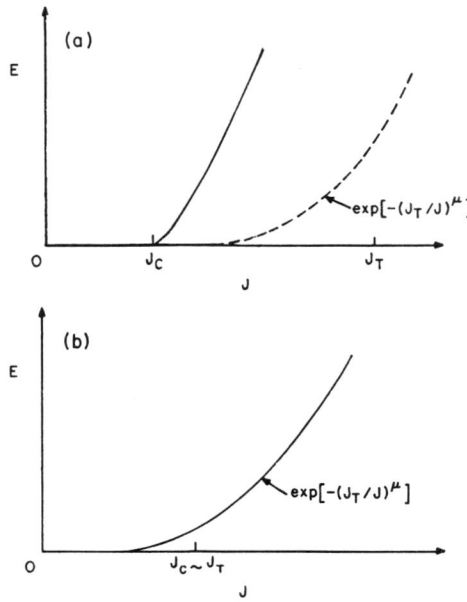

Figure 6. Qualitative current-voltage characteristics of the mixed state of dirty Type II superconductors in a magnetic field. In (a), appropriate to conventional superconductors, the electric field, E, is extremely small at small current densities, J, as illustrated by the extrapolation to higher currents shown by the dashed line. The characteristic scale of this thermally activated vortex creep is J_T which is much larger than the current, J_c at which the vortices start to move easily because the force from the current exceeds the pinning forces from the random impurities in the sample. The current J_c thus appears to be a sharp critical current. In (b), appropriate to the high temperature cuprate superconductors at intermediate temperatures not-too-far below the vortex glass transition, there is only one characteristic current scale and the "critical current" is not well-defined.

What is the evidence for a vortex glass phase? [38,60] As was mentioned above in the context of spin glasses, the best way to distinguish a true transition to a vortex glass phase from the gradual slowing down characteristic of a vortex slush is to study the behavior in a magnetic field as the temperature is lowered from the normal state. In the resulting mixed state of superconductors, the experiments [38,63-67] have focused on the non-linear (and hence, non-equilibrium) current-voltage characteristics. The predicted form near a second order vortex glass transition at $T_G(H)$ is [38,64]

$$E(J,T) \sim J|T - T_G|^{(z-1)\nu}\hat{\mathcal{E}}(J/J_\times) \qquad (48)$$

with the characteristic current scale $J_\times \sim |T - T_G|^{-2\nu}$ with ν and z the correlation length and dynamic critical exponents respectively. For $T > T_G$, $\hat{\mathcal{E}} \to$ constant as $J \to 0$ so that the resistivity vanishes as

$$\rho \sim (T - T_G)^{(z-1)\nu} \quad . \qquad (49)$$

Below T_G,

$$\hat{\mathcal{E}} \sim e^{-(J_\times/J)^\mu} \qquad (50)$$

for small J yielding the characteristic vortex glass behavior with $J_T \sim J_c \sim J_\times$. Right at T_G,

$$E \sim J^{\frac{z+1}{2}} \quad . \qquad (51)$$

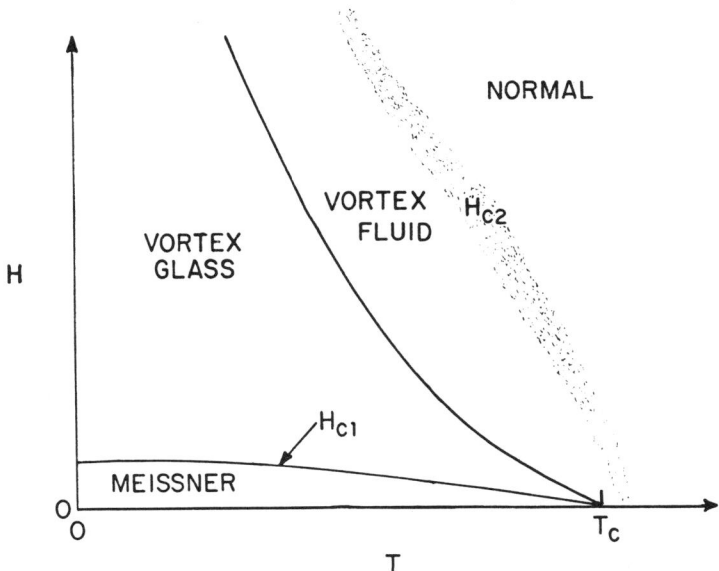

Figure 7. Schematic phase diagram of a dirty Type II superconductor with strong thermal fluctuations as a function of applied magnetic field H and temperature T. As the temperature is lowered in a field, there is a gradual crossover denoted by the fuzzy H_{c2} "line" from a normal metal to a vortex fluid phase which is not superconducting. At lower temperatures below a sharp phase transition, indicated by the solid line, a vortex glass phase forms; this phase has zero linear resistivity and is thus truly superconducting.

In Y-Ba-Cu-O films which are quite strongly disordered, Koch et al. [64] have found rather good evidence for a second-order vortex glass transition with scaling behavior fitting Eq(48) for fields in the few Tesla range where T_G's are supressed by of order 10K from the nominal upper critical field line $H_{c2}(T)$, which becomes just a crossover from normal metal to vortex fluid behavior. Associated with this crossover, the resistance starts to drop appreciably on cooling in a field. The phase diagram is shown, qualitatively, in Fig. 7. In contrast, very clean twin-free single crystals of Y-Ba-Cu-O appear to exhibit a *first order* transition [59,65] (except in very high fields) to a phase with immeasurable resistance which is presumably a vortex glass, although there is as yet no direct evidence for this. [65] The first order transition is presumably associated with the expected first order vortex lattice melting transition [54] in a clean superconductor and will occur at approximately the same temperature. [38]

Understanding the behavior of the phase diagrams, phase transitions, and low temperature properties of the cuprate superconductors as a function of magnetic field, randomness, anisotropy (Bi-Sr-Ca-Cu-O is much more anisotropic than Y-Ba-Cu-O and behaves very differently) [66,68] and macroscopic defects such as twin boundaries, [69,70] will clearly be an interesting experimental and theoretical challenge for quite some time. At this point, however, the experimental evidence does support the existence of a real phase transition in the mixed state of dirty Type II superconductors, to a new phase which is truly superconducting with vanishing linear resistivity.

As a final historical point, it is interesting to note that our current understanding of the interplay between fluctuations, correlations, randomness and linear and non-linear transport in Type-II superconductors would not have been possible without the developments of many ideas in seemingly unrelated areas during the period when the field of superconductivity was relatively quiescent. The field was thus ripe for the rapid growth in understanding of strongly fluctuating superconductors when the cuprate superconductors burst onto the scene six years ago.

Acknowledgements

The ideas and understandings summarized in this paper have been developed primarily in collaboration with David Huse and, for the work on superconductivity, with Matthew Fisher, to both of whom I am greatly indebted. I have also benefitted from fruitful interactions with many other colleagues, especially Onuttom Narayan, Terry Hwa, David Nelson, Geoff Grinstein, Alan Middleton, Peter Gammel, Aharon Kapitulnik and David Bishop. To them and others too numerous to mention, I am most grateful.

This work was supported by the A.P. Sloan Foundation, and by the National Science Foundation via grant DMR 9106237 and Harvard University's Materials Research Laboratory.

References

1 Hohenberg, P.C. and B.I. Halperin (1977), Rev. Mod. Phys. 49, 435.
2 Lifshitz, I.M. (1962), Sov. Phys.—JETP 15, 939.
3 For a pedagogical review of the theory of random magnets, see Fisher, D.S., G.M. Grinstein and A. Khurana (Dec., 1988), Phys. Today 41: 12, 56.
4 In random systems, especially spin glasses, defining ground states of infinite systems is somewhat tricky and itself illuminates some of the issues. See Huse, D.A. and D.S. Fisher (1987), J. Phys. A 20, L997.
5 Huse, D.A. and C.L. Henley (1984), Phys. Rev. Lett. 54, 2708; and D.A. Huse, C.L. Henley and D.S. Fisher (1985), Phys. Rev. Lett. 55, 2924.

6 Fisher, M.E. (1967), Physics 3, 255.
7 Polyakov, A.M. (1975), Phys. Lett. B 59, 79.
8 Widom, B. (1965), J. Chem. Phys. 43, 3898.
9 Grinstein, G. and A. Luther (1976), Phys. Rev. B 13, 1329.
10 This is analogous to the so-called "universal conductance fluctuations" in disordered metallic films. See Lee, P.A. and A.D. Stone (1985), Phys. Rev. Lett. 55, 1622.
11 Ludwig, A.W.W. (1990), Nucl. Phys. B 330, 639.
12 Fisher, D.S. and D.A. Huse (1988), Phys. Rev. B 38, 386; J. Phys. A 20, L1005 (1987); M. Aizenman and D.S. Fisher, unpublished.
13 For a recent review of spin glasses, see Binder, K. and A.P. Young (1986), Rev. Mod. Phys. 58, 801.
14 Parisi, G. (1979), Phys. Rev. Lett. 43, 1574; Mézard, G. Parisi, N. Sourlas, G. Toulouse and M. Virasoro (1984), Phys. Rev. Lett. 52, 1156, and J. Physique 45, 843.
15 Anderson, P.W. and C.M. Pond (1978), Phys. Rev. Lett. 40, 903; J.R. Banavar and M. Cieplak (1983), J. Phys. C 16, L755; W.L. McMillan (1985), Phys. Rev. B 31, 340; A.J. Bray and M.A. Moore (1985), Phys. Rev. B31, 631.
16 Fisher, D.S. and D.A. Huse (1988), Phys. Rev. B 38, 373.
17 Bray, A.J. and M.A. Moore (1987), Phys. Rev. Lett. 58, 57.
18 Balents, L. and D.S. Fisher, submitted to Phys. Rev. B.
19 Fisher, D.S., unpublished.
20 Chayes, J.T., L. Chayes, D.S. Fisher and T. Spencer (1986), Phys. Rev. Lett. 57, 299; Comm. Math. Phys. 1 20, 501 (1989).
21 Fisher, D.S. (1986), Phys. Rev. Lett. 56, 1964, and references therein.
22 Fisher, D.S. and D.A. Huse (1991), Phys. Rev. B 43, 10728, and references therein.
23 Schulz, U., J. Villain, E. Brézin and H. Orland (1988), J. Stat. Phys. 51, 1.
24 Mézard, M. (1990), J. Physique 51, 1831; M. Mézard and G. Parisi, J. Physique I1, 809 (1991); M. Mézard, private communication.
25 Kardar, M. and Y-C Zhang (1987), Phys. Rev. Lett 58, 2087.
26 See e.g., Hwa, T. and E. Frey, (1991), Phys. Rev. A 44, 7873, and references therein.
27 Huse, D.A. and D.S. Fisher (1987), Phys. Rev. B 35, 684.
28 "Stretched exponential" decay of correlations is often thought to be characteristic of "glassy" behavior or "hierarchies" of relaxation times [see e.g. Palmer, R.G., D.L. Stein, E. Abrahams and P.W. Anderson (1984), Phys. Rev. Lett. 53, 958.], but we see that it occurs in the pure 2D Ising model!
29 Villain, J. (1984), Phys. Rev. Lett. 52, 1543. G. Grinstein and J. Fernandez (1984), Phys. Rev. B 29, 6389, consider the effects of a discrete lattice. The effects of rare regions on the coarsening have been shown to be small by D.S. Fisher, unpublished.
30 Fisher, D.S. (1987), J. Appl. Phys. 61, 3672.
31 Refrigier, P.L., E. Vincent, J. Hammann and M. Ocio (1987), J. Physique 48, 1533; M. Alba, J. Hammann, M. Ocio, P.L. Refrigier and H. Bouchiat (1987), J. Appl. Phys. 61, 3683.
32 Nordblad, P., P. Svendlindh, P. Granberg and L. Lundgren (1987), Phys. Rev. B 34, 7150.
33 Nattermann, T., S. Stepanow, L-H Tang and H. Leschom (1992), J. Physique II2, 1483.
34 Narayan, O. and D.S. Fisher, Phys. Rev. B, in press.
35 Rubio, M.A., C.A. Edwards, A. Dougherty and J.P. Gollub (1989), Phys. Rev. Lett. 63, 1685; V.K. Horvath, F. Family and T. Vicsek (1991), J. Phys. A 24, L25.
36 Narayan, O. and D.S. Fisher, submitted to Phys. Rev. Lett.; O. Narayan, Ph.D. thesis, Princeton Univ. (1992).

37 For background information on Type II superconductors, see e.g. Tinkham, M. (1975) *Introduction to Superconductivity*, Krieger Publishers, Florida.
38 Fisher, D.S., M.P.A. Fisher and D.A. Huse (1991), Phys. Rev. B 43, 130, and references therein. For a pedagogical review, see Huse, D.A., M.P.A. Fisher and D.S. Fisher (1992), Nature 358, 553.
39 Narayan, O. and D.S. Fisher (1992), Phys. Rev. B 46, 11520, and references therein.
40 Fisher, D.S. (1987), in A.R. Bishop et al. (ed.), *Nonlinearity in Condensed Matter*, Springer-Verlag, New York.
41 Martys, N., M.O. Robbins and M. Cieplak (1991), Phys. Rev. B 44, 12294, and references therein.
42 Carlson, J.M. and J.S. Langer (1989), Phys. Rev. Lett. 62, 2632.
43 Bak, P., C. Tang and K. Wiesenfeld (1980), Phys. Rev. A 38, 364; L. Kadanoff, S. Nagel, L. Wu and S. Zhou (1989), Phys. Rev. A 39, 6524.
44 Sethna, J.P., K. Dahmen, S. Kartha, J.A. Krumhansl, B.W. Roberts and J.D. Shore, preprint.
45 There are exceptions to this, for example random field Ising systems which have activated dynamic scaling near their *critical* temperatures. See reference [30] and Villain, J. (1985), J. Physique 46, 1843; Fisher, D.S. (1986), Phys. Rev. Lett. 56, 416.
46 Fisher, D.S. and D.A. Huse (1987), Phys. Rev. B 36, 8937.
47 Kenning, G.G., J. Slaughter and J.A. Cowen (1987), Phys. Rev. Lett. 59, 2596.
48 Sherrington, D. and S. Kirkpatrick (1975), Phys. Rev. Lett. 35, 1972.
49 de Almeida, J.R.L. and D.J. Thouless (1978), J. Phys. A 11, 983.
50 Ocio, M., H. Bouchiat and P. Monod (1985), J. Physique Lett. 246, L647.
51 Resistance noise has recently been used as an indirect probe of the spin configurations in small samples of spin glasses; Alers, G.B., M.B. Weissman and N.E. Israeloff (1992), Phys. Rev. B 46, 507.
52 Many experiments studying the approach to the spin glass transition in zero field have been performed, for example, Monod, P. and H. Bouchiat (1982), J. Physique Lett. 43, 145; H. Bouchiat and P. Monod (1983), J. Magn. Matter 30, 175, and references in [13].
53 For the Sherrington-Kirkpatrick model, power law decay of autocorrelations in the ordered phase have been obtained by Sompolinsky, H. and A. Zippelius (1982), Phys. Rev. B 25, 6860. Experiments [31] show much slower decay, consistent with the predictions of [12] and [16].
54 Nelson, D.R. (1988), Phys. Rev. Lett. 60, 1973; A. Houghton, R.A. Pelcovits and S. Sudbo (1989), Phys. Rev. B 40, 6763.
55 Anderson, P.W. and Y.B. Kim (1964), Rev. Mod. Phys. 36, 39.
56 Larkin, A.I. and Yu. N. Ovchinikov (1979), J. Low Temp. Phys. 34, 409.
57 Kim, J.V. and J.M. Kosterlitz (1989), Phys. Rev. Lett. 62, 2289.
58 Feigel'man, M.V. and V.M. Vinokur (1990), Phys. Rev. B 41, 8986.
59 Worthington, T.K., M.P.A. Fisher, D.A. Huse, J. Toner, A.D. Marwick, T. Zabel, C.A. Field and S. Holtzberg (1992), Phys. Rev. B 46, 11854.
60 Fisher, M.P.A. (1989), Phys. Rev. Lett. 62, 1415.
61 Moore, M.A. (1988), Phys. Rev. B 39, 136, has pointed out that fluctuations of the vortex lines actually destroy the order of ψ in the Abrikosov lattice phase.
62 See, however, recent work of S. Bhattacharya and M.J. Higgins, preprint;
63 Sandvold, E. and C. Rossel (1992), Physica C 190, 309.
64 Koch, R.H., V. Foglietti, W.J. Gallagher, G. Koren, A. Gupta and M.P.A. Fisher (1989), Phys. Rev. Lett. 63, 1511.

65 Safar, H., P.L. Gammel, D.A. Huse, D.J. Bishop, J.P. Rice and D.M. Ginsberg (1992), Phys. Rev. Lett. 69, 824.
66 Palstra, T.T.M., B. Batlugg, L.F. Schneemeyer and J.V. Wasczak (1988), Phys. Rev. Lett. 61, 1662; H. Safar, P.L. Gammel, D.J. Bishop, D.B. Mitzi and A. Kapitulnik (1992), Phys. Rev. Lett. 68, 2672.
67 Gammel, P.L., L.F. Schneemeyer and D.J. Bishop (1991), Phys. Rev. Lett. 66, 953.
68 Fisher, D.S. (1991) in K. Bedell et al. (eds.), *Phenomenology & Applications of High Temperature Superconductors*, Addison-Wesley, New York.
69 Fleshler, S. W-K Kwok, U. Welp, V.M. Vinokur, M.K. Smith, J. Downey and G.W. Crabtree, preprint.
70 The effects of *columnar* defects have been considered by Nelson, D.R. and V.M. Vinokur (1992), Phys. Rev. Lett. 68, 2398.

STATICS AND DYNAMICS OF THE FLUX-LINE LATTICE IN HIGH-T_c SUPERCONDUCTORS

ERNST HELMUT BRANDT
Max-Planck-Institut für Metallforschung, Institut für Physik
Heisenbergstr. 1
W-7000 Stuttgart 80
FR Germany

ABSTRACT: As discovered by Abrikosov, a magnetic field B can penetrate a type-II superconductor in form of flux lines or vortices which carry a quantum of flux each and arrange to a more or less regular triangular lattice. The flux-line lattice has interesting elastic and fluctuation properties, in particular in the highly anisotropic high-T_c superconductors (HTSC) with layered structure. Under the action of an electric current density $J > J_c$ the flux lines move and dissipate energy, but for $J < J_c$ they are pinned by material inhomogeneites. In HTSC thermally activated depinning causes a finite resistivity ρ even at current densities $J < J_c$. At sufficiently high temperature T ohmic resistivity $\rho(T, B)$ is observed down to $J \to 0$. This indicates that the flux lines are in a "liquid state" with no shear stiffness and with small depinning energy. At lower T, $\rho(T, B, J)$ is non-linear since the pinning energy of an elastic vortex lattice or "vortex glass" increases with decreasing J. In the extremely anisotropic Bi- and Tl-based HTSC short vortex segments ("pancake vortices" in the CuO layers) can depin individually with very small activation energy.

1. The Flux-Line Lattice in Type-II Superconductors

Since the discovery of the phenomenon of superconductivity in 1911 various phenomenological theories of superconductivity have been a powerful tool for describing the electromagnetic and thermodynamic properties of superconductors [1-3]. In particular, the linear and local electromagnetic theory of Fritz and Heinz London of 1935 has proven very useful and applies to the entire temperature range $0 < T < T_c$ where T_c is the superconducting transition temperature. The London equation states a direct proportionality of the supercurrent density $\mathbf{J}(\mathbf{r})$ and the vector potential $\mathbf{A}(\mathbf{r})$ in the "London gauge" (div$\mathbf{A} = 0$, $\mathbf{A}$ parallel to all surfaces): $\mathbf{J} = -(\mu_0 \lambda^2)^{-1}\mathbf{A}$, or $\lambda^2 \text{rot rot}\,\mathbf{B} + \mathbf{B} = 0$ where $\mathbf{J} = \text{rot}\,\mathbf{H}$ and $\mathbf{B} = \mu_0 \mathbf{H} = \text{rot}\,\mathbf{A}$ and λ is the magnetic penetration depth, for which a temperature dependence $\lambda(T) = \lambda(0)/(1 - T^4/T_c^4)^{1/2}$ was observed in agreement with Gorter's two-fluid theory. Pippard extended the London equation to a nonlocal relationship. The range ξ_p of the Pippard kernel was later found close to the BCS coherence length ξ_0.

The most successful phenomenological theory was established in 1950 by Ginzburg and Landau (GL) from very general principles of second order phase transitions and the gauge invariance of the complex GL order parameter (or GL function) $\psi(\mathbf{r}) = |\psi| \exp(i\phi)$. In the GL theory, $\mathbf{B}(\mathbf{r})$ varies over the GL penetration length $\lambda(T)$ and $\psi(\mathbf{r})$ over the GL coherence length $\xi(T)$. Both lengths diverge near T_c as $\lambda \propto \xi \propto (T_c - T)^{-1/2}$. The ratio $\lambda/\xi = \kappa$ is the GL parameter. GL showed that the type of solutions qualitatively differs for $\kappa < 1/\sqrt{2}$ (type-I superconductors, wall energy between normal and superconducting domains is positive; e.g., most metals except Nb and V) and $\kappa > 1/\sqrt{2}$ (type-II superconductors, wall energy negative; e.g., Nb and its alloys and the non-metallic superconductors). Though originally derived for temperatures $T \approx T_c$, the GL theory in many cases gives qualitatively correct results valid at all temperatures $0 < T < T_c$.

Landau's student Alexei Abrikosov in 1957 published a two-dimensional (2D) periodic solution of the linearized GL equations for $\kappa \geq 1/\sqrt{2}$ and interpreted this as a regular lattice of current vortices, each carrying a quantum of magnetic flux $\Phi_0 = h/2e = 2.07 \times 10^{-15}\,\mathrm{Tm}^2$, Fig. 1. Abrikosov vortices are also called fluxons, fluxoids, flux tubes, flux lines, or vortex lines. The vortex positions in this flux-line lattice (FLL) are defined by the zeros of ψ. In 1967 Träuble and Essmann observed the FLL in an electron microscope by a Bitter decoration technique using iron microcrystallites which condensed from an evaporating wire in a helium atmosphere of a few Torr pressure ("magnetic smoke") [4]. After the discovery of high-T_c superconductors (HTSC) similar decoration experiments proved the existence of an Abrikosov FLL also in these materials [6].

When the applied magnetic field B_a is increased from zero, flux lines start to penetrate an ideal (pin-free) type-II superconductor when B_a equals the lower critical field $B_{c1}(T) = (\Phi_0/4\pi\lambda^2)(\ln\kappa + \alpha)$ ($\alpha \approx 0.5$ for $\kappa \gg 1$, $\alpha = 1 - \ln\sqrt{2} = 0.653$ at $\kappa = 1/\sqrt{2}$). The flux lines form a periodic triangular lattice of spacing $a = (\sqrt{3}\Phi_0/2B)^{1/2}$ [B = flux density = spatial average of $B(\mathbf{r})$] which becomes denser until superconductivity disappears at the upper critical field $B_{c2}(T) = \Phi_0/2\pi\xi^2$ where $a = 2.69\xi(T)$ and $B_a = B = B_{c2}$. The order

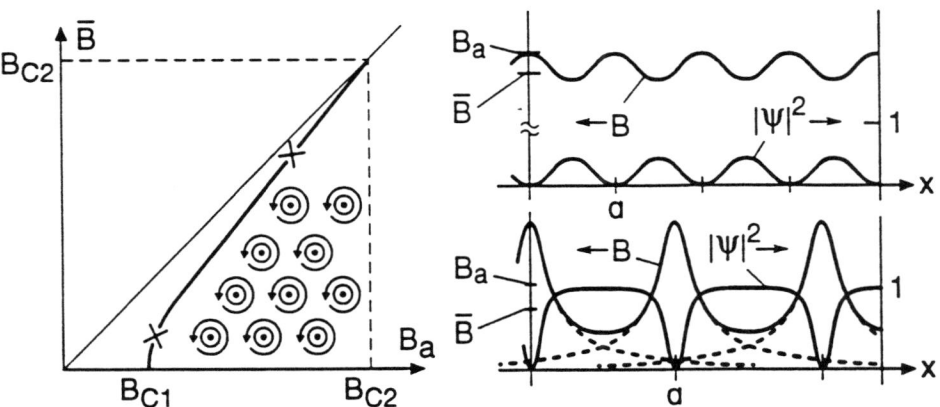

Figure 1: The flux-line lattice (FLL) in type II superconductors: *Left:* Average internal magnetic field $\bar{B}$ versus external field B_a. *Right:* Profiles of magnetic field B and order parameter $|\psi|^2$ at large (*top*) and small (*bottom*) applied fields B_a, corresponding to the two crosses in the curve $\bar{B}(B_a)$ at left. The FLL is shown schematically in the insert.

parameter $|\psi|^2$ is nearly constant except in the vortex core where it goes to zero in a tube of radius $r_c \approx \sqrt{2}\xi$. Its spatial average is $\langle|\psi|^2\rangle \approx 1 - b$ with $b = B/B_{c2}$. For $\kappa > 2$ and $B > 4B_{c1}$ and zero demagnetizing effects (long specimen in parallel field) one has a very small *reversible* magnetization, $-M = B_a - B \approx (B_{c2} - B)/2\kappa^2 \ll B$. For $\kappa > 2$ and $B < 0.25 B_{c2}$ the overlap of the vortex cores is very weak, thus $|\psi^2| \approx 1$ outside the cores, and the GL theory reduces to the London theory. In this region the magnetic field $\mathbf{B}(\mathbf{r})$ of an arbitrary arrangement of parallel vortices is the linear superposition of isolated vortex fields $B_v(r)$. The logarithmic divergence of the London $B_v(r)$ at the vortex center $r = 0$ is rounded over the GL vortex core radius, $B_v(r) \approx (\Phi_0/2\pi\lambda^2)K_0[(r^2 + 2\xi^2)^{1/2}/\lambda]$ [7] where $K_0(x)$ is a modified Bessel function. The free energy of this system is the sum over the (repulsive) pairwise interation of all vortices $V = \Phi_0 B_v(|\mathbf{r}_\mu - \mathbf{r}_\nu|)/\mu_0$, where $\mathbf{r}_\mu$, $\mathbf{r}_\nu$ are the two-dimensional (2D) vortex positions. When the vortices are curved, pairs of vortex line elements at 3D positions $\mathbf{r}_\mu$, $\mathbf{r}_\nu$ interact by a vectorial (non-scalar) potential $V = [\Phi_0^2 \exp(-r/\lambda')/(8\pi\mu_0 \lambda'^2 r)] d\mathbf{r}_\mu \cdot d\mathbf{r}_\nu$ [8] where $r = |\mathbf{r}_\mu - \mathbf{r}_\nu|$. The total energy of the vortex arrangement is obtained by integrating this potential along the flux lines and summing over all vortex pairs. These transparent London results, valid for $b < 0.25$, can be extended to higher inductions [8] by means of the GL theory. This increases the range of the magnetic repulsion $\lambda \to \lambda' = \lambda/\langle|\psi|^2\rangle^{1/2} \approx \lambda/(1-b)^{1/2}$ and adds an attractive "condensation energy term" of range $\xi' = \xi/(2 - 2b)^{1/2}$ which originates from the overlap of the vortex cores.

Besides having a higher transition temperature T_c, the HTSC differ from conventional superconductors by their short coherence length and by their pronounced anisotropy. In these notes only monocrystalline HTSC will be considered in order to avoid the additional complications connected with the granular structure of most larger-size ceramic specimens.

2. Anisotropic London Theory

HTSC to a good approximation exhibit uniaxial crystal symmetry and are characterized by magnetic penetration depths for currents in the ab-plane, λ_{ab}, and along the c-axis, λ_c, and by coherence lengths ξ_{ab} and ξ_c. The anisotropy ratio $\Gamma = \lambda_c/\lambda_{ab} = \xi_{ab}/\xi_c$ is ≈ 5 for YBCO and > 60 for Bi- and Tl-based HTSC. The London equation for the magnetic induction $\mathbf{B}(\mathbf{r})$ in a uniaxial superconductor containing arbitrarily arranged straight or curved flux lines with one quantum of flux Φ_0 and the free energy F of this system read

$$\mathbf{B} + \nabla \times [\Lambda \cdot (\nabla \times \mathbf{B})] = \Phi_0 \sum_i \int d\mathbf{r_i}\, \delta_3(\mathbf{r} - \mathbf{r_i}). \tag{1}$$

$$F = \frac{1}{2\mu_0}\int \left[\mathbf{B}^2 + (\nabla \times \mathbf{B})\Lambda(\nabla \times \mathbf{B})\right] d^3 r. \tag{2}$$

Here the tensor Λ has the components $\Lambda_{\alpha\beta} = \Lambda_1 \delta_{\alpha\beta} + \Lambda_2 c_\alpha c_\beta$ where c_α are the cartesian components of the unit vector $\hat{\mathbf{c}}$, (α, β) denote (x, y, z), and $\Lambda_1 = \lambda_{ab}^2$ and $\Lambda_2 = \lambda_c^2 - \lambda_{ab}^2$. For *isotropic* superconductors one has $\Lambda_1 = \lambda^2$, $\Lambda_2 = 0$, $\Lambda_{\alpha\beta} = \lambda^2 \delta_{\alpha\beta}$, and the l.h.s. of (1) becomes $\mathbf{B} + \lambda^2 \nabla^2 \mathbf{B}$ since $\mathrm{div}\mathbf{B} = 0$. The integral in (1) is along the position of the ith flux line which may be parametrized as $\mathbf{r}_i(z) = [x_i(z); y_i(z); z]$; $\delta_3(\mathbf{r})$ is the three-dimensional

(3D) delta function. The solution of (1) and (2) may be written as [9, 10]

$$B_\alpha(\mathbf{r}) = \Phi_0 \sum_i \int dr_i^\beta \, f_{\alpha\beta}(\mathbf{r} - \mathbf{r}_i) \tag{3}$$

$$F = \frac{\Phi_0^2}{2\mu_0} \sum_i \sum_j \int dr_i^\alpha \int dr_j^\beta \, f_{\alpha\beta}(\mathbf{r}_i - \mathbf{r}_j) \tag{4}$$

$$f_{\alpha\beta}(\mathbf{r}) = \int \exp(i\mathbf{k}\mathbf{r}) f_{\alpha\beta}(\mathbf{k}) \frac{d^3k}{8\pi^3} \tag{5}$$

$$f_{\alpha\beta}(\mathbf{k}) = \frac{\exp(-2g)}{1 + \Lambda_1 k^2} \left(\delta_{\alpha\beta} - \frac{q_\alpha \, q_\beta \, \Lambda_2}{1 + \Lambda_1 k^2 + \Lambda_2 q^2} \right) \tag{6}$$

$$g(k, q) = \xi_{ab}^2 q^2 + \xi_c^2 (k^2 - q^2). \tag{7}$$

Here $\mathbf{q} = \mathbf{k} \times \hat{\mathbf{c}}$, $\Lambda_1 = \lambda_{ab}^2$ and $\Lambda_2 = \lambda_c^2 - \lambda_{ab}^2 \geq 0$. The integrals in (3) and (4) are along the ith and jth flux lines. The integration in (5) is over all k-space. The tensor function $f_{\alpha\beta}(\mathbf{r})$ (5) gives both the interaction potential between the line elements of the flux lines in (4) and the source field generated by a vortex segment $d\mathbf{r}_i$ in (3). Due to the tensorial character of $f_{\alpha\beta}$, this source field in general is not parallel to $d\mathbf{r}_i$. The factor $\exp(-2g)$ provides an elliptical cutoff at large $\mathbf{k}$ or small $\mathbf{r}$ and can be derived from GL theory with anisotropic coherence length ($\xi_c = \xi_{ab}/\Gamma$). This cutoff originates from the finite vortex-core radius and means that the 3D delta function $\delta_3(\mathbf{r})$ in (1) is replaced by a 3D gaussian of width $\sqrt{2}\xi_{ab}$ along a and b and width $\sqrt{2}\xi_c$ along the c-axis. As shown recently by Carneiro et al. [11], the tensorial potential (6) may formally be replaced by a diagonal potential since the r.h.s. of (1) is divergence-free because vortex lines cannot end inside the bulk.

For *isotropic* superconductors one has $\lambda_{ab} = \lambda_c = \lambda$, thus $\Lambda_2 = 0$ and $f_{\alpha\beta} = \delta_{\alpha\beta} \exp(-r/\lambda)/(4\pi r \lambda^2)$; this means the source field is now spherically symmetric, and the interaction between two line elements contains the scalar product $dr_i^\alpha dr_j^\beta \delta_{\alpha\beta} = d\mathbf{r}_i \cdot d\mathbf{r}_j = dr_i \cdot dr_j \cos\alpha$ where α is the angle between these vortex segments. The interaction thus *vanishes* when $\alpha = \pi/2$ and becomes even attractive when $\alpha > \pi/2$. For anisotropic superconductors this magnetic attraction occurs only at larger angles $\alpha = 2\Theta$ close to π, when the flux lines are almost antiparallel and almost in the ab-plane. The spontaneous tilting of the vortices, which prefer to be in the ab-plane and antiparallel to each other, facilitates vortex cutting [12].

Since the discovery of HTSC numerous authors have solved the anisotropic London equations (1, 2) in particular cases to calculate energy, magnetization, shear moduli, and field profiles for the static FLL when B or B_a are arbitrarily oriented. Unexpected results were obtained, *e.g.*, field reversal or attraction of vortices when B is at an oblique angle, or multiple B_{c1} values and metastable configurations of the FLL. For some of these references see [6c, 9-11].

3. Non-Local Elasticity of the Flux-Line Lattice

The energy of small distortions of the flux-line lattice caused by its structural defects, by pinning, or by thermal fluctuations, is conveniently calculated from linear elasticity theory.

For larger deformations (*e.g.* plastic shear) the energy expressions from London theory (4) or GL theory [8] may be used. Since typically the range of the vortex interaction λ exceeds the vortex spacing a, the elastic energy caused by a local distortion of the FLL is contained in a sphere with radius of several λ. This means the elastic response of the FLL is *nonlocal;* the elastic moduli depend on the length scale of the distortion field, *e.g.*, on the wavelengths $2\pi/k$ of the Fourier components of an arbitrary strain field. This *dispersion of the elastic moduli* means that the elastic energy of compressional and tilt waves in the FLL is no longer proportional to $u^2 k^2$, *i.e.* to the strain or stress squared, but at $k \gg 1/\lambda$ becomes proportional to u^2 (u is the vortex displacement). The nonlocal elasticity of the FLL was derived from the isotropic GL and London theories in [13] (summarized in [8, 5]) and recently also from the anisotropic GL [14] and London [15-22] theories. The essential result is that the FLL is *much softer* for short-wavelength compression (modulus c_{11}) or tilt (modulus c_{44}) than it is for uniform compression or tilt. The shear modulus c_{66}, however, exhibits essentially no dispersion.

Particularly important for application is the dispersion of the tilt modulus $c_{44}(k)$ [14-17]. For short tilt wavelengths the tilt modulus is related to the self energy $J(k) \leq B_{c1}\Phi_0$ and line tension $P(k) = c_{44}(k)\Phi_0/B$ of an isolated flux line as discussed in [13b] and in [16]. For anisotropic superconductors the self energy J and the line tension P do *not* coincide as they do in isotropic superconductors. In fact, the self energy is *maximum* for flux lines along c, whereas the line tension has its *minimum* there. Such a behavior is known also for dislocations in anisotropic crystal lattices. For $k = 0$ (straight vortices) one has $P(\Theta) = J(\Theta) + J''(\Theta)$ where Θ is the vortex tilt angle away from the c-axis (or in general, away from the equilibrium direction). If one puts $J = \Phi_0^2 \ln \kappa / 4\pi \mu_0 \lambda_{ab}^2 = 1$ for $B\|c$ ($\Theta = 0$), then one has for $B\|c$ $J = 1$, $P = 1/\Gamma^2$ (since the two terms in $P = J + J''$ nearly cancel), and for $B \perp c$ one has $J = 1/\Gamma$, $P = \Gamma$ for tilt *out of* the ab-plane and $P = J = 1/\Gamma$ for tilt *in* the ab-plane.

Elasticity theory is conveniently formulated in terms of the elastic matrix $\Phi_{\alpha\beta}(\mathbf{k})$, the energy of an elastic mode per unit volume and per unit displacement squared. In general, the elastic matrix of a lattice is *periodic* in k-space. $\Phi_{\alpha\beta}(\mathbf{k})$ for the FLL may be expressed as a sum over all reciprocal lattice vectors $\mathbf{Q}$ of the FLL [13] ($B\|z$, $\alpha, \beta = x, y$)

$$\Phi_{\alpha\beta}(\mathbf{k}) = \frac{B^2}{2\mu_0} \sum_{\mathbf{Q}} [g_{\alpha\beta}(\mathbf{Q} + \mathbf{k}) - g_{\alpha\beta}(\mathbf{Q})] . \tag{8}$$

Anisotropic London theory yields [15]

$$g_{\alpha\beta}(\mathbf{k}) = k_z^2 f_{\alpha\beta}(\mathbf{k}) + k_\alpha k_\beta f_{zz}(\mathbf{k}) - 2k_z k_\beta f_{z\alpha}(\mathbf{k}) \tag{9}$$

with $f_{\alpha\beta}(\mathbf{k})$ from (6). The last term in (9), $-2k_z k_\beta f_{z\alpha}(\mathbf{k})$, was first obtained by Sardella [19-21]. This term influences the compression and shear moduli of the FLL when $\mathbf{B}$ is not along the c-axis, but it does not enter the tilt moduli.

In continuum approximation, valid for $k_\perp \ll k_{BZ}$, one has

$$\Phi_{\alpha\beta}(\mathbf{k}) = [\, c_{11}(\mathbf{k}) - c_{66} \,] k_\alpha k_\beta + \delta_{\alpha\beta} [\, c_{66}\, k_\perp^2 + c_{44}(\mathbf{k}) k_z^2 \,] \tag{10}$$

where $k_\perp^2 = k_x^2 + k_y^2$, $k^2 = k_\perp^2 + k_z^2$, and $k_{BZ} = (4\pi B/\Phi_0)^{1/2} \approx \pi/a$ is the radius of the first Brillouin zone (BZ). Eq. (10) applies to isotropic and uniaxially anisotropic superconductors with $B\|c$-axis. Combining (6) to (10) with ref. [14] one obtains the compression, tilt, and shear moduli for $B\|c\|z$, large GL parameter $\kappa = \lambda_{ab}/\xi_{ab} \gg 1$, and arbitrary reduced induction $0 < b = B/B_{c2} < 1$ [14-18]:

$$c_{11}(\mathbf{k}) = \frac{B^2}{\mu_0} \frac{1 + \lambda_c'^2 k^2}{(1 + \lambda_{ab}'^2 k^2)(1 + \lambda_c'^2 k_\perp^2 + \lambda_{ab}'^2 k_z^2)} \qquad (11)$$

$$c_{44}(\mathbf{k}) = \frac{B^2}{\mu_0}\left[\frac{1}{1 + \lambda_c'^2 k_\perp^2 + \lambda_{ab}'^2 k_z^2} + \frac{f(k_z)}{\lambda_{ab}^2 k_{BZ}^2}\right] \qquad (12)$$

$$f(k_z) = \frac{1}{2\Gamma^2}\ln\frac{\xi_c^{-2}}{\lambda_{ab}^{-2} + k_z^2 + \Gamma^2 k_0^2} + \frac{\ln[1 + k_z^2/(\lambda_{ab}^{-2} + k_0^2)]}{2k_z^2 \lambda_{ab}^2} \quad (b \ll 1)$$

$$f(k_z) = \frac{1-b}{\Gamma^2} \quad (b \approx 1)$$

$$c_{66} \approx B\Phi_0(1-b)^2/(16\pi\lambda_{ab}^2\mu_0) \qquad (13)$$

with $k_0 \approx k_{BZ}$ a cutoff radius, $b = B/B_{c2}$, and $\lambda_c' = \Gamma\lambda_{ab}' \approx \lambda_c/(1-b)^{1/2}$. The increase of the effective interaction range $\lambda \to \lambda' = \lambda/\langle|\psi|^2\rangle \approx \lambda/(1-b)^{1/2}$ follows from GL theory and reflects the increase of the magnetic penetration depth when the density of the Cooper pairs goes to zero as $B \to B_{c2}$. As stated above, in HTSC typically $b \ll 1$. Note that only when non-locality is considered do all moduli vanish at B_{c2} as it is expected from continuity arguments. Tilt and compressional moduli for $B \perp c$ are given in [15-16] and for arbitrary angle Θ in [19-21].

The c_{11} (11) and the first term in (12) originate from the vortex–vortex interaction [term $Q = 0$ in (8)]. The last term in (12) originates mainly from the interaction of each vortex line with itself and may thus be called isolated-vortex contribution [16] [integral over the $Q \neq 0$ in (6) cut off at k_0]; this term is important for $B \to 0$ or $k \approx k_{BZ}$ and was constructed such as to yield the correct limits of c_{44} for $b \to 0$, $b \to 1$, $k \to 0$, $k_z \to \infty$. Due to the k_z-dependence of this term the exact tilt modulus even of isotropic superconductors depends on $\mathbf{k}$ rather than on $k = |\mathbf{k}|$. This "correction term" equals PB/Φ_0 where $P(k_z)$ is the line tension of an isolated flux line $\|c$. The second term in (14), given first in [17, 22] and for isotropic superconductors in [13b], yields a finite flux-line tension $P = \Phi_0^2/(8\pi\lambda_{ab}^2\mu_0)$ for $k = 0$ in the limit $\Gamma \to \infty$. This coincides with the line tension of a stack of "pancake vortices" in a material with isolated superconducting layers [23, 24], see Section 5.1.

4. Thermal Fluctuation and Melting of the 3D Flux-Line Lattice

A convenient, though not generally proven criterion for the melting of a lattice is to look at the amplitude of the thermal fluctuation and to compare it with the lattice spacing a (Lindeman criterion). The thermal fluctuations $\langle u^2 \rangle$ of the flux-line positions were first calculated from local elasticity in [25] and in the extreme nonlocal limit in [26]. The correct calculation [14, 22, 27] yields a $\langle u^2 \rangle$ which practically coincides with [26] but is typically much larger than the local result [25]. This enhancement is caused by the dispersion of the tilt modulus $c_{44}(\mathbf{k})$ (12). In general,

$$\langle u^2 \rangle = k_B T \int_{BZ} \frac{d^3 k}{8\pi^3}\left[\Phi_{xx}^{-1}(\mathbf{k}) + \Phi_{yy}^{-1}(\mathbf{k})\right]. \qquad (14)$$

In the continuum approximation (10) this gives

$$\langle u^2 \rangle \approx k_B T \int_0^{k_{BZ}} dk_\perp k_\perp \int_0^\infty \frac{dk_z/2\pi^2}{c_{66}k_\perp^2 + c_{44}(\mathbf{k})k_z^2} \tag{15}$$

$$\langle u^2 \rangle \approx \left(\frac{k_B^2 T^2 \mu_0}{4\pi B \Phi_0 c_{66}}\right)^{1/2} \times \left(\frac{B\kappa^2/2}{B_{c2} - B}\right)^{1/2} \times \frac{\lambda_c}{\lambda_{ab}}. \tag{16}$$

For $B \ll B_{c2}$ with $c_{66} = B\Phi_0/(16\pi\mu_0\lambda_{ab}^2)$ inserted, (16) yields $\langle u^2 \rangle \approx k_B T \mu_0 \lambda_{ab} \lambda_c (4\pi/B\Phi_0^3)^{1/2}$. The fluctuation (16) was written in the form: Local result [25], times a typically large nonlocal correction factor $(B\kappa^2/2B_{c2})^{1/2} \approx (B\ln\kappa/4B_{c1})^{1/2} \gg 1$, times the large anisotropy ratio $\lambda_c/\lambda_{ab} = \Gamma$.

In spite of the large thermal fluctuation (15) it is not quite clear at present whether a (hypothetical) pin-free 3D FLL has a sharp melting transition and what the properties of this line-liquid would be. The problem is even more complex when pinning is included. The Lindemann criterion $\langle u^2 \rangle^{1/2}/a = c_L \approx 0.1\ldots 0.2$ and similar conditions for melting [27, 28] yield a rather low 3D "melting temperature" T_M^{3D}. Note that the Lindemann criterion does not work in the ideal 2D FLL of thin films since $\langle u^2 \rangle$ diverges there. Finite pinning removes this divergence. Formally one may write $\langle u^2 \rangle_{film} = (k_B T/4\pi^2) \int_{BZ} d^2k_\perp/(c_{66}k_\perp^2 + \alpha_L)$ where d is the film thickness and α_L is the elastic pinning restoring force density (Labusch parameter) which in general has to be determined self-consistently by a theory of collective pinning.

Recently, numerous authors have tackled the FLL melting problem. Monte Carlo simulations of FLs interacting by the correct 3D potential [9-11] with periodic boundary conditions in all three directions [29] or threading a stack of superconducting layers [30] were performed. An interesting approach to melting of a 3D lattice of point vortices in layered HTSC is based on the structure function of liquids and solids [31]. A FLL instability and melting are derived from elasticity theory by Glyde et al. [32]. Various xy-models of the FLL have been considered [33-36]. The Gaussian fluctuations around the 3D FLL were investigated in detail by Ikeda et al. [37]. For analytical calculations near B_{c2} and Monte Carlo simulations of FLL melting in 2D superconductors see [38]. The correct 3D anisotropic interaction (4) is used in recent simulations [39] and analytical calculations [40].

Recent resistivity data [41-43] as a function of T, B, and J appear to indicate a melting transition in weak-pinning YBCO. As will be discussed in Section 8, several experiments which were interpreted as evidence for a phase transition can be explained by the rather abrupt onset of thermally activated depinning when T or B are increased, in particular some ac experiments and experiments on vibrating superconductors in a magnetic field [44-46], see also the review [47]. However, other such experiments on very weak-pinning twin-boundary-free YBCO crystals show a very sharp damping peak which may be identified with FLL melting [48].

5. Layered Structure of HTSC

5.1. PANCAKE VORTICES AND JOSEPHSON VORTICES

HTSC may be considered as consisting of superconducting CuO-layers of spacing s and thickness $d \ll s$ which interact with each other by Josephson coupling [49-52]. This layered structure gives rise to two novel phenomena: pancake vortices and Josephson vortex lines, Fig. 2.

(a) When B_a is nearly along the c-axis ($\Theta = 0$) the flux lines formally consist of 2D point vortices, or *pancake vortices* [23] which have their singularity (suppressed order parameter ψ) only in one layer. The field of a single point vortex at $\mathbf{r}_\mu = 0$ (Fig. 2) is confined to a layer of thickness $\approx 2\lambda_{ab}$; its z-component is rotationally symmetric $B_z(r) = (s\Phi_0/4\pi\lambda_{ab}^2 r)\exp(-r/\lambda_{ab})$, and its in-layer (radial) component follows from div$\mathbf{B}=0$ as $B_\perp(\mathbf{r}) = (s\Phi_0 z/4\pi\lambda_{ab}^2 r_\perp)[\exp(-|z|/\lambda_{ab})/|z| - \exp(-r/\lambda_{ab})/r]$ [$\mathbf{r}_\perp = (x;y)$, $\mathbf{r} = (x;y;z)$]. In a vortex *line*, the radial components of the point-vortex fields cancel and only the z-component survives. A point vortex in plane $z = z_n$ contributes a flux $(\Phi_0 s/2\lambda_{ab})\exp(-z_n/\lambda_{ab})$ to the flux through the plane $z = 0$. This means the flux of one point vortex is $\ll \Phi_0$. The self energy of a single point vortex diverges logarithmically with the layer extension; this fact does not disturb, since the energy of a stack of such point vortices, a vortex line, has an energy per *unit length* and thus also diverges for an infinitely long vortex. What is more interesting is that the interaction between point vortices is exactly logarithmic up to arbitrarily large distances, in contrast to the interaction of point vortices in an isolated film [53], which is logarithmic only up to distances $\lambda_{film} = 2\lambda_{bulk}^2/d$ where d is the film thickness [2, 23]. Point vortices in the same layer repel each other, and those in different layers attract each other. This is the reason why a regular FLL has the lowest energy.

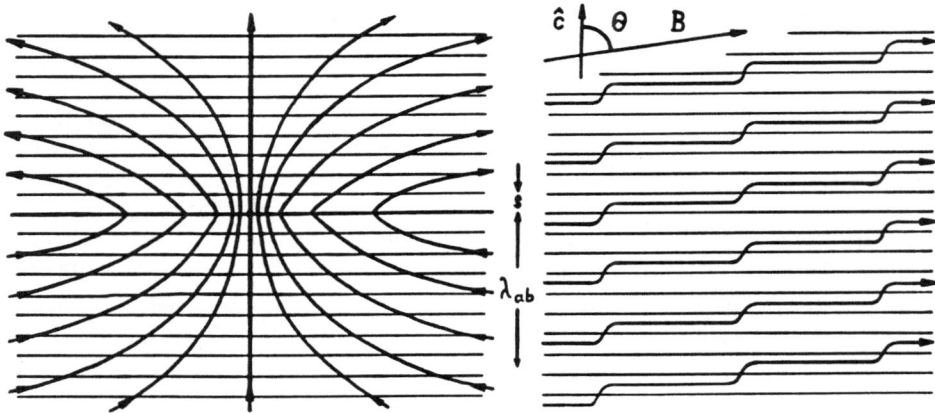

Figure 2: *Left:* The magnetic field lines of a point vortex or pancake vortex in a superconductor consisting of almost isolated superconducting layers indicated by horizontal lines. In contrast to the nearly radial field lines of a point vortex in a thin film, the field lines in a layered structure curve such that they become *parallel* to the layers at large distances from the vortex core. The field is thus confined to a disk of thickness $\approx 2\lambda_{ab}$. *Right:* When the applied field is nearly along the superconducting CuO planes the flux lines form kinks which consist of point vortices in the layers and Josephson vortices between the layers.

(b) When B_a is nearly in the ab-plane the vortex core prefers to run between the CuO layers [55]. For weak coupling between the layers the vortex lines in the ab-plane are so called *Josephson vortices*. These have no core in the usual sense of a vanishing order parameter ψ since ψ is assumed to be zero in the space between the layers anyway. Therefore, if $\Theta = \pi/2$ then ψ vanishes nowhere in the CuO planes. If $\Theta \approx \pi/2$, the flux lines form steps or kinks consisting of point vortices separated by Josephson vortices [24], Fig. 2. The width of the Josephson core is $\lambda_J = \Gamma s$ (the Josephson length) and its thickness is s. Its aspect ratio is thus $\Gamma = \lambda_J/s = \xi_{ab}/\xi_c$ like for a London vortex core in the ab-plane (width ξ_{ab}, thickness ξ_c).

5.2. LAWRENCE-DONIACH MODEL

A very useful phenomenological theory of layered superconductors is the Lawrence-Doniach (LD) theory [49, 50], which contains the anisotropic GL and London theories as limiting cases when the coherence length $\xi_c(T)$ exceeds the layer spacing s. The LD model describes the superconducting layers (*e.g.*, the CuO layers) by a two-dimensional (2D) GL parameter $\psi_n(x,y)$ (n = layer index, $z_n = ns$) and replaces the gradient along z (along $\hat{c}$) by a finite difference. The LD free energy functional reads

$$F = s\sum_n \int d^2r_\perp \left[-\alpha|\psi_n|^2 + \frac{\beta}{2}|\psi_n|^4 + \frac{\hbar^2}{2m}|(i\nabla + \frac{2e}{\hbar}\mathbf{A})\psi_n|^2 \right.$$

$$\left. + \frac{\hbar^2}{2Ms^2}|\psi_{n+1} - \psi_n \exp(iI_n)|^2 \right] + \int \frac{\mathbf{B}^2}{2\mu_0} d^3r \quad (17)$$

$$I_n(\mathbf{r}_\perp) = \frac{2e}{\hbar}\int_{ns}^{ns+s} A_z \, dz \approx \frac{2e}{\hbar}A_z s. \quad (18)$$

Here $\mathbf{r}_\perp = (x;y)$, $\alpha \propto T_c - T$ and β are the usual GL coefficients, m and M are the effective masses of Cooper pairs, $\mathbf{B} = \text{rot}\mathbf{A}$, $2e/\hbar = 2\pi/\Phi_0$, $A_z = \hat{z}\mathbf{A}$, and the integral in (18) is along a straight line. One has $\xi_{ab}^2 = \hbar^2/2m\alpha$, $\xi_c^2 = \hbar^2/2M\alpha$, $\lambda_{ab}^2 = m\beta/4\mu_0 e^2\alpha$, and $\lambda_c^2 = M\beta/4\mu_0 e^2\alpha$, thus $\lambda_c^2/\lambda_{ab}^2 = \xi_{ab}^2/\xi_c^2 = M/m = \Gamma \gg 1$ for HTSC. For $\xi_c \gg s$ the difference in (17) may be replaced by the gradient $(\partial/\partial z - 2ieA_z/\hbar)\psi(x,y,z)$; this yields the anisotropic GL theory. In the opposite limit, $\xi_c \ll s$, (17) describes weakly Josephson coupled superconducting layers.

The LD theory can be derived from a tight binding formulation of the BCS theory [49, 50] and thus is valid near T_c, but as many phenomenological theories the LD theory is more general than any microscopic theory from which it may be derived. In particular, if $\psi_n(x,y) = \text{const.} \times \exp[i\phi_n(x,y)]$ is assumed the resulting London-type LD theory applies to all temperatures $0 < T < T_c$; one now has $F = F\{\phi_n(\mathbf{r}_\perp), \mathbf{A}(\mathbf{r})\}$ [51],

$$F = \frac{s\Phi_0}{4\pi^2\mu_0\lambda_{ab}^2}\sum_n \int d^2r_\perp \left[\frac{1}{2}(\nabla\phi_n + \frac{2\pi}{\Phi_0}\mathbf{A})^2 + \frac{1-\cos\delta_n}{\Gamma^2 s^2}\right] + \int \frac{\mathbf{B}^2}{2\mu_0} d^3r \quad (19)$$

$$\delta_n(\mathbf{r}_\perp) = \phi_n - \phi_{n+1} - I_n. \quad (20)$$

The magnetic field of an arbitrary arrangement of point vortices at 3D positions $\mathbf{r}_\mu$ is in the limit of zero Josephson coupling, corresponding to the limit $\lambda_c \to \infty$ [51, 55],

$$\mathbf{B}(\mathbf{r}) = s\Phi_0 \int \frac{d^3k}{8\pi^3} \frac{\hat{\mathbf{z}} - \mathbf{k}_\perp k_z/k_\perp^2}{1 + k^2\lambda_{ab}^2} \sum_\mu \exp[i\mathbf{k}(\mathbf{r} - \mathbf{r}_\mu) - 2\xi_{ab}^2 k_\perp^2] \qquad (21)$$

where $\mathbf{k} = (\mathbf{k}_\perp; k_z)$, $\mathbf{k}_\perp = (k_x; k_y)$, $\hat{\mathbf{z}} = \hat{\mathbf{c}}$, and I have introduced the cutoff (7) (note that $q^2 = k_\perp^2$). Even for finite λ_c the field (21) is a good approximation when B_a is along z or close to it. The free energy of the point-vortex system is composed of their magnetic interaction energy F_M and the Josephson coupling energy F_J caused by the difference of the phases of ψ_n in adjacent layers, $F = F_M + F_J$ [51],

$$F_M = \frac{s^2\Phi_0^2}{2\mu_0} \int \frac{d^3k}{8\pi^3} \frac{k^2/k_\perp^2}{1 + k^2\lambda_{ab}^2} \sum_\mu \sum_\nu \exp[i\mathbf{k}(\mathbf{r}_\mu - \mathbf{r}_\nu) - 2\xi_{ab}^2 k_\perp^2] \qquad (22)$$

$$F_J = \frac{\Phi_0^2}{4\pi^2\mu_0 s\lambda_{ab}\lambda_c} \int dx \int dy \sum_n (1 - \cos\delta_n) \qquad (23)$$

In general, F_J is not easily calculated since the gauge-invariant phase differences have to be determined by minimizing $F_M + F_J$. The exact solution of the LD theory for a single straight or distorted flux line and for periodic FLL for all orientations of $\mathbf{B}$ has been obtained recently by Bulaevskii et al. [56]. For a more physical discussion see [24].

5.3. PHASE TRANSITIONS IN LAYERED SUPERCONDUCTORS

Numerous phase transitions of the vortex lattice in layered HTSC have been predicted when T or B is changed. I summarize some of these ideas in short:

1. With increasing T a transition from 2D to 3D superconductivity is expected [49, 50] when $\xi_c(T) = \xi_c(0)/(1 - T/T_c)^{1/2}$ equals the layer spacing s. For YBCO this should occur a few degrees below $T_c = 92.5$ K, and for BSCCO very close to T_c.

2. In zero field at finite temperature, vortex-antivortex pairs nucleate spontaneously in a single superconducting layer or film. These pairs dissociate at a temperature $T_{BKT} = \Phi_0^2 s/(8\pi k_B\mu_0\lambda_{ab}^2)$ since both their interaction and their entropy are proportional to $\ln(R/\xi_{ab})$ where R is the specimen size (Berezinskiĭ-Kosterlitz-Thouless transition) [51, 52, 57].

3. A vortex line threading the CuO-layers "evaporates" into single pancake vortices at the same temperature T_{BKT} [23, 48], i.e., its thermally averaged radius diverges when $T \to T_{BKT}$. This transition has recently been investigated in great detail in [59] where the coincidence of the two temperatures is explained.

4. In analogy to the 2D BKT-transition, 3D vortices at $B = 0$ may nucleate spontaneously in the bulk since their fluctuations give an entropy contribution which reduces the free energy $F = U - TS$ [22, 60]. The thermal fluctuation of the vortices influences the magnetization M such that the curves $M(B_a, T)$ for all applied fields B_a cross at the same temperature T^* close to T_c where $M = M^* = 4\pi k_B T^*/(\Phi_0 s)$ [60]. This result is also obtained from GL theory for $B_a \approx B_{c2}$ by Tešanović et al. [38].

5. A 2D vortex lattice of spacing a in a film of thickness d melts at $T_m^{2D} = a^2 dc_{66}/(4\pi k_B)$ due to the spontaneous nucleation of dislocations of energy $U = (da^2 c_{66}/4\pi)\ln(R/a)$ and

entropy $S = k_B \ln(R/a)$ (dislocation mediated melting) [61]. In layered HTSC a similar transition is expected, with the film thickness d replaced by the layer spacing s, $T_m^{2D} = a^2 s c_{66}/(4\pi k_B) = \Phi_0^2 s/(64\pi^2 k_B \mu_0 \lambda_{ab}^2)$. Thus, considering the temperature dependence of $\lambda_{ab} \propto (1 - T^2/T_c^2)^{-1/2}$, one has $T_m^{2D}/T_{BKT} = \lambda_{ab}^2(T_m^{2D})/[8\pi\lambda_{ab}^2(T_{BKT})]$ [51, 52].

6. Thermal fluctuations of the vortex positions in the layers lead to the destruction of phase coherence between adjacent layers. Glazman and Koshelev [22, 62] find that the fluctuation of the phase differences δ_n (20) becomes large, $\langle \delta_n^2 \rangle \approx 1$ (average over x, y, n, and time) when T reaches $T_0 \approx a\Phi_0^2/(4\pi \mu_0 k_B \lambda_{ab}\lambda_c) \propto B^{-1/2}$ which is typically smaller than the 2D melting temperature T_m^{2D}. They predict that fields along the ab-plane cannot be shielded for $T > T_0$.

7. Daemen et al. [63] follow this idea further and suggest that due to these phase fluctuations the Josephson coupling between neighboring layers is renormalized to a smaller value as T increases. As consequences, the effective $\lambda_{c,eff}$ and anisotropy ratio Γ_{eff} increase and the maximum possible current density along the c-axis $J_{z,max} = J_0\langle\sin\delta_n(x,y)\rangle = (\Phi_0/2\pi\mu_0 s \lambda_{c,eff}^2) \propto J_0 \exp[-B/B_D(T)]$ decreases and vanishes at a "decoupling field" $B_D(T) = [8\Phi_0^3/\mu_0 k_B T s e \lambda_c^2(T)]$ (for $\Gamma \ll \lambda_{ab}/s$) or $B_D(T) = [s\Phi_0^3/8\pi^2 \mu_0 k_B T \lambda_{ab}^4(T)] \ln(2\pi\mu_0 k_B T \lambda_{ab}^4/\Phi_0^2 s^3)$ (for $\Gamma \gg \lambda_{ab}/s$). At this *decoupling transition* the fluctuating distance of point vortices in adjacent layers becomes comparable to the flux-line distance, $\langle u^2\rangle k_{BZ}^2 \approx 1$, and therefore the notion of flux lines looses its sense, the flux lines in the FLL have evaporated.

8. In the transitions 2.-7. the vortex field was along the c-axis ($\Theta = 0$). Further phase transitions are expected when **B** is nearly parallel to the a,b-plane ($\Theta \approx \pi/2$). When Θ is very close to $\pi/2$ a "lock-in transition" [24, 56, 64-66] will switch **B** exactly into the ab-plane since the vortex lines gain energy by having their core in between the layers. This means the vortex kinks spontaneously stretch, the point vortices run out from the specimen, only Josephson vortices survive. The lock-in occurs [66] when the internal perpendicular field component B_z reaches a value $B_{lock-in} = (\Phi_0/4\pi\lambda_{ab}^2)\ln(\Gamma d/\xi_{ab})$ for $B_x < B_{cr}$ or $B_{lock-in} = (\Phi_0/2\pi\lambda_{ab}^2)\ln(\Gamma B_{c2}/B_x)$ for $B_x > B_{cr}$. Here $B_{cr} = \sqrt{3}\Phi_0/(2s^2\Gamma)$ is the field parallel to ab at which the cores of the Josephson vortices start to overlap (core width $\lambda_J = \Gamma s$). The cusp resulting from the lock-in transition in the magnetization and in the mechanical torque is difficult to observe since typical c-oriented specimens are flat and have a demagnetizing factor close to unity, which means that the angle between **B** and the ab-plane is much larger than that between $\mathbf{B}_a$ and the ab-plane.

9. The spontaneous nucleation of vortex rings in the ac or bc-planes and phase transitions for B parallel to the layers are studied by Baruch Horovitz [67] using a renormalization group technique.

10. A Berezinskiĭ-Kosterlitz-Thouless transition is predicted by Blatter et al. [68] for the "smectic vortex state" when B is along ab. Here the logarithmic interaction between the two point vortices in a double kink is mediated by two pairs of edge dislocations at a distance $> \lambda_J$. This model explains the observed [69, 70] independence of the dissipation on the angle between B and J when both B and J are in the ab-plane. For other explanations see [70-72].

11. Further transitions occur when vortex pinning is considered. Theories of dislocation-mediated 2D collective pinning and depinning transitions (cf. Section 6.4) are presented in [51, 52]. These transitions are investigated in nice recent experiments on artificially

layered materials like YBCO/PrBCO [73] and Ge/Pb [74], in which the coupling between superconducting layers can be made arbitrarily small.

6. Flux Flow and Flux Pinning

6.1. PINNING OF FLUX LINES

An electric current density $\mathbf{J}$ through the superconductor exerts a Lorentz force density $\mathbf{B} \times \mathbf{J}$ on the flux-line lattice which causes the vortices to move with mean velocity $\mathbf{v}$. This vortex drift dissipates energy and thus generates an electric field $\mathbf{E} = \mathbf{B} \times \mathbf{v}$ where $\mathbf{B}$ is the flux density or magnetic induction in the sample. The dissipation is caused by two effects which give approximately equal contributions: (a) By dipolar currents which surround each moving flux line (eddy currents) and have to pass through the normal conducting vortex core [75]. (b) By the retarded relaxation of the order parameter $\psi(\mathbf{r})$ when the vortex core moves [76]. Since at low B the dissipation of the vortices is additive and since at the upper critical field $B_{c2}(T)$ the flux-flow resistivity ρ_{FF} has to reach the normal conductivity ρ_n, one approximately gets $\rho_{FF} \approx \rho_n B/B_{c2}(T)$. A more quantitative treatment of this flux dissipation uses time dependent Ginzburg-Landau theory. For reviews of flux motion see [77-78], and for extensions to layered and anisotropic superconductors [79-81].

In real superconductors at small current densities $J < J_c$ the flux lines are pinned by inhomogeneities in the material, e.g., by dislocations, vacancies, interstitials, grain boundaries, precipitates, irradiation defects, or by a rough surface, Fig. 3. Only when J exceeds a critical value J_c do the vortices move and dissipate energy [82]. Pinning of flux lines has two important consequences:

(a) The current-voltage curve of a superconductor in a magnetic field is highly nonlinear, with $E = 0$ for $J < J_c$ and $E = \rho_{FF}J$ for $J \gg J_c$. For J slightly above J_c various shapes of $E(J)$ are observed, depending on the type of pinning and on the geometry of the

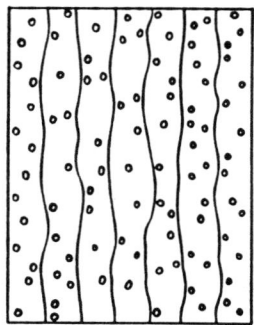
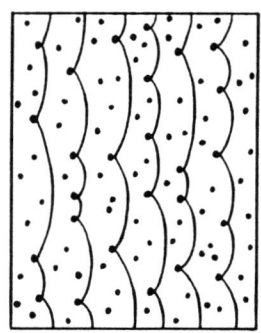
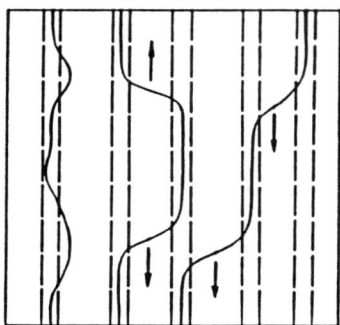

Figure 3: *Left:* Flux lines deformed by weak random pins, e.g., oxygen vacancy clusters. *Middle:* Flux lines pinned by strong point pins. The Lorentz force acts to the right. *Right:* Flux lines pinned by columnar pins can move by the formation of kinks which nucleate by thermal fluctuation or by large current density as described in [99].

sample. Often a good approximation is (with *e.g.* $p = 1$ or 2)

$$E(J) = 2\rho_{FF}[1 - (J_c/J)^p]^{1/p} \quad (J \geq J_c). \tag{24}$$

(b) The magnetization curve $M(B_a)$ exhibits a hysteresis, Fig. 4. When B_a is increased or decreased the magnetic flux enters or exits until a *critical slope* is reached like in a pile of sand, namely, a maximum and nearly constant gradient of $B = |\mathbf{B}|$. More precisely, in this *critical state* the *current density* reaches a maximum value J_c; one has $\mathbf{J} = (\partial H/\partial B)\nabla \times \mathbf{B} \approx \mu_0^{-1}\nabla \times \mathbf{B}$, where $H(B) \approx B/\mu_0$ is the (reversible) magnetic field which would be in equilibrium with the induction $\mathbf{B}$. The critical state is often well described by the "Bean model" [83], which assumes a B-independent J_c and disregards demagnetizing effects; these become important in flat superconductors in perpendicular magnetic field [84-88].

In general, the current density in type-II superconductors may have three different origins: (a) *surface currents* within the penetration depth λ, (b) a *gradient* of the flux-line density, or (c) a *curvature* of the flux lines (or field lines). The latter two contributions are easily seen by writing $\nabla \times \mathbf{B} = \nabla B \times \hat{\mathbf{B}} + B\nabla \times \hat{\mathbf{B}}$ where $\hat{\mathbf{B}} = \mathbf{B}/B$. In bulk samples typically the gradient term dominates, $J \approx \mu_0^{-1}\nabla B$, but in films the current is carried almost entirely by the *curvature* of the flux lines [84-88].

6.2. THERMALLY ACTIVATED DEPINNING

As predicted in 1962 by Anderson [82], thermally activated depinning of flux lines may

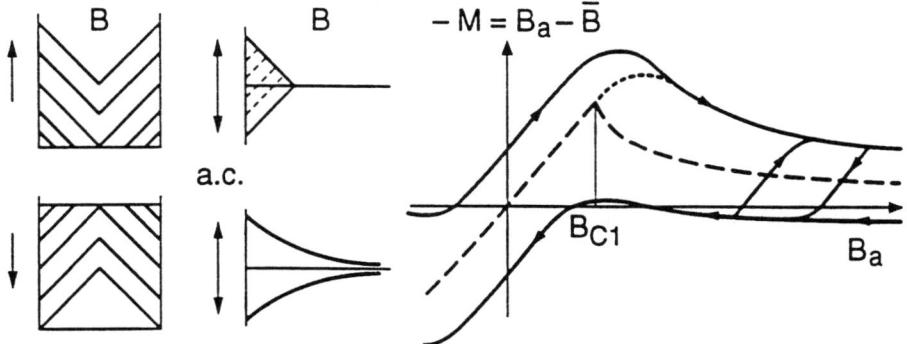

Figure 4: Field profiles in a superconducting cylinder or slab with strong pinning in increasing (*left top*) and decreasing (*left bottom*) longitudinal applied field. In this Bean critical state model the field gradient is constant. Also shown is the field profile caused by an additional weak ac field near the surface in the Bean model (*middle, top*) and in the cases of elastic pinning, viscous drag, or thermally assisted flux flow (*middle, bottom*). Strong pinning leads to hysteretic magnetization curves (*right*) with the irreversible magnetization (solid lines) lying above and below the ideal reversible curve (dashed line).

occur at finite temperatures T. In *conventional superconductors* this effect is observed only close to the transition temperature T_c as *flux creep* [89]. Flux creep occurs in the critical state after the applied magnetic field was changed. The field gradient, and the persistent currents and magnetization, then slowly *decrease with a logarithmic time law*. Formally, this flux creep is equivalent to a highly nonlinear, current dependent flux-flow resistivity, e.g., $\rho \propto \exp(J/J_1)$ or $\rho \propto \exp(-J_2/J)$. Initially, the persistent currents in a ring feel a large ρ, but as the current decays, ρ decreases rapidly and so does the decay rate $-\dot{J} \propto \rho(J)$.

In HTSC thermal depinning is observed in a large temperature interval below T_c. This "giant flux creep" [90-92] occurs mainly because (a) the superconducting coherence length ξ ($\approx$ vortex core radius) is small, (b) the magnetic penetration depth λ is large, and (c) these materials are strongly anisotropic; here $\xi = \xi_{ab}$, $\lambda = \lambda_{ab}$, and $\kappa = \lambda_{ab}/\xi_{ab}$ is used. All three properties decrease the pinning *energy* but tend to increase the pinning *force* as will be discussed now.

Small ξ means that the elementary pinning *energy* U_p of small pins (*e.g.* oxygen vacancies or clusters thereof) is small, of the order of $(B_c^2/\mu_0)\xi^3 = (\Phi_0^2/8\pi^2\mu_0\lambda^2)\xi$. The elementary pinning *force* U_p/ξ, however, is *independent* of ξ in this estimate and is thus not necessarily small in HTSC.

Large λ means that the stiffness of the FLL with respect to shear deformation and to short-wavelength tilt is small. Therefore, the flux lines can better adjust to the randomly positioned pins. This flexibility increases the average pinning force density. (The argument that a soft FLL or a flux-line liquid with vanishing shear stiffness cannot be pinned since it may flow around the pins, does not apply to the realistic situation where there are many more pins than flux lines.) The statistical summation of pinning forces at $T = 0$ [93] and $T > 0$ [94-96] requires the correct (non-local) elasticity theory of the FLL, Section 3.

Large material anisotropy effectively softens the FLL and *increases* the average pinning *force*, but it *decreases* the pinning *energy*. In particular, long columnar pins generated by high-energy (500 MeV) heavy-ion irradiation perpendicular to the CuO planes in YBCO [97] or BSCCO [98] are most effective pins at low T if the flux lines are parallel to the pins. However, at higher T, columnar pins can pin flux lines only in YBCO, but in the very anisotropic BSCCO the flux lines easily break into short segments or point vortices which then depin individually with very small activation energy [98-100], Fig. 3. As a consequence, BSCCO tapes are good superconductors only at $T = 4\,\text{K}$. In principle, if B could be kept strictly parallel to the layers, large J_c and weak thermal depinning could be achieved even at $T = 77\,\text{K}$, but this geometric condition is difficult to satisfy.

6.3. THE KIM-ANDERSON THEORY

A novel feature in HTSC is that a linear (ohmic) resistivity ρ is observed at small current densities $J \ll J_c$ in the region of thermally assisted flux flow (TAFF) [91-92]. Both effects, flux creep at $J \approx J_c$ and TAFF at $J \ll J_c$, are limiting cases of Anderson's [82] general expression for the electric field $E(B, T, J)$ caused by thermally activated flux jumps out of pinning centers, which may be written as [101]

$$E(J) = 2\rho_c J_c \exp(-U/k_B T) \sinh(JU/J_c k_B T) . \qquad (25)$$

In (25) $J_c(B)$ (the critical current density at $T = 0$), $\rho_c(B, T)$ (the resistivity at $J = J_c$), and $U(B,T)$ (the activation energy for flux jumps) are *phenomenological parameters*. The physical idea behind eqn. (25) is that the Lorentz force density $\mathbf{J} \times \mathbf{B}$ acting on the FLL *increases* the rate of thermally activated jumps of flux lines or flux-line bundels along the force, $\nu_0 \exp[-(U-W)/k_B T]$, and *reduces* the jump rate for backward jumps, $\nu_0 \exp[-(U+W)/k_B T]$. Here $U(B,T)$ is an activation energy, $W = JBVl$ the energy gain during a jump, V the jumping volume, l the jump width, and ν_0 is an attempt frequency. All these quantities depend on the microscopic model, which is still controversial, but by defining a critical current density $J_c = JU/W = U/BVl$, only measurable quantities enter. Subtracting the two jump rates to give an effective rate ν and then writing the drift velocity $v = \nu l$ and the electric field $E = vB = \rho J$ one obtains (25).

For large currents $J \approx J_c$ one has $W \approx U \gg k_B T$ and thus $E \propto \exp(J/J_1)$ with $J_1 = J_c k_B T/U$. For small currents $J \ll J_1$ one may linearize the $\sinh(W/k_B T)$ in (25) and gets *ohmic* behavior with a thermally activated linear resistivity $\rho_{TAFF} \propto \exp(-U/k_B T)$. Combining (25) with the usual flux-flow resistivity ρ_{FF} valid at $J \gg J_c$, or with the square-root result [p=2 in (24)] for a particle moving viscously across a one-dimensional sinusoidal potential (see appendix in [102]) one gets (Fig. 5)

$$\rho = (2\rho_c U/k_B T)\exp(-U/k_B T) = \rho_{TAFF} \quad \text{for } J \ll J_1 \quad \text{(TAFF)} \quad (26)$$
$$\rho = \rho_c \exp[(J/J_c - 1)U/k_B T] \propto \exp(J/J_1) \quad \text{for } J \approx J_c \quad \text{(flux creep)} \quad (27)$$
$$\rho = \rho_{FF}(1 - J_c^2/J^2)^{1/2} \approx \rho_{FF} \approx \rho_n B/B_{c2}(T) \quad \text{for } J \gg J_c \quad \text{(flux flow)}. \quad (28)$$

The existence of the linear TAFF regime (26) is confirmed by experiments if B and T are sufficiently large [41, 103-107], while at lower T and B non-linear resistivity is observed

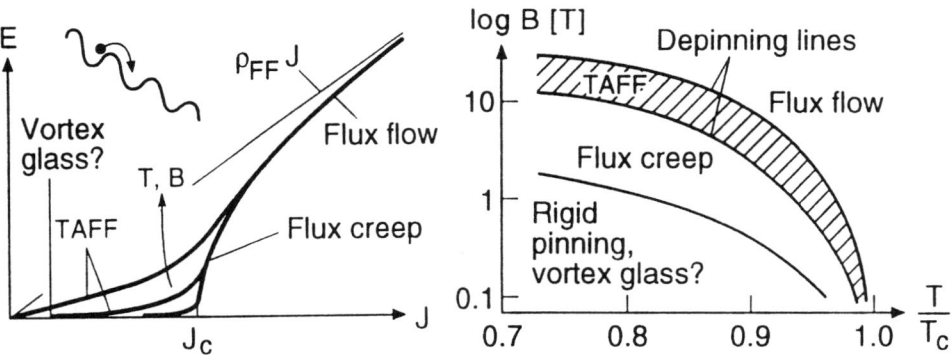

Figure 5: *Left:* Current–voltage curves. E = electric field, J = current density. $J \gg J_c$: flux flow, $E \approx J\rho_n$; $J \approx J_c$: flux creep, $E \propto \exp(J/J_1)$; $J \ll J_1 < J_c$: thermally assisted flux flow (TAFF), $E \approx J\rho_n \exp(-U/k_B T)$, or, at low T: vortex glass, $E \propto \exp[-(J_2/J)^\alpha]$. The insert shows a tilted periodic pinning potential with jumping flux-line bundle. *Right:* Depinning lines in the field–temperature plane separate the region of flux flow (with complete depinning) from the regions of flux creep (with slow logarithmic relaxation) and rigid pinning (with hysteretic behavior). Near the depinning lines (dashed area) thermally assisted flux flow (TAFF) occurs and the resistivity is linear.

[41, 42, 104-110]. It appears today that in the TAFF regime the FLL is in a "liquid" state [96], *i.e.*, it has no shear stiffness; therefore, elastic deformations of the FLL at different points are not correlated. This assumption leads to an activation energy U which does not depend on the current density.

6.4. COLLECTIVE CREEP AND VORTEX GLASS SCALING

Theories of collective pinning [94-96] going beyond Anderson's model (25) predict that the thermally jumping volume V of the FLL depends on the current density J and becomes infinitely large for $J \to 0$. As a consequence also the activation energy diverges, *e.g.*, $U \propto V \propto 1/J^\alpha$ with $\alpha > 0$, thus the resistivity becomes truly zero as $J \to 0$. This result follows for weak random pinning if the FLL is treated as an elastic medium. A diverging activation energy is also obtained in theories of depinning via a kink mechanism of vortices from the space between the CuO layers [72, 111] and from columnar pins [99-100].

A similar result is arrived at by the "vortex glass" picture [112-113]. Its basic idea is that if there is a glass-transition temperature T_G in the vortex-pin system similar to that in theories of spin glasses, then a characteristic length ξ_G (the size of the jumping volume) in the FLL should diverge as $\xi_G \propto |T - T_G|^\nu$ ($\nu \approx 1$) when T approaches T_G. The vortex-glass picture predicts scaling laws, *e.g.*, the electric field should scale as $E\xi_G^{z-1} = f_\pm(J\xi_G^{D-1})$ where $z \approx 4$, D is the spatial dimension, and $f_\pm(x)$ are scaling functions for the regions above and below T_G. For $x \to 0$ one has $f_+(x) = $ constant and $f_-(x) \to \exp(-x^{-\mu})$. At T_G, a power-law current–voltage curve is expected, $E \propto J^{(z+1)/(D-1)}$, thus

$$\rho \propto J^{(z+1)/(D-1)-1} \quad \text{for } T = T_G \tag{29}$$

$$\rho \propto \exp[-(J_2/J)^\alpha] \quad \text{for } T < T_G. \tag{30}$$

In the theory of collective pinning [94, 114] there is no explicit glass temperature, but the picture is similar since collective creep occurs only below a "melting temperature" T_m above which the FLL looses its elastic stiffness. Thus T_m has a similar meaning as T_G. A vortex glass state should not occur in 2D flux-line lattices [51, 52]. More details may be found in [113-116] and in the review paper [21].

Experiments which measure the magnetization decay or the voltage drop with high sensitivity appear to confirm this scaling law in various HTSC in an appropriate range of B and T. For example, by plotting T-dependent creep rates $\dot{M} = dM/dt$ in reduced form, $(1/J)|1 - T/T_G|^{-\nu(z-1)}\dot{M}$ versus $J|1 - T/T_G|^{-2\nu}$, van der Beek at al. [106] in BSCCO measured $T_G = 13.3\,\text{K}$, $z = 5.8 \pm 1$, and $\nu = 1.7 \pm 0.15$, Fig. 6. Very detailed curves $\rho(J)$ for three YBCO samples with different pinning (without and with irradiation with protons or Au ions) are presented by Worthington *et al.* [41] for different fields B_a with T as parameter. In the sample with intermediate pinning two transitions are seen in $\rho(J)$, a "melting transition" at T_m (*e.g.*, $T_m \approx 91.5\,\text{K}$ at $B = 0.2\,\text{T}$, $J = 10^5\,\text{Am}^{-2}$) and a "glass transition" at $T_m \approx 90\,\text{K}$ (above case) or $T_m = 84.92\,\text{K}$ (strong pinning sample, very sharp transition at $B = 4\,\text{T}$, $J \leq 4 \times 10^5\,\text{Am}^{-2}$). The FLL phase in between these two transitions was named "vortex slush".

6.5. CREEP RATES

The decay of shielding currents, or of the magnetization, in principle is completely determined by the geometry and by the resistivity $\rho(T, B, J)$. In superconducting rings and hollow cylinders in axially oriented magnetic field one simply has $\dot{M} \propto \dot{J} \propto \rho(T, B, J)$. Within the Kim-Anderson model (25) the decay of persistent currents can be calculated analytically for all times [117, 118]. A large range of electric fields $E = 10^{-13}$ to 10^{-1} V/m was measured in [109] by combining current–voltage curves with highly sensitive measurements of decaying currents in rings of YBCO films (3 mm diameter, 0.1 mm width, 200 nm thickness).

The current dependent activation energy $U(J)$ may be extracted from experiments by the method of Maley and Willis [110], Fig. 6, see e.g. [106]. The Kim-Anderson model (25) originally means an effective activation energy $\propto J_c - J$, eq. (27), and actually corresponds to a zig-zag shaped pinning potential. As shown by Beasley et al. [89] a more realistic smooth potential yields $U \propto (J_c - J)^{3/2}$. Collective creep theory yields $U \propto 1/J^\alpha$ with $\alpha > 0$ depending on B and the pinning strength. For single-vortex pinning one predicts $\alpha = 1/7$, for short jumps $l < a$ $\alpha \approx 7/9$ [94], and for long jumps $l > a$ $\alpha \approx 1/2$ [95]. Other experiments suggest a logarithmic dependence $U \propto \ln(J_2/J)$ (corresponding to the limit $\alpha \to 0$) for which the creep rate can be calculated analytically [119]. Combining the collective creep result $U(J) = U_c(J_c/J)^\alpha$ for $J \ll J_c$ with the Kim-Anderson formula $U(J) = U_c(1 - J/J_c)$ for $J \approx J_c$ one obtains for rings and cylinders the decaying current density

$$J(t) \approx J_c \left[1 + \alpha \frac{k_B T}{U_c} \ln(1 + \frac{t}{t_0})\right]^{-1/\alpha} \tag{31}$$

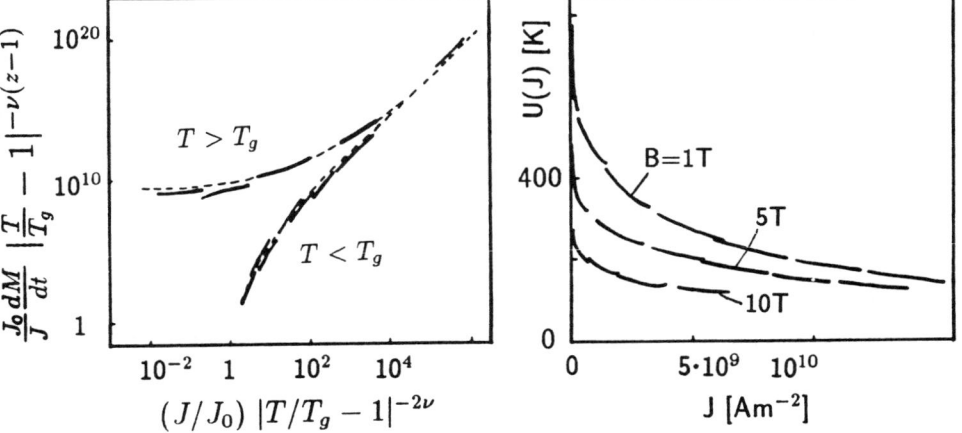

Figure 6: *Left:* Scaling plot (Section 6.4) of the decay rate (in units A/ms) versus current density (in units $J_0 = k_B T/\phi_0$), from [106]. *Right:* The (nearly temperature independent) activation energy $U(J)$ extracted from various experiments on the same sample at different temperature T for fields B of 1, 5, and 10 Tesla. Schematic from [106].

$$J(t) \approx J_c(t/t_0)^{-k_B T/U_c} \quad \text{for } \alpha \ll 1 \tag{32}$$

where t_0 is an integration constant. From (32) one obtains a temperature dependent $J(T)$ and the normalized creep rate

$$J(T) \approx J_c \exp(-T/T_0), \quad T_0 \approx U_c/[k_B \ln(t/t_0)] \tag{33}$$

$$S = -\frac{d \ln J}{d \ln t} \approx \frac{k_B T}{U_c + \alpha k_B T \ln(1 + t/t_0)} \tag{34}$$

Experiments on HTSC yield $T_0 \approx 10\,\text{K}$ [120] and $U_c/k_B \approx 100$ to $1000\,\text{K}$. Note that the decay rate S (34) decreases with increasing time and saturates at $k_B T > U_c/[\alpha \ln(t/t_0)]$ to a value $1/[\alpha \ln(t/t_0)]$. Numerical solutions to the relaxation rates for various dependencies $U(J)$ are given in [118]. Good fits to relaxation rates are also achieved by fitting a spectrum of activation energies [121-124]. The success of theories which assume only *one* activation energy U suggests that at given values of B, J, and T essentially one effective U of an entire spectrum determines the physical process under consideration.

6.6. TUNNELING OF VORTICES AT T=0

In numerous experiments the creep rate plotted versus T appears to tend to a finite value at zero temperature. Flux creep observed at low temperatures [125] in principle may be explained by usual thermal activation from smooth shallow pinning wells [126] but it may also indicate "quantum tunneling" of vortices out of the pins [114, 127-130]. The thermal depinning rate $\propto \exp(-U/k_B T)$ is now replaced by the tunneling rate $\propto \exp(-S_E/\hbar)$ where S_E is the Euklidean action of the considered tunneling process. Tunneling of vortices differs from tunneling of particles by the smallness of the inertial mass of the vortex; this means the vortex motion is overdamped, there are no oscillations or resonances. This overdamped tunneling is treated by Ivlev *et al.* [128] and Blatter *et al.* [129]. Griessen *et al.* [130] show that the dissipative quantum tunneling theory of Caldeira and Legget [131] with the usual vortex viscosity η inserted reproduces the main results of [128-129]. In a recent paper Blatter and Geshkenbein [114] give a very general theory of collective creep and (vortex-mass dominated) tunneling in anisotropic and layered superconductors. See also [21].

7. Flux Diffusion and Linear AC Response

Inserting into the induction law $\dot{\mathbf{B}} = \partial \mathbf{B}/\partial t = \nabla \times \mathbf{E}$ the electric field induced by flux flow $\mathbf{E} = \mathbf{B} \times \mathbf{v} = \rho \mathbf{J}_\perp$ where $\mathbf{J}_\perp = \hat{\mathbf{B}} \times \mathbf{J} \times \hat{\mathbf{B}}$ is the current density perpendicular to $\mathbf{B}$ and $\hat{\mathbf{B}} = \mathbf{B}/B$, one obtains an interesting equation of motion for the induction $\mathbf{B}$ in isotropic superconductors containing a FLL [101]:

$$\dot{\mathbf{B}} = \mu_0^{-1} \nabla \times \rho \hat{\mathbf{B}} \times \hat{\mathbf{B}} \times \nabla \times \mathbf{B} . \tag{35}$$

If $\mathbf{B}(\mathbf{r}, t) \approx \mathbf{B}_0 = \text{const.}$ varies little in space and time, eq. (35) may be linearized to give

$$\dot{\mathbf{B}} = D\nabla^2 \mathbf{B} + \rho \nabla \times \mathbf{J}_\parallel \tag{36}$$

plus terms of order $|\mathbf{B} - \mathbf{B}_0|^2$. In (36) $D = \rho/\mu_0 = D(B_0, T)$ means the diffusivity of flux, $\rho(B_0, T)$ equals ρ_{TAFF} (26) or ρ_{FF}. Thus, in geometries where the current component $\mathbf{J}_\parallel$ parallel to $\mathbf{B}$ vanishes, ohmic resistivity is equivalent to a *linear diffusion of the flux lines*.

Since in the TAFF region (26) $\rho = \rho_{TAFF} \propto \exp(-U/k_BT)$, one has for the thermally activated diffusion $D = \rho_{TAFF}/\mu_0 = D_0 \exp(-U/k_BT)$. For sufficiently large times, small specimens, and large T, any change of the applied magnetic field or current completely *penetrates* the HTSC, which then is in the *resistive* state. At lower T, such a change penetrates only into a thin *surface layer*, to the skin depth $\delta = (2D/\omega)$ or Campbell depth [132] $\lambda_C = (B^2/\mu_0\alpha_L)^{1/2}$ where α_L is the elastic pinning restoring force density on the FLL (Labusch parameter). The superconductor then behaves as if it were in the *Meissner* state, with almost complete expulsion of the applied field. In between these two limiting cases, the surface current penetrates more or less deeply and causes maximum dissipation when the skin depth coincides with a characteristic specimen dimension. The linear complex ac penetration depth λ_{ac} of a superconductor with a FLL may be written as [10, 133, 134]

$$\lambda_{ac}(\omega) = \left[\lambda^2 + \lambda_C^2 \frac{1 - i/\omega\tau}{1 + i\omega\tau_0}\right]^{1/2}. \tag{37}$$

The corresponding complex ac resistivity is

$$\rho_{ac}(\omega) = i\omega\mu_0\lambda_{ac} = i\omega\mu_0\lambda^2 + \rho_{TAFF}\frac{1 + i\omega\tau}{1 + i\omega\tau_0}, \tag{38}$$

and the complex ac susceptibility of a slab of thickness d in parallel ac field becomes

$$\mu(\omega) = \tanh(u)/u, \quad u = d/2\lambda_{ac} \tag{39}$$

(Fig. 7). Here $\tau_0 = \eta/\alpha_L = \lambda_C^2/D_0$ is the relaxation time of an elastically pinned FLL and $\tau = \eta_{TAFF}/\alpha_L = B^2/\rho_{TAFF}\alpha_L = \lambda_C^2/D_{TAFF} = \tau_0 \exp(U/k_BT) \gg \tau_0$ is the creep time, *i.e.* the relaxation time for linear thermal depinning. Eqs. (37) to (39) apply also when the creep is caused by *tunneling* of vortices out of the pinning wells, Section 6.6. They describe the dissipation by flux-line motion only. Near T_c, additional losses by the normal conducting electrons may become important; this has been accounted for by Coffey and Clem [133] within a two-fluid model. For concrete applications, λ_{ac}, $1/\lambda_{ac}$, λ_{ac}^2, ρ_{ac}, μ, etc., have to be decomposed into their real and imaginary parts.

Eqs. (37-39) show that for sufficiently low circular frequencies $\omega \ll 1/\tau$ one has *flux diffusion* with strongly temperature dependent diffusivity $D = D_0 \exp(-U/k_BT)$ and ohmic resistivity $\rho = \rho_{TAFF} = \rho_{FF} \exp(-U/k_BT)$, see $\mu(\omega) = \mu' + i\mu''$ in Fig. 7. One then has $\lambda_{ac} \approx \lambda_C/(i\omega\tau)^{1/2}$ (37), thus u in (39) becomes complex, $u = d/\lambda_{ac} = (1+i)(\omega d^2/8D) = (1+i)d/2\delta$ with $\delta = (2D/\omega)^{1/2}$ the skin depth; the dissipation $\mu''(\omega)$ is maximum at $\omega = 0.97/\tau_d$ with $\tau_d = d^2/\pi^2 D$. For recent experiments testing the above ac response see, *e.g.*, [135-136].

In the derivation of (37-39), after a sudden shift of the FLL an *exponential* time decay of the elastic pinning force was assumed, $\alpha_L(t) \propto \exp(-t/\tau)$. Ref. [135] observes in BSCCO ceramics an *algebraic* decay of $\alpha_L(t) \propto (1+t/\tau_0)^{-\beta}$ with $\beta = 1/(1+U/k_BT) \leq 1$, *cf.* eq. (32), a depinning limit ($\beta = 1$) near T_c, and a glass transition at $T \approx 24$ K where $\beta \approx 0.07$. Ref. [136] finds an activation energy spectrum in BSCCO crystals.

8. Depinning Lines

The flux-diffusion picture explains a wide variety of experiments which all define an irreversibility line or depinning line $T_d(B)$, Fig. 5. The irreversibility line separates in the

B–T-plane the regions of irreversible (low B, T) and reversible (large B, T) magnetic behavior and can be explained by thermally activated depinning. Above this line the pins become ineffective and the vortices can move freely, giving rise to *reversible* magnetization curves, whereas below $T_d(B)$ hysteretic behavior is observed.

An interesting feature of the depinning line is that it depends on the size and shape of the specimen and on the time scale or frequency $\nu = \omega/2\pi$ of the experiment. This sometimes overlooked effect means that $T_d(B)$ is *not* a genuine intrinsic property of the material but originates from the *diffusive character* of the flux motion and in general is *not* indicative of a sharp phase transition of any kind. Various experiments yield such depinning lines:

(a) The irreversibility (hysteresis) in magnetization curves vanishes at $T_d(B)$.

(b) The imaginary part of the ac susceptibility (39) has a maximum at $T_d(B)$.

(c) The ac penetration depth (37) measured by the screening of an ac field by a superconducting film between two coils [137] diverges at $T_d(B)$.

(d) In a particularly sensitive and interesting type of experiment, tilt vibrations in a constant magnetic field are performed by a superconducting reed or by a superconductor glued on a silicon reed. A sharp maximum in the attenuation of the vibrating HTSC and a decrease of the pinning-caused enhancement of the resonance frequency is observed at $T_d(B)$ [44–47], Fig. 7. Recently, even two or more such peaks were observed when the field was applied at an oblique angle with respect to a flat HTSC [138, 139]. These peaks may belong to different diffusion modes [140] or indicate some transition.

(e) The conduction noise in HTSC films at constant current density and at a given frequency exhibits a sharp peak as a function of B and T [141]. This noise is caused by depinning processes: each "plucking" of a vortex releases elastic energy of the FLL, which then relaxes viscously with an exponential time law. Below $T_d(B)$ the noise is small since only few depinning processes occur; at $T = T_d(B)$ the noise is maximum; and above $T_d(B)$

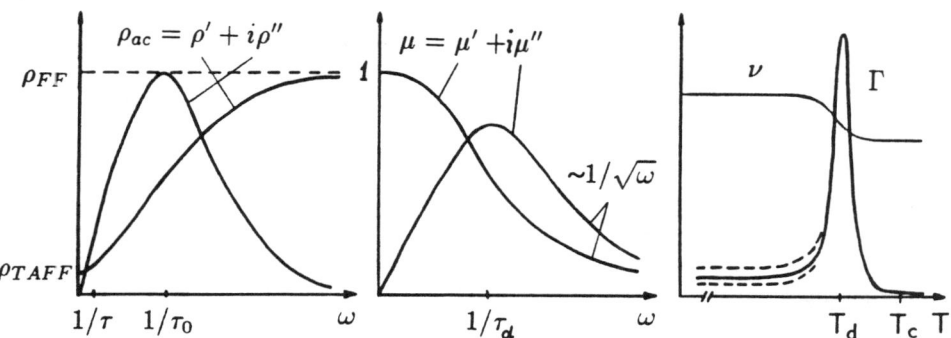

Figure 7: *Left:* Complex ac resistivity $\rho_{ac}(\omega) = \rho' + i\rho''$ (38). *Middle:* Complex ac susceptibility $\mu(\omega) = \mu' + i\mu''$ (39) of slabs in the TAFF state $\omega \ll \tau^{-1}$, see text. *Right:* Resonance frequency ν and attenuation Γ of a HTSC performing tilt vibrations in constant magnetic field. The dashed lines indicate the amplitude dependence of the hysteretic losses.

it decreases again since the viscous motion of the thermally depinned vortices is smooth. The depinning line shifts to larger T or B when a higher frequency band is selected.

(f) In the broadening of the resistive transition $\rho(T,B)$ in a magnetic field [41, 103-110] the low-ρ tail originates from TAFF but the main part of $\rho(T)$ (on a linear scale) may originate from thermal fluctuations of the order parameter near T_c, where superconducting islands nucleate and decay again. In BSCCO with its almost isolated CuO-layers, the very broad smearing of $\rho(T)$ may also be caused by the thermal nucleation of vortex–antivortex pairs in the layers, which leads to a Kosterlitz-Thouless-like transition with a power law current-voltage curve [70].

(g) An ultrasonic attenuation peak and sound velocity enhancement occur at $T_d(B)$ [142, 143]. Ultrasound probes the FLL *far inside* a superconductor, whereas in the other methods (a) – (f) the FLL interacts with the outer world (the applied field or transport current) only near its *surface*, in a layer of thickness λ where shielding currents or transport surface currents exert forces on the vortex lines or on their end points (magnetic monopoles). The resulting compression or tilt deformation of the FLL then *diffuses* into the interior as described in Section 7. At lower T and at larger amplitude, this diffusion and the corresponding resistivity become *nonlinear*, i.e., they depend on the amplitude of the ac field or current.

9. Concluding Remarks

The above outline of static and dynamic properties of the flux-line lattice cannot be complete. For more detailed information see the review papers [5, 9, 10, 21, 24, 47, 50, 87]. For lack of space I could not mention the thermo-electric (Seebeck and Nernst) effects and the thermo-magnetic (Peltier and Ettingshausen) effects [144-147] which give additional information on flux motion. A further fascinating problem is the Hall effect of moving flux lines. [146-152]. Several explanations were given for the unusual Hall angle with sign reversal just below T_c, see *e.g.*, [150-152]. Recent informative experiments on dc [41] and ac [105] resistivity and magnetic relaxation [106] of HTSC as functions of T, B, J, frequency or time require detailed discussion. A simple scaling behavior near T_c discovered for transport and thermodynamic properties of YBCO in [153] possibly may be understood in terms of the transport and fluctuation theory of Troy and Dorsey [147] generalizing the theory of Vecris and Pelcovits which combines time dependent GL theory with the distorted FLL solutions of [13d].

In the last two or three years large progress was made in understanding the electromagnetic properties of HTSC in terms of the phenomenological London, Ginzburg-Landau, and Lawrence-Doniach theories. Also statistical and scaling concepts have proven very useful as indicated in this short review. For a more detailed discussion of all these theories and concepts I refer to the forthcoming excellent review paper by Blatter *et al.* [21].

References

[1] P. G. De Gennes, *Superconductivity of Metals and Alloys* (Benjamin, New York, 1966).
[2] M. Tinkham, *Introduction to Superconductivity* (McGraw-Hill, New York, 1975).
[3] A. A. Abrikosov, *Fundamentals of the Theory of Metals* (North Holland, Amsterdam, 1988).
[4] U. Essmann and H. Träuble, *Phys. Lett. A* **24**, 526 (1967); *Scientific American* **224**, 75 (March 1971).
[5] E. H. Brandt and U. Essmann, *phys. stat. sol. (b)* **144**, 13 (1987) (review).
[6] P. L. Gammel, D. J. Bishop, G. J. Dolan, J. R. Kwo, C. A. Murray, L. F. Schneemeyer, and J. V. Waszczak, *Phys. Rev. Lett.* **59**, 2592 (1987); L. Ya. Vinnikov, L. A. Gurevich, G. A. Yemel'chenko, and Yu. A. Ossipyan, *Solid State Comm.* **67**, 421 (1988); P. L. Gammel, D. J. Bishop, J. P. Rice, and D. M. Ginsberg, *Phys. Rev. Lett.* **68**, 3343 (1992).
[7] J. R. Clem, *J. Low Temp. Phys.* **18**, 427 (1975); Z. Hao and J. R. Clem, *Phys. Rev. B* **43**, 7622 (1991).
[8] E. H. Brandt, *Phys. Rev. B* **34**, 6514 (1986).
[9] E. H. Brandt, *Physica B* **165 & 166**, 1129 (1990); *Physica B* **169**, 91 (1991); *Int. J. Mod. Phys. B* **5**, 751 (1991) (review on FLL in HTSC).
[10] E. H. Brandt, *Physica C* **195**, 1 (1992).
[11] G. Carneiro, M. M. Doria, and S. C. B. de Andrade, *Physica C* **203**, 167 (1992); see also: V. G. Kogan and J. J. Campbell, *Phys. Rev. Lett.* **62**, 1552 (1989); A. M.Grishin, A. Yu. Martynovich, and S. V. Jampol'skiĭ, *Zh. Eksp. Teor. Fiz.* **97**, 1930 (1990) [*Sov. Phys. JETP* **70**, 1089 (1990)]; *Zh. Eksp. Teor. Fiz.* **101**, 649 (1992) [*Sov. Phys. JETP* **74**, 345 (1992)]; V. G. Kogan, *Phys. Rev. Lett.* **64**, 2192 (1990); A. I. Buzdin and A. Yu. Simonov, *Pis'ma Zh. Eksp. Teor. Fiz.* **51**, 168 (1990) [*Sov. Phys. JETP Lett.* **51**, 191 (1990)].
[12] A. Sudbø and E. H. Brandt, *Phys. Rev. Lett.* **67**, 3176 (1991).
[13] E. H. Brandt, *J. Low Temp. Phys.* **26**, 709; 735 (1977); **28**, 263; 291 (1977).
[14] A. Houghton, R. A. Pelcovits, and A. Sudbø, *Phys. Rev. B* **40**, 6763 (1989).
[15] A. Sudbø and E. H. Brandt, *Phys. Rev. B* **43**, 10482 (1991); *Phys. Rev. Lett.* **66**, 1781 (1991); *Phys. Rev. Lett.* **68**, 1758 (1992).
[16] E. H. Brandt and A. Sudbø, *Physica C* **180**, 426 (1991).
[17] D. S. Fisher, in: *Phenomenology and Applications of High-Temperature Superconductors*, K. S. Bedell et al. eds., (Addison-Wesley, New York, 1992) p. 287; S. Nieber and H. Kronmüller, *Physica C* (submitted).
[18] E. H. Brandt, *Phys. Rev. B* 'Tilted vortices in anisotropic superconducting slabs' (submitted).
[19] E. Sardella, *Phys. Rev. B* **45**, 3141 (1992); *Phys. Rev. B* **44**, 5209 (1991).
[20] G. Blatter, V. B. Geshkenbein, and A. I. Larkin, *Phys. Rev. Lett.* **68**, 875 (1992).
[21] G. Blatter, M. V. Feigel'man, V. B. Geshkenbein, A. I. Larkin, and V. M. Vinokur, *Rev. Mod. Phys.* (in preparation).
[22] L. I. Glazman and A. E. Koshelev, *Phys. Rev. B* **43**, 2835 (1991).
[23] J. R. Clem, *Phys. Rev. B* **43**, 7837 (1991); S. N. Artemenko and A. N. Kruglov, *Phys. Lett. A* **143**, 485 (1990); K. H. Fischer, *Physica C* **178**, 161 (1991).

[24] D. Feinberg, *Physica C* **194**, 126 (1992).
[25] D. R. Nelson and H. S. Seung, *Phys. Rev. B* **39**, 9174 (1989).
[26] M. A. Moore, *Phys. Rev.* **B 39**, 9174 (1989); **B 45**, 7336 (1992).
[27] E. H. Brandt, *Phys. Rev. Lett.* **63**, 1106 (1989).
[28] E. H. Brandt, *Physica C* **162–164**, 1167 (1989).
[29] Hong-ru Ma and S. T. Chui, *Phys. Rev. Lett.* **68**, 2528 (1992).
[30] S. Ryu, S. Doniach, Guy Deutscher, A. Kapitulnik, *Phys. Rev. Lett.* **68**, 710 (1992).
[31] S. Sengupta, C. Dasgupta, H. R. Krishnamurty, G. I. Menon, and T. V. Ramakrishnan, *Phys. Rev. Lett.* **67**, 3444 (1991).
[32] H. R. Glyde, L. K. Molenko, and P. Findeisen, *Phys. Rev. B* **45**, 2409 (1992).
[33] Ying-Hong Li and S. Teitel, *Phys. Rev. Lett.* **66**, 3301 (1991); *Phys. Rev. B* **45**, 5718 (1992); *Phys. Rev. B* **47**, 359 (1993).
[34] P. Minnhagen and P. Olsson, *Phys. Rev. B* **44**, 4503 (1992); **45**, 5722 (1992); *Phys. Rev. Lett.* **67**, 1039 (1992).
[35] H. Weber and H. J. Jensen, *Phys. Rev. B* **44**, 454 (1991).
[36] R. E. Hetzel, A. Sudbø, and D. A. Huse, *Phys. Rev. Lett.* **69**, 518 (1992).
[37] R. Ikeda, T. Ohmi, and T. Tsuneto, *J. Phys. Soc. Japan* **61**, 254 (1992); *Physica C* **185-189**, 1563 (1991); R. Ikeda, preprint.
[38] Z. Tešanović, L. Xing, L. N. Bulaevskii, Q. Li, and M. Suenaga, *Phys. Rev. Lett.* **69**, 3563 (1992); J. A. Neill and M. A. Moore, *Phys. Rev. Lett.* **69**, 2582 (1992); Y. Kato and N. Nagaosa, *Phys. Rev. B* **47**, 2932 (1993).
[39] R. G. Carneiro, R. Cavalcanti, and A. Gartner, *Phys. Rev. B* **47**, 5263 (1993).
[40] M. V. Feigel'man, V. B.Geshkenbein, V. M. Vinokur, *Pis'ma Zh. Eksp. Teor. Fiz.* **527**, 1141 (1990) [*Sov. Phys. JETP Lett.* **52**, 546 (1990)]; and in preparation.
[41] T. K. Worthington, M. P. A. Fisher, D. A. Huse, J. Toner, A. D. Marwick, T. Zabel, C. A. Feild, and F. Holtzberg, *Phys. Rev. B* **46**, 11854 (1992).
[42] H. Safar, P. L. Gammel, D. A. Huse, D. J. Bishop, J. P. Rice, and D. M. Ginsberg, *Phys. Rev. Lett.* **69**, 824 (1992).
[43] W. K. Kwok, S.Fleshler, U. Welp, V. M. Vinokur, J. Downey, and G. W. Crabtree, *Phys. Rev. Lett.* **69**, 3370 (1992).
[44] P. L. Gammel, L. F. Schneemeyer, J. V. Waszczak, and D. J. Bishop, *Phys. Rev. Lett.* **61**, 1666 (1988); comment: E. H. Brandt, P. Esquinazi, and G. Weiss, *Phys. Rev. Lett.* **62**, 2330 (1989); reply: R. N. Kleiman, P. L. Gammel, L. F. Schneemeyer, J. V. Waszczak, and D. J. Bishop, *Phys. Rev. Lett.* **62**, 2331 (1989).
[45] A. Gupta, P. Esquinazi, and H. F. Braun, *Phys. Rev. Lett.* **63**, 1869 (1989); *Physica B* **165 & 166**, 1151 (1990).
[46] J. Kober, A. Gupta, P. Esquinazi, H. F. Braun, and E. H. Brandt, *Phys. Rev. Lett.* **66**, 2507 (1991).
[47] P. Esquinazi, *J. Low Temp. Phys.* **85**, 139 (1991) (review on vibrating superconductors). *Sol. St. Comm.* **74**, 75 (1990).
[48] D. E. Farrell, J. P. Rice, and D. M. Ginsberg, *Phys. Rev. Lett.* **67**, 1165 (1991); R. G. Beck, D. E. Farrell, J. P. Rice, D. M. Ginsberg, and V. G. Kogan, *Phys. Rev. Lett.* **68**, 1594 (1992).
[49] W. E. Lawrence and S. Doniach, Proc. 12th Internatl. Conf. of Low Temperature Physics LT12 (E.Kanda ed., Academic Press of Japan, Kyoto, 1971) p. 361.

[50] L. N. Bulaevskii, *Int. J. Mod. Phys. B* **4**, 1849 (1990) (review on LD theory).
[51] M. V. Feigel'man, V. B. Geshkenbein, A. I. Larkin, *Physica C* **167**, 177 (1990).
[52] V. M. Vinokur, P. H. Kes, and A. E. Koshelev, *Physica C* **168**, 29 (1990).
[53] J. Pearl, *J. Appl. Phys.* **37**, 4139 (1966).
[54] M. Tachiki and S. Takahashi, *Solid State Comm.* **70**, 291 (1989); *Physica B* **169**, 121 (1991); L. Schimmele, H. Kronmüller, and H. Teichler, *phys. stat. sol. (b)* **147**, 361 (1988).
[55] E. H. Brandt, *Phys. Rev. Lett.* **66**, 3213 (1991).
[56] L. N. Bulaevskii, M. Ledvij, and V. G. Kogan, *Phys. Rev. B* **46**, 366, 11807 (1992).
[57] J. M. Kosterlitz and D. J. Thouless, *J. Phys. C* **6**, 1181 (1973); **7**, 1046 (1974); B. I. Halperin and D. R. Nelson, *Phys. Rev. B* **19**, 2457 (1979).
[58] L. N. Bulaevskii, S. V. Meshkov, and D. Feinberg, *Phys. Rev. B* **43**, 3728 (1991).
[59] G. Hackenbroich and S. Scheidl, *Physica C*, **181**, 163 (1991); S. Scheidl and G. Hackenbroich, *Phys. Rev. B* **46**, 14010 (1992).
[60] L. N. Bulaevskii, M. Ledvij, and V. G. Kogan, *Phys. Rev. Lett.* **68**, 3773 (1992); V. G. Kogan, M. Ledvij, A. Yu. Simonov, J. H. Cho, and D. C. Johnston, *Phys. Rev. Lett.* **70**, 1870 (1993).
[61] B. A. Hubermann and S. Doniach, *Phys. Rev. Lett.* **43**, 950 (1979); D. S. Fisher, *Phys. Rev. B* **22**, 1190 (1980).
[62] L. I. Glazman and A. E. Koshelev, *Physica C* **173**, 180 (1991).
[63] L. L. Daemen, L. N. Bulaevskii, M. P. Maley, and J. Y. Coulter, *Phys. Rev. Lett.* **70**, 1167 (1993). *Phys. Rev. B* (submitted); see also: L. N. Bulaevskii, J. R.Clem, L. I.Glazman, and A. P. Malozemoff, *Phys. Rev. B* **45**, 2545 (1992).
[64] D. Feinberg and C. Villard, *Phys. Rev. Lett.* **65**, 919 (1990).
[65] D. Feinberg and A. M. Ettouhami, 'The lock-in transition in layered superconductors' (preprint).
[66] A. E. Koshelev, *Phys. Rev. B* 'Kink walls and Critical behavior of magnetization near the lock-in transition in layered superconductors' (submitted).
[67] B. Horovitz, *Phys. Rev. Lett.* **67**, 378 (1991); *Phys. Rev. B* **45**, 12632 (1992); *Phys. Rev. B* **47**, 5947, 5964 (1993); G. Carneiro, *Phys. Rev. B* **45**, 2391; 2403 (1992); S. E. Korshunov, *Europhys. Lett.* **11**, 757 (1990).
[68] G. Blatter, B. Ivlev, and J. Rhyner, *Phys. Rev. Lett.* **66**, 2392 (1991).
[69] Y. Iye, S. Nakamura, and T. Tamegai, *Physica C* **159**, 443 (1989); *Physica C* **174**, 4227 (1991); R, C. Budhani, D. O. Welch, M. Suenaga, and R. L. Sabatini, *Phys. Rev. Lett.* **64**, 1666 (1990); D. H. Kim, K. E. Gray, R. T. Kampwirth, and D. M. McKay, *Phys. Rev. B* **42**, 6249 (1990); H. Raffy, S. Labdi, O. Laborde, and P. Monceau, *Phys. Rev. Lett.* **65**, 2515 (1991); Y. Iye, I. Oguro, T. Tamegai, W. Datars, N. Motohira, and K. Kitazawa, *Physica C* **199**, 154 (1992).
[70] Y. Ando, N. Motohira, K. Kitazawa, J. Takeya, and S. Akita, *Phys. Rev. Lett.* **67**, 2737 (1991).
[71] P. H. Kes, J. Aarts, V. M. Vinokur, and C. J. van der Beek, *Phys. Rev. Lett.* **64**, 1063 (1990).
[72] B. I. Ivlev and N. B. Kopnin, *J. Low Temp. Phys.* **80**, 161 (1990).
[73] O. Brunner, L. Antognazza, J.-M. Triscone, L. Miéville, and Ø. Fischer, *Phys. Rev. Lett.* **67**, 1354 (1991).

[74] D. Neerink, K. Temst, M. Baert, E. Osquiguil, C. Van Haesendonck, and Y. Bruynseraede, A. Gilabert, and Ivan K. Schuller, *Phys. Rev. Lett.* **67**, 2577 (1991).

[75] J. Bardeen and M. J. Stephen, *Phys. Rev.* **140**, A1197 (1965).

[76] M. Tinkham, *Phys. Rev. Lett.* **13**, 804 (1964).

[77] L. P. Gor'kov and N. B. Kopnin, *Sov. Phys.-Uspechi* **18**, 496 (1976).

[78] A. I. Larkin and Yu. N. Ovchinnikov, in: *Nonequilibrium Superconductivity*, D. N. Langenberg and A. I. Larkin, eds. (Elsevier, Amsterdam, 1986), p. 493.

[79] B. I. Ivlev and N. B. Kopnin, *Phys. Rev. B* **42**, 10052 (1990).

[80] J. R. Clem and W. M. Coffey, *Phys. Rev. B* **42**, 6209 (1990).

[81] Z. Hao and J. R. Clem, *IEEE Trans. Magn.* **27**, 1086 (1991).

[82] P. W. Anderson, *Phys. Rev. Lett.* **9**, 309 (1962); P. W. Anderson and Y. B. Kim, *Rev. Mod. Phys.* **36**, 39 (1964).

[83] C. P. Bean, *Rev. Mod. Phys.* **36**, 31 (1964); *J. Appl. Phys.* **41**, 2482 (1970).

[84] M. Däumling and D. C. Larbalestier, *Phys. Rev. B* **40**, 9350 (1989).

[85] L. W. Connor and A. P. Malozemoff, *Phys. Rev. B* **43**, 402 (1991).

[86] H. Theuss, A. Forkl, and H. Kronmüller, *Physica C* **190**, 345 (1992).

[87] S. Senoussi, *J. Phys. (Paris) III* **2**, 1041 (1992) (review).

[88] E. H. Brandt, *Phys. Rev. B* **46**, 8628 (1992); E. H. Brandt, M. Indenbom, and A. Forkl, *Europhys. Lett.* (submitted).

[89] M. R. Beasley, R. Labusch, and W. W. Webb, *Phys. Rev.* **181**, 682 (1969); see also: C. Rossel et al., *Physica C* **165**, 233 (1990); P. Berghuis and P. H. Kes, *Physica B* **165 & 166**, 1169 (1990); P. Svedlindh et al., *Phys. Rev. B* **43**, 2735 (1991); M. Suenaga, A. K. Gosh, Y. Xu, and D. O. Welch, *Phys. Rev. Lett.* **66**, 177 (1991).

[90] Y. Yeshurun and A. P. Malozemoff, *Phys. Rev. Lett.* **60**, 2202 (1988).

[91] D. Dew-Hughes, *Cryogenics* **28**, 674 (1988).

[92] P. H. Kes, J. Aarts, J. van den Berg, C. J. van der Beek, and J. A. Mydosh, *Supercond. Sci. Technol.* **1**, 242 (1989).

[93] A. I. Larkin and Yu. N. Ovchinnikov, *J. Low Temp. Phys.* **43**, 109 (1979); E. H. Brandt, *J. Low Temp. Phys.* **64**, 375 (1986).

[94] M. V. Feigel'man, V. B. Geshkenbein, A. I. Larkin, and V. M. Vinokur, *Phys. Rev. Lett.* **63**, 2303 (1989).

[95] T. Nattermann, *Phys. Rev. Lett.* **64**, 2454 (1990); K. H. Fischer and T. Nattermann, *Phys. Rev. B* **43**, 10372 (1991).

[96] V. M. Vinokur, M. V. Feigel'man, V. B. Geshkenbein, and A. I. Larkin, *Phys. Rev. Lett.* **65**, 259 (1990); M. V. Feigel'man and V. M. Vinokur, *Phys. Rev. B* **41**, 8986 (1990); V. M. Vinokur, V. B. Geshkenbein, M. V. Feigel'man, and A. I. Larkin, *Zh. Eksp. Teor. Fiz.* **100**, 1104 (1991) [*Sov. Phys. JETP* **73**, 610 (1991)]; A. E. Koshelev, *Phys. Rev. B* **45**, 12936 (1992).

[97] L. Civale, A. D. Marwick, T. K. Worthington, M. A. Kirk, J. R. Thompson, L. Krusin-Elbaum, Y. Sun, J. R. Clem, F. Holtzberg, *Phys. Rev. Lett.* **67**, 648 (1991); M. Konczykowski et al., *Phys. Rev. B* **47**, 5531 (1993).

[98] W. Gerhäuser, G. Ries, H. W. Neumüller, W. Schmidt, O. Eibl, G. Saemann-Ischenko, and S. Klaumünzer, *Phys. Rev. Lett.* **68**, 879 (1992); D. Prost et al., *Phys. Rev. B* **47**, 3457 (1993).

[99] E. H. Brandt, *Phys. Rev. Lett.* **69**, 1105 (1992); *Europhysics Letters* **18**, 635 (1992); I. F. Lyuksyutov, *Europhysics Letters* **20**, 273 (1992).

[100] D. R. Nelson and V. M. Vinokur, *Phys. Rev. Lett.* **68**, 2398 (1992).

[101] E. H. Brandt, *Z. Physik B* **80**, 167 (1990).

[102] A. Schmid and W. Hauger, *J. Low Temp. Phys.* **11**, 667 (1973).

[103] T. T. M. Palstra, B. Battlogg, R. B. van Dover, L. F. Schneemeyer, and J. V. Waszczak, *Phys. Rev. B* **41**, 6621 (1990).

[104] R. H. Koch et al. *Phys. Rev. Lett.* **63**, 1511 (1989); C. Dekker, W. Eidelloth, and R. H. Koch, *Phys. Rev. Lett.* **68**, 3347 (1992).

[105] Ph. Seng, R. Gross, U. Baier, M. Rupp, D. Koelle, R. P. Huebener, P. Schmitt, G. Saemann-Ischenko, and L. Schultz, *Physica C* **192**, 403 (1992).

[106] J. C. van der Beek, G. J. Nieuwenhuys, P. Kes, H. G. Schnack, and R. P. Griessen, *Physica C* **197**, 320 (1992).

[107] T. K. Worthington et al., *Phys. Rev. B* **43**, 10538 (1991).

[108] P. L. Gammel, L. F. Schneemeyer, and D. J. Bishop, *Phys. Rev. Lett.* **66**, 953 (1991); N.-C. Yeh, D. S. Reed, W. Jiang, U. Kriplani, F. Holtzberg, A. Gupta, B. D. Hunt, R. P. Vasquez, M. C. Foote, and L. Bajuk, *Phys. Rev. B* **45**, 5654 (1992); H. Safar et al., *Phys. Rev. Lett.* **68**, 2672 (1992).

[109] E. Sandvold and C. Rossel, *Physica C* **190**, 309 (1992).

[110] M. P. Maley and J. O. Willis, *Phys. Rev. B* **42**, 2639 (1990).

[111] S. Chakravarty, B. I. Ivlev, and Yu. N. Ovchinnikov, *Phys. Rev. Lett.* **64**, 3187 (1990); *Phys. Rev. B* **42**, 2143 (1990).

[112] M. P. A. Fisher, *Phys. Rev. Lett.* **62**, 1415 (1989).

[113] D. S. Fisher, M. P. A. Fisher, and D. A. Huse, *Phys. Rev. B* **43**, 130 (1991).

[114] G. Blatter and V. B. Geshkenbein, *Phys. Rev. B* **47**, 2725 (1993).

[115] A. P. Malozemoff and M. P. A. Fisher, *Phys. Rev. B* **42**, 6784 (1990).

[116] M. V. Feigel'man, V. B. Geshkenbein, A. I. Larkin, and V. M. Vinokur, *Phys. Rev. B* **43**, 6263 (1991).

[117] A. A. Zhukov, *Sol. St. Comm.* **82**, 983 (1992).

[118] H. G. Schnack, R. Griessen, J. G. Lensink, C. J. van der Beek, and P. H. Kes, *Physica C* **197**, 337 (1992).

[119] V. M. Vinokur, M. V. Feigel'man, and V. B. Geshkenbein, *Phys. Rev. Lett.* **67**, 915 (1991).

[120] S. Senoussi, M. Ousséna, G. Collin, and I. A. Campbell, *Phys. Rev. B* **37**, 9792 (1988).

[121] C. W. Hagen and R. Griessen, *Phys. Rev. Lett.* **62**, 2857 (1989); R. Griessen, *Phys. Rev. Lett.* **64**, 1674 (1990).

[122] A. Gurevich, *Phys. Rev. B* **42**, 4857 (1990).

[123] L. Niel and J. Evetts, *Europhys. Lett.* **15**, 453 (1991).

[124] H. Theuss, T. Reininger, and H. Kronmüller, *J. Appl. Phys.* **72**, 1936 (1992).

[125] J. G. Lensink, C. F. J. Flipse, J. Roobeek, R. Griessen, and B. Dam, *Physica C* **162-164**, 663 (1989); A. C. Mota, G. Juri, P. Visani, and A. Pollini, *Physica C* **162-164**, 1152 (1989); R. Griessen et al., *Cryogenics* **30**, 536 (1990); A. Fruchter et al., *Phys. Rev. B* **43**, 8709 (1991); M. Lairson et al., *Phys. Rev. B* **43**, 10405 (1991); A. C. Mota et al., *Physica C* **185-189**, 343 (1991).

[126] R. Griessen, *Physica C* **172**, 441 (1991).
[127] A. V. Mitin, *Zh. Eksp. Teor. Fiz.* **93**, 590 (1987) [*Sov. Phys. JETP* **66**, 335 (1987)].
[128] B. I. Ivlev, Yu. N. Ovchinnikov, and R. S. Thompson, *Phys. Rev. B* **44**, 7023 (1991).
[129] G. Blatter, V. B. Geshkenbein, and V. M. Vinokur, *Phys. Rev. Lett.* **66**, 3297 (1991).
[130] R. Griessen, J. G. Lensink, and H. G. Schnack, *Physica C* **185-189**, 337 (1991).
[131] A. O. Caldeira and A. J. Leggett, *Phys. Rev. Lett.* **46**, 211 (1981).
[132] A. M. Campbell, *J. Phys. C* **4**, 3186 (1971); A. M. Campbell and J. E. Evetts, *Adv. Phys.* **21**, 199 (1972) (review).
[133] M. W. Coffey and J. R. Clem, *IEEE Trans. Magn.* **27**, 2136 (1991) and erratum **27**, 4396 (1991); *Phys. Rev. Lett.* **67**, 386 (1991); *Phys. Rev. B* **45**, 9872 (1992).
[134] E. H. Brandt, *Phys. Rev. Lett.* **67**, 2219 (1991); *Physica C* **185-189**, 270 (1991).
[135] R. Behr, J. Kötzler, A. Spirgatis, and M. Ziese, *Physica A* **191**, 464 (1992).
[136] D. G. Steel and J. M. Graybeal, *Phys. Rev. B* **45**, 12643 (1992).
[137] A. F. Hebard, P. L. Gammel, C. Rice, and A. Levi, *Phys. Rev. B* **40**, 5243 (1989).
[138] C. Durán, J. Yazyi, F. de la Cruz, D. Bishop, D. B. Mitzi, and A. Kapitulnik, *Phys. Rev. B* **44**, 7737 (1991); J. Yazyi, A. Arribére, C. Durán, F. de la Cruz, D. B. Mitzi, and A. Kapitulnik, *Physica C* **184**, 254 (1991).
[139] Y. Kopelevich, A. Gupta, P. Esquinazi, C. P. Heidmann, and H. Müller, *Physica C* **183**, 345 (1991).
[140] E. H. Brandt, *Phys. Rev. Lett.* **68**, 3796 (1992).
[141] A. Maeda, Y. Kato, H. Watanabe, I. Terasaki, and K. Uchinokura, *Physica B* **165 & 166**, 1363 (1990); E. S. Otabe, T. Matsushita, and K. Yamafuji, *IEEE Trans. Magn.* **27**, 1033 (1991).
[142] J. Pankert, *Physica C* **168**, 335 (1990).
[143] J. Pankert et al., *Phys. Rev. Lett.* **56**, 3052 (1990); P. Lemmens et al., *Physica C* **174**, 289 (1991); J. Pankert et al., *Physica C* **182**, 291 (1991); P. Lemmens, S. Ewert, and J. Pankert, *Physica C* **185-189**, 2271 (1991).
[144] R. P. Huebener, F. Kober, R. Gross, and H.-C. Ri, *Physica C* **185-189**, 349 (1991).
[145] K. H. Fischer, *Physica C* **200**, 23 (1992).
[146] A. Dorsey, *Phys. Rev. B* **46**, 8376 (1992).
[147] R. J.Troy and A. Dorsey, *Phys. Rev. B* **47**, 2715 (1993).
[148] G. Vecris and R. A. Pelcovits, *Phys. Rev. B* **44**, 2767 (1991).
[149] S. J. Hagen, C. J. Lobb, R. L.Greene, and M. Eddy, *Phys. Rev. B* **43**, 6246 (1991); S. J. Hagen et al., *Phys. Rev. B* **47**, 1064 (1993); J. M. Harris, Y. F. Yan, and N. P. Ong, *Phys. Rev. B* **46**, 14293 (1992).
[150] Z. D. Wang and C. S. Ting, *Phys. Rev. B* **46**, 284 (1991).
[151] R. A. Ferrell, *Phys. Rev. Lett.* **68**, 2524 (1992).
[152] N. B. Kopnin, B. I. Ivlev, and V. A. Kalatsky, *J. Low Temp. Phys.* **90**, 1 (1993).
[153] U. Welp, S. Fleshler, W. K. Kwok, R. A. Klemm, V. M. Vinokur, J. Downey, B. Veal, and G. W. Crabtree, *Phys. Rev. Lett.* **67**, 3180 (1991).

MAGNETIC FLUX LINE LATTICES IN THE CUPRATE SUPERCONDUCTORS

P. L. Gammel and D.J. Bishop
AT&T Bell Laboratories
Murray Hill, New Jersey 07974

ABSTRACT. In this article we discuss a variety of recent experiments on the static and dynamic properties of vortices and flux-line lattices in the mixed state of the copper-oxide superconductors. The experiments are of two basic types; the first images the magnetic flux patterns with magnetic decoration which gives information about static structures and the second explores the dynamics of vortices through linear and nonlinear transport. All of our results argue in favor of the existence of a phase transition in the high field mixed state from a low temperature superconducting vortex glass phase into a disordered high temperature vortex fluid phase. The vortex glass phase transition model does a good job of explaining high precision measurements of the dynamics at the transition. At low fields and temperatures very long range hexatic order in the flux-line lattice is observed.

1. Introduction

The behavior of superconductors in the presence of a magnetic field has been the subject of much scientific, as well as practical, interest over the past few decades. Superconductors can be divided into two classes: Type I superconductors which do not remain superconducting once a magnetic field penetrates the material, while Type II superconductors do remain superconducting until the penetrating field reaches a critical value for the destruction of superconductivity.

The basic theory of how type II superconductors behave in a magnetic field is due to Abrikosov[1]. He showed that in type II superconductors, the magnetic field, when greater than the lower critical field, H_{c1}, and less than the upper critical field, H_{c2}, penetrates the sample in the form of quantized flux lines, each carrying exactly one quantum, $\phi_0 = hc/2e$, of magnetic flux. The superconducting order parameter, ψ, which is a complex scalar field representing the quantum mechanical wavefunction of the paired electrons, has a vortex line for each quantized flux line. If we take $\psi = |\psi| e^{i\phi}$, the magnitude $|\psi|$ of the order parameter vanishes at the center of each vortex line, while the phase ϕ changes by 2π as one makes a full circle around a single vortex line. In the Abrikosov vortex lattice phase of a type II superconductor these vortex lines, which run parallel to the magnetic field, are arranged in a regular hexagonal crystalline array.

Soon after the discovery of the copper-oxide high-temperature superconductors it was shown that the resistivity in the temperature and magnetic field regime where the Abrikosov vortex lattice was expected to form, behaves in a qualitatively different fashion from that found in previously studied type II superconductors[2]. The reason for the difference, we now know, is that strong thermal fluctuations cause the vortex lattice to melt into a *vortex fluid* well below the temperature where the local superconducting order parameter is driven to zero[3,4]. In the fluid phase, the resistance of the material remains high, as illustrated in figure 1.

Thus the high-field phase diagram is as illustrated in fig. 2b This is to be contrasted with the

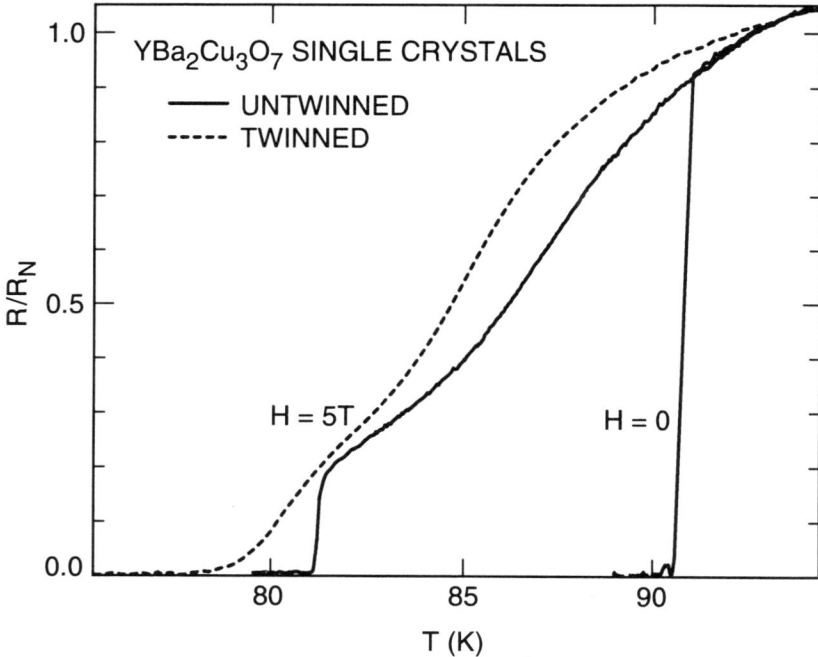

Figure 1. Resistance vs. Temperature for clean (solid) and dirty (dashed) YBCO. The mean field transition is close to the H=0 point. The high resistance below this point is evidence for a vortex fluid. The abrupt drop in the clean data is the vortex lattice melting, which turns into a continuous vortex glass transition in the presense of disorder.

phase diagram shown in figure 2a for conventional superconductors where thermal fluctuations are unimportant. For magnetic fields greater than the upper critical field, as estimated within mean-field theory ignoring thermal fluctuations, H_{c2}^{MF}, the local superconducting order parameter, ψ, is driven to zero: this is the normal state. In the absence of thermal fluctuations and pinning induced disorder, the vortex lattice would form for all fields below H_{c2}^{MF}. However, in the presence of strong thermal fluctuations, as are important in the copper-oxide superconductors, the vortices do form in the vicinity of H_{c2}^{MF}, but remain in a strongly fluctuating vortex fluid state down to significantly lower temperatures and fields before freezing. The resulting superconducting vortex glass is discussed below. Such a vortex fluid regime had been known to occur for thin-film superconductors[5]; the copper-oxide superconductors are the first bulk materials where its existence has become readily apparent. Before discussing why this is the case, and what this implies for the materials' resistivity, let us first briefly discuss what happens at much lower fields.

When the magnetic field penetrating a type II superconductor is very small, it penetrates in the form of isolated flux lines, each carrying a vortex line and a quantum, ϕ_0, of magnetic flux. A cross-section of one of these vortex/flux lines reveals two characteristic length scales of the superconductor: The magnitude of the order parameter, ψ, is significantly suppressed only in the *core* of the vortex, which is of size, ξ, the superconducting coherence length. At low

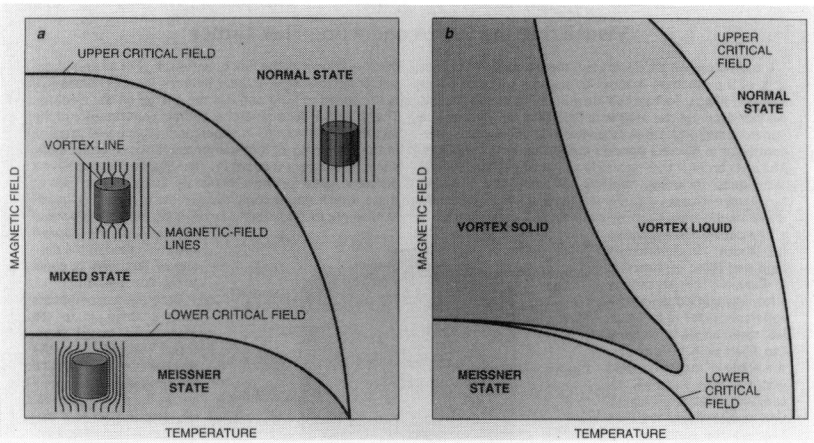

Figure 2. Schematic phase diagrams are shown for conventional superconductors where thermal fluctuations are unimportant a) and high T_c superconductors b) as functions of temperature, T, and applied magnetic field, H. Note that the field scale is highly distorted in b. For the copper oxide superconductors $H_{c1}(T=0)$ is at least several orders of magnitude smaller than $H_{c2}(T=0)$. The range of stability of the reentrant vortex liquid is uncertain. Although observed in Bitter patterns at low temperatures, it may only represent a frozen vestige of equilibrium near T_c.

temperatures ξ measures the spatial extent of the Cooper pairs of electrons and is believed to be less than 20 Å for the copper oxide superconductors. This very small coherence length is one thing that makes the copper oxide materials qualitatively different; previously known superconductors have much larger coherence lengths. The coherence length grows as the temperature is increased and diverges to infinity at the transition temperature, T_c.

The magnetic field in an isolated vortex/flux line is confined by circulating screening currents in the vicinity of the vortex, with the field intensity decaying approximately exponentially as one moves far away from the flux line, the decay length being the magnetic penetration depth, λ. This length ($\lambda \geq 1400$ Å for YBCO) far exceeds the coherence length, ξ, in the copper-oxide superconductors. A superconductor is type II when $\lambda > \xi/\sqrt{2}$; the copper-oxide superconductors are *strongly* type II, i.e. $\lambda \gg \xi$ or the Ginzburg-Landau parameter $\kappa \equiv \lambda/\xi \gg 1$. The low field regime is where the spacing between flux lines a_o is greater than λ so the tubes of flux do not strongly overlap. For $\lambda \approx 1400$ Å this corresponds to fields below 1000 gauss. When one is well within this regime, the pattern of magnetic flux emerging from the surface of a sample can be imaged by the Bitter decoration technique; results of such studies are described below. The interaction energy between flux lines also decays exponentially with decay length λ; for a perfectly clean material the vortex/flux line lattice should then melt at low fields (thus large spacing between flux lines) when this interaction energy, which stabilizes the lattice, becomes too small to withstand the thermal fluctuations of the flux lines. At these low fields the vortex lattice will also be easily disordered by random pinning. Such a result has been found in the decoration experiments and will be discussed below.

The high-field regime is where the flux lines are strongly overlapping so the spacing between them is less than the magnetic penetration length, λ. In this regime the magnetic field in the material is fairly uniform; it remains higher at the vortex cores than in between the vortices, but the difference is smaller than the average field. In the extreme Type II limit this difference is proportional to ϕ_0/λ^2. Thus, in this regime, it is not really appropriate any longer to describe the system as one of flux lines. However, the positions of the vortices and their normal cores remain well defined because $a_o \gg \xi$ where a_o is the spacing between vortices.

There are four (not completely unrelated) factors that make the vortex fluid regime in the high-field phase diagram large for the copper oxide superconductors. These are: (i) high temperatures that cause larger thermal fluctuations; (ii) small coherence lengths ξ, which allow the vortices to form at a high field $H_{c2}^{MF} \simeq \phi_0/(2\pi\xi^2)$; (iii) large penetration depths λ, since the interactions between vortices in this high field regime are proportional to $1/\lambda^2$ and therefore small; (iv) strong anisotropy: these layered materials have very anisotropic normal state conductivity and supercurrent densities, reflecting the fact that the carriers move readily within the copper-oxide layers but hop much less readily between layers. The effective mass anisotropy $\Gamma \equiv (m_c/m_a)^{1/2}$ can be at least as high as 60, as compared to 1 in conventional superconductors. This anisotropy results in a reduced interaction between vortices in different copper-oxide layers. The reduced interactions here and in (iii) allow larger thermal fluctuations of the vortices.

What do the vortices have to do with the important practical property of a superconductor, namely its electrical resistivity? For fields below H_{c2}^{MF} the primary source of resistivity is dissipation due to motion of vortices across the current[6]. Thus in order to make a type II superconductor with vortices really have zero resistivity, one must prevent all the vortices from moving. This actually does happen at low enough temperatures because the vortices get pinned to imperfections in the materials. As discussed above, the superconducting order parameter, ψ, is suppressed in the core of a vortex. The energy cost of this suppression of the superconductivity will depend on the local environment, typically being reduced near chemical or structural imperfections in the material where the superconductivity is weaker. At low temperatures, the vortex will thus tend to get pinned at such places where it has a lower energy.

What effect do such randomly placed pinning centers have on the Abrikosov vortex lattice? Larkin and Ovchinikov[7] calculated the resulting distortions of the lattice, showing that the long-range crystalline order of the lattice is destroyed by even weak pinning. The resulting vortex pattern has short-ranged crystalline order, but the crystalline order is disrupted at long distances. However, as we will show below, the lattices can still have quite long range orientational order. An important question to ask about the resulting pattern is: Are the vortices mobile, resulting in a nonzero resistivity? The answer to this question was believed for many years to be "yes". The total pinning energy for each finite region with short-range vortex lattice order is finite. Thus if each such region of vortices is assumed to be able to move without consideration of its interactions with other regions, the free energy barriers, U_o, that would have to be surmounted are finite. This assumption leads to a thermally activated resistivity proportional to $\exp(-U_o/k_B T)$ which may be very small, but remains nonzero for all positive temperatures.

However, it has recently been argued by Fisher[8] and Fisher, Fisher and Huse[9] that random pinning instead turns the vortex lattice phase into a vortex glass phase, where the vortices are frozen into a particular random pattern that is determined by the details of the pinning in the particular sample being considered. In this vortex glass phase the vortices are not mobile so the ohmic linear resistivity is strictly zero below the phase transition into this phase at a temperature T_g. The phase is named vortex glass by analogy with the spin glass phase of random magnetic materials, and was first introduced for random arrays of Josephson junctions by Shih, Ebner and Stroud[10].

In both the vortex fluid and vortex glass phases, an instantaneous snapshot of the vortex pattern shows no apparent long-range order. However, in the vortex fluid phase the vortex pattern is constantly rearranging, so the correlations between the superconducting order parameter, $\psi(\vec{r})$, at pairs of points in space decay with distance between the two points, vanishing at large distances. In the vortex glass phase, on the other hand, there are long-range correlations between $\psi(\vec{r})$ and $\psi(\vec{r}')$ even for pairs of points $\vec{r}$ and $\vec{r}'$ that are well separated. These correlations are not in a simple pattern, but rather in a static but random pattern that is determined by where all the vortices are located in their frozen configuration. As the vortex glass phase is approached from the vortex fluid phase these long-range correlations develop continuously, with the vortex-glass correlation length, $\xi_{VG} \sim (T-T_g)^{-\nu}$, diverging as a power of the temperature difference from the transition. The scaling theory of this continuous phase transition has been quite nicely confirmed by experiments by Koch et al[11] and Gammel et al[12] on samples of YBCO(123), as will be discussed below.

Having now briefly summarized some of the theoretical ideas about flux line/vortex patterns and dynamics in type II superconductors, let us now describe some of the recent experiments done on the copper-oxide high temperature superconductors. These experiments are of two basic types: the first directly probe the patterns of magnetic flux by magnetic decoration, while the second probe the dynamics by studying the linear and nonlinear electrical conductivity.

2. Static Flux Lattice Structures For Low Temperatures and $\vec{H} \| \hat{c}$

Direct information on the ordering of the magnetic vortices in the mixed state of the high T_c superconductors can be obtained through the use of the Bitter imaging technique[13] in which samples are cooled in an applied magnetic field and subsequently exposed to a smoke of ferromagnetic particles formed by evaporation into a helium buffer gas. The technique was pioneered by Trauble and Essman[14] and Sarma[15] to study individual vortices. The apparatus used in the present experiments is sketched in figure 3. The ferromagnetic particles travel down magnetic field lines outside the surface of the superconducting sample and form clusters on the surface which decorate the locations of the vortices. The particles stick to the surface with van der Waals forces. The applied field is then removed, the sample warmed to room temperature, and the clusters of particles are viewed using an electron microscope.

Direct real space imaging of the arrangement of individual vortices provides information on both the transitional and bond-orientational order of a two-dimensional slice of the vortex lattice as it pierces the sample surface. Present experiments have been limited for the most part to field-cooled samples at T = 4K, subsequently viewed by scanning electron microscopy, for which sufficient contrast is obtained when vortices are separated by roughly $a_o > 0.3$ μm (H < ~200 Oe). Below this separation (or above this field) the ferromagnetic particles have a tendency to form strings by dipole-dipole interaction and the decorated image becomes difficult to interpret. One must take into account the demagnetizing factor of the sample in order to determine the actual magnetic field in the bulk of the sample. Most of the samples studied to date are thin slabs of ~1mm extent along the $\hat{a}$, $\hat{b}$ axes and ~5–30 μm thickness along $\hat{c}$. The $\hat{a}$ and $\hat{b}$ axes span the copper oxide layers of these materials, while $\hat{c}$ is normal to the layers. For $\vec{H}$ along $\hat{c}$, vortices penetrate the sample at H~0.5 Oe, rather than the measured H_{c1} of ~150G, due

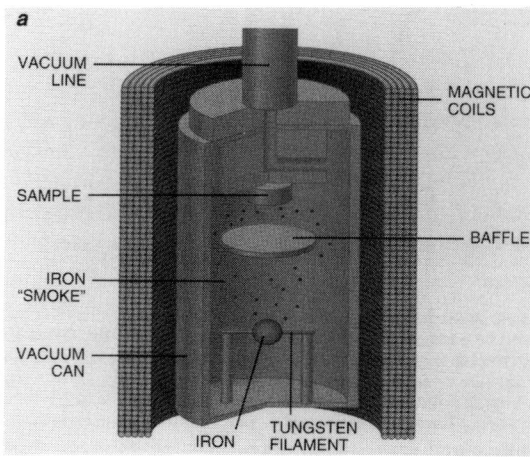

Figure 3. A schematic of the decoration apparatus. The sample to filament distance is 2-3cm.

to this demagnetizing effect. The Bitter decoration technique is limited to a static snapshot of the vortex lattice arrangement in the sample averaged over the time required to decorate — about 1 second. Also, since there is pinning and possibly entanglement of the vortices, the temperature during the sample cool down at which the microscopic arrangement of vortices goes out of equilibrium and freezes into the resulting Bitter pattern can be quite high. In general, this temperature lies somewhere near the temperature at which the bulk DC magnetization goes out of equilibrium, T_{irr} which for these low fields is quite close to T_c.

The present Bitter decoration experiments have established the following important points about the mixed state of the high T_c superconductors for $\vec{H}$ parallel to $\hat{c}$: 1) the vortices exist as hexagonally correlated, singly quantized vortices with one flux quantum, hc/2e, per vortex as in a conventional Type II superconductor[16]; 2) the vortices experience pinning at twin boundaries, crystal defects and individually at other intrinsic lattice sites in ostensibly defect-free regions[17]; 3) in twin-free samples, rather than the long range translational order expected in the crystalline state of the Abrikosov vortex lattice, the vortex arrangement has short range translational order with correlation lengths on the order of a few nearest neighbor spacings, but long range bond-orientational (hexatic) order which extends over ten to several hundred nearest neighbor spacings[18]; 4) a rather sharp transition with applied field is observed in BSCCO(2212) between isotropic disorder in the vortex arrangement for H < 20 Oe and hexatic order at higher fields; this transition occurs at considerably lower field (H~8 Oe) for samples which have been annealed in oxygen; for both types of samples the translational and bond-orientational order in the hexatic arrangement increase monotonically with field up to 100 Oe[19], 5) motion of individual vortices comparable to their separation within the 1 sec decoration time appears to occur at 15K, presumably due to thermal motion of the vortices[20] 6) the vortex lattice exhibits the expected ~10% anisotropy from a perfect hexagonal structure in the a,b plane in YBCO(123) due to the in-plane effective mass anisotropy of the electrons,[21,22] but a smaller anisotropy of ~3-10% rather than the nearly 30% expected in BSCCO(2212) from a-b mass anisotropy[18,19,21].

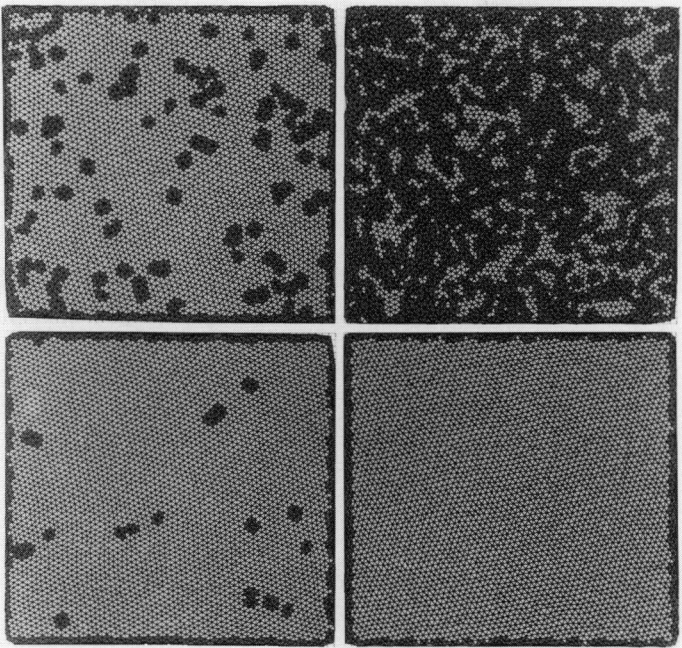

Figure 4. Shown are Delaunay triangulations for image-processed scanning electron micrographs of bitter decorated oxygen-annealed BSCCO crystals cooled to 4.2K in magnetic fields of 69 Oe, 23 Oe, 17 Oe and 8 Oe respectively. The fields of view are adjusted to keep a fixed number, roughly 4000, of vortices in the image. Shaded triangles join vertices which are not six-fold coordinated.

The advantages of using BSCCO(2212) in the decoration experiments are that excellent quality untwinned single crystals can be obtained which can be cleaved to expose a good surface layer for decoration. In addition, the Ginsburg-Landau parameter $\kappa = \lambda/\xi \approx 200$ and the a-c carrier mass anisotropy parameter $\Gamma \geq 55$ for BSCCO(2212), make it a rather exotic superconductor. For comparison, YBCO(123) has $\kappa \approx 100$ and $\Gamma \approx 5$, and a conventional Type II superconductor such as Nb_3Sn has $\kappa \approx 20$ and $\Gamma = 1$.

In Figure 4a-d are shown defect maps known as Delaunay triangulations of the arrangement of roughly 4000 vortices from three different BSCCO samples cooled to 4K in fields parallel to $\hat{c}$ of 8 Oe, 17 Oe, 23 Oe and 69 Oe, respectively[19]. These samples had been previously annealed at 600°C for ~24 hrs in 1 atm oxygen and then quenched to room temperature. Indications are that the oxygen annealing process probably does not greatly affect either κ or Γ when compared to those of the as-made samples[23], but does reduce the concentration of oxygen vacancies[24] which could serve as pinning centers for vortices. In the defect maps shown, each vortex center is represented as a vertex of nearest neighbor bonds. Non-sixfold coordinated centers, defects in a perfect hexagonal array, are shaded in the figure. The vortices are quite disordered at 8 Oe, whereas much less so at 23 Oe, and no defects in the vortex lattice are visible in the field of view at 69 Oe. The translational correlation length ξ_G, as determined from exponential fits to the

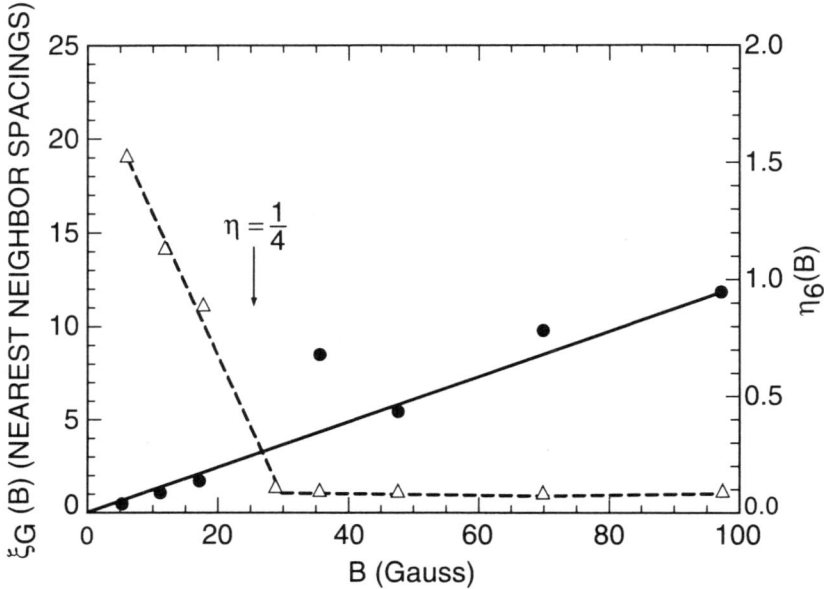

Figure 5. Shown are the translational correlation lengths ξ_G for as made samples in units of nearest neighbor spacings (closed) and the bond-orientational correlation exponents η_6(open). The points at $\eta_6 \approx 0.06$ are at our limit of experimental resolution. the arrow at $\eta = .25$ is the Hexatic phase boundary.

correlation function of the translational order parameter of the vortex lattice $\psi_G(\vec{r}) = e^{i\vec{G}\cdot\vec{r}}$, where $\vec{r}$ is a vortex position, is shown vs. H in Fig. 5 for these samples. A monotonic increase of ξ_G from ~2 a_o at 5 Oe to ~20 a_o is observed for the annealed samples, whereas the as-made samples have ξ_G roughly half that value at each field.

In the 23 Oe and 69 Oe defect maps, Fig. 4c,d one can easily sight down rows of vortices, despite the relatively small value of ξ_G compared to the size of the image. This is the signature of an hexatic, which exhibits short range translational order and long range bond-orientational order[25,26,27]. The bond-orientational order of the vortex lattice is characterized by an order parameter $\psi_6(\vec{r}) \equiv e^{i6\theta(\vec{r})}$. The correlation function of this bond-orientational order parameter measures the correlation of a bond angle $\theta(\vec{r})$ at $\vec{r}$ (modulo $2\pi/6$) with that at the origin. Assuming a power law dependence for the decay of correlations of the form $<\psi_6^*(\vec{0})\cdot\psi_6(\vec{r})> \sim r^{-\eta_6}$, one can extract a correlation exponent, η_6, from fits to computed bond-orientational correlation functions from the measured position of vortex centers[28]. The exponents are shown for the same series of decorations in Fig. 5. Immediately obvious from the figure is the abrupt change of the fitted η_6 from a relatively large value of 0.8, a rather steep decay of the bond-orientational order, to the limits of the experimental resolution, 0.06, where one can say it does not decay at all to within our experimental resolution. The change occurs rapidly in only a few Oe applied field, but for different fields for the as-made and annealed samples, presumably reflecting the change in the concentration of intrinsic pinning sites in the two types of sample. This abrupt change in the bond-orientational order with H is difficult to reconcile with theories which only include a range of pinning energies but include no phase transition. These data are consistent with the predictions of an isotropic vortex fluid-hexatic vortex glass[8,9,27] or hexatic vortex fluid phase transition[29], or a transition between a strongly pinned disordered glassy phase to a less strongly pinned hexatic near H_{c1}[30]. Experimental data on vortex mobility vs. temperature and the microscopic irreversibility temperature vs. H are needed to discriminate among predictions and will be discussed in the following section.

3. Vortex Dynamics

Up until now we have been discussing static structures of the vortex/flux-line lattices in these materials. We will now discuss the dynamics. Early evidence for unconventional behavior of the dynamics in these systems came from measurements of the decay of magnetization in ceramic LaSrCuO[31]. In those experiments, an irreversibility line was found which has been taken as evidence for thermally activated depinning of the vortex lines[32,33]. This is a fundamentally single particle view of the dynamics. The basic idea is that either individual vortex lines or bundles of small numbers of vortex lines are thermally activated over pinning barriers which are present in the sample due to disorder. In this picture, the resistivity should be thermally activated with the functional form $\exp(-U_0/k_B T)$ which is nonzero at all nonzero temperatures.

A different and more controversial point of view was put forth as a result of subsequent high Q oscillator measurements by Gammel, et al[3]. These experiments suggested that this transition was vortex lattice melting from a low temperature ordered phase into a high temperature vortex fluid. The high temperatures, short coherence lengths, and large ĉ-axis anisotropies were postulated to increase the importance of thermal fluctuations to allow a vortex lattice melting transition similar to that both predicted and found in two-dimensional films[5,34]. It was also suggested[3] that a Lindemann criterion for vortex lattice melting[35,36] could satisfactorily explain the transition temperatures found in these systems.

The melting or phase transition picture takes the point of view that many-body effects are crucially important and must be taken into account if one is to understand the statics and dynamics of the vortex arrays in these systems. In this picture there must be a low temperature ordered phase with a broken symmetry which then undergoes a phase transition into a high temperature disordered vortex fluid phase. This transition is driven by thermal fluctuations. As discussed previously, one such candidate ordered ground state with long range orientational order but short range positional order has been found in recent low field decoration experiments.

During the ensuing controversy about the melting idea, it was correctly pointed out that the simple idea of melting fails to take into account the disorder in the lattice which must be present due to random pinning of the flux lines. However, more recently theories which do include pinning disorder but also find a phase transition have evolved. Following the work of Shih, Ebner and Stroud[10], Fisher, Fisher and Huse[9] have postulated a vortex glass transition. In this picture the vortices at low temperatures are frozen into a particular random configuration as determined by the details of the pinning centers in the specific sample. In the vortex glass phase, the vortices are not free to move and so the ohmic linear resistivity is strictly zero.

We now know that both melting and vortex glass phase transitions can occur, with the distinction determined by the amount of pinning disorder in the material. The data in figure 1 already show the essential difference in the two regimes. In clean material, the sudden drop in resistance at $\sim 0.15 R_N$ is the signature of a first order melting transition[37] and is well described by a Lindemann criterion[38]. Indeed, in the clean limit, the sudden drop is actually found to

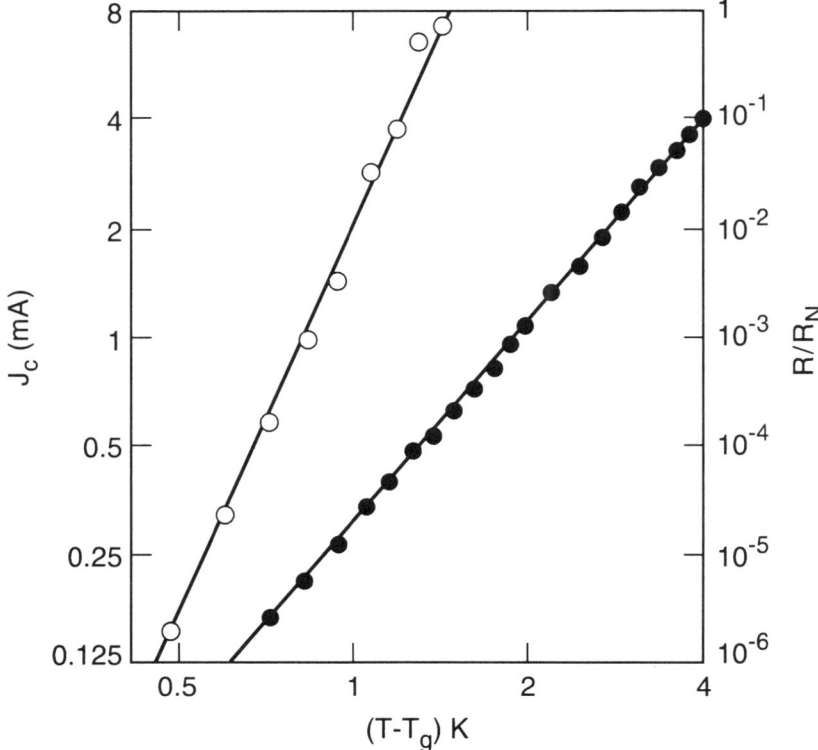

Figure 6. Shown are the linear resistivity R and the current scale for linear response J_{sc} vs. the reduced temperature $(T-T_g)$ on a log-log plot for a YBCO(123) single crystal at an applied field of 6 Tesla. The straight lines are the fits to the scaling theory for the vortex glass phase transition.

exhibit hysteretic behavior associated with a first order transition. In dirty material, the sudden drop turns into a smooth shoulder as seen in figure 1. However, the phase transition into a zero resistance state is still present, but now as a second order glass transition. Koch et al[11] have found evidence for such a transition on YBCO films. However claims were made that the data could still be fit by conventional thermally activated behavior[39,40]. However, recent measurements by Gammel, et al. [12] with picovolt sensitivity on YBCO(123) single crystals with roughly six orders of magnitude greater sensitivity than the previous measurements have provided even more compelling support for the vortex glass phase transition. Those measurements will be discussed in more detail here.

In the vortex glass model there is a true phase transition at T_g, between vortex fluid and vortex glass phases. Associated with this transition there is a diverging correlation length given by $\xi_{VG} \sim (T-T_g)^{-\nu}$ and a diverging correlation time $\tau \sim \xi_{VG}^z$. The vortex glass model makes several predictions about the resistivity as a function of temperature, magnetic field and current for a three dimensional type II superconductor near this phase transition. The first prediction is that the linear response resistivity should vanish near T_g as $R \sim (T-T_g)^{+(z-1)\nu}$. The second

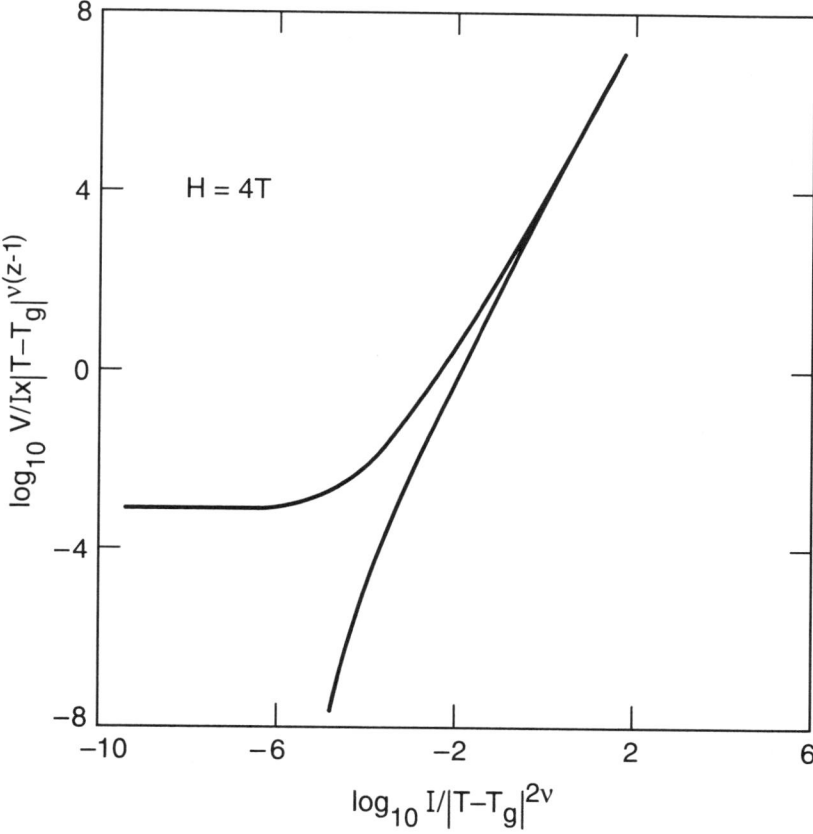

Figure 7. Scaled I-V curves at different temperatures from the thin film data[11]. The upper curves are above T_g, and the lower curves are below T_g. The scaling function is similar to that obtained using single crystals[12].

prediction is that the current scale for linear response should vary as $J_{sc} \sim (T-T_g)^{2\nu}$. The physical idea for this vanishing current scale for linear response is as follows: the measuring current defines a length scale by $L \sim (ck_B T/\phi_0 J)^{1/2}$. This is the length scale below which the effects of current on the thermal distributions of vortices are linear. In order to remain in the linear response regime, we must have $L > \xi_{VG}$. The response goes nonlinear at a current density J_{sc} where $L \sim \xi_{VG}$. As $T \to T_g$ the correlation length, ξ_{VG}, diverges to infinity and the current scale for linear response therefore vanishes.

The alternative point of view, that of thermally activated flux flow (TAFF), says that the resistance should go as $R \sim R_0 \exp(-U_0/k_B T)$ and that the current scale for linear response should be proportional to T. It was our reasoning that measurements at and near linear response at the picovolt level would provide the biggest testable differences between the various theories. To that end we have built a special squid picovoltmeter which allowed us to measure voltage vs.

current curves in magnetic fields up to 6 Tesla with sub-picovolt sensitivity. Our measurements of the temperature dependence of both the linear response resistivity and the onset of nonlinear response strongly constrain theoretical fitting parameters and have allowed us to rule out the class of models which attempt to explain the dynamics in these systems as solely due to thermally activated hopping over barriers. Shown in Fig. 6 is a log-log plot of the linear response resistivity R and the current scale for linear response J_{sc} for a YBCO(123) single crystal at a field of 6T. Both quantities vanish as powers of $(T-T_g)$ with $T_g = 74.0K$. In any TAFF model, such behavior for the linear resistivity and the current scale for linear response is impossible to obtain. In a TAFF model both quantities are only singular at $T = 0$. For example in figure 6, TAFF models would predict J_{sc} to be an essentially horizontal straight line on the scales shown here. These TAFF predictions are clearly at odds with the data. From measurements such as shown in Fig. 6 we can obtain the critical exponents for the vortex glass phase transition. From the data in figure 6 we obtain $\nu = 2.0 \pm 1$ and $z = 4.5 \pm 1.5$ for YBCO(123) single crystals.

Using the scaling of the current and the resistance, it is actually possible to collapse the entire current-voltage characteristic. Such a collapse of the Koch[11] data is shown in figure 7. Above the glass transition, the I-V curves all shown resistive behavior at low current, and collapse onto the upper curve. Below the glass transition, the downward curvature of the scaling plots shows the signature of the zero resistance state. At T_g, the I-V curve is an exact power law, whose expponent is $(z+1)/2$ in the vortex glass theory.

Shown in figure 8 is data on the BSCCO(2212) system[41]. In this case a plot of the effective activation energy from an Arrhenius plot is shown versus temperature. The low temperature increase suggests that a vortex glass transition is present in this system as well. In this type of plot the vortex glass would correspond to an activation energy diverging as $T^2\nu(z-1)/(T-T_g)$. For temperatures below T^*, this provides an excellent fit, with the glass transition temperature indicated on the plot. In both YBCO(123) and BSCCO(2212) the general phase diagram follows the form of figure 2b. The H_{c2} line that represents where the transition would be if one could neglect thermal fluctuations now only markes a crossover into the regime of a strongly fluctuating vortex fluid. The low field behavior of the phase boundaries is both theoretically and experimentally an open question. For example it is not now clear how the low field heaxtic glass phase boundary seen in decoration experiments should join the phase transition line as measured with high field probes such as the picovoltmeter. More work needs to be done.

At the moment results on YBCO(123) films[11], YBCO(123) single crystals[12] and single crystals of BSCCO(2212)[41] all show convincing proof of the type shown above for the vortex glass model. It is worthwhile pointing out why so many workers in the field find "evidence" for thermal activation models and fail to see the vortex glass phase transition. In the best of circumstances, it is experimentally quite hard to tell the difference between a large power law and an exponential law for the resistivity vs. temperature. In order to do so one needs to be able to follow the dependence over a wide range. Experiments which probe the system either at too high a current level or in ceramic materials which cut off the diverging vortex glass correlation length at the grain size, will never be able to see the transition. This does not mean that it does not exist, merely that most experiments will not be able to probe it. For example, in BSCCO(2212), in a field of 3 Tesla, the vortex glass critical region corresponds to a typical sample resistance of $10^{-7}\,\Omega$ and below. With typical measuring currents of 1 ma to avoid sample heating, one finds that voltage sensitivities of 10^{-10} volts and better are needed in order to probe the critical region. This is far outside the voltage sensitivity of most experiments. Thus many conventional experiments do not have the sensitivity to probe the vortex glass transition. However it is

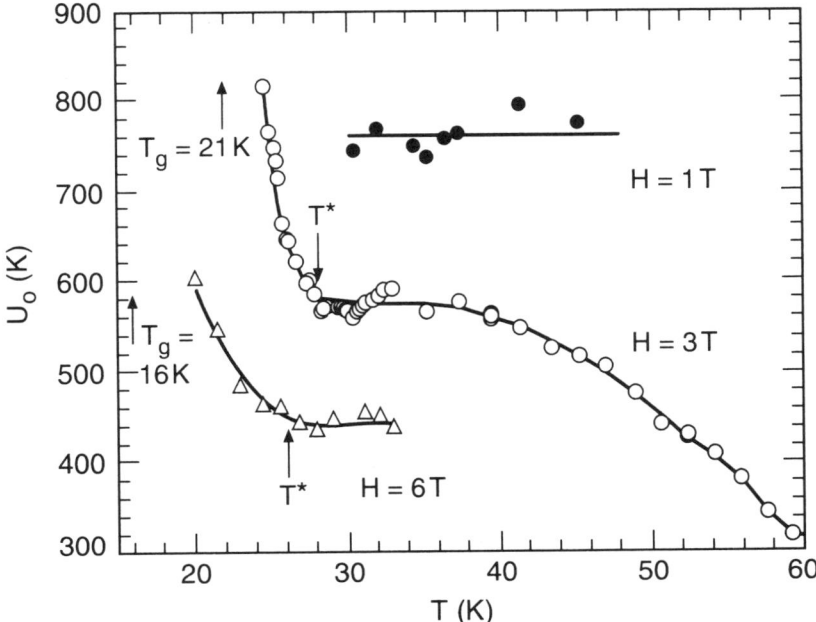

Figure 8. Effective activation energy vs temperature for BSCCO(2212). At high fields, the low temperature upturn signals the nascent vortex glass transition. The fitted T_g is shown by the arrows. At high temperatures, $T > T^8$, the vortices are two-dimensional.

important to remember that in experimental physics the absence of proof should not be confused with the proof of absence.

4. Conclusions

The implications of the data presented here are significant. For a long time it was believed that in a magnetic field the resistance of a Type II superconductor only became strictly zero at $T = 0$. We now know that there exists a finite temperature phase transition at high fields at which the resistance becomes zero. We now know in some detail what the nature is of the low temperature ordered ground state at low fields. It is an hexatic, a state not seen in many experimental systems. We are slowly making real progress in our understanding of the vortex/flux lattice statics and dynamics. The oxide superconductors have proven to be a wonderful testing ground for our understanding in this area and have forced us to re-examine, extend and in some cases discard certain theoretical models. There remain many unanswered questions. These include the role of anisotropy, a microscopic understanding of the critical currents in these systems and the behavior of the lattices in the very clean limit. There is much work to be done.

The authors would like to thank David Nelson, Daniel Fisher, Matthew Fisher, Peter Littlewood, Chandra Varma, Gerald Dolan, Tom Palstra and Betram Batlogg for many helpful discussions and acknowledge fruitful collaborations with Christian Bolle, Gerald Dolan, David Grier, Aron Kapitulnik, Raynien Kwo, David Mitzi and Lynn Schneemeyer.

5. References

1. for a review see A. L. Fetter and P. C. Hohenberg, in *Superconductivity*, edited by R. D. Parks (Dekker, New York, 1969) vol. 2.

2. B. Oh, K. Char, A.D. Kent, M. Naito, M.R. Beasley, T.H. Geballe, R.H. Hammond, A. Kapitulnik and J.M. Graybeal, Phys. Rev. B*37*, 7861 (1988).

3. P. L. Gammel, L. F. Schneemeyer, J. V. Waszczak and D. J. Bishop, Phys. Rev. Lett. *61*, 1666 (1988). D.J. Bishop, Bull. Am. Phys. Soc. *33*, 606 (1988).

4. D. R. Nelson and H. S. Seung, Phys. Rev. B*39*, 9153 (1989).

5. D. S. Fisher, Phys. Rev. B*22*, 1190 (1980); P. L. Gammel, A. F. Hebard and D. J. Bishop, Phys. Rev. Lett. *60*, 144 (1988).

6. Y. B. Kim and M. J. Stephen, in *Superconductivity*, edited by R. D. Parks (Dekker, New York, 1969) vol. 2.

7. A. I. Larkin and Yu. N. Ovchinikov, J. Low Temp. Phys. *34*, 409 (1979).

8. M. P. A. Fisher, Phys. Rev. Lett. *62*, 1415 (1989).

9. D. S. Fisher, M. P. A. Fisher and D. A. Huse, Phys. Rev. B*43*, 130 (1991).

10. W. Y. Shih, C. Ebner and D. Stroud, Phys. Rev. B*30*, 134 (1984).

11. R. H. Koch, V. Fogliette, W.J. Gallagher, G. Koren, A. Gupta and M.P.A Fisher, Phys. Rev. Lett. *63*, 1511 (1989).

12. P. L. Gammel, L. F. Schneemeyer and D. J. Bishop, Phys. Rev. Lett. *66*, 953 (1991).

13. For a review see: R. P. Huebner, "Magnetic flux structures in superconductors," Springer-Verlag (Berlin, 1979).

14. H. Trauble and U. Essmann, J. Appl. Phys. *25*, 273 (1968).

15. N. V. Sarma, Philos. Mag. *17*, 1233 (1968).

16. P. L. Gammel, D. J. Bishop, G. J. Dolan, J. R. Kwo, C. A. Murray, L. F. Schneemeyer, J. V. Waszczak, Phys. Rev. Lett. *59*, 2952 (1987).

17. G. J. Dolan, G. V. Chandrashekhar, T. R. Dinger, C. Field, and F. Holtzberg, Phys. Rev. Lett. *62*, 827 (1989).

18. C. A. Murray, P. L. Gammel, D. J. Bishop, D. B. Mitzi and A. Kapitulnik, Phys. Rev. Lett. *64*, 2312 (1990).

19. D. G. Grier, C. A. Murray, C. A. Bolle, P. L. Gammel, D. J. Bishop, D. B. Mitzi and A. Kapitulnik, Phys. Rev. Lett. *66*, 2270 (1991).

20. R. N. Kleiman, P. L. Gammel, L. F. Schneemeyer, J. V. Waszczak, and D. J. Bishop, Phys. Rev. Lett. *62*, 2331 (1989).

21. G. J. Dolan, F. Holtzberg, C. Field and T. R. Dinger, Phys. Rev. Lett. *62*, 2184 (1989).

22. L. Ya. Yinnikov, J. V. Grigoriera, L. A. Gurevich and Yu. A. Ossipyan in *High Temperature Superconductivity from Russia* (World Scientific, London, 1989), eds. A. I. Larkin and N. V. Zavaritsky.

23. S. Martin, private communication.

24. F. Parmigiani, Z. X. Shen, D. B. Mitzi, I. Lindau, W. E. Spicer and A. Kapitulnik, Phys. Rev. B*43*, 3085 (1991).

25. B. I. Halperin and D. R. Nelson, Phys. Rev. Lett. *41*, 121 (1978).

26. D. R. Nelson, M. Rubinstein and F. Spaepen, Philos. Mag. A*46*, 105 (1982).

27. E. M. Chudnovsky, Phys. Rev. B*40*, 11355 (1989).

28. For a review see: D. R. Nelson, in "Phase Transitions and Critical Phenomena," Vol. 7, ed. C. Domb and J. L. Lebowitz (Academic, London, 1983), p. 1.

29. M. C. Marchetti and D. R. Nelson, Phys. Rev. B*40*, 1910 (1990).

30. D. R. Nelson and P. LeDousal, Phys. Rev. B*42*, 10113 (1990).

31. K. A. Muller, M. Takashige and J. G. Bednorz, Phys. Rev. Lett. *58*, 1143 (1987).

32. A. P. Malozemoff, L. Krusin-Elbaum, D. C. Cronemeyer, Y. Yeshurun and F. Holtzberg, Phys. Rev. B*38*, 6490 (1988).

33. Y. Yeshurun and A. P. Malozemoff, Phys. Rev. Lett. *60*, 2202 (1988).

34. B. A. Huberman and S. Doniach, Phys. Rev. Lett. *43*, 980 (1979).

35. A. Haughton, R. A. Pelcovits and S. Sudbo, Phys. Rev. B*40*, 6763 (1989).

36. M. A. Moore, Phys. Rev. B*39*, 136 (1989).

37. H. Safar, P.L. Gammel, D.A. Huse, D.J. Bishop, J.P. Rice and D.M. Ginsberg, Phys. Rev. Lett. *69*, 824 (1992). T.K. Worthington, et al., Cryogenics *30*, 417 (1990). M. Charalambous et al., Phys. Rev. B*45*, 45 (1992).

38. W.K. Kwok, et al., Phys. Rev. lett. *69*, 3370 (1992).

39. R. Griessen, Phys. Rev. Lett. *64*, 1674 (1990).

40. S. N. Coppersmith, M. Inui and P. B. Littlewood, Phys. Rev. Lett. *64*, 2585 (1990).

41. H. Safar, P. L. Gammel, D. J. Bishop, D. B. Mitzi and A. Kapitulnik, Phys. Rev. Lett. *68*, 2672 (1992).

THEORY AND EXPERIMENT OF FLUX PINNING AND FLUX CREEP IN HIGH T_c SUPERCONDUCTORS

P.H. KES
Kamerlingh Onnes Laboratory, Leiden University
P.O. Box 9506, 2300 RA Leiden, The Netherlands

May 4, 1993

Abstract

The vortex lattice in extremely anisotropic superconductors has many intriguing properties. These peculiarities are very predominant in the oxidic cuprates, especially the Bi and Tl compounds, where they determine new regimes in the field- temperature phase diagram. They are also responsible for the weak flux pinning in extended regions of the mixed state, and cause a relatively large flux creep decay of the magnetization even at low temperatures. In the first part of this contribution several pin mechanisms in Bi:2212 and Y:123 are discussed and compared with experimental results, e.g. pinning by (oxygen) vacancies in the CuO_2 layers and by screw dislocations in Y:123 films. In the second part we concentrate on thermally activated phenomena, mainly flux creep. Numerical solutions of the creep equation by Van der Beek et al. are compared to an approximate analytical solution both for collective creep and a logarithmic creep model. Finally, experimental results of magnetic relaxation in Bi:2212 single crystals are analyzed in terms of the models just described.

1 Introduction

Recent experiments by Farrel et al. [1] proved the large anisotropy of the superconducting state in the Bi and Tl- compound high-temperature superconductors (HTS). Though less extreme as the resistivity anisotropy in the normal state which has recently been reviewed by Batlogg [2], the large values of the anisotropy parameter Γ are responsible for a total renovation of our view on the vortex-lattice (VL) properties in layered superconductors. The superconductivity is purely located in the conducting planes which are weakly coupled via Josephson tunneling [3]. The conducting planes consist of double or triple CuO_2 layers (with Ca layers in between) seperated by non-conducting BiO, SrO, or TlO and BaO layers. It is appealing to contribute the much smaller anisotropy of $YBa_2Cu_3O_7$ (YBCO) to the conducting CuO chain layers.

The former two compounds therefore actually form a stacking of superconducting-insulating layers (SISI) with Josephson coupling, whereas the latter compound rather is a stacking of superconducting-normal layers (SNSN) with proximity coupling.

In this contribution we will consider the phenomena occuring when a field is applied along the crystallographic c axis. For large effective mass ratios $m_c/m_{ab} \equiv \Gamma$, the field in the cores of the flux line is only screened off by supercurrents in the S layers which means that the flux lines are segmented into pancake vortices [4]. The "pancakes" in adjacent S layers are weakly coupled giving rise to a strong reduction (by a factor Γ^{-1}) of the bending modulus of a flux line as predicted by Houghton et al. [5] and experimentally verified by Koorevaar et al. [6] in the conventional layered superconductor 2H-NbSe$_2$. For the HTS the parameter Γ can be as large as 3 x 10^4 for Bi:2212 and 10^5 for Tl:2212. The interaction between the pancake vortices in adjacent S layers drops exponentially at a distance equal to the Josephson length $R_J = \Gamma^{\frac{1}{2}} s$ along the layers [7], where s is the periodicity of the SI multilayer, i.e. $s = 1.5$ nm for Bi and Tl:2212. Consequently, positional fluctuations of the pancake VL within a length scale R_J are effectively decoupled which leads to quasi two-dimensional (2D) behavior. When $a_0 \simeq (\Phi_0/B)^{\frac{1}{2}} < R_j$, i.e. $B > B_{2D}$ with $B_{2D} \equiv \Phi_0/\Gamma s^2$, the VL attains 2D properties comparable to a description valid for a thin film superconductor of thickness s [7,8,9]. We compute $B_{2D} \simeq 30$ mT for Bi:2212 and 10 mT vor Tl:2212, while it is 50 T for YBCO.

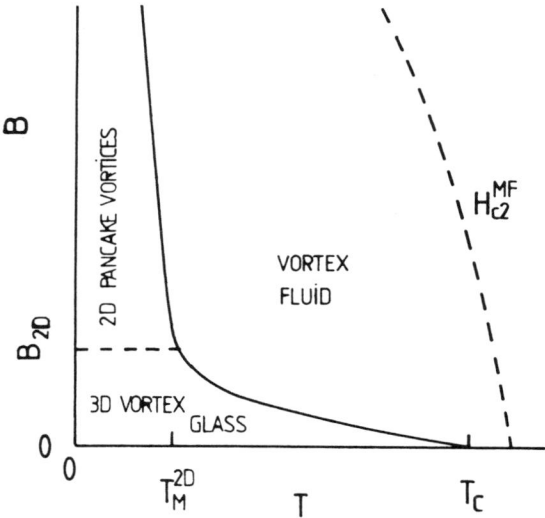

Figure 1: Phase diagram of a layered quasi- twodimensional superconductor.

Some features of the resulting $B - T$ phase diagram are given in Fig. 1 in case of the presence of pinning centres. The role of pinning is most predominant at low B and T. It destroys the long range order in the VL and leads to the 3D vortex-glass state. At higher fields and low temperatures we enter the regime of a 2D disordered (by pinning) VL. By increase of the temperature thermal fluctuations become important giving rise to both thermal depinning [10] and melting [5,11,12] of the VL. In case of weak pinning of a 2D VL melting preceeds thermal depinning [7,13]. The melting line is then given by the Kosterlitz-Thouless temperature T_M^{2D} for unbinding of vortex-dislocation pairs. The value of T_M^{2D} for Bi:2212 is estimated to be about 30 K. Melting of the 3D VL in presence of pinning has not yet been worked out in detail [9], but it is believed that it will closely follow the predictions of the Lindeman criterion $<u^2>_T \approx (a_0/6)^2$ without pinning [5,11,12], where $<u^2>_T$ is the average-squared thermal displacement of the VL. In the 3D regime thermal depinning and melting are also closely related. Raising the temperature leads to a transition to a vortex liquid state which is expected to show reversible magnetic properties away from the transition line. This vortex liquid regime, as sketched in Fig. 1, is very extended in Bi:2212.

We will now address the consequences for pinning and creep.

2 Pin mechanisms in HTS

The elementary pin interaction of a single defect with a single vortex has been computed for several pin mechanisms, e.g. for twin planes in Y:123 and grain boundaries [14], for oxygen vacancies [15] and for dislocation loops or stacking faults [16]. These are among the strongest pinning centers in bulk single crystals. In sintered bulk materials small precipitates of Y_2BaCuO_5 [17] or CuO [18] may be effective. However, their elementary interaction is only roughly known. Recent STM studies of c-oriented Y:123 films revealed a large density of screw dislocations along the growth direction [19]. We want to investigate pinning by (oxygen) vacancies and screw dislocations in some detail here. The field is always supposed to be in the c direction and $\ll B_{c2}$.

2.1 Oxygen vacancies

(Oxygen) vacancies in the CuO_2 double layers pin most effectively because the pin interaction is caused by the local change of the GL coherence length ξ and it is proportional to the maximum value of the order parameter. In the HTS the order parameter between the superconducting layers is either much smaller (Y:123) or almost zero and uniform (Bi:2212) [20]. Therefore, pinning must be due to defects in the superconducting layers. The pinning force between a single vacancy of diameter D and a single vortex is computed in [15] using Thuneberg's electron scattering formalism [21]. At $t = T/T_c < 0.6$ one gets

$$f_p \approx [10/(1+t)^4](B_c^2/\mu_0\xi)\sigma_{tr}\xi_0 \qquad (1)$$

where $\sigma_{tr} = \pi D^2/4$, ξ_0 is the BCS coherence length and B_c the thermodynamic critical field. Using $D = 0.29$ nm, the diameter of O^{2-}, $B_c = 0.3 - 0.8$ T, a maximum force at $T = 0$ is computed between 0.6×10^{-13} and 5×10^{-13} N. Note that this value does not depend on $\xi(0)$ ($= 0.74 \xi_0$), but only on $B_c(0)$. Assuming that the distance between the defects is smaller than the range of the pinning force r_f (for isolated vortices $r_f = \xi$), the collective effect of all vacancies in the vortex core should be estimated.
i. For the Bi compound in fields $B_{2D} < B < 0.1B_{c2}$ (B_{2D} is the field where the

interaction between the pancake vortices in the same layer is stronger than between pancake vortices in adjacent layers, about 0.03 T in Bi:2212) the 2D vortices are individually pinned and one finds for the total force F_v per 2D vortex [15]

$$F_v \approx (0.9 n_\Box \pi \xi^2 f_p^2)^{\frac{1}{2}} \tag{2}$$

where $n_\Box$ denotes the density of vacancies in the double layer. The pin energy follows from $U_v \approx F_v \xi$. Substituting number for the parameters [15] one obtains $U_v \approx 35$ K. The bulk pinning force F_p is computed by deviding F_v by the volume of the vortex segment, i.e. $a_0^2 s$. This gives $J_c = F_p/B \approx 5 \times 10^{10}$ Am^{-2}.

ii. In case of Y:123 the 2D vortices are not decoupled and one should make an estimate for a vortex line of length L_c given by [14]

$$L_c \approx [\Phi_0^3 B r_f^2 / (\Gamma^2 \mu_0^2 \lambda^4 W)]^{\frac{1}{2}} \tag{3}$$

with $W = 0.5 n_\Box f_p^2 \pi \xi^2 / (a_0^2 s)$ and λ the penetration depth. Substituting $n_\Box = 3.5 \times 10^{17}$ m^{-2} (this number was found for Bi:2212 [15] and means there is one vacancy per 80 oxygen atoms in each CuO$_2$ layer), $s = 1.2$ nm, $\xi = 1.5$ nm, $\lambda = 140$ nm, $\Gamma = 29$ and $f_p = 5 \times 10^{-13}$ N one gets $W \approx 0.12 B/[T]$ N^{2}m^{-3} and $L_c \approx 7$ nm. Using $F_p = (W/L_c a_0^2)^{\frac{1}{2}}$ and $J_c = F_p/B$ one obtains $J_c \approx 9 \times 10^{10}$ Am^{-2} in reasonable agreement with experimental values at low temperatures. For the pin energy one finds $U_p = F_p L_c a_0^2 \xi \approx 140$ K. In Section 3 the implication of these results for flux creep will be further discussed.

The justification for taking the limit of single-vortex pinning follows from an estimate of the elastic shear energy $U_{el} \approx 3 c_{66} (\xi / a_0)^2 a_0^2 L$ for a line element L and shear modulus c_{66}, this can be expressed as $U_{el} \approx 0.12(\epsilon_0 s B / B_{c2} \ln(\kappa))$; $\epsilon_0 \equiv (\Phi_0^2 / 4\pi \mu_0 \lambda^2)$ $\ln(\kappa)$ is the line energy. It turns out that for the HTS $\epsilon_0 = 10^3 \ln(\kappa)$ K/nm. Substituting $L = s$ and $L = L_c$ one gets $U_{el} = 180$ B/B_{c2} K and $U_{el} \approx 8.8 \times 10^2 B/B_{c2}$ K. A comparison with the pin energies U_v and U_p shows that they both exceed U_{el} in the field regime we consider.

Finally, it should be mentioned that other non-conducting defects with $D < \xi$ act equivalent to vacancies. The pinning force of conducting defects, however, will be much smaller because of the proximity effect [22]. One may thus use the above considerations to estimate an optimal pinning force by taking $D \approx \xi(0)$ and $n_\Box \approx 0.1 \xi(0)^{-2}$. With this density of defects one may hope that T_c does not decrease due to proximity coupling, but that an increase of J_c by an order of magnitude is still possible. More important is that U will increase by roughly the same amount which would reduce the creep process at 4.2 K considerably.

2.2 Screw dislocations

The pin interaction of a screw dislocation can be estimated by considering the dislocation core as a non-conducting cilinder of diameter s. This assumption is supported by the fact that the resistance and T_c of Y:123 are very sensitive to disorder. Further, it seems reasonable to take for the pitch of the screw dislocation the diameter of the disordered core region. For parallel vortices the interaction can be estimated from the cross section for electron scattering making use of Thuneberg's formalism [21], see also [14]. One obtains per unit length in the single vortex limit

$$f_{pl} \approx (B_c^2 \mu_0) p D \xi_0 / \xi \qquad (4)$$

where p is the scattering probability and D the diameter of the dislocation core, $D \approx s$.

An other pin mechanism one should consider, is the surface roughness created by the screw dislocations. If one models the roughness by $\delta d.\sin(2\pi x/L_s)$ with L_s the average distance between the dislocations and $2\delta d$ the thickness modulation, the pinning force per unit length roughly is

$$f_{pl} \approx 4\pi \epsilon_0 (\delta d/d)/L_s \qquad (5)$$

with d the film thickness. Note that f_{pl} represents the maximum force.

To obtain the bulk pinning force one generally has to sum over the actual forces which depend on the mutual positions of the defects and the vortices taking into account the fortex deformations. However, strong pins may disrupt the vortex lattice (VL) giving rise to a granular VL consisting of bundless seperated by edge dislocations. This might be the case for the Y:123 films. In such a situation the direct summation applies. For each of the pin mechanisms considered in the preceeding paragraph one estimates

$$F_p \approx f_{pl}/L_s^2 \qquad (6)$$

A quite different mechanism that might play a role here, is VL shear. This means that weakly pinned regions of the VL flow along strongly pinned areas. When both the screw dislocations and the valeys between them act as such strong pins, this effect may occur. Upon increase of the driving force percolation paths of weak pinning regions develop bridging the width of the film. The force to be exceeded is the flow stress. Suppose the percolation path-length is L_p, the width of this path is $(q/2)a_0$ (being q VL planes), W the width of the film, and $\tau = \alpha C_{66}$ the flow stress, the resulting critical shear force density $F_s = J_s B$ is [41]

$$F_s = (4\alpha L_p/qW) C_{66}/a_0 \qquad (7)$$

In practice the prefactor will be of order unity yielding $J_s \approx C_{66}/(a_0 B) = (\Phi_0 B)^{\frac{1}{2}}/(16\pi\mu_0\lambda^2)$. Thus, we estimate $J_s \approx 3.7 \times 10^{10} (B/[T])^{\frac{1}{2}}$ Am^{-2}. Since this value is an order of magnitude smaller than the experimental J_c [19], and, in addition, has a field dependence which is not observed, the shear mechanism seems to be ruled out.

We still should compare these ideas with the possible effect of oxygen vacancies keeping in mind though that the actual value of $n_\square$ has been obtained indirectly. Oxygen vacancies would give a line force on each vortex of 1.9×10^{-13} N/nm and a J_c of 9×10^{10} Am^{-2}. This should be compared to $f_{pl} = 4.2 \times 10^{-13}$ N/nm for the line force of a screw dislocation as determined from Eq.(4). From Eq.(6) taking $L_s = 300$ nm, $J_c = 4.7 \times 10^9 /(B/[T])$ Am^{-2} is obtained. For the case of thickness variations taking $d = 100$ nm and $\delta d = 5$ nm [19], we get $f_{pl} = 1.3 \times 10^{-13}$ N/nm and $J_c = 1.6 \times 10^9 /(B/[T])$ Am^{-2}.

In [19] it is suggested that the grain boundaries caused by the screw dislocations are responsible for F_p. Using the expressions derived in [14] we estimate this contribution to be almost two orders of magnitude too small.

In conclusion, the direct core interaction of the screw dislocations, the effects of surface corrugation and oxygen vacancies are all of the same order of magnitude. At small fields the two former mechanisms are probably predominant, while oxygen vacancies may take over at large fields. The combined effect provides a good explanation for the large J_c's reported in the literature.

3 High temperature properties

3.1 Thermally activated depinning

So far, it has been assumed that thermal fluctuations are not important. However, close to T_c this assumption is not justified and one has to consider the effect of thermal depinning [10]. The reason is that both the elastic and the pinning energies scale with B_c^2/μ_0 so that near T_c they will be of order $k_B T$. Moreover, also a scaling factor $(1-b)^p$, with $b = B/B_{c2}$ and p either 1 or 2 related to the reduction of the order parameter, occurs which broadens the thermal depinning regime to far below the $B_{c2}(T)$ transition line, see Fig. 1. Thermal fluctuations of both the order parameter and its phase effectively lead to fluctuations of the vortices about their equilibrium positions. The influence on the pinning can be illustrated by considering a single vortex and a point pin. The pin potential has a typical range ξ. The thermal displacements then lead to an averaging of the pin potential over an area of order $<u^2>_T$. When this area is of order ξ^2 the pinning force decays $\propto \exp((-T/T_d)^3)$ in 3D [10] and $\propto (T/T_d)^{-\frac{5}{2}}$ in 2D at the depinning temperature T_d [7]. In larger fields similar considerations apply for a VL, but T_d is now determined by $<u^2>_T \approx a_0^2$.

3.2 Depinning or melting?

The above condition for depinning is very similar to the condition for melting. In the single vortex limit for which $\xi \ll a_0$, depinning may take place below the melting temperature, whereas for large fields probably the opposite happens. It should be mentioned however, that strong pinning may still change these results quantitatively. When the pinning energy is larger than the elastic energy, the vortices move independently in their pin potentials and the mean square displacement is inversely proportional to the curvature of the pin potential, the Labusch parameter α_L [23,24]. Depinning now occurs when $k_B T_c^s \approx U_p$ and $U_p = F_p a_0^2 L_c^s \xi \leq (B_c^2/\mu_0)\xi^2 L_c^s$. (Superscripts s and w denote the strong and weak-pinning case). When the pinning is weak the VL has short range order over distances large compared to a_0 and the displacements are determined by $k_B T/a_0 C_{ef}$ where $C_{ef} \approx (C_{44}C_{66})^{\frac{1}{2}}$ is the effective elastic constant of the VL. The depinning condition in this case can be written as $k_B T_d^w \approx c_{ef}\xi^2 L_c^w$. One can further analyse these conditions in order to compare the depinning temperatures. In case of strong pinning a temperature increase weakens the effect of the pins so that the correlation in the VL growths and eventually the configuration goes over into the weak pinning case. The melting temperatures in both cases are expected to be almost equal.

3.3 Flux creep in the vortex-glass phase

In the vortex-glass phase thermal fluctuations in presence of a driving force give rise to hopping of flux bundles over a distance u_h from one metastable configuration to an other separated by an energy barrier U. The hopping frequency is given by $\nu_h = \nu_0 exp(-U/k_B T)$ with ν_0 a microscopic attempt frequency. The net effect is a diffusion-like motion of the VL with an average velocity $v = \nu u_h$ in the direction of the driving force which is either caused by a transport current, a flux density gradient, or a flux-line curvature. The resistance generated by the creep process is given by

$$\rho \approx \rho_f exp(-U/k_B T) \qquad (8)$$

where ρ_f is the flux-flow resistivity. The activation barrier depends on B and T, and generally also on J. For $\vec{J} = \nabla \times \vec{B}/\mu_0$, one obtains with $\nabla \times \vec{E} = -\partial \vec{B}/\partial t$ and using $\vec{E} = \rho(J)\vec{J}$ a nonlinear diffusion equation in $\vec{B}$ [25,26]

$$\partial \vec{B}/\partial t = -\mu_0^{-1} \nabla \times \{\rho(J)\nabla \times \vec{B}\} \qquad (9)$$

or $\vec{J}$

$$\partial \vec{J}/\partial t = \mu_0^{-1}(\nabla \cdot \nabla)(\rho(J)\vec{J}) \qquad (10)$$

results which describes the flux-density decay with time and position inside the superconductor. For simple geometries, e.g. a slab in parallel field, and $\rho \neq \rho(J)$ a linear equation describing the small-current behavior results [27]

$$\partial B/\partial t = D_0 \partial^2 B/dx^2 \qquad (11)$$

with $D_0 = \rho/\mu_0$ and ρ from Eq.(8). In this thermally-assisted-flux-flow (TAFF) limit a simple relation is obtained for the irreversibility lines as measured by several techniques [28,29], see below. Note, one should solve for $B(x,t)$ and average over the sample cross section in order to obtain the decay of the magnetization with time or the ac susceptibility before one can compare with experiments. The decay in the TAFF limit is exponential. The general behavior of χ' and χ'' can be expressed in terms of one parameter a/λ_{ac}, where λ_{ac} is the penetration depth of the ac field [15], and a the half- thickness or the radius of the sample. The peak in χ'' will occur when $\lambda_{ac} \approx a$. Since

$$\lambda_{ac} = (D_0/\pi \nu)^{\frac{1}{2}} \qquad (12)$$

this condition leads to a relation between field, temperature, and frequency ν of the ac field which defines the irreversibility line. Below this line the VL oscillates in its pin potential which offers a method to measure the Labusch parameter α_L and the range of reversible collective motion of the VL in the pin potential [30,31]. The combination with TAFF is worked out in [15].

In the limit of large current densities $J \leq J_c$ and for a simple geometry Eq.(10) leads to the well-known flux creep result [32,33] with a logarithmic time decay for J in constant field [34,35]:

$$J(t) = J_c[1 - (k_BT/U)\ln(t/t_0)], \quad t_0 \ll t \ll t_{cr} \tag{13}$$

where $t_{cr} \approx t_0 \exp(U/k_BT)$ is a cross-over time to exponential decay, and t_0 a macroscopic scaling time determined by the condition that the normal-state skin depth equals the sample size, i.e. $t_0 \approx (\mu_0 a^2/\rho_n)$ rather then ν_0^{-1} as is often assumed [36]. From Eq.(13) it follows that

$$S \equiv -d\ln(J)/d\ln(t) = [\ln(t_{cr}/t)]^{-1} \simeq k_BT/U, \quad t \ll t_{cr} \tag{14}$$

Note, however, that S is not equal to the experimental decay parameter

$$S_{exp} \equiv -M_0^{-1} dM(t)/d\ln(t) = [\ln(t_{cr}/t_i)]^{-1} = (k_BT/U)/\ln(t_0/t_i) \tag{15}$$

where $M_0 = M(t_i)$ and t_i is the time at which the measurement is started. Here it has been assumed that J is uniform so that $M \propto J$. This model can be extended to cases where U depends on J via the shape of the pin potential [33,37] and one may try to deconvolute for a distribution of energy barriers [38].

In the preceding paragraphs the hopping distance and the size of the flux bundle L_b were assumed to be independent of J. This might be justified for $J \approx J_c$, or in the special case that the TAFF models holds (see below), but for smaller values of J both u_h and L_b have to grow in order to conserve energy in the hopping process, i.e. the energy gained from the external force should be equal to the elastic deformation energy of the VL. The thermally activated motion of an elastic object in a random potential has been treated for a VL in the collective creep theory [10,8,7]. The key result of this theory is

$$U(J) = U_c(J_c/J)^\alpha \tag{16}$$

which leads to

$$J(t) \simeq J_c/[1 + (\alpha k_B T/U_c)\ln(t/t_0)]^{\frac{1}{\alpha}} \tag{17}$$

with $\alpha = (2\zeta - 1)/(2 - \zeta)$ and ζ the wandering exponent, determined by the dimensionality of the medium and the hopping vector. Typically, for a VL $\alpha \approx 1$. As L_b determines whether local or nonlocal elastic moduli are relevant, α also depends on the current regime leading to a more general result $U(J) = U_i(J_i/J)^{\alpha_i}$ for $J_{i-1} < J < J_i \leq J_c$. In a 3D VL α changes from 1/7 to 3/2 to 7/9 with decreasing current density [39,40]. In 2D the values are 9/8 and 1/2, but in 2D, plastic creep of VL defects takes ofer at low J [7,8]. For plastic or dislocation mediated creep the energy barrier does not depend on J any more which means that the TAFF limit would apply. This is also expected for a 3D vortex-liquid just above the glass-transition temperature where it probably behaves as a plasto-elastic medium.

The logarithmic decay rate S which follows from both Eqs.(13) and (15), will at low temperature increase with T since both U and U_c will be almost constant. Closer to the melting temperature the pinning and thus t_{cr} will decrease. At this point S will level off and eventually decrease for still higher T. The relation between U_c and the collective pinning potential $U_p = F_p V_c \xi$ depends on the size of the hopping bundle which consists of about $(C_{11}/C_{66})^{\frac{1}{2}}$ correlated regions of volume V_c [10]. Hence, $U_c \gg U_p$. Finally, by measuring ρ and using Eq.(8) it seems straight-foward to determine the activation barrier. However, U at high temperature strongly depends on T, which may show up as a very large prefactor "ρ_f".

3.4 Collective creep experiments and the activation barrier

Magnetic relaxation measurements at low temperatures provide an experimental tool to investigate the validity of the collective creep model discussed in the previous section. Especially, the prediction for the exponent α in Eq.(16) is of interest. The collective creep theory assumes that the VL is mainly elastically distorted. Therefore, it seems appropriate to concentrate the experiments on a 2D system, for in 3D plastic disorder dominates [14]. So we choose to perform the experiments on Bi:2212 single crystals in fields above B_{2D}. An extensive report of the results and a detailed analysis has been given by Van der Beek et al. [41]. Remaining questions about the influence of geometry effects, e.g. the demagnetization of the single crystalline platelet, and the possible field dependence of J_c are dealt with in a seperate paper [42]. The conclusion is that these effects can be seperated out from the time dependent effects. A brief account of Ref. [42] is given in Section 3.5.

3.4.1 Experimental

The experiments were performed on three BSCCO crystals prepared in a mirror furnace using the travelling-solvent floating-zone technique [43]. Laue X-ray diffraction confirmed the crystal structure and HREM evidenced the high sample quality, crystals containing $Bi_2Sr_2CuO_6$-intergrowths in a concentration of 2 per 1000 CuO_2 double layers. Most results were obtained on a crystal with square dimensions $4 \times 4 \times 0.11$ mm^3. It weighed 11.4 mg and had a T_c of 88.1 K with a transition width of 1.5 K, as determined from AC susceptibility.

Magnetization experiments were carried out with the field H applied along the sample c-axis (perpendicular to the sample plane) in a temperature range of 5.0 to 26 K. Most measurements were carried out using a commercial SQUID magnetometer. Further experiments in magnetic fields up to 12 T were carried out in a vibrating sample magnetometer (VSM) which is a significantly faster apparatus than a SQUID: the measurement of a single data point takes approximately 5 s, against 100 s in a SQUID.

The procedure for relaxation measurements was to zero-field cool the sample to the measurement temperature. The field was then ramped at a fixed rate, $\mu_0 \dot{H}_a = 1 \times 10^{-3}$ T s^{-1}, to the measurement value. In every case, it was cycled in such a way as to ensure full penetration of the sample by a shielding current of uniform direction. For the measurements at the lowest fields of 0.1 and 1 T, this meant ramping the field up to 5 T, and then reducing it to the respective measurement value. The sample moment is measured during the field cycle. At temperatures below 5 K, the field ramp rate had to be reduced in order to prevent flux jumps. This resulted in an upper limit to the current densities that could be obtained, $J < 1.5 \times 10^{10}$ A m^{-2}. After the field ramp was stopped, the decay with time of the magnetic moment was monitored. Relaxation measurements were carried out at field values 0.1, 1.0, 2.0, 5.0 and 12.0 T.

From the measured magnetization, the current density at the sample surface is determined using the Bean model, $J_s \approx 6(M - M_{eq})/a$. The equilibrium magnetization M_{eq} was determined from a measurement of $M(H)$ at $T = 80$ K and subsequent scaling according to the Abrikosov prediction to account for the temperature dependence.

3.4.2 Results and analysis

The results of the above experiments have among others been analyzed according to the method suggested by Maley et al. [44]. For the square platelet the time dependence of the magnetization is related to the average electric field E_s on the sample circumference by [42]

$$-\frac{\partial M}{\partial t} = \frac{aE_s}{6\Lambda} = \frac{a}{6}\frac{\rho_{f,s}J_s}{\Lambda}exp\left(-\frac{U(J_s)}{k_BT}\right) \tag{18}$$

where a is the lateral sample dimension and the subscripts s denote evaluation at the sample edge. The "self-inductance" $\Lambda \equiv \frac{1}{6}a\,\partial B^*/\partial J_s$, B^* is the self field generated by a uniform shielding current J_s and ρ_s is roughly the flux-flow resistivity. In Ref. [44], $\rho_f J$ was identified with $Bw(J)\nu_0$, where $w(J)$ is the average jump length at current density J and ν_0 is a microscopic "attempt" frequency. Because to first order the magnetization is proportional to the current density at the sample surface, $M \approx \frac{1}{6}J_s a$, a plot of $k_B T(c - \ln(|\partial M/\partial t|))$ versus $|M - M_{eq}|$, with $c \approx ln\left(\frac{a}{6}\frac{\rho_{f,s}J_s}{\Lambda}\right)$ represents the dependence of the activation barrier $U(J)$ on the current density J. An alternative approach is to conjecture that the prefactor $\rho_{f,s}J_s$ in Eq.(18) is independent of current density. It may then be identified with $\rho_{f,s}J_c$ [45]: at current densities $J \gg J_c$, activation barriers $U(J)$ vanish and the resistivity of the sample should equal the flux-flow resistivity, $\rho_f \approx \rho_n B/B_{c2}$.

Figure 2(a) shows typical results of the time dependence of the magnetization at $\mu_0 H_a = 2.0$ T. Although M is seemingly logarithmic in time, the logarithmic creep rate S depicted in fig. 2(b) reveals that this is not so. Not only is S non-constant, it is even non-monotonic in time. This means that experiments are *not* carried out in the long time limit of relaxation, as is often assumed [46]. It can be shown explicitly that the maximum in $S(\ln t)$ is a consequence of transient effects [42]. It was derived that if $U_c \gg k_B T$, a maximum should occur at $t \approx B^*/\mu_0 \dot{H}_a$. Taking the results for $\mu_0 H_a = 2$ T and $T=5$ K as an example, one expects the maximum in S at $t \approx 1.8 \times 10^3$s, of the same order of magnitude as seen in experiment. This result means that not only an analysis in terms of S, but also the direct comparison of the data to functional forms as Eqs.(15) and (17) is inappropriate, except at the highest temperatures.

The result of applying the method of Maley et al. is shown in Fig. 3. Measurements at different temperatures were combined for each measurement field in order to increase the current density window. For each field value, it was possible to fit results for temperatures $T \leq 11$ K onto a common curve using a single constant $c \approx 27$ (the unit of M is taken to be A m^{-1}). At temperatures above 11 K, the $U(J)$-data could still be fitted onto the common curve when divided by a factor ranging between 1 and 0.77. The curves thus obtained are shown in Fig. 4. The correction factors systematically decrease with temperature, and are most obviously interpreted as measuring the temperature dependence of $U(J_c)$. Hence, the conclusion may be drawn that below 11 K, both the current-independent prefactor U_c and the critical current density J_c in Eq.(16) are temperature-independent.

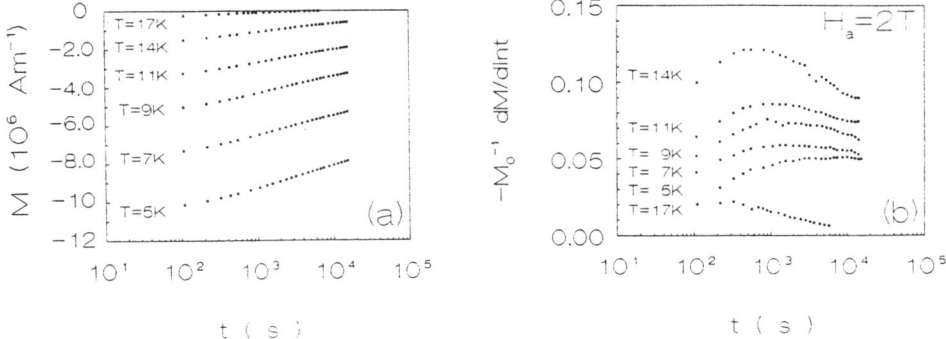

Figure 2: (a) Magnetization vs. time in a field $H_a = 2.0$ T. Measurement temperatures are indicated. (b) The relaxation rate as function of time, as derived from (a).

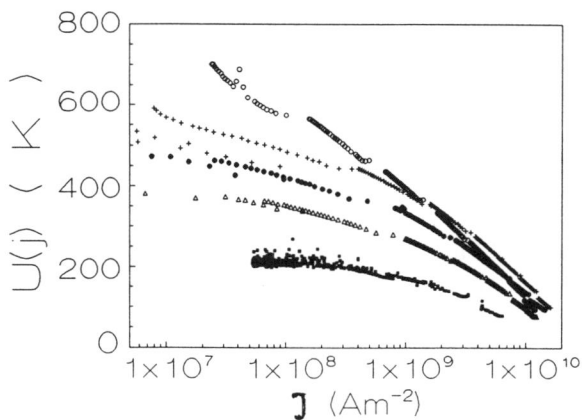

Figure 3: Flux creep activation barriers as determined by the method of Ref. [44] from magnetic relaxation data. Symbols denote the field values: 0.1 T (○), 1.0 T (+), 2.0 T (●), 5.0 T (△), and 12.0 T (■)

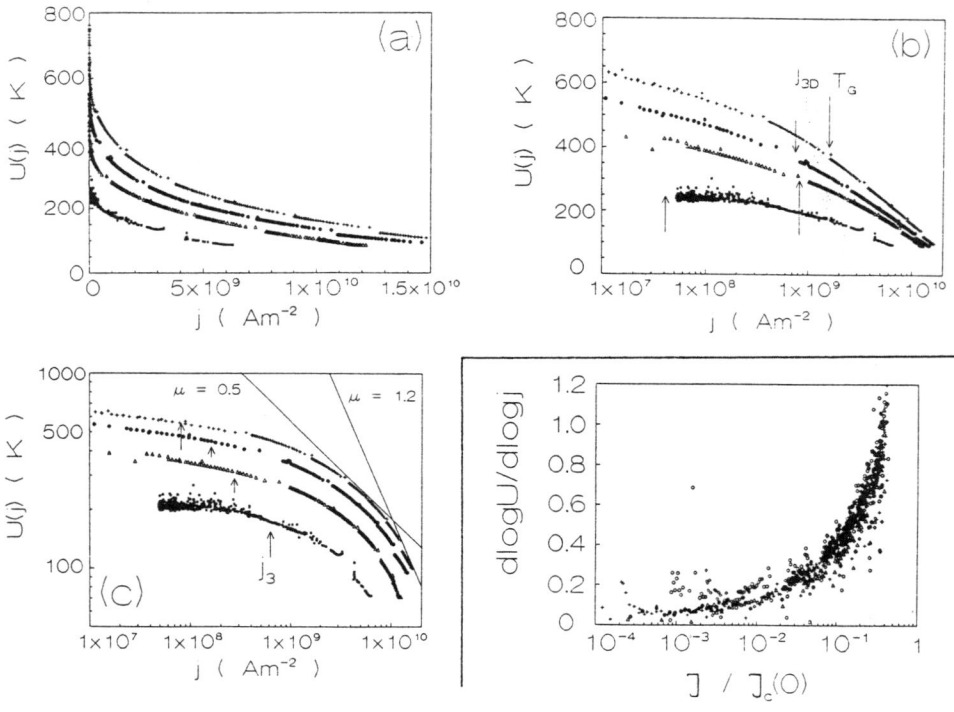

Figure 4: Corrected flux creep activation energies vs. current density in fields of 1.0 T (+), 2.0 T (•), 5.0 T (△), and 12.0 T (■). (b) The same data on a semi-logarithmic scale. Solid arrows separate data sets measured above and below the glass transition line $T_G(B)$, dotted arrows indicate the expected crossover to 3D flux-creep, see [41]. (c) Data on a double-logarithmic scale. The straight lines denote the procedures used to obtain U_c (label $\mu = 1.2$, corresponding to the high current asymptotic μ-value) and J_2 (label $\mu = 0.5$). Arrows denote the estimated current density values J_3 where the activation barrier should become current-independent due to the activation of VL point defects in the absence of a vortex-glass transition.

Figure 5: Scaling plot dlog U/dlog $J \equiv \mu$ for all measurements on sample A, $H_a = 0.1$ T (○), 1.0 T (+), 2.0 T (•), 5.0 T (△), 12.0 T (■) and sample C, $H_a = 2.0$ T (◇).

The method of analysis was applied to the data under both assumptions for the prefactor $\rho_{f,s}J_s$ outlined above, namely, that it is either current-independent or linear in J. The latter assumption means that the additive factor c should be taken as $\tilde{c}M$. Although the different assumptions for the preexponential factor $\rho_{f,s}J_s$ do not seriously affect the data at the highest current densities, there is a difference at lower J. the procedure that accounts for a J-dependent preexponential factor required a larger correction for the temperature dependence of U_c. It was therefore supposed that replacement of J_s, in the preexponential factor by a constant current density in Eq.(18) is more appropriate, and all subsequent data analysis was carried out in this way. The following quantities are obtained from experiment.

Critical current density. The semi-logarithmic plots of the experimentally deduced $U(J)$, Fig 4(b), show a quasi- logarithmic dependence that allows for the extraction of the critical current density at zero temperature, $J_c(0)$, by simple extrapolation to $U(J) = 0$. This extrapolation does not imply that $U(J_c(0))$ is necessarily zero. A compilation of experimentally determined critical current density values shows that $J_c(0) \approx 3.5 \times 10^{10}$ A m^{-2}, and is roughly field-independent.

Activation energies. The free energy barrier for thermally activated vortex motion is a strongly decreasing function of current density, which means that flux creep is due to the thermally activated motion of elastic objects. At high current densities, $U(J) \approx U_c \ln(J_c/J)$, i.e. the (I, V)-characteristic roughly follows a power-law, as reported by several groups [44,47]. An activation barrier of the form (16) with a "current-dependent exponent μ", as reported by Dekker et al. [48], may also result in such behavior.

It is important to realize that the results in Fig. 4 depict $U(J)$ evaluated on the sample *circumference*. This means that the field at which $U(J)$ is evaluated is *not* equal to the applied field H_a, but to H_a plus the self-field component in the z-direction. Because the magnitude of the shielding current decreases with time, the field component $B_{z,s}$ changes during the measurement. Estimates of the total field on the sample circumference show that for $J = 1 \times 10^{10}$ A m^{-2}, $B_{z,s} \approx \mu_0 H_a \pm 0.25$ T, depending on whether the magnetic moment is negative or positive with respect to the applied field. The fact that data for the lowest applied field value, $\mu_0 H_a = 0.1$ T, did not comply to the general trend in Fig. 3 thus finds a natural explanation when one recognizes that at this field the induction on the sample side edge changes polarity when $J \approx 2 \times 10^9$ A m^{-2}. Analysis of the 0.1 T-data is therefore potentially complicated by the possibility of vortex annihilation *within* the sample and is therefore not shown in Fig. 4.

In spite of these complications, it can be shown, however, that the flux creep mechanism is the same at all applied field values. Namely, in that case the activation barrier is of a form $U(J, B) = \tilde{U}(B) \mathcal{F}(J/J_c)$, with $\mathcal{F}$ the same function in the whole field regime. The result of differentiation of the experimental $\log(U(J))$ with respect to $\log(J)$ for *all* field values, depicted in Fig. 5 shows that this is indeed the case. This is the more striking if it is remembered that the field is non-constant during each measurement of $M(t)$. From Fig. 4(b) it can be seen directly that an attempt to carry out a similar type of scaling over the whole current range for a hypothetical dependence $U(J) \propto \ln(J)$ will be unsuccessful.

The dependence of dlog U/dlog $J \equiv "\mu"$ on J can be compared to that found in an YBCO thin film [48]. There, "μ" increases from 0.19 to a saturation value 0.94 at current densities $J \approx 8 \times 10^9$ A m^{-2}. The data in Fig. 5, however, show no sign of such saturation.

Collective pinning analysis. With the values $J_c(0)$ know, one is in a position to make a number of estimates as regards the object that is thermally activated, i.e. a bundle of point vortices, a vortex line (loop) or a vortex line bundle. Using CP theory and $J_c(0) \approx 4 \times 10^{10}$ A m^{-2}, it is found that L_c is always smaller than the interlayer spacing, and the critical current density is determined by the pinning of vortex segments within the CuO_2-layers. With $L_c \sim s$, this means that the correlated volume consists of one point vortex of dimensions $a_0^2 s$, i.e. point vortices in the CuO_2-double layers are pinned independently. In the 2D individual pinning regime, the elastic energy is not important, and the critical current density is obtained from the fluctuation of the elementary pinning force over the point vortex core. In Ref. [15], J_c in the 2D individual pinning regime was determined under the assumption that pinning is due to randomly distributed oxygen vacancies. The result, a field-independent $J_c \approx 5.0 \times 10^{10}$ A m^{-2}, is in very good agreement with the present results.

Comparison with 2D collective creep. The twodimensional nature of pinning found above suggests an interpretation in terms of 2D collective creep [7,8], i.e. the motion of vortices in 2D bundles. Such a description is meaningful only when the bundle size in the jump direction $R(J) \approx R_c(J_c/J)^{5/8}$ is larger than a_0, which corresponds to a current criterion $J < J_1 \approx \frac{8}{3} J_c (R_c/a_0)^{8/5}$. At high fields, $R_c \approx a_0$, so that $J_1 > J_c(0)$; at lower fields, $R_c \approx 0.2 a_0$, and $J_1 \approx 0.4 J_c(0)$. This means that the condition $J < J_1$ is fulfilled in the experiment and 2D collective flux creep is a relevant flux transfer mechanism.

Near $J = J_1$, the theory of 2D collective expects $U(J)$ to scale as in Eq.(16) with $\mu = \frac{9}{8}$. In view of the above results, the activation barrier U_c should be the pinning energy of a point vortex. The data plotted on a double-logarithmic scale in Fig. 4(c) show that the high-current asymptote is close to a power-law with exponent $\frac{9}{8}$ for fields below 5.0 T. Fig. 5 shows that the behaviour is in fact the same for all applied fields and the exponent at high currents can be directly read, $\mu \approx 1.2$.

A fit of the high-current asymptotic behavior of $U(J)$ to a power-law yields the pinning energy when $J \approx J_c(0)$, U_c. This is found to be field independent, in agreement with the limit of single vortex pinning. The magnitude of U_c is found to be 35-42 K, the same as the result of Ref. [15], where a value $U_v = 35$ K was found for the pinning energy of a single vortex disc.

When the current density has decreased to $J_2 = J_1(a_0/\lambda)^{8/5}$, 2D collective creep expects the bundle size $R(J)$ to become of the order λ, and a crossover to $\mu = \frac{1}{2}$ should be found. J_2 is the current density at which the non-locality of the VL compression modulus C_{11} becomes unimportant. Substituting in favour of J_c, one finds that $J_2 \approx J_c(R_c/\lambda)^{8/5}$. With the results for J_c and R_c found above and $\lambda \approx$ 290 nm, $J_2 \approx 2 \times 10^{-3} J_c$, independent of magnetic field. If $\lambda \approx 140$nm is taken, $J_2 \approx 0.02 J_c$ is expected. The activation barrier for $J \approx J_2$ should be described by $U(J) = U_2(J_2/J)^{1/2}$ [7], where $U_2 = U_c(\lambda/a_0)^{9/5}$. Because U_c was found to be field-independent, U_2 is expected to increase with field.

The crossover current densities J_2 were determined as the J-values for which the tangent to the $\log(U(J))$ versus $\log(J)$-curves has a slope $\frac{1}{2}$. The experimental J_2/J_c-

values are approximately field-independent, which agrees nicely with the expectation. The obtained magnitude, $J_2 \approx 0.15 J_c$, is much larger than the expected value. The experimental activation barriers $U_2 = U(J = J_2)$, are seen to decrease with applied field, which is contrary to expectation.

In Ref. [7], it was proposed that in two dimensions, the activation barrier $U(J)$ should eventually rise above the nucleation energy of 2D VL dislocation pairs (point defects). As a consequence, the measured activation energy should become current-independent at a value $J_3 \approx J_c(\xi/a_0)^2$. Flux creep mediated by such 2D VL point defects could successfully explain the AC susceptibility in fields above 0.4 T, see Ref. [15]. The values of the current density J_3 were computed for the different measurement fields and are indicated in Fig. 4(c). It is seen that the activation barrier does not become J-independent at J_3.

Summarizing, it is found that at current densities $J \geq J_2 \approx 0.15 J_c$, 2D collective creep theory describes the data well.

3.5 Numerical and approximate solution to the creep equation

3.5.1 Approximate solutions

The analogue of Eq.(18) for an infinite slab of thickness d in a parallel field is

$$\frac{\partial J_s}{\partial t} = -\frac{\rho_{f,s} J_s(t)}{\Lambda} exp\left(-\frac{U(J_s(t))}{k_B T}\right) \tag{19}$$

where the self-inductance takes the form $\Lambda = \frac{1}{2}\mu_o d\partial M/\partial J_s$. We now assume that Λ may be treated as a constant. This approximation, to be justified below, provides a way to take the influence of sample geometry and field-dependence of J_c into account. It is equivalent to the assumption that the flux profile can be described by the product of a space- and a time-dependent function, that is, $M(t) \propto J_s(t)$.

Upon multiplication of both sides of Eq.(19) with $(\partial U/\partial J_s)\exp(U(J_s)/k_B T)$, it may be directly integrated with respect to time. In an actual experiment, the time origin can be chosen to correspond to the moment that the field ramp is stopped. Denoting the initial average current density $J_s(0)$ as J_i, one obtains

$$\exp\left(\frac{U(J_s(t))}{k_B T}\right) = exp\left(\frac{U(J_i)}{k_B T}\right) - \left[\frac{1}{t}\int_0^t \frac{\rho_{f,s} J_s(t')}{\Lambda k_B T}\left(\frac{\partial U}{\partial J_s}\right) dt'\right] t \equiv \frac{\tau_i + t}{\tau} \tag{20}$$

Thus, in general, an approximate solution for J_s can be found from

$$U(J_s(t)) = U(J_s(0)) + k_B T \ln\left(1 + \frac{t}{\tau_i}\right) \tag{21}$$

which we write as

$$U(J_s(t)) = k_B T \ln\left(\frac{\tau_i + t}{\tau}\right) \tag{22}$$

The normalization time

$$\tau \equiv \tilde{\tau} \left[\frac{1}{t} \int_0^t \frac{J_s}{U_c} \left(-\frac{\partial U(J_s)}{\partial J_s} \right) dt' \right]^{-1} \quad (23)$$

describes the time scale of relaxation and

$$\tau_i \equiv \tau \exp\left(\frac{U(J_i)}{k_B T} \right) \quad (24)$$

is a measure of the time regime in which transient contributions to $J_s(t)$ are important. The quantity $\tilde{\tau}$ introduced in Eq.(23) is given by

$$\tilde{\tau} = \frac{\Lambda}{\rho_{f,s}} \frac{k_B T}{U_c} \quad (25)$$

where the approximation that Λ is constant has been used. It can be shown that for the potentials U the integrand in Eq.(23) $J_s(\partial U(J_s)/\partial J_s)$, depends only very weakly on time. Thus, the normalization time τ can be treated as constant during the relaxation. The application of Eq.(22) to some flux-creep models yields the following results:

(1) $U(J) = U_c \ln(J_c/J)$: in this case the solution to Eq.(19) is exact [49], J_s decays according to a power law,

$$J_s(t) = J_c \left(\frac{\tau_i + t}{\tau} \right)^{-k_B T/U_c} \quad (26)$$

and the normalization time is given by $\tau = \tilde{\tau}$ of Eq.(25).

(2) *Collective creep model,* $U(J) = U_c((J_c/J)^\mu - 1)$: no exact solution is known, but to logarithmic accuracy the current decays as

$$J_s(t) = J_c \left(1 + \frac{k_B T}{U_c} \ln\left(\frac{\tau_i + t}{\tau} \right) \right)^{-1/\mu} \quad (27)$$

where $\tau \simeq \tilde{\tau}/\mu$.

The approximation that is generally made in deriving the time dependence of the magnetization [9] is that $M \propto J_s$, i.e., Λ is a constant. The simplest current distribution which yields this result is that assumed by the Bean model, namely that J is constant over the sample cross-section. The magnetization of an infinite slab of thickness d in a constant field H in the case of a constant current of magnitude J_c is [50]

$$M = \mp \frac{1}{4} J_c d \quad (H > J_c d/2) \quad (28)$$

During the initial stage of a field ramp, magnetix flux only penetrates the sample to a depth x_f, equal to HJ_c in the case of field-independent J_c. Then, $J=J_c$ only in the penetrated region leading to a partial critical state. The magnetization is given by

$$M = \frac{H^2}{J_c d} - H \quad (H < J_c d/2) \tag{29}$$

The choice of sign in Eq.(28) corresponds to situations of increasing and decreasing induction, respectively. A comparison of exact numercial results for the magnetization M with Eq.(28), with J_c substituted by the form for $J_s(t)$ appropriate to the flux-creep model under consideration, measures the validity of the approximation $M \propto J_s$.

From Eqs.(29) and (28), the self-inductances are easily obtained. For partial penetration, $\Lambda = \frac{1}{2}\mu_0 H^2/J_c^2 = \frac{1}{2}\mu_0 x_f^2$; for fullpenetration, $\Lambda = \frac{1}{8}\mu_0 d^2$. The flux creep normalization times $\tilde{\tau}$ become

$$\tilde{\tau} = \frac{\mu_0 x_f^2}{2\rho_{f,s}} \frac{k_B T}{U_c} \quad (t < t^*), \tag{30}$$

$$\tilde{\tau} = \frac{\mu_0 d^2}{8\rho_{f,s}} \frac{k_B T}{U_c} \quad (t > t^*). \tag{31}$$

The time of full flux penetration t^* is that at which the current density has decreased to a vlaue $J_s = 2H/d$.

In the absence of flux-creep, a dependence of J_c on the induction B leads to a non-constant shielding current over the sample cross-section. If, for example,

$$J_c = J_0 \frac{B_0}{B_0 + B} \tag{32}$$

as suggested by Kim, Hempstead and Strnad [51], the critical state (again defined as $J=J_c$ over the whole region of flux penetration) in the considered slab geometry then results in a simple parabolic flux-profile. For further details of the "Kim model" we refer to [42].

3.5.2 Relaxation of the critical state

The Bean model. For the simulation of magnetic relaxation experiments, see [42] for details, it is assumed that an initial flux distribution is established at $t = t_0$. The magnetic relaxation was calculated starting from the following initial states:
(1) a full critical state $J_i = \mp J_c$ over $0 < x < d/2$, where $\mp$ corresponds to the situation arrived at after either increasing or decreasing the field, respectively;
(2) a full subcritical state, in which J_i is constant over $0 < x < d/2$ but smaller than J_c;
(3) a partial critical state;
(4) a partial subcritical state.

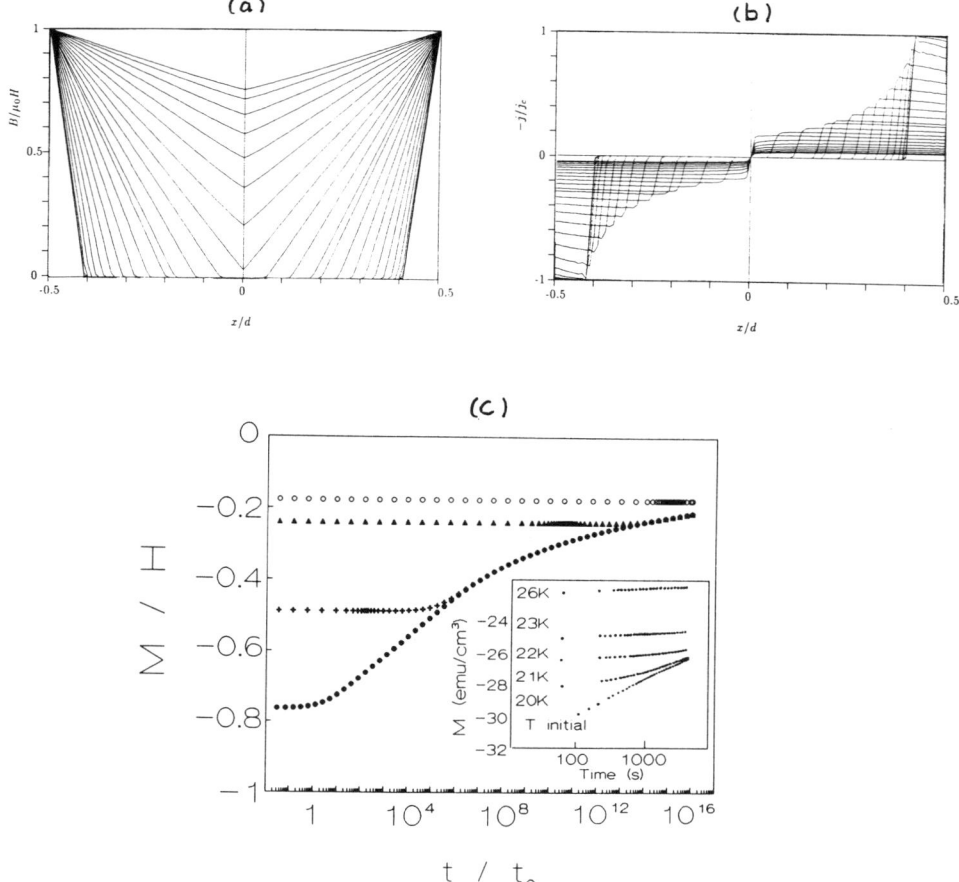

Figure 6: Flux-density (a) and current-density (b) profiles in an infinite superconducting slab of thickness $d = 4$ mm for 2D collective creep, $U = U_c(J_c/J)^{9/8} - 1)$, and field independent J_c. Parameter values were $U_c/k_BT = 7$, $J_c = 4.5 \times 10^9 Am^{-2}$, $\rho_n = 50\mu\Omega$ cm, $B_{c2} = 70$ T. A field $\mu_0 H = 5$ T is applied along the slab surface. Snapshots of the induction and current density distributions are shown, taken at exponentially spaced time intervals between $0.01\ t_0$ and $1 \times 10^{12} t_0$. (c) The time evolution of the magnetization of a superconducting slab of width 4 mm when $U = U_c(J_c/J)^{9/8} - 1)$ for different initial flux distributions in the sample. In all cases J_i was constant in a sheet $d/2 > x > \max(d/2 - H/J_i, 0)$, with values $J_i = J_c(\bullet)$, $0.4 J_c$ (+) (partially penetrated initial flux profiles), $0.2 J_c(\blacktriangle)$, and $0.1 J_c(\circ)$ (fully penetrated initial flux profiles). Arrows denote the threshold time τ_i. Inset to the figure are data on grain-oriented $YBa_2Cu_3O_{7-\delta}$ by Maley et al. [44]. In this experiment, different initial current-densities were obtained by cycling the sample to temperatures labeled $T_{initial}$ before the relaxation measurement was begun at $T_{meas} = 20$ K.

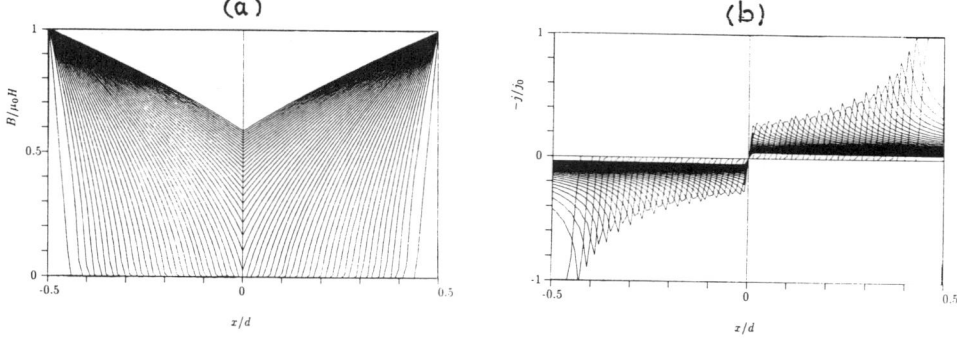
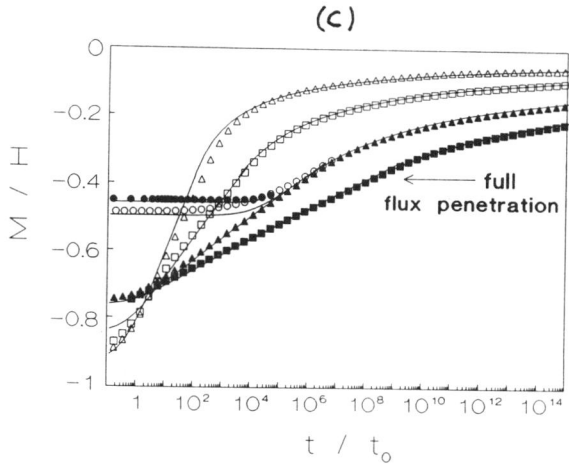

Figure 7: Flux-density (a) and current-density (b) profiles obtained in the twodimensional collective creep model, $U = U_c(J_c/J)^{9/8} - 1$, using a field-dependent critical current- density of the form (32). Parameter values were $U_c/k_BT = 7$, $J_0 = 7 \times 10^{10}$ A m^{-2}, $\rho_n = 50\mu\Omega$ cm and $B_{c2} = 70$ T. (c) The time evolution of the magnetization in an infinite superconducting slab for forms of $U(J)$ and $J_c(B)$ as in (a). Parameter values, chosen to represent Bi:2212, are $U_c/k_BT = 2(\triangle)$, 4 ($\square$), 7 ($\circ, \bullet, \blacktriangle$), and 10 ($\blacksquare$), remaining parameter values were the same as in (a). Open and closed symbols represent calculations started from linear and parabolic flux distributions, respectively. Different initial values of M correspond to different initial flux-profiles, lines depict the expected time evalution of the magnetization as obtained by substituting $(J_0/J_c)J_s(t)$, with $J_s(t)$, as given in Eq.(27).

Any field dependence of U and J_c is initially disregarded. Although this approximation is only valid in the case of individually pinned isolated vortices, such an approach brings out the unobscured features following from the chosen $U(J)$ dependence.

Figure 6(a,b) shows the time development of the flux and current-density profiles for $U = U_c(J_c/J)^{9/8} - 1)$ starting from a partial critical state. The corresponding time decay of the magnetization is depicted in Fig. 6(c), along with results for different initial states. In general, changes in J and B only become appreciable at times $t = \mu_0 d^2/4\rho_n$, provided that J is of the order of J_c. At lower initial current-densities J_i, the decay of magnetization remains extremely slow until t exceeds the threshold time $\tau_i = \tau \exp(U(J_i)/k_B T)$. At shorter times, the probability that vortices move from their respective metastable configurations is exponentially small. At times greater than the threshold time, the solution tends to the same asymptote regardless of the initial flux profile. Such behaviour was found by several other methods [52].

As can be seen from Fig. 6(b), the current-density distribution initially penetrates the sample as a sharp front, the value of J being almost constant and the flux-profile linear in the region of penetration. The profiles reflect the characteristics of the Bean model even when J is less than J_c. Moreover, the flux- and current-density distributions were found to be stable with respect to small perturbations, introduced by suddenly ramping $B = \mu_0 H$ at the surface up or down. Such behaviour was found for all models involving a current-dependent activation barrier. An analytic treatment of the logarithmic $U(J)$-dependence [49] yields the same result. Thus, in all considered creep models, there exists, for a given set of parameters, a single, time-dependent current distribution to which the solution $J(x,t)$ evolves, indicative of self-organized critically in the stage of incomplete penetration.

At the time of full penetration, $t = t^*$, the solution changes owing to the extra condition that $J \equiv 0$ in the sample centre. In contrast to the expectation in Ref. [49], a kink in the magnetization relaxation curve was not found at t^*. Instead, the time-dependent logarithmic creep rate $S(t)$ has a maximum. However, when the time decay of magnetization after a small field step was calculated, as in [49], a kink was in fact found at t^*.

At times $t > t^*$, we can group results into two classes. If the activation barrier $U(J)$ diverges for $J \to 0$, as in Fig. 6, the flux profile remains (almost) linear, i.e., scales with the (Bean) critical state at all times and the magnetization still tends to the same asymptotic curve independent of the initial flux profile. Note that the requirement that J be continuously differentiable at the centre of the sample prohibits the formal Bean state from being realized physically. In the case of finite $U(J \to 0)$ TAFF behaviour is eventually observed. The flux- and current-density profiles become hyperbolic functions of the spatial coordinate, the magnetization decays exponentially with time, and the insensitivity to the initial conditions is no longer observed.

Field-dependent J_c. The next step is the introduction of an explicit field dependence of the critical current density. For illustrative purposes, we chose the field dependence of Eq.(32). It must be stressed that the above form is taken only because of analytical simplicity and not because it is expected to represent the actual behaviour of high-temperature superconductors.

Figure 7(a,b) illustrates the flux and current profiles found for the two-dimensional collective creep model, $U = U_c((J_c/J)^{9/8} - 1)$, and parameter values representing the Bi:2212 high-temperature superconductor. Magnetization curves for a variety of U_c/k_BT ratios and initial states are shown in Fig. 7(c). Again the asymptotic shape of the flux-profile is independent of the initial conditions, and scales with the shape expected from the field dependence of the critical current density. This is vividly illustrated by the figure. An initial Bean state (linear flux profile) was expressly chosen, but after a few steps in the calculation the flux-profile has become parabolic, in correspondence with the input $J_c(B)$-dependence. The calculations for $U_c/k_BT = 7$ depicted in Fig. 7(c) were started from either linear or parabolic critical or subcritical states. Nevertheless, the magnetization at sufficiently long times is the same.

The scaling of the $J(x)$-profile with the current distribution in the critical state means that the time-dependent magnetization M should be well described by the appropriate expression for the magnetization in the critical state, but with J_c replaced by the time-dependent current density $J_s(t)$ as derived in Section 3.5.1. In [42] this proposition is further substantiated for the flux-creep models under consideration. In the situation of decreasing induction, possible transient contributions arising from the time regime of incomplete flux penetration are absent and the assessment whether $M \propto J_s$ can be made over the longest possible time interval. In the case of partial penetration M is a (nearly) linear function of the flux penetration depth x_f which is in turn linear with $H/J(t)$.

Acknowledgements

Stimulating conversations with many colleagues are gratefully acknowledged. In particular, I would like to thank Valerii Vinokur and Kees van der Beek for many helpful, clarifying and pleasant discussions. Sections 3.4 and 3.5 are based on the Ph.D. thesis of Van der Beek. This work was financed in part by the Netherlands Foundation for Research on Matter (FOM).

References

[1] E.E. Farrell et al., Phys. Rev. Lett. **63**, 782 (1989) and Phys. Rev. **B42**, 6758 (1990).

[2] B. Battlog in "High Temperature Superconductivity, Los Alamos Symposium 1989", edited by K.S. Bedell et al. (Addison Wesley, Redwood City, 1990) p. 37.

[3] W.E. Lawrence and S. Doniach in Proceedings of the Twelfth International Conference on Low Temperature Physics, Kyoto, 1970, edited by E. Kanda (Kigaku, Tokyo, 1971) p. 361.

[4] J.R. Clem, Bull. Am. Phys. Soc. **35**, 260 (1990).

[5] A. Houghton, R.A. Pelcovits and A. Sudbø, Phys. Rev. **B40**, 6763 (1989).

[6] P. Koorevaar, J. Aarts, P. Berghuis and P.H. Kes, Phys. Rev. **B42**, 1004 (1990).

[7] V.M. Vinokur, P.H. Kes and A.E. Koshelev, Physica **C168**, 29 (1990).

[8] M.V. Feigelman, V.B. Geshkenbein and A.I. Larkin, Physica **C167**, 177 (1990).

[9] D.S. Fisher, M.P.A. Fisher and D. Huse, Phys. Rev. **B43**, 130 (1991).

[10] M.V. Feigelman and V.M. Vinokur, Phys. Rev. **B41**, 8986 (1990).

[11] D. Nelson, Phys. Rev. Lett. **60**, 1973 (1988).

[12] E.H. Brandt, Phys. Rev. Lett. **63**, 1106 (1989).

[13] P. Berghuis and P.H. Kes, Phys. Rev. **B47**, 262 (1993).

[14] P.H. Kes and J. van den Berg in: Studies in High Temperature Superconductors, Vol. 5, A. Narlikar, ed. (NOVA Science Publishers, New York, 1990) p. 83.

[15] C.J. van der Beek and P.H. Kes, Phys. Rev. **B43**, 13032 (1991).

[16] E.J. Kramer, Phil. Mag. **33**, 331 (1976); C.S. Pande, Appl. Phys. Lett. **28**, 462 (1976).

[17] M. Murakami, M. Morita, K. Doi, K. Miyamoto, Jpn. J. Appl. Phys. **28**, 1189 (1989).

[18] S. Jin, T.H. Tiefel, S. Nakahera, J.E. Graebner, H.M. O'Bryan, R.A. Fastnacht and G.W. Kammlott, Appl. Phys. Lett. **56**, 1287 (1990).

[19] M. Hawley, J.D. Raistrick, J.G. Beery and R.J. Houlton, Science **251**, 1587 (1991); Ch. Gerber, D. Anselmetti, J.G. Bednorz, J. Mannhart and D.G. Schlom, Nature **350**, 279 (1991).

[20] L.N. Bulaevski, Zh. Eksp. Teor. Fiz. **64**, 2241 (1973) [Sov. Phys. JETP **37**, 1133 (1973)].

[21] E.V. Thuneberg, Cryogenics **29**, 236 (1989); J. Low Temp. Phys. **57**, 415 (1984).

[22] E.J. Kramer and H.C. Freyhardt, J. Appl. Phys. **51**, 4930 (1980); T. Matsushita, J. Appl. Phys. **54**, 281 (1983); A.A. Golub et al., Fiz. Nizk. Temp. **10** [Sov. J. Low Temp. Phys. **10**, 133 (1984).

[23] R. Labusch, Crystal Lattice Defects **1**, 1 (1969).

[24] A.M. Campbell and J.E. Evetts, Adv. Phys. **21**, 199 (1972).

[25] E.H. Brandt, Z. Phys. **B80**, 167 (1990).

[26] R. Griessen, in Concise Encyclopedia of Magnetic and Superconducting Materials, ed. J.E. Evetts (Pergamon, Oxford 1991), p. 144.

[27] D. Dew Hughes, Cryogenics **28**, 675 (1988).

[28] P.H. Kes, J. Aarts, J. van den Berg and C.J. van der Beek, Superc. Sci. Technol. **1**, 242 (1989).

[29] A. Gupta, P. Esquinazi and H.F. Braun, Phys. Rev. Lett. **63**, 1869 (1989).

[30] K. Yamafuji, F. Fujiyoshi, K. Toko and T. Matsushita, Physica **C159**, 743 (1989).

[31] A.M. Campbell, J. Phys. C: Solid St. Phys. **4**, 3186 (1971).

[32] P.W. Anderson and Y.B. Kim, Rev. Mod. Phys. **36**, 39 (1964).

[33] M.R. Beasley, R. Labusch and W.W. Webb, Phys. Rev. **181**, 682 (1969).

[34] C.W. Hagen, R. Griessen and E. Salomons, Physica **C157**, 199 (1989).

[35] P. Esquinazi, Solid State Commun. **74**, 75 (1990).

[36] C.J. van der Beek, G.J. Nieuwenhuys and P.H. Kes, Physica **C185-189**, 2241 (1991).

[37] Y. Xu, M. Suenaga, A.R. Moodenbaugh and D.O. Welch, Phys. Rev. **B40**, 10882 (1989); R. Griessen, Physica **C172**, 441 (1991).

[38] C.W. Hagen and R. Griessen, Phys. Rev. Lett. **62**, 2857 (1989).

[39] T. Nattermann, Phys. Rev. Lett. **20**, 2454 (1989).

[40] K.H. Fischer and T. Nattermann, Phys. Rev. **B43**, 10372 (1991).

[41] C.J. van der Beek, P.H. Kes, M.P. Maley, M.J.V. Menken and A.A. Menovsky, Physica **C195**, 307 (1992).

[42] C.J. van der Beek, G.J. Nieuwenhuys, P.H. Kes, H.G. Schnack and R.P. Griessen, Physica C197, 320 (1992).

[43] M.J.V. Menken, Ph.D. thesis, University of Amsterdam, 1991.

[44] M.P. Maley, J.O. Willis, H. Lessure and M.E. McHenry, Phys. Rev. B42, 2639 (1990).

[45] M.V. Feigelman, V.B. Geshkenbein and V.M. Vinokur, Phys. Rev. B43, 6263 (1991); E.H. Brandt, Z. Phys. B80, 167 (1990).

[46] A.P. Malozemoff and M.P.A. Fisher, Phys. Rev. B42, 6784 (1990).

[47] B.M. Lairson, J.Z. Sun, T.H. Geballe, M.R. Beasley and J.C. Bravman, Phys. Rev. B43, 10405 (1991); G. Ries, H.W. Neumüller and W. Schmidt, Superc. Sci. Technol. 5, S81 (1992).

[48] C. Dekker, W. Eidelloth and R.H. Koch, Phys. Rev. Lett. 68, 3347 (1992).

[49] V.M. Vinokur, M.V. Feigelman and V.B. Geshkenbein, Phys. Rev. Lett. 67, 915 (1991).

[50] C.P. Bean, Phys. Rev. Lett. 8, 250 (1962); C.P. Bean, Rev. Mod. Phys. 36, 31 (1964).

[51] Y.B. Kim, C.F. Hempstead and A.R. Strnad, Phys. Rev. Lett. 9, 306 (1962).

[52] R. Griessen, J.G. Lensink, T.A.M. Schröder and B. Dam, Cryogenics 30, 563 (1990).

Lectures presented at the NATO Advanced Study Institute: "Phase Transitions and Relaxation in Systems with Competing Energy Scales," April 13–23, 1993, Geilo, Norway.

Vortex Line Fluctuations in Superconductors from Elementary Quantum Mechanics

David R. Nelson
Lyman Laboratory of Physics
Harvard University
Cambridge, Massachusetts 02138

ABSTRACT. Concepts from elementary quantum mechanics can be used to understand vortex line fluctuations in high-temperature superconductors. Flux lines are essentially classical objects, described by a string tension, their mutual repulsion, and interactions with pinning centers. The classical partition function, however, is isomorphic to the imaginary time path integral description of quantum mechanics. This observation is used to determine the thermal renormalization of critical currents, the decoupling field, the flux lattice melting temperature at low and moderate inductions, and to estimate the degree of entanglement in dense flux liquids. The consequences of the "polymer glass" freezing scenario, which assumes that the kinetic constraints of entanglement prevent field cooled flux liquids from crystallizing, are reviewed.

1. Introduction

The past five years have been a time of great ferment in the study of high-temperature superconductors. Although no general consensus about the microscopic mechanism for superconductivity in the cuprate materials has emerged, there is now considerable understanding of the remarkable behavior of these materials in a magnetic field. The theoretical analysis of vortex fluctuations in these extreme Type II materials requires only an underlying Ginzburg-Landau theory with a BCS-like order parameter. The basic conclusions are *independent* of the precise microscopic mechanism, and rely instead on the remarkably different Ginzburg-Landau parameters (coherence length, temperature range and anisotropy) which distinguish the cuprates from their low T_c counterparts.

Figure 1 shows a schematic temperature-magnetic field phase diagram for cuprate superconductors subjected only to weak point disorder in the form of oxygen vacancies. Throughout this paper we assume for simplicity a field oriented along the c axis, perpendicular to the copper-oxide planes. The magnetic field B (proportional to the vortex density) is plotted because demagnetizing corrections in the usual slab-like experimental geometry insure that B rather than the magnetic induction H is held fixed in most experimental situations. The famous Abrikosov flux lattice, which exists for all temperatures below the upper critical field $B_{c2}(T)$ in mean field theory [1], appears here only below a much lower "melting line" $T_m(B)$. Above this line, melted vortex arrays entangle in a novel "flux liquid." The line $B_{c2}^0(T)$ marks the onset of enhanced diamagnetism but is not

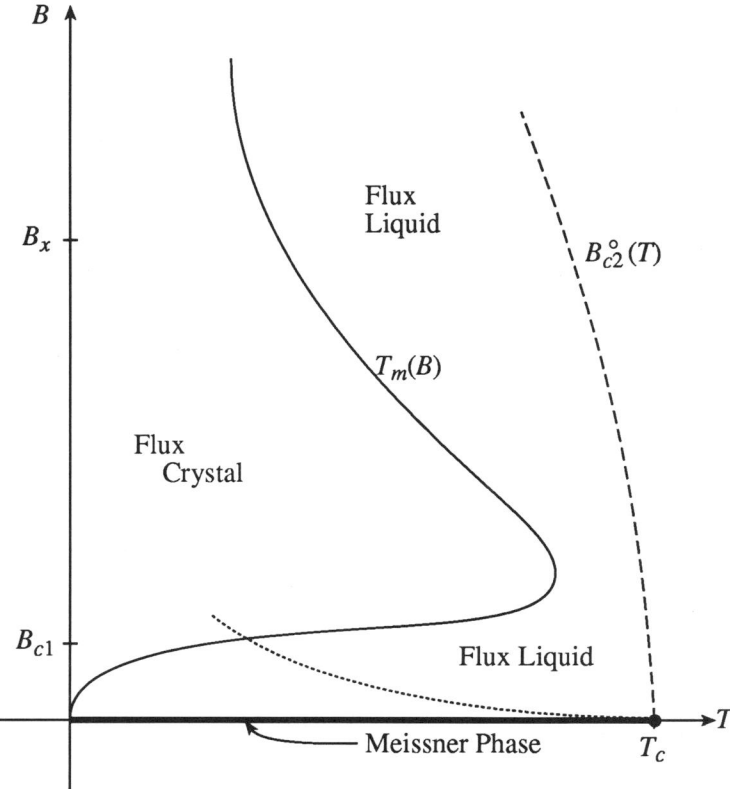

Figure 1. Schematic phase diagram of a high temperature superconductor in the clean limit. Although quantized vortex lines appear below $B_{c2}^0(T)$, the Abrikosov crystal phase appears only below $T_m(B)$. The Meissner phase collapses to the heavy line along the temperature axis in this representation. Point disorder only affects the flux liquid below the dotted line. Ref. [8] presents arguments that the maximum in the melting curve actually bends down closer to T_c than indicated here.

expected to be a sharp phase boundary. The crossover fields B_{c1} and B_x are discussed in Sec. 3. There is now considerable evidence [2–5] that the Abrikosov lattice in clean (twin-free) single crystals of yttrium barium copper oxide (YBCO) melts at $T_m(B)$ via a first-order phase transition [6] into a flux liquid in which the quantized vortex filaments presumably wander and entangle in a complicated fashion. Weak point disorder significantly alters the properties of the flux liquid only below the dotted line [7]. Note that a sliver of flux liquid may exist above this line and below the melting curve down to quite low temperatures. It is still unclear whether point disorder alone is sufficient to produce a distinct thermodynamic "vortex glass" phase [8] below the dotted line or within the crystalline region.

Figure 2 shows another striking development—a remarkable shift in the "irreversibility

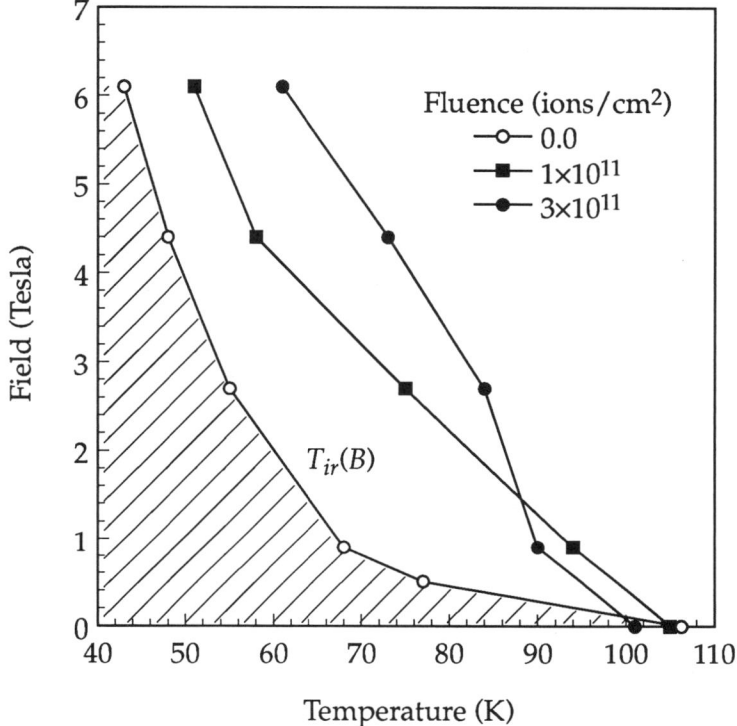

Figure 2. Effect of columnar pins on the irreversibility line of $Tl_2Ba_2Ca_2Cu_3O_{10}$ [9]. In the absence of irradiation, the resistivity only becomes unmeasurably small in the shaded region below $T_{ir}(B)$. The reentrant low field vortex liquid regime of Fig. 1 appears for $B \lesssim 10^{-2}$ T, and is hence not visible on this scale. Figure courtesy of R. Budhani, Brookhaven National Laboratory.

line" of a highly anisotropic thallium-based compound engineered via the deliberate introduction of columnar pins [9]. Similar pinning centers have been injected by many groups via heavy ion irradiation [10–13] with sufficient energy to produce long tracks of damaged material. The "irreversibility line" $T_{ir}(B)$ is the boundary below which the dynamics of field cooled materials slows down drastically [14]. In samples *without* correlated disorder the melting curve and irreversibility line may in fact be almost identical [8].

Since $T_{ir}(B)$ coincides approximately with the temperature at which the resistivity becomes unmeasurably small, the large upward shift in Fig. 2 is of considerable technological as well as intellectual interest. Note for example that the effective critical temperature for superconductivity at $B = 2.7$ tesla shifts from 53 K to 87 K upon irradiation. The critical current at liquid nitrogen temperature (77 K) increases by three orders of magnitude [9]. It is believed that after irradiation the irreversibility line becomes a locus of thermodynamically sharp "bose glass" transitions, below which the flux lines are localized on the columnar defects [15]. Other forms of correlated disorder may be important even in unirradiated samples. Twin boundaries, for example, appear to be responsible for the apparently continuous transition observed in twinned YBCO samples with the field aligned parallel to the c axis [3].

In these lectures, we highlight these developments by describing how flux lines interact with correlated pinning centers and with each other. To account for thermal fluctuations, one must average over vortex configurations. Because the configuration sums bear a strong resemblance to the imaginary time path integral formulation of quantum mechanics [16], wave functions and binding energies which appear in simple quantum problems (such as the square well and the harmonic oscillator) have important implications for flux lines. With this identification we can quickly compute the free energy and localization length of vortices in the bose glass phase, the corresponding critical currents, and determine as well the decoupling field and line of melting temperatures in clean systems [15]. In the flux liquid, we can readily estimate the degree of entanglement of the wandering vortex lines. If entanglement occurs in temperature and field ranges where flux cutting barriers are high, it may prevent crystallization and lead to a polymer-like glass transition even in the absence of strong pinning disorder [17]. The experimental consequences of this scenario are reviewed at the end of this brief, elementary review.

The idea of using Schroedinger's equation to solve problems in classical statistical mechanics is not new. It was used by S.F. Edwards, P.G. deGennes and others starting in the 1960s to treat the conformations of flexible polymer chains [18]. One can go further with this analogy for flux lines, however, because these are equivalent to a system of *directed* polymers, with a common average orientation. As a result, distant self-interactions along the filaments can usually be neglected, in contrast to isotropic self-avoiding polymer solutions. The essential physics is captured if one introduces a line tension to control vortex wandering and allows interactions between *different* polymers as well as interactions with the relevant pinning centers. Although we focus here primarily on simple one and two-line problems, many flux line statistical mechanics (with multiple pins) is in fact equivalent to the many body quantum mechanics of bosons in two dimensions [15, 17]. This system differs from the otherwise closely related problem of helium films on disordered substrates [19] because vortices behave dynamically like *charged* bosons, and are easily manipulated in experiments by the injection of supercurrents. A simple explanation of why vortex probability distributions are boson-like in a thick sample can be found in Sec. 2.3. The richness and complexity of the *many* flux line problem rivals the physics of correlated electrons in semiconductors and metals. For details, readers are referred to Refs. 15 and 17, and a recent review [20].

2. Correlated Pinning and Quantum Bound States

2.1. MODEL-FREE ENERGY

We start with a model-free energy F_N for N flux lines in a sample of thickness L, defined by their trajectories $\{\vec{r}_j(z)\}$ as they traverse a sample with both columnar pins and external magnetic field aligned with the z-axis, i.e., in the direction perpendicular to the CuO_2 planes,

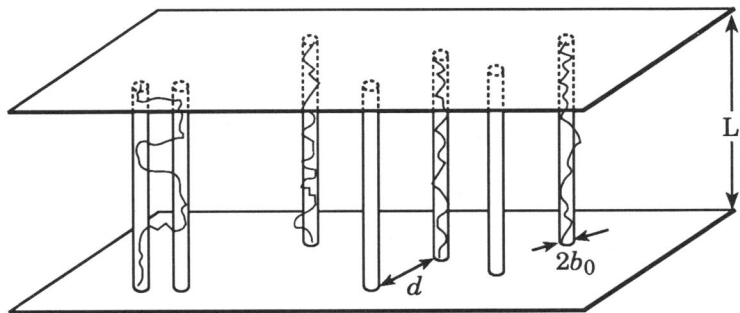

Figure 3. Schematic of columnar pins and pinned vortex lines.

$$F_N = \frac{1}{2}\tilde{\epsilon}_1 \sum_{j=1}^{N} \int_0^L \left|\frac{d\vec{r}_j(z)}{dz}\right|^2 dz + \frac{1}{2}\sum_{i\neq j}\int_0^L V(|\vec{r}_i(z) - \vec{r}_j(z)|)\,dz$$
$$+ \sum_{j=1}^{N}\int_0^L V_D[\vec{r}_j(z)]\,dz \qquad (2.1a)$$

with

$$V_D(\vec{r}) = \sum_{k=1}^{M} V_1(\vec{r} - \vec{R}_k)\,. \qquad (2.1b)$$

Here $V(|\vec{r}_i - \vec{r}_j|) = 2\epsilon_0 K_0(r/\lambda_{ab})$, is the interaction potential between lines with in-plane London penetration depth λ_{ab} and the random potential $V_D(\vec{r})$ arises from a z-independent set of M disorder-induced columnar pinning potentials $V_1(\vec{r})$ centered on sites $\{\vec{R}_k\}$. The tilt modulus $\tilde{\epsilon}_1 \approx (M_\perp/M_z)\epsilon_0 \ln(\lambda_{ab}/\xi_{ab})$, where the material anisotropy is embodied in the effective mass ratio $M_\perp/M_z \ll 1$, and $\epsilon_0 \approx (\phi_0/4\pi\lambda_{ab})^2$ is the energy scale for the interactions. The potential $V_D(\vec{r})$ arises from identical cylindrical traps assumed for simplicity to pass completely through the sample with well depth per unit length U_0 and effective radius b_0. The parameter $b_0 \approx \max\{c_0, \xi_{ab}\}$ where $c_0 \approx 25 - 40$ Å is the radius of the columnar pins and ξ_{ab} is the superconducting coherence length in the ab-plane.

A complete analysis of the many-line statistical mechanics associated with Eqs. (2.1) requires multiple path integrals over vortex trajectories weighted by $e^{-F_N/T}$ and subject to a complicated random pinning potential. See Fig. 3. Our goal here is to illuminate the essential physics by studying a few simple problems involving one or two flux lines and only a few columnar pins.

2.2. ONE FLUX LINE AND ONE COLUMNAR PIN

Consider a vortex line trapped near a single columnar pin with well depth U_0 and radius b_0 parallel to z in an otherwise defect-free sample of thickness L, as shown in Fig. 4. A quantity of considerable physical interest is the binding *free* energy per unit length

$$U(T) = U_0 - TS\,, \qquad (2.2)$$

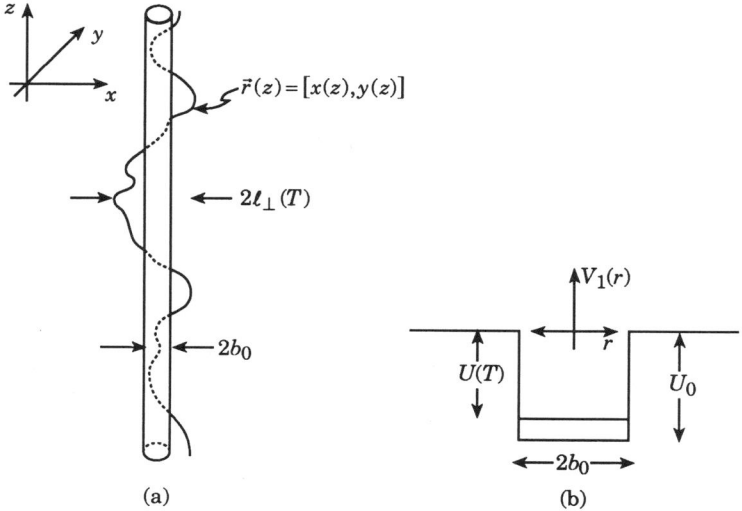

Figure 4. One columnar pin and one vortex line, indicating (a) the localization length $\ell_\perp(T)$ and (b) the thermally renormalized pinning potential $U(T)$.

where S is the entropy reduction due to confinement. This free energy is given by a path integral,

$$e^{U(T)L/T} = \frac{\int \mathcal{D}\vec{r}(z)\exp\left[-\frac{\tilde{\epsilon}_1}{2T}\int_0^L \left(\frac{d\vec{r}}{dz}\right)^2 dz - \frac{1}{T}\int_0^L V_1[\vec{r}(z)]\,dz\right]}{\int \mathcal{D}r(z)\exp\left[-\frac{\tilde{\epsilon}_1}{2T}\int_0^L \left(\frac{d\vec{r}}{dz}\right)^2 dz\right]}, \qquad (2.3)$$

where the denominator is required to subtract off the entropy of an unconfined line far from the pin. The cylindrically symmetric confining potential $V_1(r)$, indicated in Fig. 4b, tends to zero as $|\vec{r}| \to \infty$. The path integrals in (2.3) follow from standard statistical methods which express them in terms of the eigenvalues of a transfer matrix. In the limit $L \to \infty$, the smallest eigenvalue dominates, and $U(T) = -E_0(T)$, where $E_0(T)$ is the ground state energy of a two-dimensional "Schroedinger equation" (see Appendix A),

$$\left[-\frac{T^2}{2\tilde{\epsilon}_1}\nabla_\perp^2 + V_1(r)\right]\psi_0(\vec{r}) = E_0\psi_0(\vec{r}). \qquad (2.4)$$

Here and henceforth, all vectors $\vec{r}$ will refer to positions in the plane perpendicular to $\hat{z}$. Note that T plays the role of the Planck parameter $\hbar$ and $\tilde{\epsilon}_1$ plays the role of mass m in this quantum mechanical analogy.

The ground state wave function $\psi_0(\vec{r})$ determines the localization length $\ell_\perp(T)$ displayed in Fig. 3a. As shown in Appendix B, the probability $P(\vec{r})$ of finding a point on the vortex at transverse displacement $\vec{r}$ relative to the center of the pin is independent of z and given by the square of $\psi_0(\vec{r})$, just as in elementary quantum mechanics,

$$\mathcal{P}(\vec{r}) = \psi_0^2(\vec{r}) \bigg/ \int d^2r\,\psi_0^2(\vec{r}). \qquad (2.5)$$

Because (2.4) is unchanged under complex conjugation, $\psi_0(\vec{r})$ can always be chosen to be real. We then define the localization length as

$$\ell_\perp^2 = \int d^2r \, r^2 \psi_0^2(r) \bigg/ \int d^2r \, \psi_0^2(r) \,. \tag{2.6}$$

The properties of a vortex near a columnar pin now follow from standard results for a quantum particle in a cylindrical potential [21]. If we assume for simplicity a cylindrical square well, $(V_1(r) \equiv -U_0, \, r < b_0,$ and $V_1(r) = 0, \, r > b_0)$ the binding-free energy $U(T) = -E_0$ takes the form [20],

$$U(T) = U_0 f(T/T^*) \tag{2.7}$$

where T^* is an important characteristic temperature defined by

$$T^* = \sqrt{\tilde{\epsilon}_1 U_o} \, b_0 \,. \tag{2.8}$$

When $T \ll T^*$, the well depth is effectively infinite, and we find the usual particle-in-a-box result,

$$U(T) \approx U_0 - c_1 \frac{T^2}{2\tilde{\epsilon}_1 b_0^2} \tag{2.9}$$

where c_1 is a constant related to the first zero of the Bessel function $J_0(x)$ which solves Eq. (2.4) in this limit. The localization length is then

$$\ell_\perp(T) \approx b_0[1 + \mathcal{O}(1/\kappa b_o)] \,, \tag{2.10}$$

where $\kappa^{-1} \approx T/\sqrt{2\tilde{\epsilon}_1 U(T)} \ll b_0$ is the distance the "particle" penetrates into the classically forbidden region. The low-temperature correction in Eq. (2.9) represents the entropy lost each time a wandering flux line is reflected off the confining walls of the binding potential [22]. At the crossover temperature T^*, this "zero point energy" of confinement becomes comparable to the well depth.

When $T \gg T^*$, the flux line is only weakly bound, although a strictly localized ground state *always* exists in this effectively two-dimensional problem [21]. In the limit, one finds [15,20]

$$U(T) \approx \frac{1}{2} U_0 \left(\frac{T}{T^*}\right)^2 e^{-2(T/T^*)^2} \,, \tag{2.11}$$

and a localization length $\ell_\perp(T) \approx \kappa^{-1} = T/\sqrt{2\tilde{\epsilon}_1 U(T)}$, so that

$$\ell_\perp(T) \approx b_0 e^{(T/T^*)^2} \,. \tag{2.12}$$

The flux line now "diffuses" within a confining tube with radius of order $\ell_\perp(T)$ as it crosses the sample. The length along $\hat{z}$ required to "diffuse" across this tube is $\ell_z \approx \ell_\perp^2/(T/\tilde{\epsilon}_1)$, i.e.,

$$\ell_z = (b_0^2 \tilde{\epsilon}_1/T) e^{2(T/T^*)^2} \,. \tag{2.13}$$

These results imply a strong thermal renormalization of the critical current $J_c(T)$ [15]. $J_c(T)$ is the current necessary to produce a Lorentz force $f_c = J_c\phi_0/c$ so strong that thermal activation is unnecessary to tear a flux line away from its columnar pin. At low temperatures, one would expect that $f_c \approx U_0/b_0$, i.e.,

$$J_c \approx cU_0/\phi_0 b_0 . \tag{2.14}$$

Here we assume that the confining potential $V_1(r)$ does not really jump abruptly at $r = b_0$, but instead rises smoothly to zero in a distance of order b_0. Results such as (2.11–2.13) are in any case independent of the precise form of the microscopic potential [21]. To account for line wandering, we should replace U_0 by $U(T)$ and b_0 by $\ell_\perp(T)$. For $T \gg T^*$, this leads to

$$J_c(T) \approx J_c(0) e^{-3(T/T^*)^2} . \tag{2.15}$$

2.3. ONE FLUX LINE AND TWO COLUMNAR PINS

The localization length discussed above is like the Bohr radius of an isolated "atom" consisting of one columnar pin and one vortex line. Now consider a vortex line which is able to hop between *two* nearby identical columnar pinning sites at $\vec{R}_1$ and $\vec{R}_2$ (analogous to an H_2^+ molecule) as it traverses the sample—see Fig. 5. For now, we ignore the additional dashed flux line. The binding potential $V_1(r)$ in Eq. (2.4) in this case should be replaced by

$$V_2(\vec{r}) = V_1(\vec{r} - \vec{R}_1) + V_1(\vec{r} - \vec{R}_2) , \tag{2.16}$$

where identical wells of depth U_0 and radius b_0 separated by d are assumed. As in a quantum double well, the dominant configurations are those in which the particle delocalizes further by "tunneling" from one well to another. In the subspace spanned by the isolated pin ground state wave functions $\psi_0(\vec{r} - \vec{R}_1)$ and $\psi_0(\vec{r} - \vec{R}_2)$, the effective Hamiltonian takes the form

$$\mathcal{H}_2 = \begin{pmatrix} E_0, & t \\ t, & E_0 \end{pmatrix} , \tag{2.17}$$

where E_0 is the ground state energy determined in the previous subsection, and $t > 0$ is a tunneling matrix element. The two lowest energy eigenvalues for the double well are then

$$E_\pm = E_0 \pm t . \tag{2.18}$$

The partition function in this approximation is $Z_{tot} = \sum_{i,j=1}^{2} \langle i| e^{-\mathcal{H}L/T} |j\rangle$, where $|1\rangle$ and $|2\rangle$ are states localized on pin 1 or 2.

To determine t, we can proceed variationally and minimize

$$E(\psi) \equiv \frac{\int d^2r \left[\frac{T^2}{2\tilde{\epsilon}_1} |\nabla \psi(\vec{r})|^2 + V_2(\vec{r}) |\psi(\vec{r})|^2 \right]}{\int d^2r |\psi(\vec{r})|^2} , \tag{2.19}$$

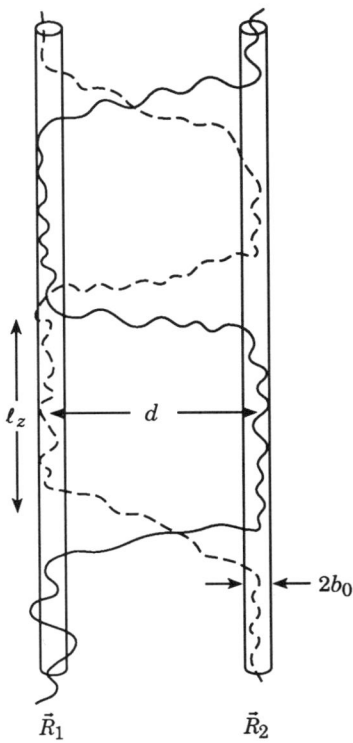

Figure 5. Two columnar pins with a single vortex (solid line) tunneling between them. The dashed line shows a second vortex which requires the introduction of a simple Hubbard model.

with the trial function

$$\psi(\vec{r}) = \alpha \psi_0(\vec{r} - \vec{R}_1) + \beta \psi_0(\vec{r} - \vec{R}_2), \qquad (2.20)$$

where $\psi_0(\vec{r})$ is the ground state for an isolated well. We assume widely separated wells, i.e., $d \gg \ell_\perp(T)$. The minimum occurs for the symmetric case $\alpha = \beta$ and has energy $E = E_0 - t$, with

$$t \approx \text{const.} \times \frac{U(T)}{\sqrt{E_k/T}} e^{-E_k/T}, \qquad (2.21)$$

where $U(T) = -E_0(T) > 0$, and

$$E_k = \sqrt{2\tilde{\epsilon}_1 U(T)}\, d \qquad (2.22)$$

is similar to a WKB tunneling exponent.

The flux line has now delocalized a distance $\sim d$ in the transverse direction. Delocalization proceeds via wandering in a tube of radius $\ell_\perp(T)$, and occasional tunneling across to a neighboring tube, as indicated in Fig. 5. When $b_0 \ll \ell_\perp \ll d$ the spacing between such tunneling events along the z axis is of order

$$\ell_z(T) \approx \left[\frac{b_0^2 \tilde{\epsilon}_1}{T} e^{2(T/T^*)^2}\right] e^{E_k/T}, \qquad (2.23)$$

where the prefactor (an inverse "attempt frequency" in imaginary time) comes from the isolated pin result Eq. (2.13). Note the close analogy between Fig. 5 and the configurations of a classical one-dimensional Ising model with exchange constant $J = E_k$ disrupted by kinks along the z axis.

The flux line will not be delocalized much further by adding a third, more distant columnar pin to the problem, because the new available state is not in resonance with the double well ground state energy calculated above. One isolated flux line will always be localized by a random array of columnar pins [15].

2.4. A HUBBARD MODEL: TWO FLUX LINES AND TWO COLUMNAR PINS

Interactions are crucial for determining vortex configurations when correlated pinning is present. In absence of a repulsive pair potential all vortices would pile up at $T = 0$ in a deep minimum of the random pinning potential produced, for example, by an unusually dense region of columnar pins. By allowing *two* vortices to wander simultaneously between a pair of columnar pins, we can study interactions in a particularly simple context. This elementary model also illustrates why flux lines behave like bosons in thick samples. A related treatment describes the H_2 molecule in real quantum mechanics [23].

Figure 5 shows two vortices (solid and dashed lines) hopping back and forth between two columnar pins as we trace their trajectories along the z axis. In the absence of interactions, the probability distribution of each vortex would be described by the symmetric double well ground state wave function discussed in the previous section. The vortices would find themselves on the same columnar pin approximately half the time. Introducing a repulsive energy between vortices will lead to *correlated* hopping as the fluxons exchange places.

To model this situation, we introduce a tight-binding Hamiltonian similar to Eq. (2.17) which operates on a set of four normalized orthogonal basis states, $|12\rangle$, $|21\rangle$, $|11\rangle$, and $|22\rangle$. The state $|12\rangle$ means that the first fluxon line occupies pin 1 and the second occupies pin 2, while $|21\rangle$ is the state where the vortices have exchanged places. Both fluxons are localized on pin 1 or on pin 2 in the states $|11\rangle$ and $|22\rangle$, respectively. The classical partition function which gives the probability of making a transition from one of these four states $|a\rangle$ to a final state $|b\rangle$ across a slab of thickness L is then

$$\mathcal{Z}(a,b;L) = \langle b| e^{-\mathcal{H}L/T} |a\rangle, \qquad (2.24)$$

where the tight-binding model is defined by the 4×4 matrix Hamiltonian

$$\mathcal{H} = \begin{array}{c} \\ |12\rangle \\ |21\rangle \\ |11\rangle \\ |22\rangle \end{array} \begin{pmatrix} |12\rangle & |21\rangle & |11\rangle & |22\rangle \\ 2E_0 & 0 & t & t \\ 0 & 2E_0 & t & t \\ t & t & 2E_0 + V_{\text{int}} & 0 \\ t & t & 0 & 2E_0 + V_{\text{int}} \end{pmatrix}. \qquad (2.25)$$

The diagonal terms include the one vortex binding energy E_0 discussed in Sec. 2.2., and a "Hubbard repulsion" term V_{int} which represents the energy which arises when two vortices occupy the same columnar pin. A reasonable estimate of V_{int} when $d \ll \lambda_{ab}$ is [15]

$$V_{int} \approx 2\epsilon_0 \ln(d/\xi_{ab}) , \qquad (2.26)$$

reflecting the energy cost of one doubly quantized vortex as opposed to two singly quantized vortices separated by d. For $d \gg \lambda_{ab}$, $V_{int} \approx 2\epsilon_0 \ln(\lambda_{ab}/\xi_{ab})$. The off-diagonal terms in (2.25) reflect transitions between the four basis states caused by hops of a single vortex. The Hamiltonian breaks into symmetric and antisymmetric subspaces,

$$\mathcal{H} = \begin{array}{c} |-\rangle \\ |+\rangle \\ |11\rangle \\ |22\rangle \end{array} \begin{pmatrix} 2E_0 & \vdots & 0 & 0 & 0 \\ \cdots & \cdots & \cdots & \cdots & \cdots \\ 0 & \vdots & 2E_0 & \sqrt{2}t & \sqrt{2}t \\ 0 & \vdots & \sqrt{2}t & 2E_0 + V_{int} & 0 \\ 0 & \vdots & \sqrt{2}t & 0 & 2E_0 + V_{int} \end{pmatrix} \qquad (2.27)$$

when $|12\rangle$ and $|21\rangle$ are eliminated in favor of the symmetrized states

$$|-\rangle = \frac{1}{\sqrt{2}}(|12\rangle - |21\rangle) , \qquad (2.28a)$$

and

$$|+\rangle = \frac{1}{\sqrt{2}}(|12\rangle + |21\rangle) . \qquad (2.28b)$$

The one dimensional subspace of states antisymmetric under vortex interchange would describe spinless fermions in conventional quantum mechanics. The three dimensional symmetric subspace is appropriate to boson quantum mechanics. As $L \to \infty$, partition functions such as (2.24) will be dominated by the smallest eigenvalue of $\mathcal{H}$. The four eigenvalues of (2.27) are easily found to be

$$\lambda^* = 2E_0$$
$$\lambda_0 = 2E_0 + V_{int}$$
$$\lambda_+ = 2E_0 + \tfrac{1}{2}V_{int} + \tfrac{1}{2}\sqrt{V_{int}^2 + 16t^2}$$
$$\lambda_- = 2E_0 + \tfrac{1}{2}V_{int} - \tfrac{1}{2}\sqrt{V_{int}^2 + 16t^2} . \qquad (2.29)$$

The eigenvalue λ^* belongs to the fermion subspace, and is unaffected by interactions. The remaining eigenvalues are bosonic. As shown in Fig. 6, the lowest eigenvalue is always λ_-, in the boson subspace. It differs from the fermion eigenvalue $2E_0$ by an amount proportional to t^2/V_{int} as $V_{int} \to \infty$.

The boson character of the ground state is a general feature of Hamiltonians which are symmetric under particle interchange [16]. It arises in the present context because repeated

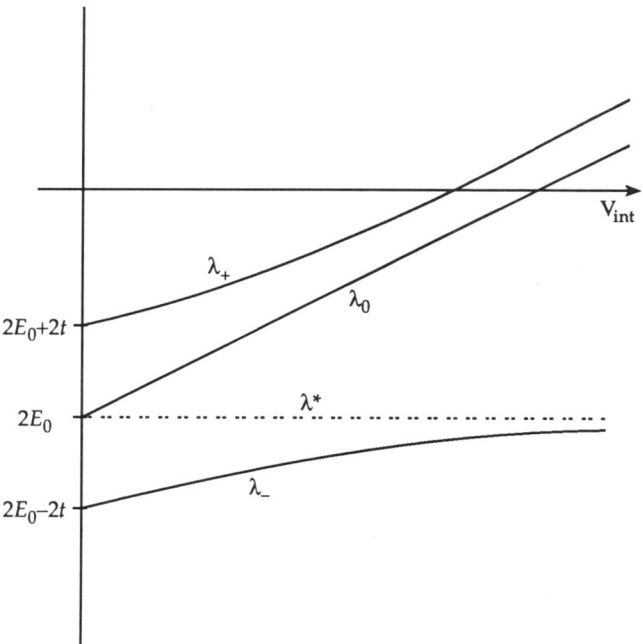

Figure 6. Energy eigenvalues for the four state Hubbard model. Solid lines represent the symmetric "boson" subspace, while the dashed line corresponds to the antisymmetric "fermion" excitation.

hopping between the columnar pins leads to a probability distribution symmetric in the two vortex coordinates for sufficiently thick samples. The pair of flux lines in Fig. 6 will behave like bosons for sample thicknesses $L \gg \ell_z$, where $\ell_z(T)$ is the spacing between these tunneling events.

The eigenvector of the ground state is

$$|0\rangle = \text{const.} \times \left[\frac{1}{\sqrt{2}} |+\rangle + \frac{1}{2} \left(\sqrt{1 + \left(\frac{V_{\text{int}}}{4t}\right)^2} - \frac{V_{\text{int}}}{4t} \right) (|11\rangle + |22\rangle) \right] . \quad (2.30)$$

Note that the probability of double occupancy of a columnar pin vanishes like $(t/V_{\text{int}})^2$ as $V_{\text{int}} \to \infty$.

Although the smallest eigenvalue of the transfer matrix V dominates the partition function as $L \to \infty$, the excited states in Fig. 6 are important for correlation functions connecting different values of z and when $L \lesssim \ell_z$. Only states in the *bosonic* subspace contribute even in this case, however. To understand this, recall that Eq. (2.24) must be summed over states $|a\rangle$ and $|b\rangle$ describing the entry and exit points of the vortices. The total partition function is thus

$$Z_{\text{tot}} = \langle \alpha | e^{-\mathcal{H}L/T} | \alpha \rangle , \quad (2.31)$$

where $|\alpha\rangle$ is some linear combination of $|11\rangle$, $|22\rangle$, $|12\rangle$ and $|21\rangle$. Because the flux lines are indistinguishable, the initial state must take the form

$$|\alpha\rangle = a|11\rangle + b|22\rangle + c(|12\rangle + |21\rangle), \tag{2.32}$$

where a, b, and c are constants. We allow for $a \neq b$ because the columnar pins could in fact have slightly different binding energies both at the surface or in the interior, as in the Anderson model of localization [15]. Even in this case, the initial state $|\alpha\rangle$ lies completely within the *boson* subspace of Eq. (2.27). Thus the fermion eigenstate never contributes to the statistical mechanics. Similar arguments show that only boson excitations are relevant to arbitrarily large assemblies of interacting flux lines [17,20].

3. Flux Melting and the Quantum Harmonic Oscillator

Consider one representative fluxon in the confining potential "cage" provided by its surrounding vortices in a triangular lattice. The partition function for a fixed entry point $\vec{0}$ and exit point $\vec{r}_\perp$ in a sample of thickness L is

$$\mathcal{Z}_1(\vec{r}_\perp, \vec{0}; L) = \int_{\vec{r}(0)=\vec{0}}^{\vec{r}(L)=\vec{r}_\perp} \mathcal{D}\vec{r}(z) \exp\left\{-\frac{1}{T}\int_0^L \left[\frac{1}{2}\tilde{\epsilon}_1\left(\frac{d\vec{r}(z)}{dz}\right)^2 + V_1[\vec{r}(z)]\right] dz\right\}, \tag{3.1}$$

where $V_1[\vec{r}]$ is now a one-body potential chosen to mimic the interactions in Eq. (2.1a). We assume clean samples and high temperatures so that both correlated and point disorder can be neglected.

Three important field regimes for fluctuations in vortex crystals are easily extracted from this simplified model. Following the approach in the previous section, we rewrite this imaginary time path integral as a quantum mechanical matrix element,

$$\mathcal{Z}(\vec{r}_\perp, \vec{0}; L) = \langle \vec{r}_\perp | e^{-L\mathcal{H}/T} | \vec{0} \rangle, \tag{3.2}$$

where $|\vec{0}\rangle$ is an initial state localized at $\vec{0}$, $\langle \vec{r}_\perp |$ is a final state localized at $\vec{r}_\perp$, and the "Hamiltonian" $\mathcal{H}$ is the operator which appears in Eq. (2.4),

$$\mathcal{H} = -\frac{T^2}{2\tilde{\epsilon}_1}\nabla_\perp^2 + V_1(\vec{r}). \tag{3.3}$$

Recall that the probability of finding the flux line at transverse position $\vec{r}$ within the crystal is $\psi_0^2(\vec{r})$, where $\psi_0(\vec{r})$ is the normalized ground state eigenfunction of (3.3).

When $B \gg B_{c1} \equiv \phi_0/\lambda_{ab}^2$, the pair potential $V(r_{ij}) = 2\epsilon_0 K_0(r_{ij}/\lambda_{ab})$ is logarithmic, $K_0(x) \approx \ln x$, and we expand $V_1(\vec{r}_\perp)$ about its minimum at $\vec{r}_\perp = 0$ to find

$$\left[-\frac{T^2}{2\tilde{\epsilon}_1}\nabla_\perp^2 + \frac{1}{2}kr_\perp^2\right]\psi_0 = E_0\psi_0 \tag{3.4}$$

where (neglecting logarithmic corrections to ϵ_0 and constants of order unity)

$$k \approx \left.\frac{d^2V}{dr^2}\right|_{r=a_0}$$
$$\approx \frac{\epsilon_0}{a_0^2}, \tag{3.5}$$

and a_0 is the mean vortex spacing. Equation (3.4) is the Schroedinger equation for a two-dimensional quantum oscillator, with $\hbar \to T$ and mass $m \to \tilde{\epsilon}_1$. The ground state wave function is

$$\psi_0(r_\perp) = \frac{1}{\sqrt{2\pi}\, r_*}\, e^{-r^2/4r_*^2} \tag{3.6}$$

with spatial extent

$$r_* = \left(\frac{T^2 a_0^2}{\epsilon_0 \tilde{\epsilon}_1}\right)^{1/4}. \tag{3.7}$$

Melting occurs when $r_* = c_L a_0$, where c_L is the Lindemann constant, so the melting temperature is

$$T_m = c_L^2 \sqrt{\epsilon_0 \tilde{\epsilon}_1}\, a_0, \qquad (B_{c1} < B \lesssim B_x) \tag{3.8}$$

in agreement with other estimates [24]. Vortices in the crystalline phase will travel across their confining tube of radius $r_\perp^*$ in a "time" along the z axis of order ℓ_z^0, where [17]

$$\ell_z^0 \approx r_*^2/(T/\tilde{\epsilon}_1)$$
$$\approx \sqrt{\frac{\tilde{\epsilon}_1}{\epsilon_0}}\, a_0. \tag{3.9}$$

A new high field regime arises when $\ell_z^0 \lesssim d_0$, where d_0 is the average spacing of the copper-oxide planes, i.e., for $B \gtrsim B_x$, with decoupling field

$$B_x \approx \frac{\tilde{\epsilon}_1}{\epsilon_0}\, \frac{\phi_0}{d_0^2}, \tag{3.10}$$

again in agreement with earlier work [8, 24]. Above this field, the planes are approximately decoupled, and T_m may be estimated from the theory of two-dimensional dislocation mediated melting [8, 24]

$$T_m \approx \frac{\epsilon_0 d_0}{8\pi\sqrt{3}}, \qquad (B \gtrsim B_x). \tag{3.11}$$

The estimate (3.8) also breaks down at low fields $B \lesssim B_{c1}$ where the logarithmic interaction potential must be replaced by an exponential repulsion. The two-dimensional harmonic oscillator model again applies, with the replacement

$$k \to \frac{\epsilon_0}{\lambda_{ab}^2}\, e^{-a_0/\lambda_{ab}}. \tag{3.12}$$

The transverse wandering distance is now

$$r_* \approx \left(\frac{T^2 \lambda_{ab}^2}{\epsilon_0 \tilde{\epsilon}_1}\right)^{1/4} e^{a_0/4\lambda_{ab}}. \tag{3.13}$$

Table 1. Estimates for the flux lattice melting temperature determined for the three regimes discussed in the text

Regime	$T_m(B)$	
$B_x \lesssim B$	$\epsilon_0 d_0/8\pi\sqrt{3}$	
$B_{c1} \lesssim B \lesssim B_x$	$c_L^2\sqrt{\epsilon_0\tilde{\epsilon}_1}(\phi_0/B)^{1/2}$	$B_x \approx \frac{\tilde{\epsilon}_1}{\epsilon_0}\frac{\phi_0}{d_0^2}$
$B \lesssim B_{c1}$	$c_L^2\sqrt{\epsilon_0\tilde{\epsilon}_1}\lambda_{ab}\left(\frac{B_{c1}}{B}\right)e^{-\frac{1}{2}(B_{c1}/B)^{1/2}}$	$B_{c1} \approx \phi_0/\lambda_{ab}^2$

and takes place over a longitudinal distance

$$\ell_z^0 = \sqrt{\frac{\tilde{\epsilon}_1}{\epsilon_0}}\,\lambda_{ab}e^{a_0/2\lambda_{ab}}. \tag{3.14}$$

The low field melting temperature becomes

$$T_m \approx c_L^2\sqrt{\epsilon_0\tilde{\epsilon}_1}\,\frac{a_0^2}{\lambda_{ab}}e^{-a_0/2\lambda_{ab}} \qquad (B \lesssim B_{c1})\,, \tag{3.15}$$

consistent with earlier predictions [17,8]. Although we have retained the distinction between $\tilde{\epsilon}_1$ and ϵ_0 in these formulas, note that $\tilde{\epsilon}_1 \approx \epsilon_0$ in this regime [8].

The predictions (3.8), (3.11) and (3.15) are combined to give the reentrant phase diagram for melting shown in Fig. 1. Analytic estimates and boundaries for melting in the various regimes are summarized in Table 1.

4. Vortex Entanglement in the Liquid Phase

Above the melting line in Fig. 1, weak point disorder due to oxygen vacancies can usually be neglected [7] and we must consider a liquid of wandering, essentially unconfined lines. To estimate the degree of entanglement, we consider a *single* flux line $\mathbf{r}(z)$ and determine how far it wanders perpendicular to the z axis as it traverses the sample. The relevant path integral is

$$\langle |\mathbf{r}(z)-\mathbf{r}(0)|^2\rangle = \frac{\int \mathcal{D}\mathbf{r}(s)|\mathbf{r}(z)-\mathbf{r}(0)|^2\exp\left[-\frac{\tilde{\epsilon}_1}{2T}\int_0^L\left(\frac{d\mathbf{r}}{ds}\right)^2 ds\right]}{\int \mathcal{D}\mathbf{r}(s)\exp\left[-\frac{\tilde{\epsilon}_1}{2T}\int_0^L\left(\frac{d\mathbf{r}}{ds}\right)^2 ds\right]},$$

which, when discretized as in Appendix A, yields

$$\langle |\mathbf{r}(z)-\mathbf{r}(0)|^2\rangle = \frac{2T}{\tilde{\epsilon}_1}|z|\,, \tag{4.1}$$

which shows that the vortex "diffuses" as a function of the timelike variable z,

$$\langle |\mathbf{r}(z)-\mathbf{r}(0)|^2\rangle^{1/2} = (2Dz)^{1/2}\,, \tag{4.2}$$

with diffusion constant

$$D = \frac{T}{\tilde{\epsilon}_1} = \frac{M_z}{M_\perp} \frac{4\pi T}{\phi_0 H_{c1}} . \tag{4.3}$$

At $T = 77$K, we take $H_{c1} \approx 10^2$ G and $M_z/M_\perp \approx 10^2$ and find $D = 10^{-6}$ cm, so that vortex lines wander a distance of order 1 μm while traversing a sample of thickness 0.01 cm.

Close encounters between neighboring vortex lines will thus occur quite frequently in fields of order 1 T or more, where vortices are separated by distances of order 500 Å or less. The "entanglement length" ℓ_z is defined as the distance along the z axis such that $\langle |\vec{r}(\ell_z) - \vec{r}(0)|^2 \rangle = a_0^2 = B/\phi_0$, i.e. [17]

$$\ell_z = \frac{a_0^2}{2D} = \frac{\tilde{\epsilon}_1 a_0^2}{2T} . \tag{4.4}$$

Collisions and entanglement of vortex lines wil be important for flux liquids whenever

$$L > \ell_z , \tag{4.5}$$

i.e., for $B > B_{x1} \approx (M_\perp/M_z)(\phi_0 \epsilon_0/LT)$.

The discrete flux filaments which comprise the flux liquid first form when a superconductor is cooled through the mean field transition line $B_{c2}^0(T)$. These fluxons are "phantom vortices" for $B \lesssim B_{c2}^0(T)$ in the sense that the barriers to line crossing are expected to be negligible. The barriers will increase, however, as the flux liquid is cooled toward the fluctuation induced melting temperature of the Abrikosov flux crystal. Entanglement can then cause the flux liquid to become very viscous [17,26].

The consequences for transport of vortices in the vicinity of a few strong pinning centers such as twin boundaries can be quite striking. Assume that a supercurrent $\vec{j}_s$ flows along $\hat{\mathbf{y}}$, $\vec{j}_s = j_s \hat{\mathbf{y}}$, and that the magnetic field, as usual, is parallel to $\hat{\mathbf{z}}$. Vortices will then be subjected to a constant driving Lorentz force,

$$\vec{f}_L = \frac{1}{c} n_0 \phi_0 \hat{z} \times \vec{j}_s , \tag{4.6}$$

where $n_0 \approx a_0^{-2}$ is the vortex density. The equation of motion which describes a z-independent flux liquid velocity field in the vicinity of, say, a twin boundary in the xz plane, is then [26]

$$-\gamma \vec{v} + \eta \nabla_\perp^2 \vec{v} + \vec{f}_L = 0 . \tag{4.7}$$

Eq. (4.7) is a hydrodynamic description of flux flow, valid on large scales compared to the intervortex spacing. The Bardeen-Stephen parameter γ is a "friction" coefficient which represents the resistance encountered to vortex motion by the normal electrons in the core. The combination of the drag and viscous terms in Eq. (4.7) introduces an important new length scale into the problem,

$$\delta = \sqrt{\eta/\gamma} . \tag{4.8}$$

The length δ is the scale over which the velocity rises to its bulk value from the center of the twin boundary where it is small or vanishes entirely.

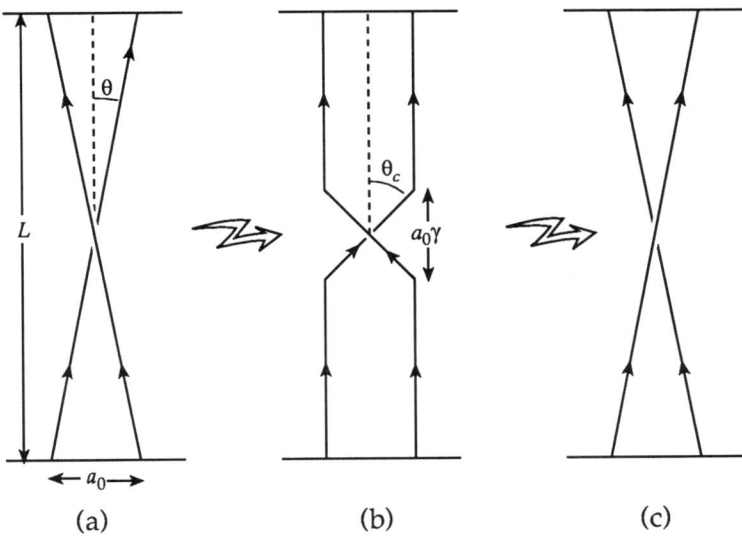

Figure 7. A mechanism for vortex line crossing. Vortices in the initial (a) and final (c) configurations are displaced out of the plane of the figure by a distance $\sim a_0$. An intermediate saddle point configuration connecting these states is shown in (b). Here $\gamma = (M_\perp/M_z)^{1/2}$.

The flux line viscosity has been estimated, e.g., by Cates [27], who finds a remarkably simple formula for δ

$$\delta \approx a_0 e^{U_\times/T} \qquad (4.9)$$

where $U_\times$ is the barrier to line crossing. We estimate this barrier as shown in Fig. 7, following the treatment of Obukhov and Rubinstein [28]. The projections of two vortices (displaced initially by $\sim a_0$ perpendicular to the plane of the figure) intersect near the center of the sample and approach the intervortex spacing at $z = \pm L/2$. Crossing is very difficult at low angles θ, because of the extra energy associated with a doubly quantized filament formed near the crossing region [17]. Crossing is easier at high angles, however. In an isotropic superconductor, the crossing energy goes to zero when $2\theta = 90°$, for example [29]. A simple rescaling argument shows that this critical angle for zero crossing energy becomes

$$\theta_c \approx \tan^{-1}(\sqrt{M_z/M_\perp}) \qquad (4.10)$$

for anisotropic materials in the symmetrical situation shown in Fig. 7. One way to exchange the lines shown in Fig. 7a is to pass to the intermediate configuration shown schematically in Fig. 7b, where the vortices bend and acquire extra length so they are inclined at the critical angle. The filaments can then pass through each other remaining crossed, relaxing finally to the configuration shown in Fig. 7c.

The line energy of a wandering vortex can be parameterized in terms of its arc length s as

$$\int_0^L \epsilon_1(\theta) \frac{ds}{dz} dz, \qquad (4.11)$$

where $\theta(z)$ is the local angle of the inclination relative to z and [30]

$$\epsilon_1(\theta) = \epsilon_1 \sqrt{\cos^2 \theta + \frac{M_\perp}{M_z} \sin^2 \theta} \qquad (4.12)$$

with $\epsilon_1 = \epsilon_0 \ln(\lambda_{ab}/\xi_{ab})$. For the configuration in Fig. 7a, this formula leads immediately to an energy

$$E_0 = 2\epsilon_1 L[1 + \mathcal{O}(a_0^2/L^2)]. \qquad (4.13)$$

The saddle point energy shown in Fig. 7b, on the other hand, has energy

$$E_0' = 2\epsilon_1 \left(L - \sqrt{\frac{M_\perp}{M_z}} a_0 \right) + 2\epsilon_1 \sqrt{\cos^2 \theta_c + \frac{M_\perp}{M_z} \sin^2 \theta_c} \sqrt{a_0^2 + (\cot^2 \theta_c) a_0^2} \qquad (4.14)$$

which simplifies using Eq. (4.10) to

$$E_0' = 2\epsilon_1 L + 2(\sqrt{2} - 1) \sqrt{\frac{M_\perp}{M_z}} a_0 \epsilon_1. \qquad (4.15)$$

The change in the interaction energy between the lines due to the sharp bends and nonzero tilt will add a correction to the coefficient of the second term, which we neglect. Note also that the terms neglected in (4.13) are higher order in a_0/L than the terms kept in (4.15). The crossing energy $U_\times \approx E_0' - E_0$ is thus approximately [28]

$$U_\times \approx 2(\sqrt{2} - 1) \sqrt{\frac{M_\perp}{M_z}} a_0 \epsilon_1. \qquad (4.16)$$

We see that the crossing energy tends to zero as $M_\perp/M_z \to 0$, and has essentially the same functional form as the intermediate field melting temperature displayed in Eq. (3.8). The key parameter which determines the viscous length scale (4.9) at the melting temperature is then

$$\frac{U_\times}{T_m} \approx \frac{c_\times}{c_L^2}, \qquad (4.17)$$

where $c_\times$ is a dimensionless constant, $c_\times = (2\sqrt{2} - 1) \ln(\lambda_{ab}/\xi_{ab})$, for the simple model discussed above. This dimensionless ratio should be independent of field strength in the range $B_{c1} \ll B \ll B_x$. If $U_\times/T_m \lesssim 1$, the crossing barriers associated with entanglement will not interfere with crystallization into an Abrikosov flux lattice. (This should *always* be the case for $B \gg B_x$.) Note, however, that it is the Lindemann ratio squared which enters the denominator of (4.17). Assume for concreteness that $c_\times = 0.75$. If $c_L = 0.3$, then $U_\times/T_m \approx 8$ and $\delta = a_0 e^{U_\times/T_m} \approx 4 \times 10^3 a_0$. If $c_L = .15$, as indicated in the most recent experiments on untwinned YBCO [3], then $U_\times/T_m = 33$, and $\delta \approx 3 \times 10^{14} a_0$! If $U_\times/T_m$ is really this large, the kinetic barriers associated with entanglement will *preclude* crystallization on experimental time scales in samples thick enough or fields high enough to allow multiple entanglements of the vortex lines. The flux liquid will instead form a "polymeric glass" phase upon cooling [17,28]. It is interesting to note for comparison purposes that $U_\times/T_m \approx 75$ in a *real* polymer like polyethyene. More accurate estimates of the numerator of (4.17) would, of course, be highly desirable.

Some evidence for the "polymer glass" scenario already exists. Recent experiments by Safar *et al.* find that the first order melting transition in untwinned YBCO goes away in sufficiently high magnetic fields, $B \gtrsim 10T$. One possibility is that point disorder somehow becomes more important and causes a "vortex glass" transition [8] at high fields [31]. An alternative explanation, however, is that the flux liquid simply becomes more entangled and viscous as its density increases with increasing field, leading eventually to undercooling and a polymer glass transition. Recent simulations of a lattice superconductor have revealed large crossing barriers and are never able to recover the crystalline phase upon cooling once the flux lattice has melted [32], in agreement with this picture. In real experiments, the hysterisis loops in the resistivity associated with first order freezing would slowly go away as crystallization became more difficult for a fixed cooling rate. The low temperature dynamics of this polymer glass in the presence of point disorder should be similar to that predicted by the collective pinning theory [33], with a *polymeric* shear modulus replacing the usual crystalline flux lattice elastic constant c_{66}.

Acknowledgements

This work was supported by the National Science Foundation, through Grant No. DMR 91-15491 and through the Harvard Materials Research Laboratory. Much of this work is the result of stimulating collaboration with V.M. Vinokur. See Ref. 15 for a more complete treatment of correlated pinning. Discussions with D. Bishop, R. Budhani, L. Civale, G. Crabtree, D.S. Fisher, P.L. Gammel, T. Hwa, P. Le Doussal and M.C. Marchetti, are also gratefully acknowledged.

Appendix A: Transfer Matrix Representation of the Partition Function

We review here how path integrals like those represented in Eq. (2.3) can be rewritten in terms of a transfer matrix, which is the exponential of the Schroedinger operator which appears in elementary quantum mechanics [16].

We first consider

$$Z(\vec{r}_\perp, \vec{0}; \ell) = \int_{\vec{r}(0)=\vec{0}}^{\vec{r}(\ell)=\vec{r}_\perp} \mathcal{D}\vec{r}(z) \exp\left[-\frac{\tilde{\epsilon}_1}{2T}\int_0^\ell \left(\frac{d\vec{r}}{dz}\right)^2 dz - \frac{1}{T}\int_0^\ell V_1[r(z)]\, dz\right] \tag{A1}$$

and discretize this path integral as indicated in Fig. 8, where the planes of constant z are separated by a small parameter δ. This is precisely the situation which arises in the high

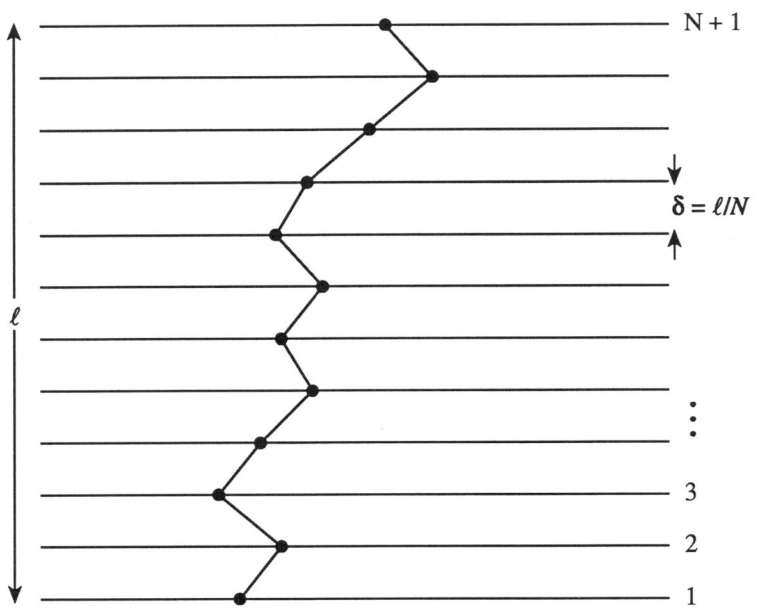

Figure 8. Discretized path integral representation of vortex line passing through $N+1$ CuO_2 planes with average spacing $\delta = \ell/N$.

T_c superconductors with field parallel to the c axis, provided δ represents the mean spacing between CuO_2 planes. The discretized path integral reads

$$Z(\vec{r}_\perp, \vec{0}, \ell) \approx \left(\prod_{j=2}^{N} \frac{\tilde{\epsilon}_1}{2\pi T \delta} \int d^2 r_j \right) \exp\left[\frac{-\tilde{\epsilon}_1}{2T\delta} \sum_{j=2}^{N+1} (\vec{r}_j - \vec{r}_{j-1})^2 - \frac{\delta}{T} \sum_{j=1}^{N+1} V_1(\vec{r}_j) \right], \quad (A2)$$

where it is understood that $\vec{r}_1 = \vec{0}$, $\vec{r}_{N+1} = \vec{r}_\perp$, and the normalization of the integrals is chosen so that $Z(r_\perp, 0; \ell) = 1$ when $V_1(\vec{r}) = 0$.

As usual in statistical mechanics, the effect of adding one copper oxide plane to the system can be represented in terms of a transfer matrix

$$Z(\vec{r}, \vec{0}; \ell + \delta) = \int d^2 r' T(\vec{r}, \vec{r}') Z(\vec{r}', \vec{0}; \ell) \quad (A3)$$

where (neglecting small edge effects at the top and bottom of the sample)

$$T(r, \vec{r}') = \frac{\tilde{\epsilon}_1}{2\pi T \delta} \exp\left\{ \frac{-\tilde{\epsilon}_1}{2T\delta} |\vec{r} - \vec{r}'|^2 - \frac{\delta}{2T} [V_1(\vec{r}) + V_1(\vec{r}')] \right\}. \quad (A4)$$

Ryu et al. have studied the spectrum of $T(\vec{r}, \vec{r}')$ for finite δ [34]. Here, we shall instead consider the limit of small δ. We can then expand the potential term and derive a differential equation for Z,

$$T\partial_\ell Z(\vec{r},\vec{0};\ell) = \left[\frac{T^2}{2\tilde{\epsilon}_1}\nabla_\perp^2 - V_1(\vec{r})\right] Z(\vec{r},0;\ell). \tag{A5}$$

A formal expression for the partition function results from integrating Eq. (A5) across a sample thickness L with initial state $|\vec{0}\rangle$ and final state $|\vec{r}_\perp\rangle$,

$$Z(\vec{r}_\perp,0;\ell) = \langle \vec{r}_\perp | e^{-\mathcal{H}L/T} | \vec{0} \rangle, \tag{A6}$$

where $\mathcal{H} = \frac{-T^2}{2\tilde{\epsilon}_1}\nabla_\perp^2 + V(\vec{r})$. The partition function $Z(\vec{0},\vec{r}_\perp;L)$ has been defined to be the ratio of the path integrals which appear in Eq. (2.3). Upon inserting a complete set of normalized eigenfunctions $\{\psi_n(\vec{r})\}$ of $\mathcal{H}$ with energies $\{E_n\}$ into (A6), we obtain

$$\begin{aligned} e^{U(T)L/T} &= Z(\vec{r}_\perp,0;L) \\ &= \sum_n \psi(\vec{r}_\perp)\psi(\vec{0})e^{-E_n L/T} \end{aligned} \tag{A7}$$

which leads when $L \to \infty$ to the result

$$U(T) = -E_0(T) \tag{A8}$$

quoted in the text.

Appendix B: Vortex Probability Distributions

We show here that the probability of finding an individual vortex line at height z with position $\vec{r}_\perp$ in an arbitrary binding potential $V_1(\vec{r}_\perp)$ is related to the ground state wave function of the corresponding Schroedinger equation. The probability distribution at a free surface is proportional to the wave function itself, while the probability far from the surface is proportional to the wave function squared.

Consider first a fluxon which starts at the origin $\vec{0}$ and wanders across a sample of thickness L to position $\vec{r}_\perp$. As discussed in Appendix A, the partition function associated with this constrained path integral may be written as a quantum mechanical matrix element

$$\begin{aligned} \mathcal{Z}(\vec{r}_\perp,\vec{0};L) &= \int_{\vec{r}(0)=\vec{0}}^{\vec{r}(L)=\vec{r}_\perp} \mathcal{D}r(z) \exp\left[-\frac{\tilde{\epsilon}_1}{2T}\int_0^L \left|\frac{d\vec{r}}{dz}\right|^2 dz - \frac{1}{T}\int_0^L V_1[\vec{r}(z)]dz\right] \\ &\equiv \langle \vec{r}_\perp | e^{-L\mathcal{H}/T} | 0 \rangle \end{aligned} \tag{B1}$$

where $|\vec{0}\rangle$ is an initial state localized at $\vec{0}$ while $\langle \vec{r}_\perp|$ is a final state localized at $\vec{r}_\perp$. The "Hamiltonian" $\mathcal{H}$ appearing in (B1) is the Schroedinger operator,

$$\mathcal{H} = -\frac{T^2}{2\tilde{\epsilon}_1}\nabla_\perp^2 + V_1(\vec{r}). \tag{B2}$$

The probability distribution $\mathcal{P}(\vec{r}_\perp)$ for the vortex tip position at the upper surface is then

$$\mathcal{P}(\vec{r}_\perp) = \mathcal{Z}(\vec{r}_\perp,\vec{0};L) \Big/ \int d^2r_\perp \mathcal{Z}(\vec{r}_\perp,\vec{0};L). \tag{B3}$$

Upon inserting a complete set of (real) energy eigenstates $|n\rangle$ with eigenvalues E_n into Eq. (B1), we have

$$\mathcal{P}(\vec{r}_\perp) = \frac{\sum_n \psi_n(\vec{0})\psi_n(\vec{r}_\perp)e^{-E_n L/T}}{\sum_n \psi_n(0)\int d^2 r_\perp \psi_n(\vec{r}_\perp)e^{-E_n L/T}} . \tag{B4}$$

In the limit $L \to \infty$ the ground state dominates, the probability $\mathcal{P}(\vec{r}_\perp)$ becomes

$$\mathcal{P}(\vec{r}_\perp) \approx \frac{\psi_0(\vec{r}_\perp)}{\int d^2 r_\perp \psi_0(\vec{r}_\perp)} \left[1 + \mathcal{O}\left(e^{-(E_1-E_0)L/T}\right)\right] , \tag{B5}$$

where E_1 is the energy of the first excited state. Because the ground wave function is nodeless [21], $\mathcal{P}(r_\perp)$ is always positive and well defined.

Consider now a more general problem of a vortex which enters the sample at $\vec{r}_i$, exits at $\vec{r}_f$, and passes through $\vec{r}$ at a height z which is far from the boundaries. The normalized probability distribution is now

$$\tilde{\mathcal{P}}(\vec{r}; L) = \tilde{\mathcal{Z}}(\vec{r}; L) / \int d^2 r \tilde{\mathcal{Z}}(\vec{r}; L) \tag{B6}$$

where

$$\tilde{\mathcal{Z}}(\vec{r}; L) = \int d^2 r_i \int d^2 r_f \mathcal{Z}(\vec{r}_f, \vec{r}; L-z)\mathcal{Z}(\vec{r}, \vec{r}_i; z) \tag{B7}$$

and $Z(\vec{r}_2, \vec{r}_1; L)$ is given by Eq. (B1). Upon inserting complete sets of states as before, we find that

$$\tilde{\mathcal{P}}(\vec{r}; L) = \frac{\psi_0^2(r)}{\int d^2 r_\perp \psi_0^2(r)} \left[1 + \mathcal{O}\left(e^{-L(E_1-E_0)/2T}\right)\right] , \tag{B8}$$

where the correction assumes $\vec{r}$ is at the midplane of the sample.

References

1. A.A. Abrikosov, *Zh. Eksperim. i Theor. Fiz.* **32**, 1442 (1957) [*Sov. Phys. JETP* **5**, 1174 (1957)].
2. M. Charalmabous *et al.*, *Phys. Rev.* **B45**, 45 (1992).
3. W.K. Kwok, S. Fleshler, U. Welp, V.M. Vinokur, J. Downey, G.W. Crabtree and M.M. Miller, *Phys. Rev. Lett.* **69**, 3370 (1992).
4. D.E. Farrell, J.P. Rice and D.M. Ginsberg, *Phys. Rev. Lett.* **67**, 1165 (1991).
5. H. Safar, P.L. Gammel, D.A. Huse, D.J. Bishop, J.P. Rice and D.M. Ginsberg, *Phys. Rev. Lett.* **69**, 824 (1992).
6. E. Brezin, D.R. Nelson and A. Thiaville, *Phys. Rev.* **B31**, 7124 (1985).
7. D.R. Nelson and P. Le Doussal, *Phys. Rev.* **B42**, 10112 (1990).
8. D.S. Fisher, M.P.A. Fisher and D.A. Huse, *Phys. Rev.* **B43**, 130 (1991).
9. R.C. Budhani, M. Suenaga and H.S. Liou, *Phys. Rev. Lett.* **69**, 3816 (1992).
10. M. Konczykowski *et al.*, *Phys. Rev.* **B44**, 7167 (1991).

11. L. Civale, A.D. Marwich, T.K. Worthington, M.A. Kirk, J.R. Thompson, L. Krusin-Elbaum, Y. Sun, J.R. Clem and F. Holtzberg, *Phys. Rev. Lett.* **67**, 648 (1991).
12. W. Gerhauser et al., *Phys. Rev. Lett.* **68**, 879 (1992).
13. V. Hardy et al., *Nucl. Instr. and Meth.* **B54**, 472 (1991).
14. A.P. Malozemoff, T.K. Worthington, Y. Yeshurun and F. Holtzberg, *Phys. Rev.* **B38**, 7203 (1988).
15. D.R. Nelson and V.M. Vinokur, *Phys. Rev. Lett.* **68**, 2392 (1992), and Harvard University preprint.
16. R.P. Feynman and A.R. Hibbs, *Quantum Mechanics and Path Integrals* (McGraw-Hill, New York, 1965); R.P. Feynman, *Statistical Mechanics* (Benjamin, Reading, MA, 1972).
17. D.R. Nelson, *Phys. Rev. Lett.* **60**, 1973 (1988); D.R. Nelson and S. Seung, *Phys. Rev.* **B39**, 9153 (1989).
18. S.F. Edwards, *Proc. Phys. Soc.* **85**, 613 (1965); P.G. de Gennes, *Rep. Prog. Phys.* **32**, 187 (1969).
19. M.P.A. Fisher, P.B. Weichman, G. Grinstein and D.S. Fisher, *Phys. Rev.* **B40**, 546 (1989), and references therein.
20. D.R. Nelson, in *Phenomenology and Applications of High Temperature Semiconductors*, edited by K. Bedell, M. Inui, D. Meltzer, J.R. Schrieffer, and S. Doniach (Addison-Wesley, New York, 1991).
21. L.D. Landau and E.M. Lifshitz, *Quantum Mechanics*, 2nd. Edition (Pergammon, New York, 1965).
22. See, e.g., D.R. Nelson, *J. Stat. Phys.* **57**, 511 (1989).
23. N.W. Ashcroft and N.D. Mermin, *Solid State Physics* (Sanders College, Philadelphia, 1976), Chapter 32.
24. See, e.g., L.I. Glazman and A.E. Koshelev, *Phys. Rev.* **B43**, 2835 (1991).
25. D.S. Fisher, *Phys. Rev.* **B22**, 1190 (1980).
26. M.C. Marchetti and D.R. Nelson, *Phys. Rev.* **B42**, 9938 (1990); *Physica* **C174**, 40 (1991).
27. M. Cates, *Phys. Rev.* **B45**, 12415 (1992).
28. S. Obukhov and M. Rubinstein, *Phys. Rev. Lett.* **66**, 2279 (1991); see also S. Obukhov and M. Rubinstein, *Phys. Rev. Lett.* **65**, 1279 (1990).
29. E.H. Brandt, J.R. Clem and D.G. Walmsley, *J. Low Temp. Phys.* **37**, 43 (1979).
30. V.G. Kogan, *Phys. Rev.* **B24**, 1572 (1981).
31. H. Safar, P.L. Gammel, D.A. Huse, D.J. Bishop, W.C. Lee, J. Giapintzakis and D.M. Ginsberg, AT&T Laboratories preprint.
32. S. Teitel, private communication.
33. M.V. Feigel'man, V.B. Geshkenbein, A.I. Larkin and V.M. Vinokur, *Phys. Rev. Lett.* **63**, 2303 (1989).
34. S. Ryu, A. Kapitulnik and S. Doniach, Stanford University preprint.

ANISOTROPY AND STRONG PINNING IN $YBa_2Cu_3O_7$ WITH Y_2BaCuO_5 INCLUSIONS

M. G. Karkut, L. K. Heill, M. Slaski[*], L. T. Sagdahl and K. Fossheim
Division of physics, The Norwegian Institute of Technology and SINTEF Applied Physics
N-7034 Trondheim, Norway
[*]*School of Physics, University of Birmingham*
Birmingham, B15 2TT, UK

ABSTRACT. We have measured the ac magnetic permeability μ in $YBa_2Cu_3O_7$ with Y_2BaCuO_5 inclusions and in single crystal $YBa_2Cu_3O_7$ in applied magnetic fields parallel and perpendicular to the c-axis. We find that the inclusions act as strong pinning sites when H∥c, but have a deleterious effect on the superconductor when H⊥c.

1. Introduction

The melt-process-melt-growth material (MPMG), $YBa_2Cu_3O_7$ (Y123) with insulating Y_2BaCuO_5 (Y211) inclusions, is a technologically promising material due to its enhanced levitation force[1]. Surprisingly, it has not received much attention from the theoretical community. The nearly spherical Y211 inclusions of about 0.1–0.2 μm in size are randomly distributed throughout the sample and act as strong pinning sites[2]. This arrangement presents us with a superconducting system which is intermediate between the randomly distributed point defects which collectively pin the vortex system and for which the vortex glass theory has been put forward[3], and the long columnar defects, produced by heavy ion irradiation, which pin individual vortex lines and for which the Bose glass theory has been advanced[4]. The mechanism by which the Y211 inclusions act as strong pinning sites is yet unknown but that they do so is clear. Here we present ac permeability measurements on a MPMG sample and compare with results on a single crystal of $YBa_2Cu_3O_7$ (SC) in order to extract information about the nature of the flux pinning in MPMG.

2. Experimental

The MPMG sample was made at ISTEC, SRL in Tokyo, with nominal stochiometry Y:Ba:Cu = 1.8:2.4:3.4, and is almost cubic; 1.38×1.29×1.25 mm^3. The Y211 inclusions make up about 20% of the otherwise single crystal like sample volume. The SC sample, made at the University of Birmingham, is also almost cubic; 1.100×1.065×1.015 mm^3. The similar shapes and dimensions of the two samples imply similar demagnetization factors and allow direct comparison of data. The principal difference between the two samples is the presence/absence of Y211 material.

The complex ac permeability $μ = μ' + iμ''$ was measured with a susceptometer built in-house. The excitation and pick-up coils are oriented parallel to the dc magnetic field (≤8 T). Data were taken during heating, after an initial phase adjustment zeroing $μ''$ in the normal state (100 K).

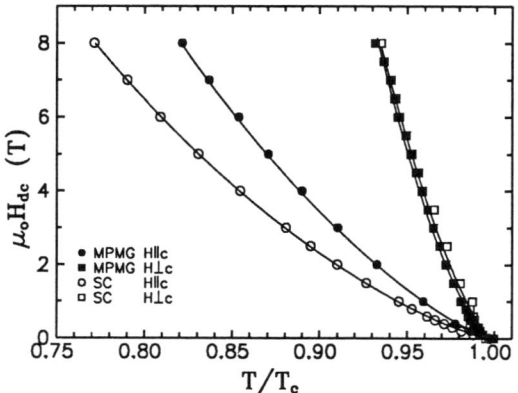

Figure 1. The position of the peak of μ" plotted vs reduced temperature T/T_c as a function of applied dc magnetic field for the MPMG and the SC samples for two orientations of the field, H∥c and H⊥c. The amplitude and frequency of the excitation field are 1 G and 121 Hz.

3. Results and Discussion

We define the dynamically determined irreversibility line here as the loci of μ" maxima in the H-T plane. Figure 1 presents the peak position of μ" as a function of reduced temperature T/T_c and applied dc field, for the MPMG and SC samples for two applied field directions, H∥c and H⊥c. Since these lines are close to the true irreversibility lines[2] we can immediately conclude that for both samples the irreversibility temperature $T_{irr}(H_{dc})$ is considerably higher for H⊥c than for H∥c. This is in good agreement with the expected melting of the vortex lattice due to thermal fluctuations when the crystalline anisotropy is taken into account[5]. Qualitatively, to explain the difference in melting temperature, we would argue that for the magnetic field parallel to the CuO_2 planes (H⊥c) the flux lines are largely confined between CuO_2 planes because it is energetically more favorable to do so. This removes a degree of freedom of the fluxons. Although these flux lines can easily move between the CuO_2 planes by external magnetic pressure, they should be quite well ordered in a long-range pattern since they repel each other over a long distance. Only close to T_c can one expect this ordering to be disrupted by thermal agitation, to a large extent by production of thermally activated kinks in the vortex lines. For H normal to the CuO_2 planes, on the other hand, the vortices are rather free to move in both sideways directions. Furthermore they are soft since they can be considered as embryonic pancake vortices due to mass anisotropy. Thermal disconnection and reconnection as well as large amplitude thermal motion in the a,b directions are possible. This H∥c vortex structure melts more easily than the H⊥c vortex structure.

The second point of interest in Figure 1 is that for H∥c, the irreversibility line of the MPMG sample is clearly at a higher temperature and field than for that of the SC sample. The Y211 inclusions in the MPMG sample therefore have the effect of increasing the melting temperature of the vortex structure. This raises the question of the effect of weak and strong pinning on the irreversibility line. In a series of experiments an IBM group has probed this question using ac on YBCO with the applied dc field parallel to the crystalline c-axis. They demonstrated that the irreversibility line does not shift upon proton irradiation of single crystal YBCO even though there is an order of magnitude increase in J_c at 77 K of the irradiated sample [6]. They show that the irreversibility line for YBCO films thicker than 1000 Å was the same as the irreversibility line for single crystal YBCO even though J_c (at 77 K) is about two orders of magnitude higher in the thin film[7]. These results imply a largely pinning independent nature of the irreversibility line when the additional pinning sites are weak or point-like. The increase of J_c is due to the increase in the total number of weak pinning sites. However, the same group also reports that heavy ion

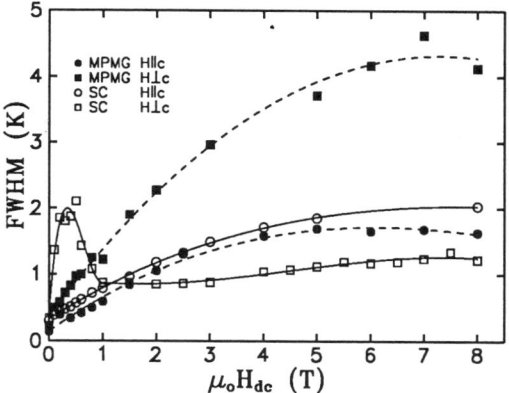

Figure 2. The FWHM (full width at half maximum) vs the applied dc field of $\mu''(H,T)$ for the MPMG and SC samples for both field orientations. The excitation frequency and field are 121 Hz and 1 G.

radiation produces, in addition to an increase in J_c, a shift of the irreversibility line to higher fields and temperatures due to the strong pinning nature of the created columnar defects [8]. Our results in Figure 1 demonstrate that the presence of ~0.2 μm inclusions in YBCO similarly has the effect of raising the irreversibility line to higher fields and temperatures than that of single crystal YBCO for the same H∥c orientation. The detailed nature of pinning by the Y211 inclusions in the MPMG sample is not known, but the implication here is that it provides strong pinning when H∥c and that these strong pins seem to provide additional stability to the vortex structure resulting in a higher vortex melting temperature. These strong pins are of course superimposed upon the existing random distribution of weak pins which provide the collective pinning.

The unresolved question is how do the Y211 inclusions act as strong pinning sites for H∥c. We do not believe that the Y211 acts, for instance, to pump oxygen from the Y123 surroundings. This would increase the number of point defects, and experiments up to now have shown that this does not increase $T_{irr}(H_{dc})$. For strong pinning, i.e. an increase in $T_{irr}(H_{dc})$, to occur it would seem that the density of the inclusions must be large enough so that the average distance between them must be less than the longitudinal characteristic length of a flux line or bundle of flux lines. Otherwise it will be the collective pinning of the point defects controlling the dissipation between the Y211 inclusions, and hence no increase in $T_{irr}(H_{dc})$ should take place.

It is reasonable to assume that pinning occur at the Y123/Y211 interface. In our cubic ~1 mm^3 MPMG sample, the inclusions of radii ~0.15 μm make up ~20% of the sample volume. The average distance between neighbouring interfaces is about 0.1 μm and the areal density is ~10^{13} m^{-2}, matching the flux line density B/Φ_0 at 120 G. With a coherence length ~20 Å and assuming that in each horizontal slice all flux lines must be accomodated by a strong pinning inclusion, each inclusion can host roughly 300 flux lines along its circumference. We have measured the MPMG samples in magnetic fields of 8 T, corresponding to ~400 flux lines per inclusion. In principle, if on increasing H_{dc} the number of flux lines exceeds the number of available sites, the extra flux lines cannot profit by the strong pinning and the irreversibility line determined by the loss peak should decrease in temperature until it resembles that of YBCO weak pinning crystals. We do not see evidence of this at 8 T. This is treated more thoroughly in Reference [9].

We now consider the effect of the Y211 inclusions for the H⊥c orientation. On a reduced temperature scale, the irreversibility lines for both samples are the same. However, by plotting the full width at half maximum of μ", FWHM, vs H_{dc}, we get a fuller measure of the dissipation since this probes the overall loss with temperature. In Figure 2 we plot the FWHM for the two samples for both H∥c and H⊥c. For H∥c, the FWHM is about the same for both samples. So even though the irreversibility line is raised to a higher temperature in Figure 1, the losses for the two samples are similar. For the SC sample, the FWHM for the two field orientations is roughly similar (except for the anomaly at low fields which we discuss in Reference [9]). In complete contrast to this is the large dissipation measured in the MPMG sample with H⊥c. It is not only considerably greater than the FWHM for the two H∥c orientations but it is also much larger than the SC H⊥c orientation (again, except for low fields). Thus, even though the irreversibility lines determined by the position of the loss peak, as in Figure 1, are similar for the H⊥c orientation, the FWHM implies that the dissipation is much greater for the MPMG sample. Clearly, the strong pinning effects of the 211 inclusions are effective only for H∥c. They seem to be deleterious when H⊥c. One possible way to look at the situation is to consider the strong pinning energies. If we let the characteristic length of the pinning at the Y123/Y211 interface be d, and that this length be the same for fields both parallel and perpendicular to the c-axis – not unreasonable since these inclusions are essentially spherical – then for H∥c, the pinning energy is given by $U_{p\parallel} = (\mu_o/2)H_c^2(\pi\xi_{ab}^2 d)$ whereas for H⊥c, it is given by $U_{p\perp} = (\mu_o/2)H_c^2(\pi\xi_{ab}\xi_c d)$ and so the pinning energies differ by a factor of ξ_{ab}/ξ_c, the anisotropy of the material. So, although with H∥c the pinning due to the Y211 is strong, with H⊥c it is far weaker as seen by the greater ac loss in this orientation. In addition to the much weaker strong pinning for H⊥c, the removal due to the presence of the Y211 inclusions of weak pinning sites such as stacking faults, cracks etc. in the Y123 material produces an **enhanced** dissipation in MPMG.

4. Conclusions

We have measured and compared the ac response of a Y123 crystal containing Y211 inclusions with a single crystal Y123 sample. Several interesting conclusions can be drawn from this study. 1) the irreversibility temperature $T_{irr}(H_\perp) > T_{irr}(H_\parallel)$ for both samples. This is in accord with vortex melting predictions. 2) For H∥c, T_{irr}(MPMG) > T_{irr}(SC). This implies that the Y211 inclusions act as strong pins. Assuming that the pinning takes place at Y123/Y211 interfaces, and given the Y211 density and size, then above a certain applied dc field the Y211 sites will no longer be effective. 3) For H⊥c, the losses are **increased** by the presence of the Y211 inclusions.

5. References

1. M. Murakami, Mod.Phys.Lett. **B4** (1990) 163; T.H. Johansen, these proceedings.
2. K. Fossheim, M.G. Karkut, L.K. Heill, M. Slaski, and L.T. Sagdahl, Physica Scripta **T42** (1992) 20. The irreversibility line separates the H-T plane into pinning and non-pinning regions. We note here that the temperature position of the loss peak will depend on excitation frequency, field amplitude and orientation with respect to the c-axis.
3. D.S. Fisher, these proceedings.
4. D.R. Nelson, these proceedings.
5. G. Blatter, V.B. Geshkenbein, and A.I. Larkin, Phys.Rev.Lett. **68** (1992) 875.
6. L. Civale, A.D. Marwick, M.W. McElfresh, T.K. Worthington, A.P. Malozemoff, F.H. Holtzberg, J.R. Thompson, and M.A. Kirk, Phys.Rev.Lett **65** (1990) 1164.
7. L. Civale, T.K. Worthington, and A. Gupta, Phys.Rev. **B43** (1991) 5425.
8. L. Civale, A.D. Marwick, T.K. Worthington, M.A. Kirk, J.R. Thompson, L. Krusin-Elbaum, Y. Sun, J.R. Clem, and F. Holtzberg, Phys.Rev.Lett. **67** (1991) 648.
9. M.G. Karkut, M. Slaski, L.K. Heill, L.T. Sagdahl, and K. Fossheim, submitted to Physica C.

Interstitial and Vacancy Proliferation in Flux Line Lattices

ERWIN FREY,
Physik-Department der Technischen Universität München,
D-85747 Garching, Germany.
and
DAVID R. NELSON and DANIEL S. FISHER
Department of Physics, Harvard University,
Cambridge, Massachusetts 02138, USA.

ABSTRACT. We study the formation of interstitials and vacancies in the Abrikosov phase of clean Type II superconductors. The defects are line imperfections, which cannot extend across macroscopic samples at low temperatures. We argue that the entropy associated with line wandering nevertheless causes these defects to proliferate at a sharp transition which can occur below the melting temperature of the vortex crystal. Flux lines are both entangled and crystalline in the resulting "supersolid" phase, which is closely related to a two-dimensional quantum crystal with interstitials or vacancies incorporated in its ground state. We find that the supersolid phase *must* occur above the thermal decoupling field B_x. Using molecular dynamics simulation we calculate the formation energies of various types of defect lines. We find that interstitials rather than vacancies are the preferred types of defects for $B >> \phi_0/\lambda_\perp^2$.

Fluctuations, especially in high-temperature superconductors, play a prominent role in determining vortex configurations in Type II materials in an external field [1]. There is now experimental evidence that clean single crystal samples of YBCO (in the absence of twin boundary pinning) melt via a first-order phase transition [2] at a temperature $T_m(H)$ well below the upper critical field line $H_{c2}(T)$ predicted by mean field theory [3]. Point disorder, in the form of oxygen vacancies, does not seem to affect this phase transition strongly. It is quite possible that the disorder-induced translational correlation length [4] R_a greatly exceeds the vortex spacing for $T < T_m$ in the field range for which the transition is first order.

Here we consider the thermal excitations about the crystalline state on scales less than R_a. We assume for simplicity vortices which are perpendicular on average to the CuO_2 planes. Then there are several possible deviations of the vortex crystal from perfect periodicity. The most important kinds of lattice imperfections are phonons, vacancies and interstitials, and dislocations. A discussion of the effect of *phonons* on the Debye-Waller factor $\rho_{\vec{G}}(T)$ associated with the translational order parameter can be found in Ref. [5]. *Dislocation loops* are a topologically distinct excitation which, when allowed to proliferate at a melting transition, can lead to a hexatic flux liquid with residual bond orientation order [6]. Our emphasis here is on vacancies and interstitials. In conventional crystals (of point particles) the point-like nature of those defects ensures that they are present in equilibrium at all finite temperatures for entropic reasons. However, since the number of flux lines is conserved these defects are *lines* instead of points in vortex crystals. Therefore, they have an energy proportional to their *length*, which prevents them from extending completely across an equilibrated macroscopic sample at low temperatures. It is nevertheless possible for these defects to "proliferate" (i.e., to become infinitely long) at high temperatures for entropic reasons. Consider, for example, a vacancy wandering across a macroscopic sample of thickness L, as in Fig. 1.a. To estimate the free energy of this defect, we describe its trajectory along the z axis by a function $\vec{r}_d(z)$ and write its partition function as

$$\mathcal{Z}_d = e^{-\epsilon_d L/T} \int \mathcal{D}\vec{r}_d(z) \exp\left\{-\frac{1}{T}\int_0^L dz \left[\tfrac{1}{2}\tilde{\varepsilon}_d \left(\frac{d\vec{r}_d}{dz}\right)^2 + U_\ell(\vec{r}_d)\right]\right\}. \quad (1)$$

Here ε_d is the energy of an isolated defect and $\tilde{\varepsilon}_d$ is its tilt energy per unit length, defined in analogy with similar quantities ε_1 and $\tilde{\varepsilon}_1$ for isolated flux lines near H_{c1} [5]. $U_\ell(\vec{r}_d)$ is a periodic lattice potential with minima at the preferred sites of the considered defect. Implicit in the path integral (1) is a length scale ℓ_z which is the average distance along z between hops of the vacancy from one lattice position to another. As a crude estimate of the path integral we replace it by $\exp(-\varepsilon_d L/T) 6^{L/\ell_z}$, since the vacancy has six directions in which to hop on a triangular lattice. The free energy $F_d = -T \ln \mathcal{Z}_d$ is thus $F_d \approx [\varepsilon_d L - \frac{T}{\ell_z} L \ln 6]$, which becomes negative for $T > T_d$, where

$$T_d = \varepsilon_d \ell_z / \ln 6. \quad (2)$$

Above this temperature (provided the crystal does not melt first), vacancies (or interstitials) will proliferate in a crystalline phase. Provided these defects do not become pinned by point disorder, one might then expect a *linear* contribution to the resistivity due to defect motion within pinned bulk crystallites. A related phenomenon was suggested by Feigel'man *et al.* [7], who predicted the unbinding of "quarters" (i.e., quartets) of dislocations above a "decoupling" field $B_x = (\phi_0/d_0^2)(M_\perp/M_z)$, where d_0 is the layer spacing and $(M_\perp/M_z)$ the effective mass anisotropy in highly anisotropic superconductors.

The phase in which defects such as interstitials and vacancies proliferate is in fact both crystalline and entangled. Regarded in light of the analogy between thermally excited flux lines and two-dimensional bosons [5], it represents a "supersolid" quantum crystal, in which vacancies and interstitials are incorporated into the ground state [8]. The possibility of a supersolid phase for flux lines in Type II superconductors was first noted on the basis of the boson analogy by Fisher and Lee [9]. Entanglement of vortex lines in the crystal will be catalyzed by the proliferation of vacancies and interstitials, since these allow fluxons to easily move perpendicular to the z axis, resulting in a nonzero "boson" order parameter $\psi_0^B(T)$.

There are two distinct scenarios for vortex crystal melting. If the temperature for defect line proliferation is larger than the melting temperature (Type I melting) a first-order transition separates a line crystal with $\rho_{\vec{G}} \neq 0$ from a flux liquid with $\psi_0^B \neq 0$. In Type II melting, *both* order parameters are nonzero in an intermediate "supersolid" phase. Vacancies or interstitials enter the Abrikosov flux lattice at T_d in much the same way as the flux lines penetrate the Meissner phase at H_{c1}.

Although this transition can be continuous, even in the presence of strong thermal fluctuations, the melting of the supersolid into a liquid is likely to remain a first-order transition.

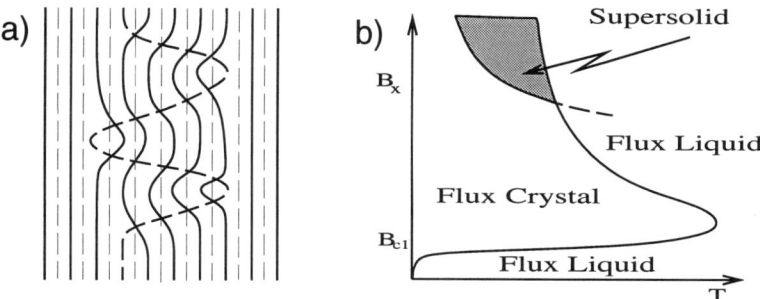

Fig. 1 a) Vacancy line (dashed) meandering through a vortex crystal. b) Schematic phase diagram of a clean Type II superconductor. The supersolid phase is shown as the shaded region.

Next we derive simple estimates for the defect unbinding temperature in various regimes, and compare them with the melting temperature. We also present numerical calculations of the formation energies for straight vortex lines in the high field regime.

For magnetic inductions in the regime $\phi_0/\lambda_\perp^2 \sim B_{c1} \ll B \leq B_x$ the pair potential is logarithmic with well-coupled CuO_2 planes. As in the related problem of electrons interacting via a $1/r$ potential [10], we find that the defect line energy is then proportional to the characteristic energy scale per unit length $\varepsilon_0 = (\phi_0/4\pi\lambda_\perp)^2$ of the pair potential

$$\varepsilon_d = \text{const.} \times \varepsilon_0 \ . \tag{3}$$

We have determined the coefficient for various types of defects via a molecular dynamics computation. Whereas there is a clear gap between the formation energies of a vacancy $\varepsilon_d^V = 0.2506\varepsilon_0$ and interstitials, the interstitial energies are comparable. We find $\varepsilon_d^{EI} = 0.1503\varepsilon_0$ and $\varepsilon_d^{CI} = 0.1482\varepsilon_0$ for an interstitial located at the edge and center of a triangle in the hexagonal vortex lattice, respectively. The center interstitial is a stable configuration, but the edge interstitial is unstable with respect to "buckling" modes, similar to the one found in Wigner crystals [11]. This shows that in fact center *interstitials*, rather than vacancies are favored energetically for $B \gg B_{c1}$.

The energy associated with a "kink" in a defect line where is moves one lattice constant (or "tunneling" event) can be estimated from the "quantum mechanical" analogy using the WKB tunneling expression $E_k \sim \sqrt{\varepsilon_0 \tilde{\varepsilon}_1}\, a_0$, where $\tilde{\varepsilon}_1 \approx \varepsilon_0 \frac{M_\perp}{M_z}$.

The spacing between kinks is then of order $\ell_z = n_{\text{kink}}^{-1}$, or $\ell_z \approx \ell_z^0 e^{E_k/T}$, where $\ell_z^0 \approx \sqrt{\tilde{\varepsilon}_1/\varepsilon_0}$ can be regarded as an inverse "attempt" frequency. Upon substituting the above expressions into Eq. (2), we find

$$T_d = c_d \sqrt{\varepsilon_0 \tilde{\varepsilon}_1}\, a_0 \ , \tag{4}$$

where c_d is a numerical constant which depends on the type of defect. Comparison with the estimate [12] for the melting temperature $T_m = c_L^2 \sqrt{\varepsilon_0 \tilde{\varepsilon}_1} \, a_0$, where c_L is the Lindemann constant, shows that supersolid order will occur whenever $c_d < c_L^2$.

Above the decoupling field B_x the defects can hop one lattice constant or more every copper-oxide plane spacing d_0, and the defect path integral (1) must be evaluated in a different limit. We can now neglect the lattice potential $U_\ell(\vec{r}_d)$ entirely and evaluate the remaining functional integral by discretizing the path integral along z in units of d_0. The result is $F_d = \varepsilon_d L - T \ln(2\pi B/B_x) L/d_0$, which leads to a defect unbinding temperature

$$T_d(B) \approx \text{const.} \times \varepsilon_0 d_0 / \ln[2\pi B/B_x] \,. \tag{5}$$

Since $T_d \sim 1/\ln(2\pi B/B_x)$ in this regime, while $T_m \approx 0.01 \varepsilon_0 d_0$ (estimated from two-dimensional dislocation mediated melting) is B-independent, defects *must* proliferate for $B \geq B_x$. Equation (5) agrees with the unbinding transition of dislocation pairs from bound dislocation quartets estimated in Ref. [7].

Finally, in the low density regime, $B \leq B_{c1}$, the lowest energy defect is presumably a vacancy, as is usual for solids with short-range interactions. The characteristic defect energy is now $\varepsilon_d = \text{const.} \times \varepsilon_0 e^{-a_0/\lambda_\perp}$. The barrier to produce a kink in the trajectory of a defect is also approximately $\varepsilon_0 e^{-\lambda/a_0}$, so the resulting kink energy is $E_k \sim \sqrt{\varepsilon_0 \tilde{\varepsilon}_1} \, a_0 e^{-a_0/2\lambda_\perp}$. Upon using Eq. (2) we find that the free energy $F_d = \varepsilon_d L \left[1 - \text{const.} \times \frac{T}{E_k} e^{-a_0/\lambda_\perp} \frac{a_0}{\lambda_\perp} e^{-E_k/T} \right]$ changes sign for

$$T_d \approx \text{const.} \times \sqrt{\varepsilon_0 \tilde{\varepsilon}_1} \, \lambda_\perp e^{a_0/2\lambda_\perp} \,. \tag{6}$$

Since T_d grows exponentially with lattice constant while $T_m \approx c_L^2 \sqrt{\varepsilon_0 \tilde{\varepsilon}_1} \frac{a_0^2}{\lambda_\perp} \lambda_\perp e^{-a_0/2\lambda_\perp}$ is exponentially small in this regime, we conclude that supersolid order is not possible in the regime of small fields.

Combining the above estimates for T_d and T_m leads to the schematic phase diagram shown in Fig. 1.b.

Acknowledgements

This work has been supported by the Deutsche Forschungsgemeinschaft (DFG) under Contracts No. Fr. 850/2-1,2 (ef), by the Harvard Materials Research Laboratory and NSF contracts DMR 91–15491 (drn) and DMR 91–06237 (dsf).

References

[1] See, e.g., the reviews in *Phenomenology and Applications of High-Temperature Superconductors*, edited by K. Bedell et al. (Addison-Wesley, New York, 1991).
[2] E. Brezin, D.R. Nelson, and A. Thiaville, Phys. Rev. B*31*, 7124 (1985).
[3] M. Charalambous, J. Chaussy and P. Lejay, Phys. Rev. B*45*, 509 (1992); H. Safar, P.L. Gammel, D.A. Huse, D.J. Bishop, J.P. Rice, and D.M. Ginsberg, Phys. Rev. Lett. *69*, 824 (1992); K.W. Kwok, S. Fleshler, U. Welp, V.M. Vinokur, J. Downey, G.W. Crabtree and M.M. Miller, Phys. Rev. Lett. *69*, 3370 (1992).
[4] A.I. Larkin and Y.M. Ovchinnikov, J. Low Temp. Phys. *34*, 409 (1979).
[5] D.R. Nelson and H.S. Seung, Phys. Rev. B*39*, 9153 (1989); D.R. Nelson, Phys. Rev. Lett. *60*, 1973 (1988); D.R. Nelson, in Ref. 1.
[6] M.C. Marchetti and D.R. Nelson, Phys. Rev. B*42*, 9938 (1990).

[7] M. Feigel'man, V.B. Geshkenbein, and A.I. Larkin, Physica C*167*, 177 (1990).
[8] J.H. Hetherton, Phys. Rev. *176*, 231 (1968); A.F. Andreev and I.M. Lifshitz, Sov. Phys. JETP *29*, 1107 (1969); G. Chester, Phys. Rev. A*2*, 256 (1970); A.J. Leggett, Phys. Rev. Lett. *25*, 1543 (1970); I.E. Dzyaloshinskii, P.S. Kondratenko, and V.S. Levchenkov, Sov. Phys. JETP *35*, 823, 1213 (1972); K. Liu and M.E. Fisher, J. Low Temp. Phys. *10*, 655 (1973).
[9] M.P.A. Fisher and D.H. Lee, Phys. Rev. B*39*, 2756 (1989).
[10] D.S. Fisher, B.I. Halperin and R. Morf, Phys. Rev. B*20*, 4692 (1979).
[11] E. Cockayne and V. Elser, Phys. Rev. B*43*, 623 (1991).
[12] See, e.g. , Ref.[7] and D.S. Fisher, M.P.A. Fisher, and D.A. Huse, Phys. Rev. B*43*, 130 (1991).

Simulations of Relaxation, Pinning, and Melting in Flux Lattices.

Henrik Jeldtoft Jensen
Department of Mathematics
Imperial College
180 Queen's Gate
London SW7 2BZ
United Kingdom

ABSTRACT: Fundamental aspects of the physics of an elastic medium in a static random background potential are discussed by use of one – and two dimensional computer simulations. The role of non-linear *elastic* instabilities and *plastic* deformations are treated in detail. Elastic instabilities are needed for the existence of a pinning force. In the limit of infinite system size plastic deformations always occur.

Contents

1	**Introduction**	**2**
	1.1 Instabilities and the Pinning Force	2
	1.2 Content of the Paper	5
2	**One Dimensional Chain**	**6**
	2.1 Model	6
	2.2 Generation of Instabilities	7
	2.3 Dynamics	11
	2.4 Statistics of discontinuities	16
	2.5 Power spectrum of pinning force	17
3	**Static Response**	**19**
	3.1 Adiabatic Relaxation	19
	3.2 Model	20
	3.2.1 Pinning Energy, and Translational Order	21
	3.3 Softening Induced by the Random Potential	25
	3.3.1 Simulation of softening	25
	3.3.2 Analytic Study of Softening	26
4	**Quasistatic Response**	**31**
	4.1 Simulation of the quasistatic response	31

	4.2	Deformations due to the quasistatic response	31
	4.2.1	The elastic region	33
	4.2.2	The elastic instability region	33
	4.2.3	Plastic flow region	34

5 Above-threshold Driving Force — 35
5.1 Simulation of Flux Flow — 35
5.2 Voltage-Current Characteristics — 38

6 Onset of Difussion — 40
6.1 Model — 41
6.2 $T - A_p$ Phase Diagram — 44

7 Critical Dynamics — 48
7.1 Model — 49
7.2 Monte Carlo Simulations — 50
7.3 $1/f$ from the Driven Diffusion Equation — 52

8 Conclusions — 53

9 Acknowledgements — 54

1 Introduction

The problem of the repsonse of an elastic medium to a set of random pinning centers possess some general features which are probably independent of the specific type of system considered. It has been the attempt to study some of these general aspects by the simulations described in the following. Thus, although the title contains the term flux lattices, I belive that many of our findings are of relevance to other pinned systems such as charge density waves, Wigner crystals, triple lines, pinning of growing interfaces, etc..

We have chosen to study particles in one and two dimensions. In one dimension we consider a chain of harmonically coupled particles. In two dimensions we use a repulsive Gausian pair potential. The motivation for these interactions is that they are simple and numerically convenient. The repulsive interaction was inspired by the repulsive interaction between parallel flux lines. But beside of that we do not try to use realistic interactions (except in Chap. 6). I expect that the qualitative phenomenology developed from our similations are independent of the specific interaction between the particles which constitute the elastic medium. The important point is to have some elastically and plastically deformable structure interacting with a quenched static random pinning potential. The quantitative details will, of course, differ from one system to another.

1.1 Instabilities and the Pinning Force

Consider a collection of particles with some kind of mutual interaction. At zero temperature the particle-particle interaction ensures that the particles organize themselves in a lattice structure. In the absence of external forces the lattice will possess infinitely long range order. External forces applied to the lattice will induce deformations away from the the ideal lattice configuration. The corresponding increase of the energy of the system will for small external forces be described by the elastic constants, or moduli, of the lattice. For small external load the deformations will be smooth and reversible. When the forces become too large the material starts to break. We induce plastic deformations which do not disappear when we remove the external load.

Even before the external forces become so large that they induce plastic deformations they may lead to elastic instabilities. In this case the deformation is reversible, that is, the initial ideal lattice configuartion is recovered when the external forces are removed. However, the deformation of the lattice undergo a *discontinuous* jump as the external forces reach the threshold at which the instability occur. The Euler buckling transition is the prototype of such elastic instabilities.[1] Consider a straight rod of length l. We apply a compression force F to the ends of the rod. As long as the force is smaller than a certain critical value F_{cr} the rod remains straight. At $F = F_{cr}$ the straight configuration becomes unstable and the rod will discontinuously jump to a configuration with a finite curvature. The transition is discontinuous in the sense that the curvature has a discontinuity at $F = F_{cr}$. It is probably intuitively clear that the value F_{cr} becomes smaller as the rod is made longer. One find that $F_{cr} \propto 1/l^2$.[1]

Let us again consider our system of interacting particles embedded in a random potential. The response of the elastic particle lattice to the action of the random potential is controled by elastic instabilities similar to the Euler buckling instability. The importance of these instabilities is most easily seen when we consider the force we have to apply to the lattice in order to move it through the random potential. We know from experience that when we place an elastic medium on a rough surface, say a metal block on a table, we need to work against the friction force in order to move the metal. The force we have to apply on average to keep the block moving is given by the instabilities which occur as the two surfaces are moved relative to each other.

We will in a moment try to make this point more clear. First some comments about flux lines. Since we have flux pinning in mind we shall often call the particles for vortices. For three dimensional systems we should consider vortex *lines* and not point objects as we are going to do below. Although the difference between line objects and point objects will be important for many aspects of pinning it is, however, of no significance for the discussion of the fundamental role played by instabilities. Moreover, one can of course think of many different types of realizations of the random background potential: planar defects (say, twinning planes), columnar defects (say, radiation tracks) or point pinning centers (say, oxygene vacancies). Again, despite the fact that the structure of the random potential will influence the value of the pinning force and be of importance for the specific way the lattice is deformed, the basic importance of the instabilities does not depend on the type of pinning.

Now back to the origin of the pinning force. For simplicity we discuss point objects.[2] Consider a system of N_v vortices and N_p pinning centers. The potential energy of the system is given by

$$U = U_{vv} + U_{vp}$$
$$U_{vv} = \frac{1}{2}\sum_{i \neq j} u_{vv}(|\mathbf{r}_i - \mathbf{r}_j|) \qquad (1)$$

$$U_{vp} = \sum_{i=1}^{N_v}\sum_{j=1}^{N_p} u_{vp}(|\mathbf{r}_i - \mathbf{r}_j^p|)$$

where U_{vv} is the contribution coming from the vortex-vortex interaction, U_{vp} describes the interaction between the vortices and the pinning centers and $\mathbf{r}_i$ and $\mathbf{r}_j^p$ are the positions of the vortices and the pinning centers, respectively.

According to Eqs. 1 the forces acting on the ith vortex are; the force due to the other vortices

$$\mathbf{f}_v^i = -\frac{\partial U_{vv}}{\partial \mathbf{r}_i}, \tag{2}$$

the force coming from the random potential

$$\mathbf{f}_p^i = -\frac{\partial U_{vp}}{\partial \mathbf{r}_i}, \tag{3}$$

and the applied homogeneous driving force $\mathbf{f}_{dr}$. We assume the vortices to follow a diffusive equation of motion, as is expected to be the relevant dynamics for flux lines in type II superconductors.[3] Thus

$$\eta \mathbf{v}_i = \mathbf{f}_{dr} + \mathbf{f}_v^i + \mathbf{f}_p^i, \tag{4}$$

where η is the damping coefficient. Multiplying Eq. 4 by $\mathbf{v}_i$ and summing over all the vortices gives

$$\mathbf{f}_{dr} \cdot <\mathbf{v}> = \eta <\mathbf{v}^2> + \frac{1}{N_v}\frac{dU}{dt}, \tag{5}$$

where $<\cdots>$ denotes averaging over the vortices. A driving force is chosen which is just above the threshold for motion at any instant

$$\mathbf{f}_{dr} = \epsilon - <\mathbf{f}_p>. \tag{6}$$

Using Eq. 4 the center of mass velocity becomes

$$<\mathbf{v}> = \frac{1}{\eta}\epsilon. \tag{7}$$

Substituting the values for $\mathbf{f}_{dr}$ and ϵ from Eq. 6 and 7 into Eq. 5 we obtain

$$-<\mathbf{f}_p> \cdot <\mathbf{v}> -\eta(<\mathbf{v}^2> - <\mathbf{v}>^2) = \frac{1}{N_v}\frac{dU}{dt}. \tag{8}$$

As long as the internal reorganization of the vortices induced by the applied force is small, the second term of the left hand side of Eq. 8 will be small compared to the first term. If, say, N_f of the N_v vortices are moving with a speed v, and $N_v - N_f$ are trapped, one has

$$<v^2> - <v>^2 = \frac{N_f}{N_v}v^2 - \left(\frac{N_f}{N_v}v\right)^2 = \left(\frac{N_v}{N_f} - 1\right)<v>^2. \tag{9}$$

In the limit of quasistatic displacement the center of mass velocity goes to zero. The first term on the left hand side of Eq. 8 goes to zero $<v>$ whereas the second term goes to zero as $<v>^2$ and is therefore negligible in this limit. Let X denote the center of mass position. Since

$$\frac{dU}{dt} = \frac{dU}{dX}\frac{dX}{dt} = \frac{dU}{dX}\langle v \rangle \tag{10}$$

we obtain from Eq. 8
$$F_p \equiv -N_v <f_p> = \frac{dU}{dx}. \tag{11}$$

The average force $\bar{F}$ which is needed to quasistatically move the vortex system through the background potential is calculated as

$$\begin{aligned}
\bar{F} &= \lim_{L \to \infty} \frac{1}{L} \int_0^L F_p(X) dX \\
&= \lim_{L \to \infty} \left\{ \frac{1}{L}[U(L) - U(0)] + \frac{1}{L} \sum_{X_k \in [0,L]} Disc[U(X_k)] \right\} \\
&= \lim_{L \to \infty} \frac{1}{L} \sum_{X_k \in [0,L]} Disc[U(X_k)], \tag{12}
\end{aligned}$$

where the sum is over the discontinuities of U;

$$Disc[U(X)] = \lim_{\epsilon \to 0^+} [U(X + \epsilon) - U(X - \epsilon)]. \tag{13}$$

The discontinuities in the potential energy occur when the vortex system jumps from one metastable configuration to an other. These jumps are caused by elastic (and plastic) instabilities.[4, 5, 6, 7] It follows from Eq. 12 that there are *no pinning force if there are no discontinuities* in the potential energy as functic of the center of mass position of the system.

This result is after all not so surprising. Think of a cyclist in the Norwegian mountains. Sometimes she will have to struggle up steep slopes for in the next moment to gain all the energy again, as she swiftly pedals down the next hill. As you travel through a random potential you go up and down, though on average you stay at the same level in the potential. The cyclist analogy is of course only valid as long as the internal degrees of freedom of the many body vortex system can be neglected. The analogy breaks down when elastic energy, which has been gradually increased as the lattice deforms in responce to the pinning potential, all as a sudden is released during an instability.

The force defined in Eq. 12 can be thought of as the dynamical pinning force in the limit of zero velocity. This force has the very important feature that it only depends on inherent elastic, as well as plastic, properties of the vortex-pin system. It is automatically averaged over different metastable configurations. The connection to the static frition force, which is the force one have to exert on the system to make it move, is not obvious, although, one would expect for a large system the two forces to be of the same order of magnitude.

So far we have tried to make it clear that instabilities are of crucial importance for the pinning force experinced as the elastic lattice is moved through the random potential. We shall see below that even when the elastic medium is relaxed statically, i.e. whithout any motion of the center of mass position, elastic instabilities occur and controls the relaxation down into the random potential.

1.2 Content of the Paper

As just explained, elastic (and as we will see later also plastic) instabilities are essential for the understanding of the response of an elastic medium to a random background potential.

We shall devote much attention to the study of the nature of the instabilities. The next four chapters deals with the study of the response at *zero* temperature. In Chap. 2. we discuss the characteristics of instabilities in a one dimensional chain draged longitudinally through a random pinning potential. In Chap. 3. we consider the static relaxation of a two dimensional lattice to a set of identical but randomly poisitioned pinning centers. Our starting configuration is the ideal lattice and zero strength of the pinning centers. We then study the instabilitites and distortions induced as the strength of the pinning centers are gradually increased. In Chap. 4. we consider the situation in which we quasistatically shifts the center of mass of the lattice through the pinning potential. This amounts to study the deformations of the lattice in the presence of an external driving force, which in every instance precisely balance the pinning force. In Chap. 5. we apply a driving force, $\langle f_{dr} \rangle$ which exceeds the threshold pinning force, f_{thr} by a finite amount. This makes the lattice move with a finite center of mass velocity, $\langle v \rangle$. We study the relation between $\langle v \rangle$ and $f_{dr} - f_{thr}$. In Chap. 6 we turn to finite temperature effects. We consider the combined effect of the random potential together with thermal fluctuations. We describe how the pinning potential influences the temperature at which vortices of the lattice begin to diffuse around. In Chap. 7 we use a simple lattice gas to study the density fluctuations in a vortex system in the region just above the depining transition. Chap. 8 contains the conclusion.

2 One Dimensional Chain

In this chapter we study a disordered version of the Frenkel-Kontorova model in one dimension.[8] More specifically, we consider a homogeneous elastic chain moving in a random potential. We use the model for a detailed study of the many-body aspects of the instabilities. As the chain is pulled from one end, stick-slip processes occur. The restriction to one dimension enables us to determine the statistical properties of the instabilities. Energy discontinuities connected with these processes are found to exhibit a power law distribution. Moreover, these jumps induce a finite pinning force. We also study the fluctuations of the force which is required to pull the system. Its power spectrum has the $1/f^\alpha$ form. The procedure of pulling the chain selects a small subset of "physical states" among the exponentially large number of possible metastable states. Pushing the chain back and forth through the random potential doesn't generate the same sequence of states.

2.1 Model

We consider a harmonic chain consisting of N particles in a random potential.[9] The potential energy of the system is defined by

$$E = \sum_{i=1}^{N-1} [\frac{k}{2}(x_{i+1} - x_i - a)^2 + V(x_i)] \qquad (14)$$

Here k is the spring constant, which we set equal to unity, x_i denotes the position of the particles, a is the lattice parameter, and $V(x_i)$ is the random potential of the following form

$$V(x) = -A_p \sum_p \exp(-[(x - x_p)/R_p]^2) \qquad (15)$$

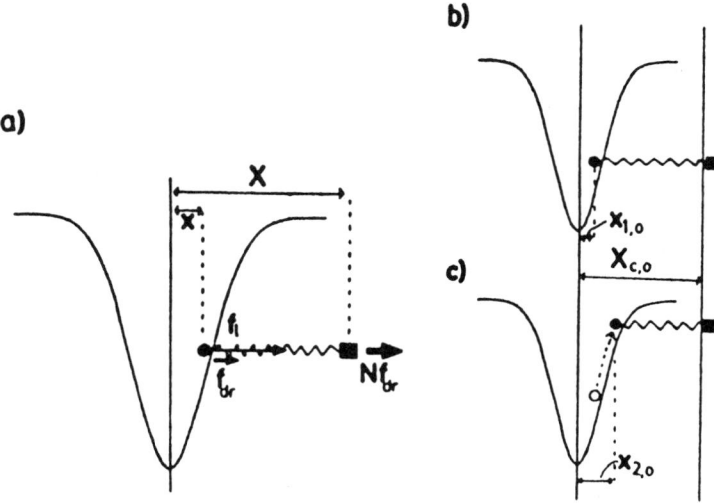

Figure 1: a) Harmonic oscillator coupled to a gaussian potential. b) and c) Show the configuration just before and just after the jump out of the well. Note that the position $X_{c,0}$ does not change during the jump.

The positions x_p of the pinning centers are randomly distributed with a uniform density n_p. The strength and the range of the individual pinning centers are respectively A_p and R_p.

The minimization of the potential energy is performed iteratively by solving the equilibrium condition $\partial E/\partial x_i = 0$ for all i. We have

$$-k(x_2 - x_1 - a) + V'(x_1) = 0 \qquad (16)$$
$$k(x_2 - x_1 - a) - k(x_3 - x_2 - a) + V'(x_2) = 0$$
$$\ldots$$
$$k(x_N - x_{N-1} - a) - k(x_{N-1} - x_{N-2} - a) + V'(x_{N-1}) = 0$$

No equation is written for the Nth particle since we assumed that an additional force applied on the Nth particle is pulling the system in such a way that this N^{th} particle always experiences a zero net force. We prescribe the position x_1 of the first particle and express all physical quantities as a function of x_1. The physical situation in which particle number N is forced to move with constant velocity (as in the experiments described in Ref. [10]) can be obtained by inverting the relation between x_1 and x_N. Even for weak potential $V(x_i)$ this relation is found to be extremely non-monotonous provided the chain is sufficiently long.

2.2 Generation of Instabilities

Consider a single vortex in a potential well, see Fig. 1. The coupling to the surrounding

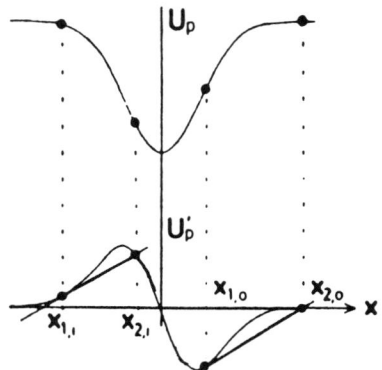

Figure 2: A Gaussian potential, U_p, and its derivative U'_p. The slope of the straight lines is k. The intersection between the straight lines and U'_p determine the solution of Eq. 17.

vortices is represented by an harmonic oscillator. The equlibrium condition is defined by

$$\frac{\partial U_p}{\partial x} = k(X - x) \tag{17}$$

The solution of this equation is shown graphically in Fig. 2. The values of the positions $x_{i,1}$, $x_{i,2}$ (or $x_{o,1}$, $x_{o,2}$), at which the vortex in the well jumps in (or jumps out), and the corresponding values of the center of mass shift X_i (or X_o), can be calculated numerically. From these positions we can compute the discontinuity ΔU in the potential energy associated with a sequence "jump in – jump out". We find that ΔU is well approximated by the following simple expression

$$\Delta U = A_p \Phi(\frac{A_p^*}{A_p}), \qquad \Phi(t) = (\frac{1-t}{t})^{\frac{5}{7}} \tag{13}$$

where $A_p^* = (e^{3/2}/4)kR_p^2$. For a given stiffness k and range R_p there will not be any instability unless the strength A_p is larger than A_p^*.

Let us now consider the chain. In order to gain some understanding of the complex behavior of the large N chain, it is helpful to start with the simplest $N = 2$ case. In Fig. 3, we show x_2 as a function of x_1, for two different values of A_p. There is a critical value A_{thr} of A_p below which x_2 is a strictly increasing function of x_1. When A_p is larger than A_{thr}, some parts of the curve correspond to unstable regions. This leads to jumps in x_1 when the control variable x_2 is smoothly increased. These jumps are the one dimensional equivalent of the elastic instabilities.[4] The evolution of the center of mass position, and the total energy versus x_2 are plotted in Figs. 4 and 5 respectively, for values of A_p smaller or larger than the threshold value A_{thr}. The presence of non monotonous regions in the x_2 versus x_1 leads to multivalued functions of the control variable x_2. In Figs. 4b and 5b, dotted lines represent metastable states, and full lines the ones which are actually reached

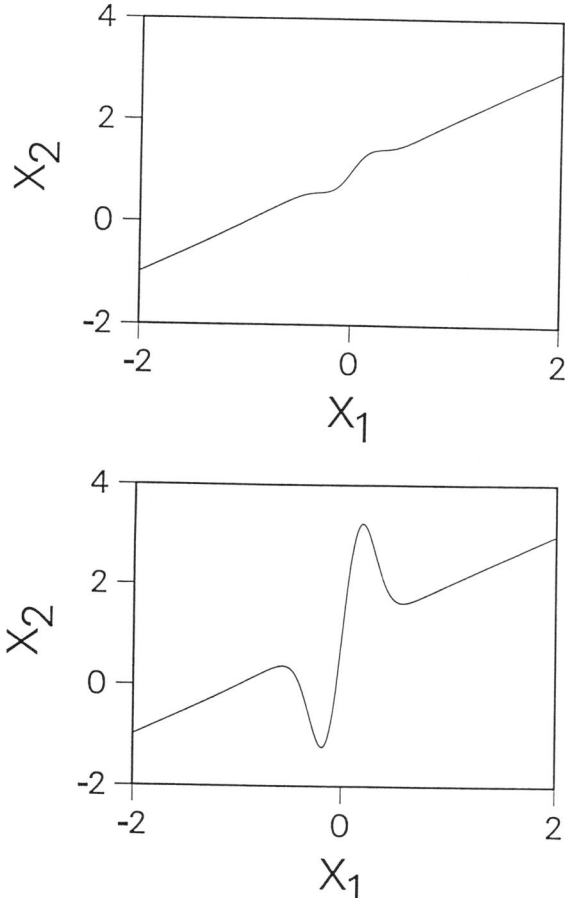

Figure 3: Case of the two particle chain ($k = 1$) with a single pinning center ($R_p = 0.25$): graph of the position x_2 of the second particle versus the position x_1 of the first particle, for different values of the pinning strength A_p a) $A_p = 0.06 < A_{thr}$: under the threshold for elastic instabilities. b) $A_p = 0.6 > A_{thr}$: above the threshold for elastic instabilities. x_1 is a multivalued function of x_2

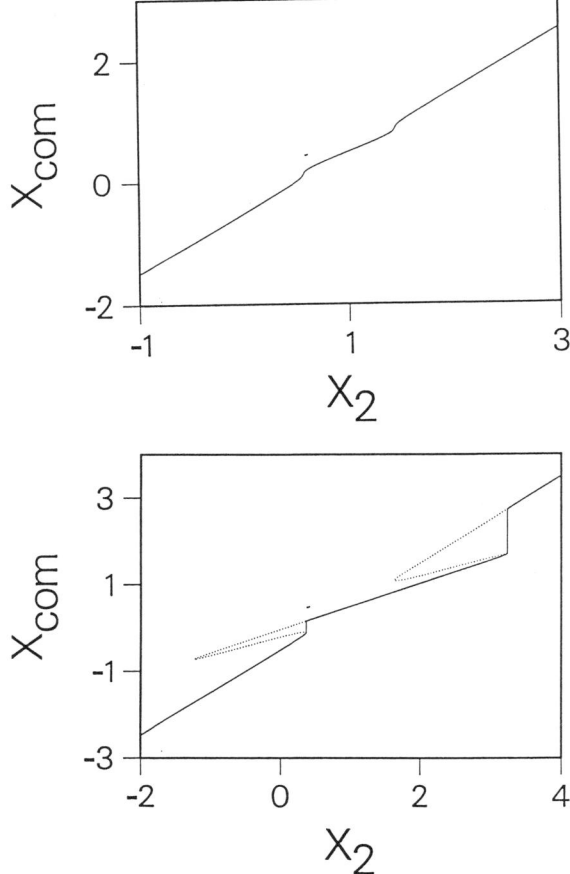

Figure 4: Case of the two particle chain ($k = 1$) with a single pinning center ($R_p = 0.25$): graph of the center of mass position x_{com} versus the position x_2 of the second particle, for different values of the pinning strength A_p a) $A_p = 0.06 < A_{thr}$: under the threshold for elastic instabilities. b) $A_p = 0.6 > A_{thr}$: above the threshold for elastic instabilities. The dotted line corresponds to all the possible metastable states. The full line shows the succession of states actually reached upon increasing x_2. This illustrates the existence of jumps in the center of mass position.

as x_2 is increased. Upon decreasing x_2, a different path in configuration space is followed. The pattern of Fig. 5b is quite generic, and will be pictorially referred to as the "cat" instability.

In the case of a longer chain, the multivalued nature of the energy and center of mass position as a function of x_N becomes more dramatic with a very large number of folds. However, this intricate pattern can be decomposed in a series of nested elementary cat instabilities. A qualitative understanding of this behavior comes from the fact that if say x_1 lies within the range of a pinning center, it generates displacements in x_p which grow linearly with p. As a result, the probability for particle number p to sweep through a pinning center increases with p. In order to compensate the pinning force acting on particle number p, x_N oscillates with an amplitude proportional to $N - p$. Furthermore, the corresponding displacement in x_1 required for particle number p to sweep through the given pinning center goes as $1/p$. After inverting the x_N versus x_1 curve, this leads to small and numerous cat instabilities induced by particles close to particle number N, superimposed on larger and less frequent instabilities due to particles closer to particle number 1. This phenomenon is well illustrated on Fig. 6, where x_N and total energy are plotted as as function of x_1. The oscillations in Figs. 6a and 6b lead to the multivalued energy function $E(x_N)$ shown in Fig. 6c.

As shown on Fig. 6c, because of these jumps the energy, E_{tot}, as function of x_N (dotted line) has always an upward curvature locally. Hence, a finite average force is required in order to induce a global displacement of the chain. This is the microscopic origin of a finite friction force between the chain and the disordered substrate. The threshold for onset of these instabilities decreases as $N^{-\sigma}$, where $\sigma = 3/2$ in agreement with the following general argument. Instabilities occur when dx_N/dx_1 becomes negative. We have

$$\frac{dx_N}{dx_1} = 1 + k^{-1} \sum_{p=1}^{N-1} pV''(x_{N-p})\frac{dx_{N-p}}{dx_1} \qquad (18)$$

dx_N/dx_1 is able to change sign when the variance of the second term in the right hand side of Eq. 18 is equal to 1. In weak disorder, we may assume the derivatives dx_{N-p}/dx_1 to be of the order of 1. Furthermore, we assume the different $V''(x_p)$ to be independent random variables for different values of p. This leads to a variance proportional to $N^3 n_p R_p^{-3} A_p^2 k^{-2}$. Hence,

$$A_{thr} \sim n_p^{-1/2} R_p^{3/2} k N^{-3/2}. \qquad (19)$$

This scaling for A_{thr} can also be derived from a criterium on the magnitude of the square displacements of the individual particles (see Ref. [11] and Chap. 3 below). Fig. 7 shows an example of this scaling of A_{thr} with the chain length N.

2.3 Dynamics

In this section, we study the motion of the chain as x_N is gradually increased by applying an external force on particle number N so that the system is always in equilibrium. The previous discussion has shown that the evolution of the system cannot always be smooth. Jumps have to occur, corresponding to cusps in the energy as a function of x_N. However, because of the large number of metastable states, an additional prescription is required in order to determine the final state after a jump. It certainly depends on the specific

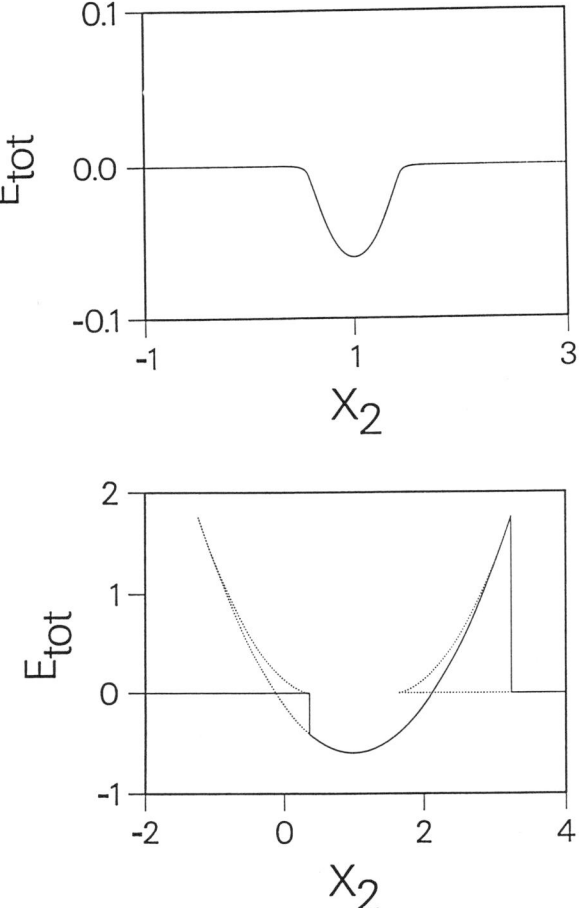

Figure 5: Case of the two particle chain ($k = 1$) with a single pinning center ($R_p = 0.25$): graph of the total energy E_{tot} versus the position x_2 of the second particle, for different values of the pinning strength A_p a) $A_p = 0.06 < A_{thr}$: under the threshold for elastic instabilities. b) $A_p = 0.6 > A_{thr}$: above the threshold for elastic instabilities. The dotted line corresponds to all the possible metastable states. The full line shows the succession of states actually reached upon increasing x_2. This has been labelled as the "cat instability" because of the shape of the $E_{tot}(x_2)$ curve.

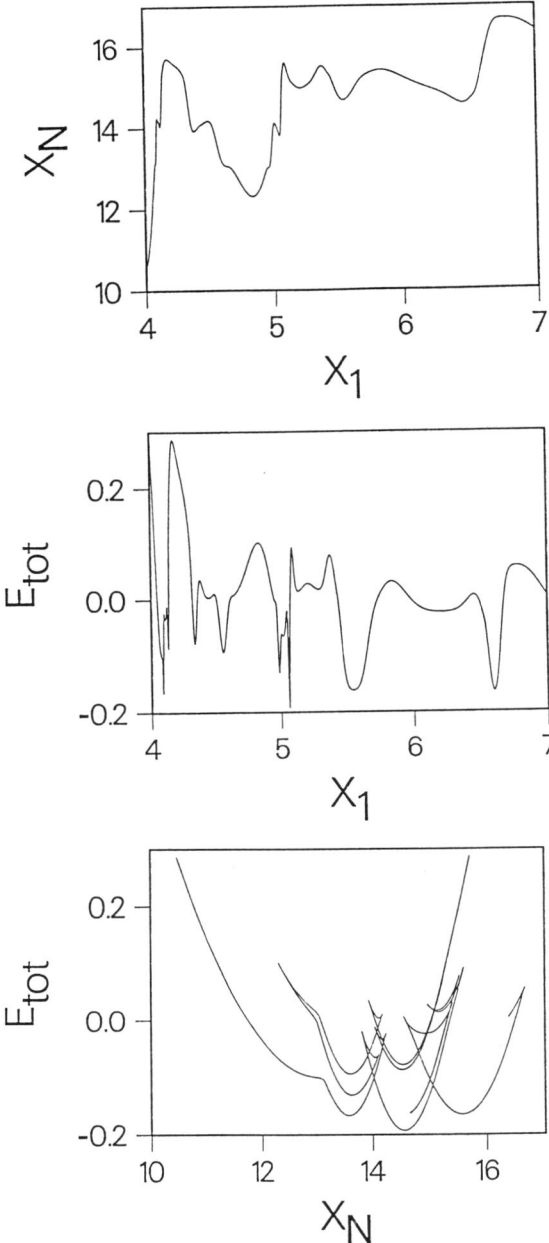

Figure 6: General case: chain of length $N = 10$ with a density $n_p = 0.5$ of pinning centers with $R_p = 0.25$ and $A_p = 0.05$ above the instability threshold A_{thr}. a) Position of the Nth particle x_N versus position of the first particle x_1. b) Total energy E_{tot} versus x_1. c) E_{tot} versus x_N from combining the two previous graphs. The rather complicated shape of this curve can be analyzed as a superposition of "elementary cats".

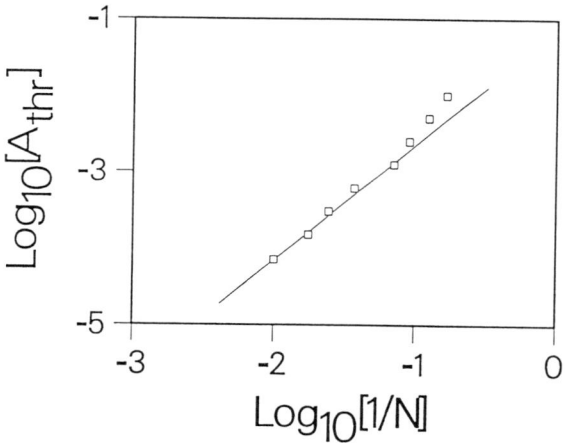

Figure 7: Dependence of the threshold for elastic instabilities A_{thr} on chain length N. Parameters are: $n_p = 0.5, R_p = 0.25, k = 1$. This log-log plot shows the scaling $A_{thr} \sim N^{-3/2}$

choice of dynamics. For the sake of simplicity, we want to construct a dynamics which can be directly related to the knowledge of the various equilibrium states of the system. Our criterion of selection is to pick the states which minimize the jump in the center of mass position. This simple criterion corresponds to choosing a situation where the center of mass dynamics is over damped. This a priori is not equivalent to assuming that each particle motion is itself over damped. It would be interesting to study the influence of the chosen dynamics on the behavior of the chain. However, this requires a full simulation and we would loose the simplicity of the present approach. The implementation of our criterion is illustrated on Fig. 8. Fig. 8a shows the center of mass position X_{com} as a function of x_N. The corresponding energy branch is displayed as the full drawn line in Fig. 8b, where the dotted line refers to the full set of metastable states.

We observed that the states which are selected by this procedure differ qualitatively from the typical metastable states. For instance, we have checked that in these selected states x_p is an increasing function of p. However most typical states do not satisfy this property. The fact that the states selected in different dynamics differ has already been observed in other complex systems such as charge density waves.[12] However, in the work by Tang et al, the preparation method is quite different from ours. It would be interesting to investigate if our models exhibits this phenomenon of pulse memory.

As already stated for the two particle case, the motion of the chain is strongly hysteretic. If the control variable x_N is cycled in a fixed interval around a given initial value, the system evolves in the following way. After the first cycle, it does not return in general to its initial state. However, it afterwards evolves along a closed loop in configuration space. We show an example of this behavior in Fig. 9. We should also note that these periodic cycles depend on the minimal and maximal values of x_N.

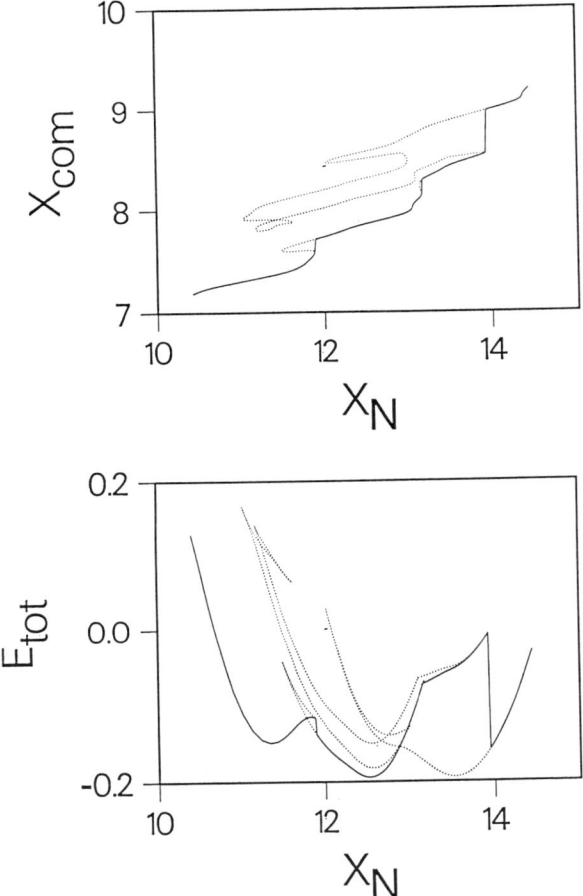

Figure 8: Selection of states in the over damped center of mass dynamics. The parameters are the same as in Fig. 6. a) Position of the center of mass x_{com} versus x_N. When the system jumps, it is assumed to select the closest value for x_{com} associated to the same x_N. b) Total energy E_{tot} versus x_N. Note that the selected states after a jump are neither the closest nor the lowest ones in energy.

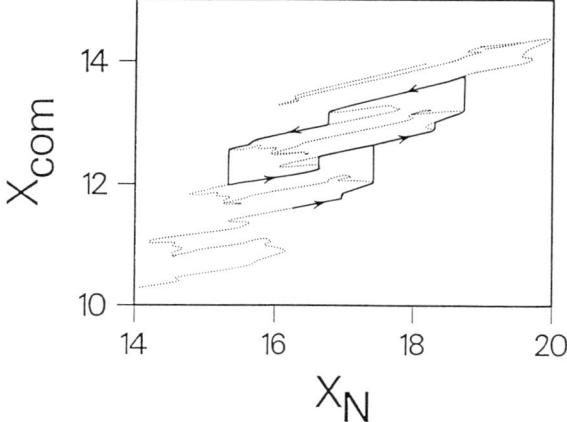

Figure 9: Hysteretic behavior. Center of mass position x_{com} as a function of x_N, with the same parameters as in Fig. 6. Note that after a first cycle in x_N, the final state is in general different from the initial state. However, a second identical cycle in x_N, starting from this new state leads to a closed hysteresis loop.

2.4 Statistics of discontinuities

In the present section, we discuss the statistical properties of the motion of the chain. In particular, we are concerned with the discontinuities in the total energy and center of mass position which occur as the system jumps from one metastable state to another. The distribution of energy discontinuities is shown on Fig. 10 for different chain lengths.

Power law behavior is observed except in the high energy region. The exponent of the scaling region, $b \approx 0.8$, is within the numerical accuracy independent of A_p, n_p, R_p and N (see Fig. 10). The crossover from scaling for higher energies seems to be related to depinning events involving a single or a small number of pinning centers. To check this idea, we have calculated the distribution of displacements of individual particles during jumps. This shows that high energy events are connected with a broad distribution of displacements centered around zero and with a width of order several times a. The low energy discontinuities are connected with a narrow displacement distribution centered around zero. Intuitively, displacements induced by few pinning centers can propagate far away and grow linearly with the distance to the relevant pinning centers. This corresponds to the high energy events, which resemble the elementary cats discussed above. By contrast, low energy events involve many pinning centers, so that induced displacements cannot propagate very far. A more systematic study of this crossover from scaling to high energy regimes is in progress. One trend, among others, is the increase in the crossover energy upon increasing A_p, as can be seen in Fig. 10.

The corresponding distribution of discontinuities in the center of mass, $D(\Delta X_{com})$, is nearly uniform (over a range from about $10^{-4}a$ up to about $a/2$) for the events in the scaling region of $D(\Delta E)$. An example of this behavior is given in Fig. 11. The local peak in $D(\Delta E)$ is connected with a peak in $D(\Delta X_{com})$ at about a, thus corresponding to large

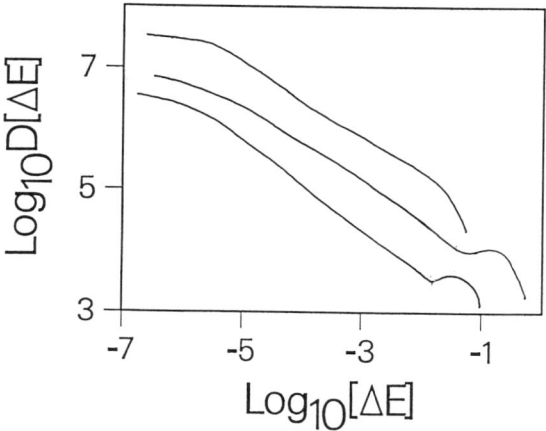

Figure 10: Distribution of energy discontinuities. Parameters are: $n_p = 0.5, R_p = 0.25, k = 1$. From top to bottom: $N = 10, A_p = 0.01$; $N = 10, A_p = 0.05$; and $N = 20, A_p = 0.01$. The three curves have been arbitrarily shifted vertically for the sake of clarity. This log-log plot shows the power law behavior in the low energy region.

energy events.

2.5 Power spectrum of pinning force

The actual pinning force $F_p(x_N)$ is obtained as

$$F_p(x_N) = \sum_{i=1}^{N-1} V'(x_i) = -k(x_N - x_{N-1} - a). \qquad (20)$$

We have found that $F_p(x_N)$ exhibits a similar saw-toothed behavior (see Fig. 12) as was observed in the experiments on solid friction.[10] It is then illuminating to calculate the power spectrum as shown in Fig. 13. This is done with the assumption that the chain is pulled at a constant velocity, so that x_N is proportional to time. We find a $1/f^{1.5}$ behavior for frequencies *smaller* than the characteristic frequency, f_2, connected with the duration of the sawtooth, and larger than a frequency f_1 which decreases as N increases. The exponent 1.5 in this scaling regime appears to be independent of the system size N and the pinning strength A_p. For f larger than f_2, the power spectrum follows a power law decay, with an exponent between 2 and 2.5. This exponent seems to decrease when N is increased, and to saturate at the value 2 in the large N limit. For f smaller than f_1, we find a white noise, i.e. a power spectrum independent of f.

It is tempting to compare our results for this simple model to experiments on solid friction. It is first useful to note that the $1/f^2$ behavior observed at high frequencies in our model is a simple consequence of the fact that the pining force is a discontinuous function of time. Indeed, the power spectrum generated by a single kink is $1/f^2$, and at high frequencies, we can neglect interferences coming from different kinks in the calculation of the power spectrum. In the experiment of Ref. [10], such a $1/f^2$ is observed at high

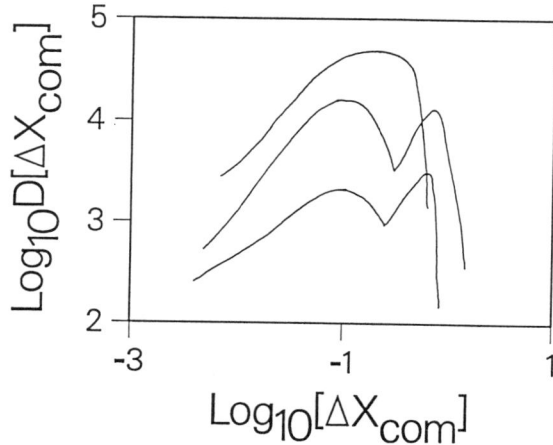

Figure 11: Distribution of discontinuities in the center of mass position x_{com}. The parameters are the same as in Fig. 10. Again, the three curves have been arbitrarily shifted vertically for the sake of clarity.

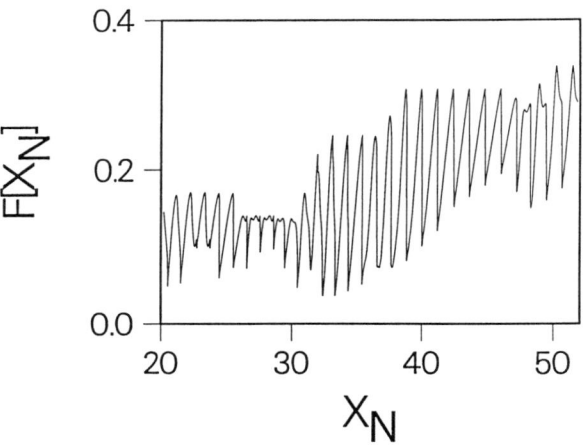

Figure 12: Sawtooth shape of the pinning force as x_N is increased gradually.

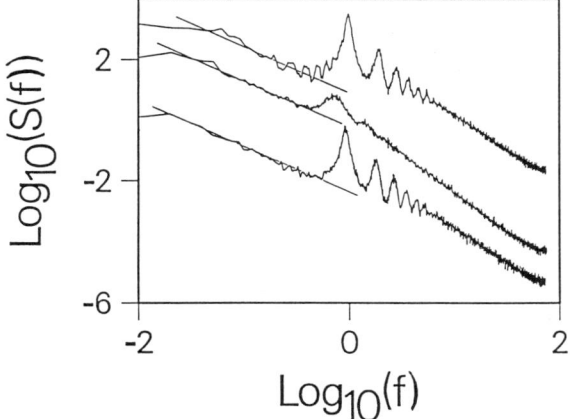

Figure 13: Power spectrum for the pinning force. The parameters are the same as in Fig. 10. The chain is assumed to be pulled with a constant velocity, so that the frequency f can be interpreted as an inverse length scale for x_N. This log-log plot shows a $1/f^{1.5}$ in an intermediate frequency range.

frequencies. This is likely to stem from the fact that in the actual experiment, the jumps occur on distances much larger than the substrate roughness. We think that observation of a $1/f^2$ behavior at high frequencies for both our model and the experiment is not relevant in order to probe the collective nature of solid friction. In the intermediate scaling regime $f_1 < f < f_2$, the power law reflects correlations between different avalanches, due to the fact that the same impurity configuration give rise to many metastable states which are sampled during the system evolution. This may be observed in a solid friction experiment similar to the one described in Ref.[10], provided the force measuring apparatus is sensitive enough to measure small jumps so that the memory of the substrate configuration is not lost after each jump.

3 Static Response

In this chapter we discuss the static response of a two dimensional lattice at zero temperature to a random background potential. We describe first simulations which show how elastic instabilities make the response to even the weakest random potential a non-pertubative effect. Next we consider the softening of the lattice induced by the background potential.

3.1 Adiabatic Relaxation

Let us now focus on the breakdown of adiabatic continuation between the ideal lattice and the ground state as the random potential is gradually switched on. We consider a two dimensional system a zero temperature and with out any external applied driving force.[11] For a finite system, the resulting deformed lattice is smoothly related to the ideal lattice configuration for random potentials weaker than a critical strength. This strength vanishes as $1/\sqrt{N}$, where N is the number of particles. When the amplitude of the pinning potential is increased non-linear elastic instabilities occur as particles jump towards local minima of the background potential. As soon as the elastic instabilities occur the sequence of ground states obtained by the gradual change of the pinning strength is no longer reversible. The probability for a particle to undergo an instability grows rapidly with A_p, the strength of the random potential. As A_p is further increased above a certain threshold topological defects are nucleated.

The pinning potential is switched on adiabatically. We study the spatial correlations in the ground state of the lattice as it responds to the pinning potential. Although the random background induces elastic instabilities as well as miss-coordinated particles the long range order decays only very weakly, probably algebraically, when the random potential is switched on adiabatically. This finding contrasts previous investigations.[13, 14] We also find that the system exhibits strong history dependence.

The instabilities observed during the adiabatic relaxation controls the efficiency of the pinning potential. We show below that the gain in pinning energy is directly determined by the density of particles which have experienced an instability.

3.2 Model

The system consist of N mobile particles of density n_v and positions $\mathbf{r}_i$. Periodic boundary conditions are used. The particles interact mutually through a repulsive pair potential and with fixed pinning centers at randomly distributed positions $\mathbf{r}_p$. The density of the pinning centers is n_p. The potential energy is given by[4, 6]

$$H = \sum_{i,j} u(r_{ij}/R_v) - A_p \sum_{i,p} u(r_{ip}/R_p) \quad (21)$$

Here $r_{\alpha\beta} = |\mathbf{r}_\alpha - \mathbf{r}_\beta|$. The pair potential is $u(r) = \exp(-r^2)$ and A_p is the amplitude of the attractive pinning centers. The amplitude of the particle-particle potential is used as our energy unit. We study the ground state of this system. For $A_p = 0$ the ground state configuration is a triangular lattice, we use the lattice spacing as our unit of length. Starting from the ideal lattice we gradually switch on the random potential by increasing A_p stepwise in small increments ($\Delta A_p = 0.002$). The subject of our study is the deformation of the particle lattice away from the ideal lattice configuration as A_p is adiabatically increased. The relaxation of the particles to the random potential is done by use of a molecular dynamics annealing method.[4] The procedure is as follows. We ascribe a unit mass to each of the particles. Ordinary molecular dynamics is used to integrate Newton's equation for the particles with potential energy given by Eq. 1. In order to find the minimum configuration of the potential energy kinetic energy is constantly extracted from the particle system. This is done by rescaling all velocities such as to keep the mean kinetic energy, E_{kin}, fixed at a *very* small value. One could suspect this method not to lead to the true ground state. We checked this by comparing the energy of the configurations obtained through the adiabatic relaxation with the energy of configurations resulting from a quench at fixed A_p from high temperature. The quench gives higher or equal energies for the small values of A_p considered here. The adiabatic procedure will eventually fail to lead to the true ground state when a significant number of miscoordinated particles (or topological defects) are induced at higher

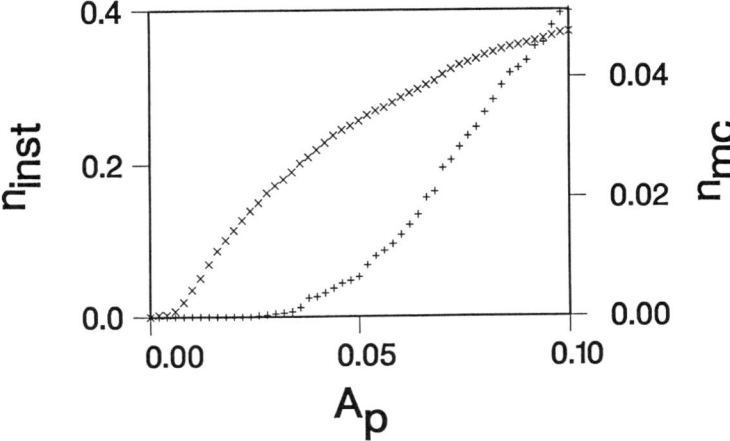

Figure 14: The accumulated density of instabilities n_{inst} and the density of miss-coordinated particles n_{mc} both of as function of A_p. The system consist of 672 particles and the same number of pinning centers of range $R_p = 0.125$ in unites of the lattice constant of the ideal particle lattice. The range of the particle pair potential is $R_v = 0.6$. For details see Ref. [4]

values of A_p. This does not concern us right now. We are only interested in the limit of very small A_p. Our goal is to illuminate how the analytic connection between the ground state of the system in the presence of the pinning potential and the ideall lattice configuration of the pure system is lost even in the limit of small A_p.

3.2.1 Pinning Energy, and Translational Order

We used the following set of system parameters in the simulations: $E_{kin} = 10^{-5}$, $R_v = 0.6$, which corresponds to a shear modulus $C_{66} = 0.27$, $R_p = 0.125$. These are rather short ranged pinning potentials, although, they might still be relatively smooth compared to the case of flux pinning. The simulations were run with equal number, N, of pinning centers and particles. In order to study size scaling we used $N = 56, 340, 418, 672, 1020$.

As A_p is increased from A_p to $A_p + \Delta A_p$ we measure the resulting change in distance, Δr_{ip}, between the particles and their nearest pinning center. The increase ΔA_p can cause some of the particles to jump discontinuously. We say that a particle has performed a jump if $\Delta r_{ip} > R_p/2$. We call these jumps elastic instabilities. They start to occur at a value of A_p which is much smaller than the one needed to create topological defects.

In Fig. 14 we show an example of the accumulated number of such instabilities per area, $n_{inst}(A_p)$, (see the × symbols) together with the areal density of the number of miss-coordinated particles (determined by the usual Voronoi polygon construction[15]), $n_{mc}(A_p)$, (see the + symbols). An averaged over 58 realizations of the pinning potential has been performed. One notices that the elastic instabilities continue to occur in the regime where topological defects are nucleated. Furthermore, the onset of the production of defects has no apparent influence on the number of elastic instabilities. The behavior of n_{inst} as well as of n_{mc} can be fitted well to the functional form $\exp(-\text{cst.}/A_p)$

The threshold for creation of elastic instabilities exhibits a dependence on system size which is consistent with a decrease as the inverse square root of the number of particles in the lattice. This behavior can be understood in the following way.

Assume that the instabilities sets in when when the root mean square of the displacements, s, of the particles away from the ideal lattice has reached a value of the order of R_p. The energy of the distorted lattice is (in a short hand notation were we suppress all summations) given by $E = \frac{1}{2}s_i\Phi_{ij}s_j - f_i s_i$, were f_i is the pinning force on particle i. The minimum configuration is obtained from the criterion $\delta E = 0$, which leads to $\Phi_{ij}s_j = f_i$. Introducing the inverse matrix χ_{ij} of the elastic matrix Φ_{ij} (see [16]) we obtain in d dimensions $\langle s^2 \rangle = \langle (\sum_{ij} \chi_{ij} f_j)^2 \rangle \sim (A_p/R_p)^2 n_p R_p^d \int_{1/L}^{1/a}(k^{d-1}/k^4)dk \sim A_p^2 n_p R_p^{d-2} N^{(4-d)/d}$

The threshold value for the pinning amplitude is determined by the condition $\langle s^2 \rangle^{1/2} \sim R_p$ and scales as

$$A_{thr} \sim R_p^{(4-d)/2} n_p^{-1/2} N^{-(4-d)/2d} \qquad (22)$$

Hence, for dimensions less than four the threshold for elastic instabilities decreases with increasing system size as $N^{-\sigma}$ with $\sigma = (4-d)/2d$. This is in agreement with the result that there is no threshold for pinning for dimensions less than four.[17] The dependence of A_{thr} on R_p and n_p given by Eq. 22 is in qualitative agreement with our two dimensional simulations (we could only afford to check a few different R_p and n_p values). The dependence of σ on dimension is in agreement with the one dimensional simulations described in Chap. 2.

The threshold for creation of miss-coordinated particles decreases with increasing system size in a way consistent with one over the logarithm of the number of particles. Similar instabilities and defects with the same size dependence of the thresholds were found in the study of the depinning transition in Ref. [4], see Chap. 4, where quasi-static depinning were investigated. This suggests that the depinning transition is closely related to the nature of the ground state of the static system. It is important to obtain a better understanding of the observed size dependence of these thresholds.

There is a direct connection between the pinning energy gained during the relaxation of the lattice to the random potential, as A_p is turned on, and the accumulated number of elastic instabilities which has occurred. The pinning energy can be thought of as composed of two terms. An average contribution coming from the set of pinning centers which has not been strongly correlated with the particles through instabilities plus a contribution from those pinning centers which have trapped a particle through the process of an instability. The pinning energy is thus given by

$$E_p = -n_v \pi R_p^2 A_p [n_p - n_{inst}(A_p)] - \alpha A_p n_{inst}(A_p) \qquad (23)$$

where α is a constant of order unity.

In Fig. 15 we show the actual measured pinning energy (as the solid line) together with the estimate given by Eq. 23 (dotted line). Here we have used n_{inst} from Fig. 14 and $\alpha = 0.7$ was chosen to get the best agreement between the two curves. This demonstrates that the elastic instabilities are crucial for the gain of pinning energy.

We have also studied the energies and the instabilities as A_p is cycled from $A_p = 0$ up to $A_p = A_{max}$ and then down again to $A_p = 0$. As soon as A_{max} becomes so large that elastic instabilities are induced, cycling of A_p becomes irreversible in the following sense. The total potential energy in Eq. 1 $E_{tot} = E_{par} + E_{pin}$ is the sum of the particle-particle

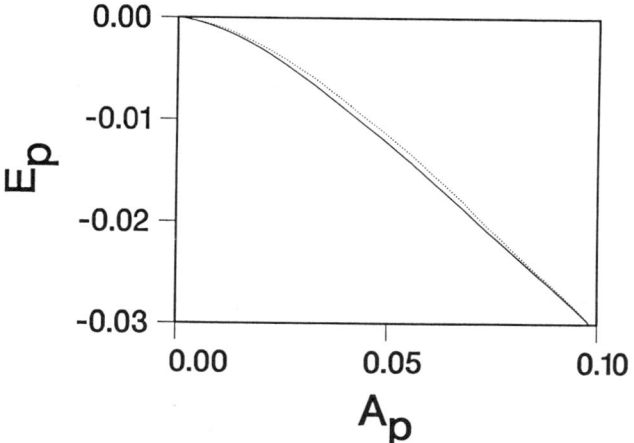

Figure 15: The system is the same as the one in Fig. 14. The solid line is the actual measured pinning energy as function of A_p the dotted line is obtained by use of n_{inst} from Fig. 14 together with Eq. 23 with $\alpha = 0.7$.

energy, E_{par}, and the particle-pinning energy, E_{pin}. The set of values assumed by E_{tot} are the same upon increasing A_p from zero to A_{max} as those obtained when A_p is decreased from A_{max} to zero. This is not the case for neither E_{par} nor E_{pin}. The values assumed by E_{par} upon increasing A_p are all smaller than the corresponding values obtained as A_p is decreased. The situation is opposite for E_{pin}. The reason is that the instability in which a particle jumps *off* a pinning center in general takes place for a different value of A_p than the one at which a particle jumps *onto* a pinning center.[2] When A_{max} is smaller than the threshold for production of miss-coordinated particles the system return to the ideal lattice configuration as A_p is lowered back to zero.

As A_{max} becomes larger than the threshold for creating miss-coordinated particles metastable configurations are created as A_p is decreased from A_{max}. These configurations contain frozen-in topological defects. The ideal lattice configuration is not recovered in this case when A_p is lowered back to zero.

In order to quantify the order in the distorted lattice we have measured the decay of the translational order parameter $\psi(\mathbf{r}) = e^{i\mathbf{G}\cdot\mathbf{r}}$ (where $\mathbf{G}$ is a reciprocal-lattice vector $\mathbf{G} = (4\pi/a_0\sqrt{3}, 0)$) for increasing strength of the random potential. This correlation function, $g_G(r) = \langle \psi(\mathbf{r})\psi(\mathbf{0}) \rangle$, is the same as the one measured in resent experiments on the order of flux line lattices.[18] The simulations show in contrast to what has been the general belief[13, 14, 19] that the dense sharp pinning centers are not very effective in destroying the translational order. In Fig. 16 we show a log-log plot of $g_G(r)$ for different values of A_p. One observe that the decay of $g_G(r)$ is consistent with an algebraic behavior $g_G(r) \sim r^{-a}$. The exponent a depends on A_p. The exponent extracted from plots like those in Fig. 16 is shown in Fig. 17. The behavior of a is consistent with the functional form $\exp(-\text{cst.}/A_p)$ as indicated by the dotted line in Fig. 17.

The weak decay of the long range order is connected with the fact that the elastic matrix, Φ_{ij}, acquires a non-zero mass[20] in the presence of a random potential or phrased

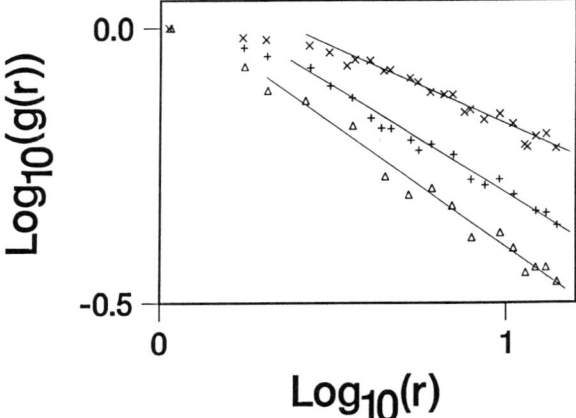

Figure 16: Double logarithmic plot of the translational correlation function for a system of size 30×34 for three different values of A_p, namely $A_p =$, 0.018, 0.038, and 0.096 from top to bottom. The slopes of the straight lines determine the exponent a. The actual slopes are -0.28, -0.39, and -0.45.

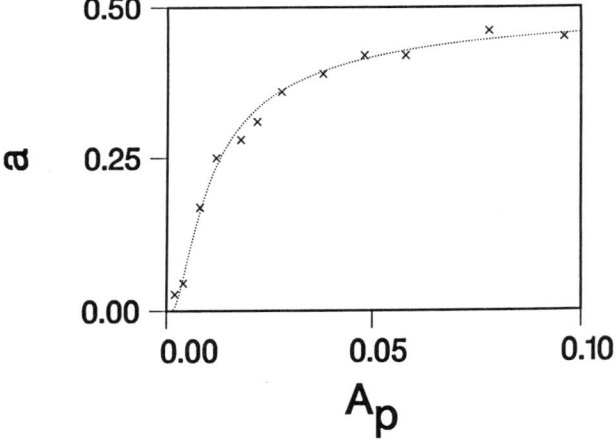

Figure 17: The exponent a of the correlation function for the same system as describe in Fig. 16 for a set of A_p values. The dotted line is given by $a = 0.5 \exp(-0.009/A_p)$.

in a different way: the random potential generates a non-zero Labusch parameter.[16] This means that the elastic response to an external stress is screened beyond a finite (phonon) localization length.[21] This prevents the formation of long range displacements which are caused by unbound topological defects. Such defects would cause an exponential decay of the translational order. Our numerical finding is consistent with the result of Ref. [22] which found that the translational order decreases slower than exponentially in two dimensions.

This behavior of the correlation functions strongly depends on the procedure used to generate the considered state. Indeed the quasi-static shift of the center of mass investigated previously[4] was able to destroy the long range order of the lattice. Moreover, the simulations show that a quench from high temperature at constant A_p easily leads to destruction of the long range translational order suggesting that in this case free dislocations are present. The energy of the quenched state is typically only slightly *higher* than the energy of the configurations obtained through the adiabatic relaxation. This might explain the observed destruction of the order in decoration experiments.[18, 23]

The strong dependence of the correlations on the history of the system makes it very interesting to compare the resulting decoration patterns obtained from different routes through the temperature and field phase-diagram of type II superconductors.[23]

3.3 Softening Induced by the Random Potential

3.3.1 Simulation of softening

The random potential makes the vortex lattice more susceptible to shear. Fig. 18 shows the measured shear modulus C_{66} as function of the pinning strength.[24]

The shear modulus is measured by simulating a shear experiment (in a mechaninical sense) in the following way. The ideal triangular lattice is relaxed to the random potential using the above described MD technique. Then the vortex system is divided into three regions A, B and C (See Fig. 19). A periodic shear wave is induced by gradually increasing the y-coordinate of all the vortices in region B at the same time as the y-coordinate of all the vortices in region C is decreased with an equal and opposite amount. The vortices in region A are always kept well relaxed by the MD method. By this procedure a strain field is induced in region A. This strain field depends only weakly on the y-coordinate, which we check by inspection of the actual displacements of the vortices in region A. The vortex-vortex energy is calculated as function of the displacement of the vortices in the boundaries B and C. An effective C_{66} is identified through

$$U_{vv} = \frac{1}{2}C_{66}\int dx \int dy \left[\frac{\partial u}{\partial x}\right]^2 = \frac{1}{2}C_{66}L_y \int dx \left[\langle\frac{\partial u}{\partial x}\rangle_y\right]^2, \qquad (24)$$

where U_{vv} denotes the total vortex-vortex energy and L_y the length of the system in the y-direction. $\partial u/\partial x$ is the derivative along the x-axis of the displacement field, which is averaged along the y-axis before it is summed along the x-direction.

As long as $A_p \leq A_{plas}$ the strain induced in region A varies slowly and an unambiguous C_{66} coefficient can be identified. The situation is different for $A_p > A_{plas}$ as the presence of miscoordinated vortices in this regime makes it difficult to measure a well defined C_{66}. The induced displacements due to the shear of the boundaries B and C varies rapidly around the defective vortices. This makes it difficult to define a strain field to be used in Eq. 24. Moreover, as A_p is increased above A_{plas}, more and more miscoordinated vortices are

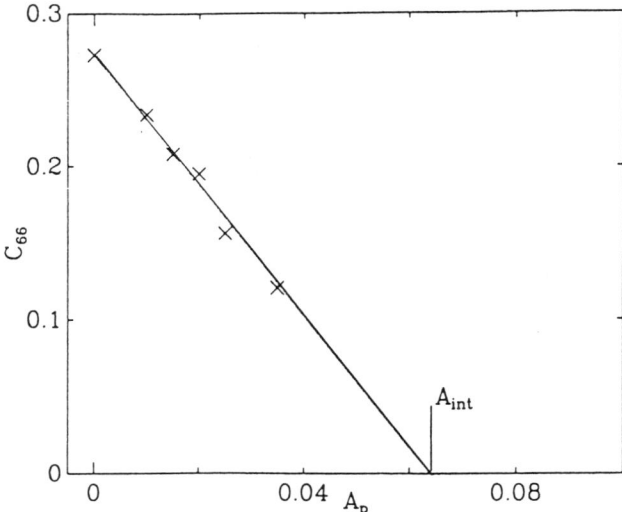

Figure 18: The effective shear modulus determined by eq. (4) as function of the pinning strength. The theoretical ideal lattice shear modulus is $C_{66} = 0.27$. The system parameters are $N_v = 1020, R_v = 0.25, N_p = 438$ and $R_p = 0.25$. The value of A_{plas} is for this specific system $\simeq 0.035$. The straight line is a best fit extrapolation.

created, which leads to a decoupling of regions B and C from the vortices in region A. No strain penetrates into region A, all the induced strain becomes localized to the interface between the regions.

As is seen on Fig. 18 the measured C_{66} values fall on a straight line. The extapolation of this line into the plastic regime above A_{plas} intersects the A_p-axis at $A_{int} \simeq 0.06$. This value of A_p coincides with the value about which the long range order vanishes in the radial distribution function of the vortex system (See Fig. 20). In this sense the vortex system begins to resemble a fluid when A_p becomes larger than A_{int}.

The behavior described above is qualitatively similar to the elastic behavior recently observed in experiments on irradiation induced amorphization[25]. In this case the measured shear modulus decreases linearly as the volume increases with the amount of disorder induced by the irradiation. The extrapolation of the linearly decreasing shear modulus intersects the volume axis at a value corresponding to the amorphous phase.

3.3.2 Analytic Study of Softening

The non-linear discontinuous behavior described so far make it clear that a perturbative calculation of the static elastic response is problematic. Nevertheless, it is very illuminating to try to cary out a first principle weak disorder perturbative calculation. We will see that up to a certain length scale perturbation theory may work. The calculation shows explicitly how the perturbation theory fails at long length scales.

The calculation of the observed softening turns out to be closely related to the generation of a finite pinning force. Below we give a brief summary of the perturbation theory.[16]

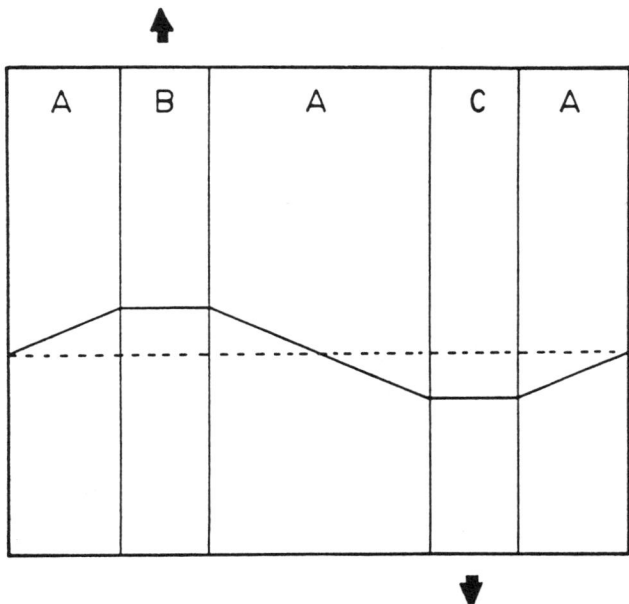

Figure 19: Schematics of the induced shear deformation. All vortices in region B are shifted in the positive y-direction while all vortices in region C are shifted in the opposite direction. The result is a periodic shear deformation since periodic boundary conditions are assumed.

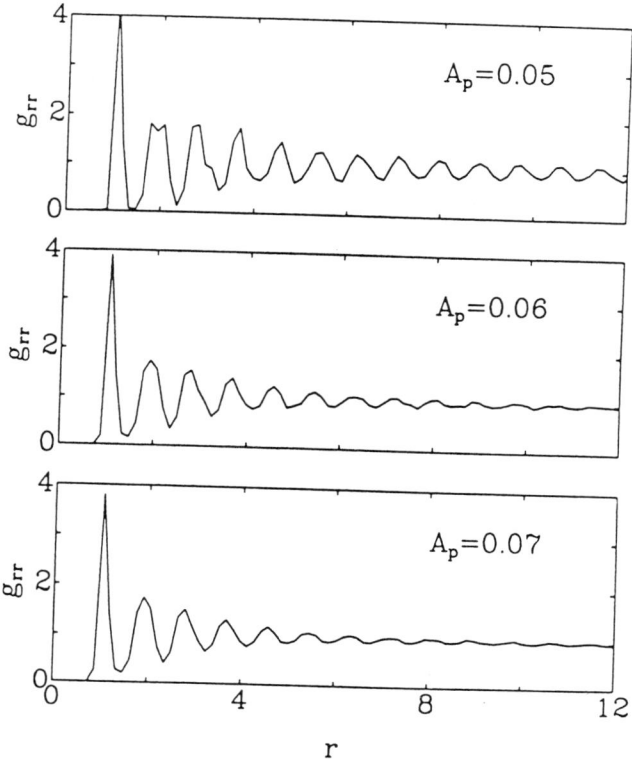

Figure 20: Radial distribution functions for pinning strengths around the value at which the extrapolated fit in Fig. 18 intersects the A_p-axis. The system parameters are as in Fig. 18.

Our starting point is the Hamiltonian

$$H = \frac{1}{2}\sum_{i,j} u_{vv}(\mathbf{r}_i - \mathbf{r}_j) + \sum_{i=1}^{N_v} V(\mathbf{r}_i) \qquad (25)$$

Here $\mathbf{r}_i$ denote the position of the N_v vortices, $u_{vv}(\mathbf{r}_i - \mathbf{r}_j)$ is the vortex-vortex interaction, and $V(\mathbf{r}_i)$ is the random pinning potential. The intention is to determine the distorted lattice configuration $\mathbf{r}_i = \mathbf{r}_i^0 + \mathbf{s}_i$ by expanding in powers of the weak random potential. This amounts to expanding the Hamiltonian in powers of the displacement $\mathbf{s}_i$ away from the ideal lattice configuration $\mathbf{r}_i^0$. It turns out that one has to expand to third order to obtain non-zero results for the Labusch parameter as well as for the correction to the shear modulus. In shorthanded notation we have in Fourier space (see Ref. [16] for details)[26]

$$\begin{aligned} H &= \int d\mathbf{q}\, \mathbf{s}\Phi\mathbf{s} + \int d\mathbf{q}\, \mathbf{V}'(\mathbf{q})\mathbf{s}(-\mathbf{q}) + \int d\mathbf{q}_1 d\mathbf{q}_2 \mathbf{V}''(-\mathbf{q}_1 - \mathbf{q}_2)\mathbf{s}(\mathbf{q}_1)\mathbf{s}(\mathbf{q}_2) + \\ &\int d\mathbf{q}_1 d\mathbf{q}_2 d\mathbf{q}_3 \mathbf{V}'''(-\mathbf{q}_1 - \mathbf{q}_2 - \mathbf{q}_3)\mathbf{s}(\mathbf{q}_1)\mathbf{s}(\mathbf{q}_2)\mathbf{s}(\mathbf{q}_3). \end{aligned} \qquad (26)$$

Here Φ is the elastic matrix of the vortex system and the tensors $\mathbf{V}', \mathbf{V}''$, and $\mathbf{V}'''$ denote the derivatives of the random potential. The displacement field $\mathbf{s}$ is determined from the condition $\partial H/\partial \mathbf{s} = 0$. The elastic properties of the disordered lattice are obtained from

$$\Phi_{dis}(\mathbf{q}) = \partial^2 H/\partial \mathbf{s}(\mathbf{q})\partial \mathbf{s}(-\mathbf{q}), \qquad (27)$$

where the right hand side is calculated for the new equilibrium configuration. The $\mathbf{V}'''$ term in equation 26 produces a finite contribution to $\Phi_{dis}(\mathbf{q})$ at $\mathbf{q} = 0$. This show explicitly that the translational invariance has been broken by the random potential and a restoring pinning force $\mathbf{F}$ has been generated:

$$\mathbf{F} = \Phi_{dis}(\mathbf{q} = 0)\mathbf{s}(\mathbf{q} = 0) \equiv \alpha_L \mathbf{s}, \qquad (28)$$

Where α_L is the Labusch parameter.[27] In our perturbative approach the averaging over the randomness allows only terms with even powers of the random potential. Accordingly we find to lowest order the Labusch parameter to be given by

$$\alpha_L = \frac{1}{2C_{66}} \frac{n_p A_p^2 n_v}{R_p^2} \ln\left[\frac{L}{a_0}\right]. \qquad (29)$$

Here C_{66} is the shear modulus of the ideal vortex lattice, n_p, A_p, and R_p the density, amplitude and range, respectively, of the pinning centers. L is a large distance cutoff, which seems natural to define as $L = min\{V^{1/2}, V_c^{1/2}\}$, V being the system size and V_c some kind of correlated volume of the vortex system.[14] From Eq. 29 we can estimate the pinning force F_p by assuming that the maximum restoring force is obtained when the correlated volume has been translated by a distance of the order of R_p. The pinning force per unit surface is then given by $F_p = \alpha_L R_p$. The estimate of F_p obtained this way agrees with the Larkin-Ovchinnikov[14] expression for the two dimensional case (except for a different exponent of the logarithm of the system size[16]).

It is important to point out that a finite Labusch parameter is only produced if one goes beyond second order in the expansion in the randomness. This is consistent with the fact that only such anharmonicities allow the existence of elastic instabilities which is a necessary condition for a nonzero pinning force.[2] On the other hand the puzzling fact is that no instabilities or discontinuities enter in any explicit way the perturbative calculation described here. Even more so, since the finite discontinuity of the displacements induced by an instability cannot be reached perturbatively.

The procedure outlined above can also be used to calculate the energy increase of the lattice relaxed to the pinning potential as a shear deformation is imposed on the boundaries of the system. We make a relative displacement of size Δ of the boundaries, say, in the y-direction. The displacement field is written as

$$s^\alpha(\mathbf{r}) = s_0^\alpha(\mathbf{r}) + \Delta[\frac{x\delta^{\alpha y}}{L} + s_1^\alpha(\mathbf{r})] + \Delta^2 s_2^\alpha(\mathbf{r}), \tag{30}$$

s_0^α, s_1^α, s_2^α are chosen with vanishing boundary conditions. Here s_o^α is the displacement coming from the relaxation to the random potential before the shear is applied. The displacement field in the presence of the shear of the relaxed lattice is calculated by minimizing the energy. The effective shear modulus is given by the coefficient to the term in the vortex-vortex energy quadratic in Δ. The result is

$$C_{66}^{eff} = C_{66} - \frac{1}{3}\alpha_L L^2. \tag{31}$$

This result calls for several comments. First, the result depends on an even power of A_p, namely A_p^2, as mentioned above this will be the case to any finite order in the random potential. However, the numerically observed dependence is linear in A_p. Secondly, C_{66}^{eff} has a very strong size dependence. The result has only meaning when C_{66}/α_L is much larger than the system size. The Larkin-Ovchinnikov correlated volume[14], V_c, can be expressed as $V_c \simeq C_{66}/\alpha_L$, this implies that perturbation theory breaks down as the correlated volume becomes smaller than the system size. The effective shear modulus can then be calculated from the following argument.[16] Assume that the system breaks into correlated volumes of size V_c separated by thin boundary regions only one lattice distance a_0 thick. The respective volume fraction of the boundary regions and the correlated volumes are f and $1-f$, respectively, where

$$f \sim \frac{a_0 V_c^{1/2}}{V_c} = \frac{a_0}{V_c^{1/2}}. \tag{32}$$

In this picture the whole system is like a composite consisting of rather rigid regions (the correlated regions) with a shear modulus equal to C_{66}, separated by soft regions (the boundary layers) with a shear modulus equal to C_{66}^b. The effective shear modulus of the composite is[28]

$$C_{66}^{eff} = fC_{66}^b + (1-f)C_{66}. \tag{33}$$

Since we assume $C_{66} \gg C_{66}^b$, we have

$$C_{66}^{eff} \approx (1 - a_0 V_c^{-1/2})C_{66} = (1 - a_0[\frac{\alpha_L}{C_{66}}]^{1/2})C_{66} \tag{34}$$

where α_L is given in Eq. 29. This expression does depend linearly on the absolute value of A_p. The coefficient of A_p obtaned from Eq. 34 after substitution of α_L from Eq. 29 is equal to 4.0 for the parameners used for the system described in Fig. 18. The slope of the straight line in Fig. 18 is 4.3. This good agreement is ammusing, although, probably fortuitous.

4 Quasistatic Response

4.1 Simulation of the quasistatic response

In this section we discuss the effect of the random potential as the we move the lattice through the potential right at threshhold for motion. Hence in every instance we have $f_{dr} = F_p$. Again we consider the system at zero temperature.[4, 5]

When the relaxed state has been reached as described in the previous chapter, we shift the center of mass (COM) of the vortices through the pins in the $\langle 10 \rangle$ direction by shifting all the vortices by a small increment $dx = 10^{-4}$ to 10^{-3} (the smaller increment is used for the strongest pins). We then let the vortices relax by the MD procedure for 10 time steps, keeping the temperature fixed at the low temperature reached in the initial cooling. The COM is kept fixed during this relaxation by subtracting the COM velocity at the end of each time step. This procedure amounts to applying an external homogeneous driving force so as to precisely balance out the pinning force at any instant. The procedure is therefore equivalent to a qusistatic shift of the vortex system. We continue the procedure of successively shifting and relaxing the lattice until the COM has been shifted a distance between 0.25 to 1.2 lattice spacings.

The changes in the potential energy, caused by a shift, are given by

$$\Delta U = \sum_i \frac{\partial U}{\partial \vec{r}_i} \cdot d\vec{x}_i = -\sum_i \frac{\partial V_p}{\partial \vec{r}_i^{(p)}} d\vec{x} = -\vec{F} \cdot d\vec{x} \tag{35}$$

($\vec{F}$ being the total pinning force) since all vortices are initially subject to the same incremental shift dx. For sufficiently complete relaxation, we always obtain a pinning force equal to the derivative of the total potential energy $U(X)$ with respect to the COM position, X. We consider this to be a check of our numerical method.

4.2 Deformations due to the quasistatic response

The vortex lattice is found to exhibit three qualitatively different types of behavior as the center of mass of the vortices is shifted over the pins. For weak pins the response of the vortex lattice is entirely elastic, the average pinning force over a shift of one lattice spacing being several orders of magnitude smaller than the maximum pinning force measured over the shift. For stronger pins (or higher pinning density, or a higher value of R_v, which is equivalent to a smaller value of C_{66}, or a larger system) the average pinning force becomes comparable with the maximum pinning force. Although none of the vortices become trapped on the pins for the entire shift of the center of mass by a lattice spacing, some vortices are briefly trapped causing elastic instabilities. Finnaly, for even stronger pins, some vortices are permanently trapped on the pinning centers. With some of the vortices fixed, the remaining vortices must flow plastically around them as the center of mass of the vortex lattice is shifted.

Fig. 21 shows graphs of $\bar{F}$ and $F_{max}/\bar{F}$ as a function of A_p, where $\bar{F}$ and F_{max} are the average and maximum pinning force measured over a shift of the center of mass of the vortices by one lattice spacing. From these graphs it can be seen that the value of A_p at which $F_{max}/\bar{F}$ falls to a constant is lower than the value of A_p at which the value of $\bar{F}$ begins to rise rapidly. By examining the vortex configurations it is found that the rapid rise

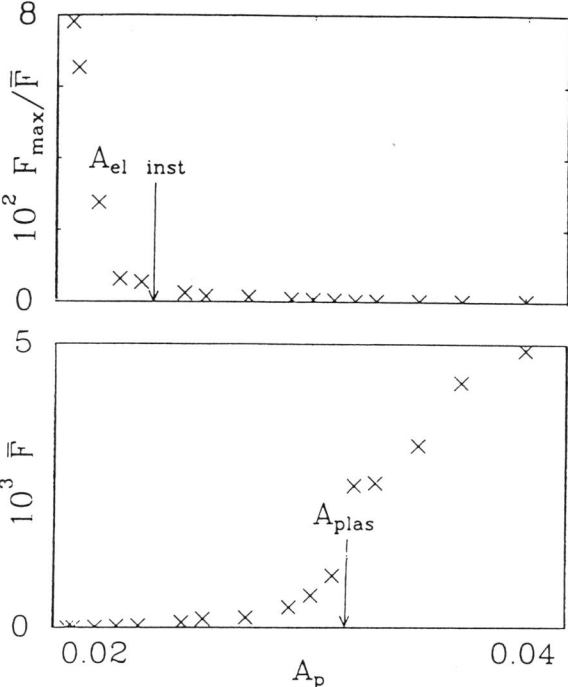

Figure 21: $\bar{F}$ and $F_{max}/\bar{F}$ as function of A_p for a system of 340 vortices with $R_v = 0.6$, $R_p = 0.25$, and $N_p = 146$.

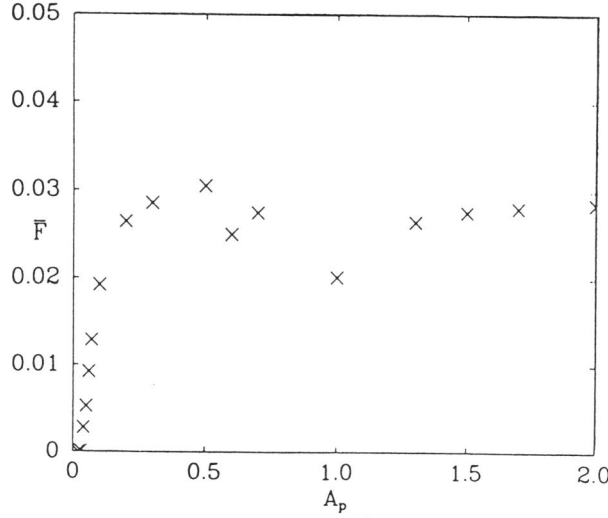

Figure 22: $\bar{F}$ as a function of A_p for a larger range of A_p than in Fig. 21.

in $\bar{F}$ is associated with the onset of plastic flow and the trapping of vortices (for $A_p > 0.032$ at least one vortex was trapped for the entire shift of the lattice). Similarly the rapid fall in the value of $F_{max}/\bar{F}$ can be associated with the crossover from the elastic regime into the elastic instability regime. For a given set of values R_p, R_v, the pinning density and the system size, we can therefore define two quantities: $A_{el.inst}$, which labels the values of A_p at which the lattice crosses over from the elastic regime to the elastic instability regime, and A_{plas}, which labels the value of A_p at the crossover between the elastic instabilitiy region and the plastic regime. Fig. 22 shows how the pinning force varies with A_p for a large range of values of A_p.

Below we briefly discuss the characteristica of the three different regimes.

4.2.1 The elastic region

For weak pinning strength $A_p \leq A_{el.ins}$ the distortion of the vortex lattice is elastic and reversible. All quantities vary smoothly and periodically as functions of the COM position with a period equal to the ideal lattice spacing, and thus the pinning force averaged over shift, $\bar{F}$, is zero. The value of $A_{el.ins}$ decreases with increasing system size, approximately as the square root of the system size, and will eventually vanish for an infinite system.

4.2.2 The elastic instability region

For somewhat stronger pins $A_{el.ins.} \leq A_p \leq A_{plas}$, we find a narrow region where the pins are strong enough to trap a vortex temporarily. As the strain builds up locally in the lattice around the trapped vortex, an elastic instability[29] occurs when the pinned vortex suddenly jumps off the pinning center and a new vortex jumps onto the pin. Even though the instability occurs on one pinning center it is collectively induced since $A_{el.ins.}$ depends

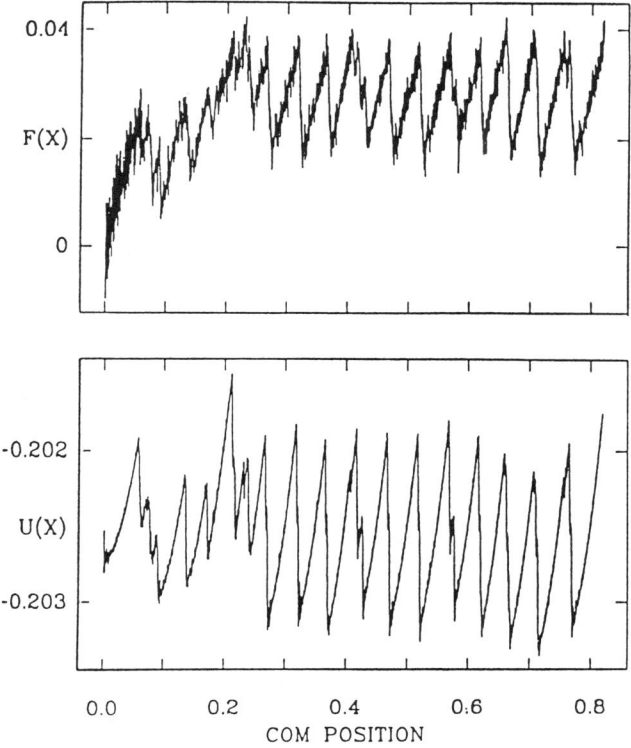

Figure 23: The potential energy $U(X)$ and the pinning force $F(X)$ normalised by the number of vortices for the following set of parameters: $N_v = 340, R_v = 0.60, N_p = 146, A_p = 1.0$ and $R_p = 0.25$. The high frequency noise is caused by the thermal fluctuations.

on the density of the pinning centers as $A_{el.ins} \sim 1/\sqrt{n_p}$ Due to these instabilities the potential energy, although it is still periodic with periodicity equal to the lattice spacing, contains discontinuties and $\bar{F}$ becomes finite in this region[5].

4.2.3 Plastic flow region

For $A_p \geq A_{plas}$ the pins are sufficiently strong to trap individual vortices permanently, and plastic flow of the lattice around the fixed vortices becomes important. The value of A_{plas} decreases inversly proportionaly to the square root of the density of the pinning centers and decreases logarithmically with increasing system size. Thus, A_{plas} is also a collective effect of the pinning centers. Recently Coppersmith has argued that plastic deformations or phase slips always will occur in (infinitely large) pinned charge density wave systems.(See the paper by Coppersmith in Ref. number [13] and her contribution in the present proceeding) The characteristic features of this region are precipitous drops in the potential energy $U(X)$ connected with a dramatic reordering of the vortex lattice. As the COM of the vortices is shifted, energy is pumped into the system, and the total

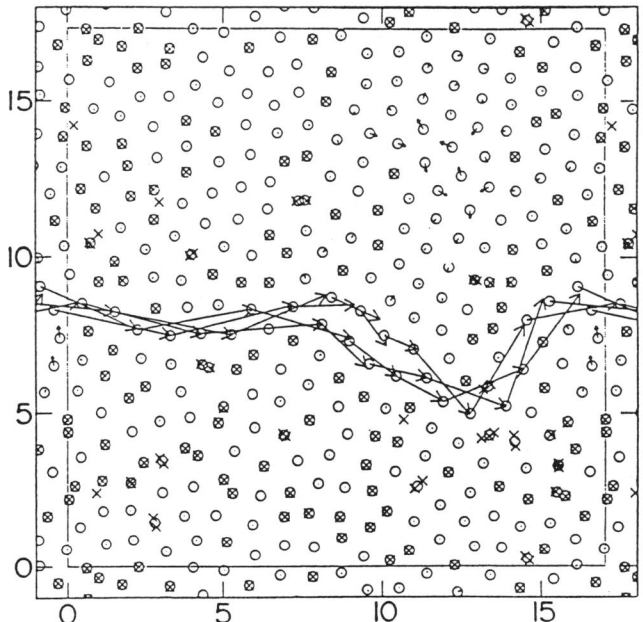

Figure 24: Flow channels through the vortex lattice. Crosses are the fixed pins, circles are vortices. The arrows show how the vortices have moved while the COM has been shifted from $X = 0.24$ to 0.42. The configuration corresponds to the plots in Fig. 23.

potential energy increases approximately quadratically, with a corresponding linear increase in the force. See Fig. 23 The energy is released when some of the vortices suddenly start to flow plastically relative to the rest of the lattice (see Fig. 24). This flow takes place through *channel*-like paths, in between pinned regions of the lattice. Fig. 25 and Fig. 26 show how the width of the channel depends on the strenght of the pins A_p.

We have analyzed the defects of the vortex lattice by calculating the the coordination number of all the vortices[15]. Defects in the vortex lattice show up as vortices with coordination numbers different from 6. For $A_p \leq A_{plas}$ no defects are observed whereas for $A_p \geq A_{plas}$ we always find a *highly* defective lattice (See Fig. 27). In Fig. 28 is shown an example of how the miscoordinated vortices are distributed through the lattice.

5 Above-threshold Driving Force

5.1 Simulation of Flux Flow

In order to simulate flux flow we add an external driving force $\mathbf{f}_{dr}$ to all the vortices. It is belived that flux lines in type II superconductors are subject to diffusive dynamics,[3] hence we assume the following equation of motion

$$\eta \frac{d\mathbf{r}_i}{dt} = -\frac{\partial U}{\partial \mathbf{r}_i} + \mathbf{f}_{dr}. \tag{36}$$

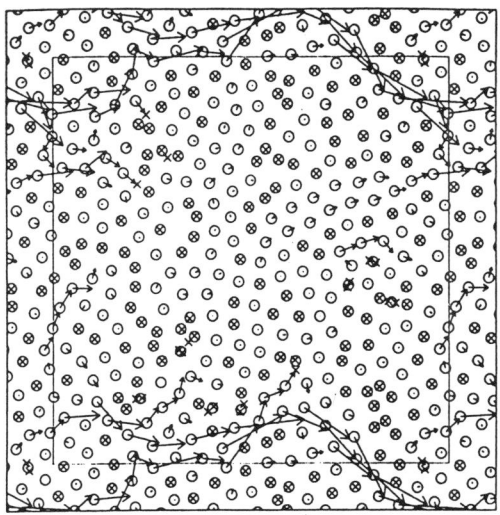

Figure 25: Vortex motion as the lattice is shifted. Crosses mark the fixed pins and cirkle the vortices. The arrows sjow the motion of the vortices as X is increased from 0 to 0.125. The system parameters are $N_v = 340$, $R_v = 0.6$, $N_p = 146$, $A_p = 1.0$, and $R_p = 0.25$.

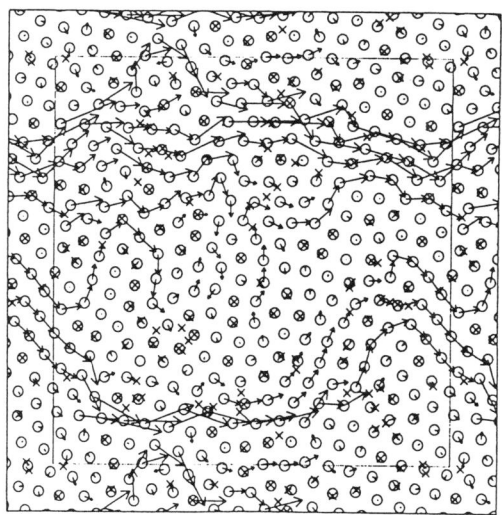

Figure 26: The vortex flow pattern for a system with the same parameters as in Fig. 26 except that $A_p = 0.07$.

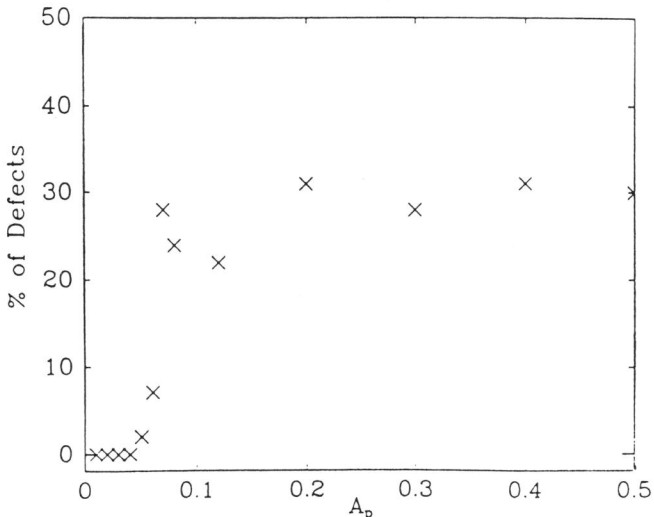

Figure 27: % of defective vortices as a function of A_p for a system of 108 vortices with $R_v = 0.6$, $R_p = 0.25$ and $N_p = 46$.

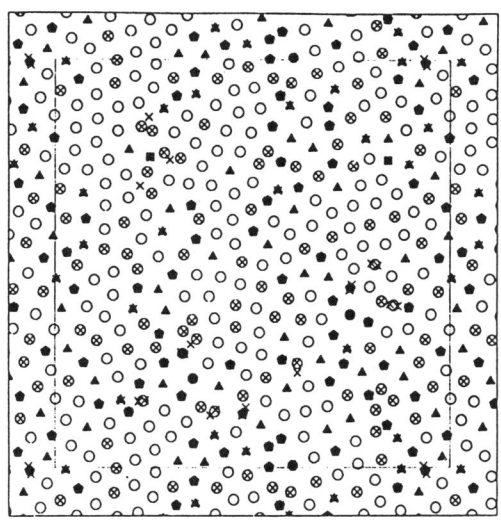

Figure 28: The initial relaxed configuration used in the run desrcibed by Fig. 25. Crosses show the fixed pins, circles the six coordinated vortices, filled pentagons the five coordinated vortices, filled triagles the seven coordinated vortices, filled squares the four coordinated vortices, and filled octagons the eight coordinated vortices.

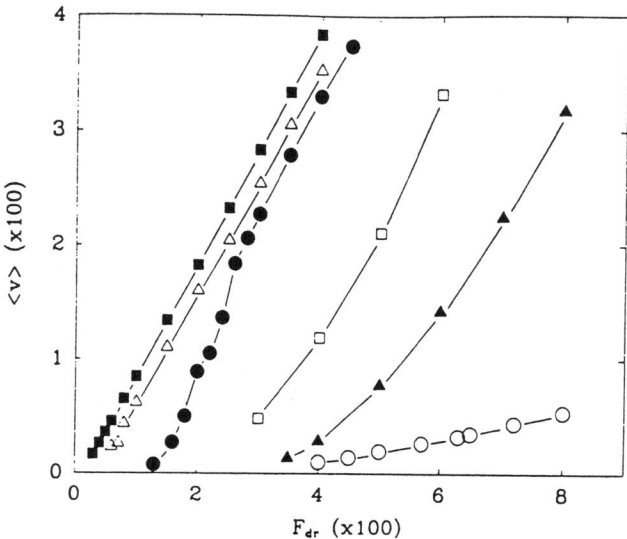

Figure 29: COM velocity versus applied driving force for different values of the pinning strength: ■ $A_p = 0.025$, △ $A_p = 0.04$, ● $A_p = 0.05$, □ $A_p = 0.07$, ▲ $A_p = 0.1$ and ○ $A_p = 0.5$. The system parameters are $N_v = 1020, R_v = 0.6, N_p = 438$ and $R_p = 0.25$.

The time scale is set by choosing the friction coefficient $\eta = 1$.

We first set the driving force to zero and relax an ideal triangular lattice to the random potential by using the above described MD annealing technique. We then apply a constant homogeneous driving force larger than the threshold force, F_T, and follow the vortex system as it evolves according to Eq. 36. The system is followed until the center of mass has shifted at least three lattice spacings. The center of mass velocity, $<v>$, time averaged over the run, is then measured. This corresponds to measuring the I-V characteristics of a type II superconductor in which case the average center of mass velocity of the flux lines is proportional to the voltage and the driving force is proportional to the current.[30]

5.2 Voltage-Current Characteristics

We have summarized our results in Fig. 29. The velocity versus driving force are shown for a range of values of the amplitude of the pinning centers. One can clearly see how a non-linear region starts to develop for $A_p \geq A_{cr} \simeq 0.04$. The value of A_{cr} is indistinguishable from A_{plas}. Above A_{plas}, some vortices remain trapped on pins as the rest of the vortex system begins to flow. The cross-over at A_{cr} also coincides with a steep increase in the threshold pinning force.(See Chap. 4 and ref. [5] In the linear regions of the $<v> -F_{dr}$ curves, i.e., for all driving forces for $A_p \leq A_{cr}$ and for large driving forces in the regime $A_p \geq A_{cr}$, the moving vortices form a well defined lattice. A fluid-like form of the radial distribution function $G(r)$ is observed in the non-linear regions of the $<v> -F_{dr}$ curves, whereas $G(r)$ is always solid-like in the linear regions (See Fig. 30). These findings agree well with neutron scattering experiments on moving FL lattices in type II superconductors[31]. All

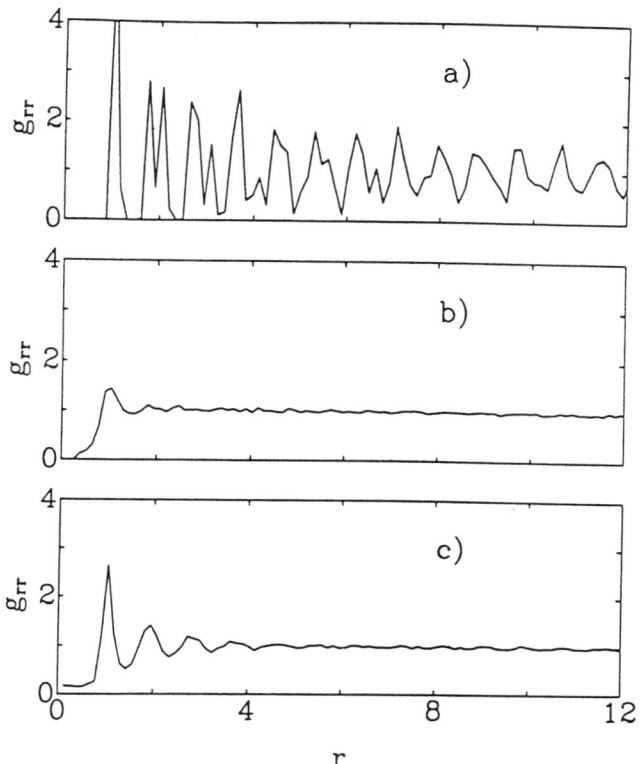

Figure 30: Radial distribution functions for the moving vortex system for different pinning strengths and driving forces. a) $A_p = 0.025$ and $F_{dr} = 0.01$ corresponds to a $<v> - F_{dr}$ curve (se Fig. 29) with no non-linearity. b) $A_p = 0.5$ and $F_{dr} = 0.5$ is in the non-linear regime of a $<v> - F_{dr}$ curve above A_{plas}. c) $A_p = 0.5$ and $F_{dr} = 50$ corresponds to the same $<v> - F_{dr}$ curve but in the linear regime at high driving force.

this strongly indicates that the random potential induces a considerable softening in the vortex lattice when the potential reaches a certain strength, $A_{cr} \simeq A_{plas}$. It should be emphasized that the amplitude A_{cr} is much smaller than the amplitude needed for a single pin to be able to induce elastic instabilities or trapping of vortices[5].

It is worth noticing the similarity between the curves in Fig. 29 for which $A_p \geq A_{cr}$ and the I-V characteristics for type II superconductors. An example of the latter[32] is shown in Fig. 31. The field dependence of the I-V characteristics is consistent with the A_p dependence of the curves in Fig. 7. The pinning potential in a type II superconductor scales with magnetic field as $1 - b$, where $b = B/B_{c_2}$ is the reduced field, B the internal field, and B_{c_2} the upper critical field.[33] The shear modulus C_{66} of the flux line lattice depends on the magnetic field as $b(1-b)^2$ (see Ref. [34]). So the pinning strength becomes relatively weaker as b is increased for small b.

It has been suggested by Fisher[35, 17] that the non-linear behavior connected with the

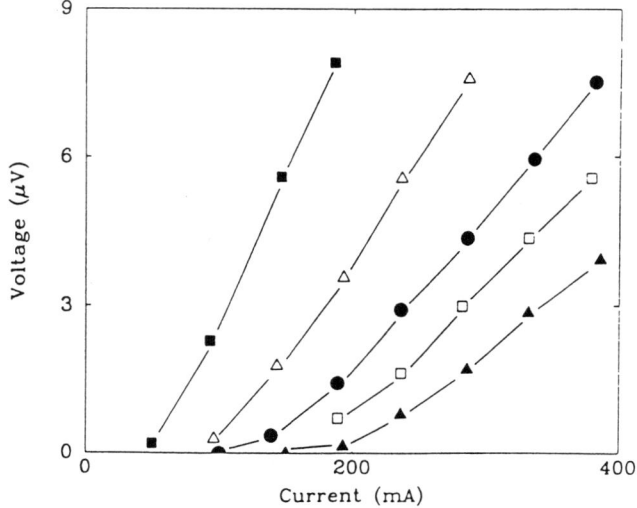

Figure 31: Experimental I-V characteristics for different values of the magnetic field. The signature is as follows: ■ $b = 0.74$, △ $b = 0.55$, ● $b = 0.46$, □ $b = 0.41$ and ▲ $b = 0.37$.

depinning phenomena in the case of weak pinning, at least in the case of charge density waves, might be an example of a dynamical critical phenomenon, and experiments on non-linear I-V characteristics in type II superconductors have been interpreted along these lines[36]. If such an interpretation is correct one would expect the correlation length of the velocity-velocity correlation function to diverge as the threshold force is approached from above[35, 17]. We have not been able to find any sign of an increase in the velocity-velocity correlation length. The absence of any critical behavior in this regime might not be surprising, given the fluid-like flow in the non-linear regions.[37] Nor did we find any critical behavior in the weak pinning regime below A_{cr}. We would therefore rather interpret the experimentally observed non-linear I-V characteristics as an indication of the presence of pinning which is strong in the sense that the flux line lattice is plastically deformed and flows in a fluid-like manner with some vortices remaining pinned.

6 Onset of Difussion

So far we have only considered zero temperature. In this section we study the combined effect of thermal fluctuations and the random pinning potential. To make the model slightly more realistic with respect to flux systems we leave our repulsive Gausian particles. Instead we consider particles which interacts with the potential pertinent to straight parallel flux lines. In place of the molecular-dynamics annealing used in the previous sections we use Langevin dynamics to simulate the over-damped dynamics at finite temperature relevant to vortex motion.[38]

6.1 Model

There are a multitude of motivations for finite temperature simulations. It has become apparent that the electrical properties of high temperature superconductors in magnetic fields are strongly influenced by the finite temperature behavior of the Abrikosov flux line system in these materials. Many interesting features have been observed. The resistive transition becomes very broad when a magnetic field is applied [39, 40]. An irreversibility line is found in the field-temperature phase diagram [41, 42]. Flux creep rates are unusually high [42], and thermally assisted flux flow has been proposed as an explanation for these high rates [43]. Aging and memory effects also seem to have been observed [44]. Although the experimental evidence for melting of the flux line lattice [45] is not conclusive [46], theoretical considerations suggest that melting might take place at temperatures low compared to the superconducting transition temperature [47].

All the above–mentioned phenomena are related to the mobility of the flux lines. This problem is complicated by the fact that due to inhomogeneities in the superconducting material, the flux lines always experience a random potential background. The mobility is thus determined by the combined effect of the random potential and thermal fluctuations.

We have simulated the onset of diffusion in the flux line lattice in a thin film superconductor. We prefer to discuss our results in terms of mobility rather than in terms of melting. The reason is that, as discussed in Chap. 3, even a weak random potential in an infinite system at zero temperature can destroy the order of the lattice, resulting in an amorphous glass–liquid like (static) structure.

In this section, the temperature, T_D, at which vortices start to become mobile is studied as function of applied magnetic field and strength of the random pinning potential. We find that *weak* randomness *reduces* T_D significantly, and that the diffusion takes place in channels along grain boundaries in the flux line lattice. The random potential facilitates the creation of these grain boundaries and thereby assists the thermal fluctuations in making the vortices diffuse.

Details of the simulation method have been published previously in [48]. We consider a set of vortices with velocities $\mathbf{v}$ determined by an over–damped diffusive equation of motion

$$\mathbf{v} = \frac{1}{\eta}\mathbf{F} + \chi, \tag{37}$$

where η is the friction (or viscosity) coefficient, and $\mathbf{F}$ is the total force on a vortex due to the other vortices and the random potential. The function χ is a Gaussian white-noise velocity which models the coupling to a heat bath at a given temperature T.

The stochastic term in Eq. 37 modelling the interaction with a¡ heat bath needs to be calibrated in such a way that we reach the correct equilibrium distribution at long times. As in the usual case of a Langevin equation for particles with an inertial mass,[49] this is done using the Fokker-Planck equation associated with Eq. 37. Let

$$P(x, t|x_0, t_0) = P(x, t - t_0|x_0) \tag{38}$$

denote the probability that a vortex is found at position x at time t if it were at position x_0 at time t_0 (for notational simplicity we discuss the one dimensional version of Eq. 37). P will obey the differential equation:

$$\frac{\partial P}{\partial t} = \sum_{n=1}^{\infty} \frac{(-1)^n}{n!} \frac{\partial^n}{\partial x^n}[M_n P(x,t|x_0)]. \tag{39}$$

where
$$M_n = \frac{1}{\tau} \int \zeta^n P(x+\zeta,\tau|x) d\zeta, \qquad (40)$$

We can make the connection between Eqs. 39 and 37 by replacing the M_n in Eq. 39 by the moments as calculated directly from Eq. 37 to $O(dt^2)$. In order to calculate these moments we assume that χ_i can be expressed as a Gaussian white noise function:

$$<\chi_i(t)> = 0, \qquad (41)$$

$$<\chi_i(t_1)\chi_k(t_2)> = A\delta(t_1-t_2)\delta(i-k), \qquad (42)$$

where A is a constant which will be evaluated later. Calculating the moments from Eq. 37,

$$M_1 \equiv \frac{1}{\tau}<\delta x> = \frac{1}{\eta}F \text{ and } M_2 \equiv \frac{1}{\tau}<[\delta x]^2> = A, \qquad (43)$$

and substituting them into Wq. 39 we obtain the following Fokker-Planck equation:

$$\frac{\partial P}{\partial t} = -\frac{\partial}{\partial x}\left[\frac{1}{\eta}F\, P(x,t|x_0)\right] + \frac{1}{2}\frac{\partial^2}{\partial x^2}\left[A\, P(x,t|x_0)\right]. \qquad (44)$$

The value of A can be determined by demanding that

$$P(x,\infty|x_0) \propto exp[-(1/kT)U(x)], \qquad (45)$$

where $U(x)$ is the total potential energy of the vortices at $t=\infty$. Substituting Eq. 45 into Eq. 44 we finally obtain:

$$A = 2kT/\eta. \qquad (46)$$

Special care must be taken in deriving the discrete form of the Gaussian white noise term in Eq. 37. This can be seen by Considering the second moment $\langle(\Delta x)^2\rangle$ where the average is over a small time interval Δ. We have according to Eq. 37

$$\langle(\Delta x)^2\rangle = \int_0^\Delta dt_1 \int_0^\Delta dt_2 [\frac{1}{\eta}F(t_1) + \chi(t1)][\frac{1}{\eta}F(t_2) + \chi(t_2)] \qquad (47)$$

$$\approx (F(0)/\eta)^2\Delta^2 + \int_0^\Delta dt_1 \int_0^\Delta dt_2 \langle\chi(t_1)\chi(t_2)\rangle$$

According to Eq. 42 we obtain

$$\langle(\Delta x)^2\rangle = (F/\eta)^2\Delta^2 + A\Delta \qquad (48)$$

Thus, the stochastic part of Eq. 37 dominates as the time step Δ is becoming small. The discrete version of the Langevin equation has, accordingly, a tendency to be completly dominated by the noise term. This problem can be circumvented in the following way.

We assume that the noise term acts with a constant average rate, $1/\tau$, on each particle.[50] The probability that a given vortex will have been acted on by the noise term during a time step of length Δ is $p \approx \Delta/\tau$ for $\Delta \ll \tau$. As we are using a discrete time variable we need to find a representation for the noise term which satisfies Eqs. 42 and 43 (with A replaced by $2kT/\eta$). It is also important that the equilibrium temperature be independent of τ. The following prescription[50] for the noise term satisfies all the above requirements:

$$\chi(t) = B \sum_j \delta(t - t^j) \gamma(t^j) \theta(p - q^j), \qquad (49)$$

where j labels the time step, $\gamma(t^j)$ is a random number chosen from a Gaussian distribution of mean 0 and width 1, q^j is a random number uniformly distributed on $[0, 1]$, and $\theta(x)$ is defined by:

$$\theta(x) = \begin{cases} 1, & \text{if } x > 0; \\ 0 & \text{if } x < 0. \end{cases} \qquad (50)$$

Substituting Eq. 49 into Eq. 42 gives us that:

$$B = \left(\frac{2\Delta kT}{\eta p}\right)^{\frac{1}{2}}. \qquad (51)$$

To make the simulation computationally manageable we consider straight, parallel flux lines, in which case the vortex-vortex force is determined by the potential [6]

$$U_{vv}(\mathbf{r}_i) = \epsilon_0(0)(1-b)(1-t^4) \sum_{j \neq i} [K_0(r_{ij}\sqrt{(1-b)}/\lambda) - K_0(r_{ij}\sqrt{(2-2b)}/\xi)] \qquad (52)$$

where the energy scale $\epsilon_0(0) = \phi_0^2/(8\pi^2 \lambda(0)^2)$ (ϕ_0 is the flux quantum and $\lambda(0)$ is the zero temperature penetration depth). $b = B/B_{c_2}(T)$ is the reduced magnetic field at the temperature T, $t = T/T_c$ is the reduced temperature (T_c is the superconducting transition temperature), K_0 is a modified Bessel function and ξ the Ginzburg-Landau coherence length. The random potential is modeled by a set of randomly positioned pinning centers of density n_p at position $\mathbf{R}_p$. For concreteness we use a Gaussian form for the individual pinning wells

$$U_{vp}(\mathbf{r}_i) = -A_p \sum_p \exp(-\left[\frac{|\mathbf{r}_i - \mathbf{R}_p|}{\xi}\right]^2). \qquad (53)$$

We have chosen the radius ($= \xi$) of the flux line core as the range of the pinning centers (point pinning) and for the amplitude we take a fraction, σ, of the condensation energy stored per length in a cylinder of the size of the flux line core [51]

$$A_p = \sigma \frac{B_{c_2}^2}{16\pi\kappa^2}(1-b)\pi\xi^2 \qquad (54)$$

where $\kappa = \lambda/\xi$. It should be emphasized that the temperature enters both through the stochastic term in Eq. 37 and through the phenomenological temperature dependance of all the superconducting parameters.[48, 52]

The final stage in the derivation of the numerical integration algorithm is to decide on the sizes of Δ and τ. This can be done by ensuring that the moments calculated numerically agree with those given in Eq. 43 (to $O(\Delta^2)$). Two conditions have to be fulfilled[48]

$$\frac{\tau}{\tau_0(0)} \gg \frac{2kT_c}{d\epsilon_0(0)} \left(\frac{f_0(0)}{\bar{f}}\right)^2 t \qquad (55)$$

$$\frac{\Delta}{\tau_0(0)} \ll \frac{2kT_c}{d\epsilon_0(0)} \left(\frac{f_0(0)}{\bar{f}}\right)^2 t$$

where $\bar{f}$ denotes the average net deterministic force on a vortex. Here the unit of time, $\tau_0(0)$, is given by

$$\tau_0(t) = 8\pi\lambda^4\eta/\phi_0^2 = \tau_0(0)/(1-t^4)^2 \tag{56}$$

and the unit of force, $f_0(0)$, is define according to

$$f_0(t) = \phi^2/(8\pi^2\lambda^3) = f_0(0)(1-t^4)^{3/2}. \tag{57}$$

In the program we explicitly calculate the second moment of the Fokker-Planck and check that it has the expected equilibrium value.

The important energy scales in the problem are the vortex-vortex interaction energy and the superconducting transition temperature $k_B T_c$ (k_B is Boltzmann's constant). We consider a system with [53] $k_B T_c/d\epsilon_0(0) = 2.5 \cdot 10^{-3}$, d being the thickness of the sample, and $\kappa(0) = 2$. This choice of parameters is dictated by numerical requirements: small number of interacting neighbors [54], and reasonable speed of the stochastic dynamics [48]. A typical high temperature superconducting sample with $d = \lambda$ should be described by $k_B T_c/d\epsilon_0(0) \simeq 2 \cdot 10^{-4}$ and $\kappa(0) \simeq 100$. A conventional superconductor like Niobium-Germanium has $k_B T_c/d\epsilon_0(0) \simeq 7 \cdot 10^{-6}$ and $\kappa(0) \simeq 60$. Although, the parameter set we consider does not directly describe a specific material, we find it useful as a model for the study of the effect of the random potential on the mobility of the vortices.

6.2 $T - A_p$ Phase Diagram

We show in Fig. 32 the onset of diffusion for a flux line lattice, consisting of 340 flux tubes in the presence of 170 randomly placed pins, as a function of temperature and strength of the random potential. The external magnetic field is kept constant [55] at $b(0) = B/B_{c_2}(0) = 0.1$. The criteria of the onset of diffusion are that the mean square displacement approaches the normal diffusive behavior (Fig. 33); and that the correlated string–like motion occurs (Fig. 34).

One sees clearly that even a weak random potential has a dramatic effect on the mobility of the vortices. The flux line lattice is softened by the disorder. The same effect has been observed at zero temperature (see Chap. 3 and Ref. [24], where numerical measurements of the shear modulus of a model system showed a linear decrease of the shear modulus as the amplitude of the random potential was increased. This decrease in the shear modulus reduces the barrier for diffusion. The diffusion onset temperature $T_D(\sigma)$ saturates as the pinning strength σ exceeds about 0.4.

One might have expected that the strong random potential, since it consist of attractive centers, would have increased the diffusion temperature due to its pinning effect. The saturation behavior of T_D may be due to the fact that the density of pins used in this particular simulation is less that the density of the vortices. Since each pinning center can trap only one vortex, the number of mobil vortices approaches a constant value for strong pinning potentials. These mobil vortices diffuse by overcoming an energy barrier which is determined by the density of the pinned vortices. Therefore for low pinning density and strong pinning strength, one expects that the diffusion onset temperature T_D is a function of the pinning density only, and that $T_D(n_p)$ increases with increasing n_p. For $n_p > n_v$ each pin will trap a vortex in the strong pinning limit and $T_D(\sigma) \sim \sigma$ for large σ.

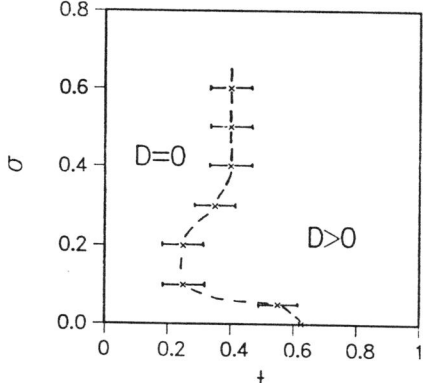

Figure 32: Diagram of the diffusive and non-diffusive regions for a fixed external magnetic field $b(0) = 0.1$. The bars indicate the temperature T_D at which the vortices start to diffuse for a given value of the amplitude of the random potential. The dotted line is a guide to the eye. The system consists of 340 vortices and 170 pinning centers. Periodic boundary conditions are used.

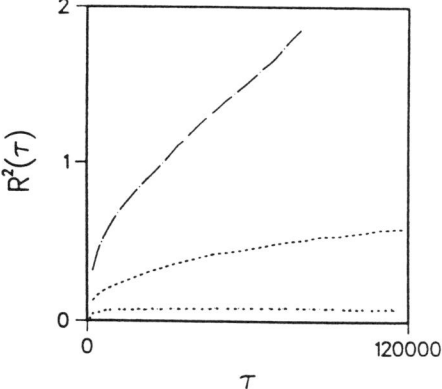

Figure 33: The mean square displacement as function of time for three different values of temperatures ($t = 0.1, 0.3$, and 0.7 from bottom to top) at a given amplitude of the random potential $\sigma = 0.3$. The magnetic field is $b(0) = 0.1$.

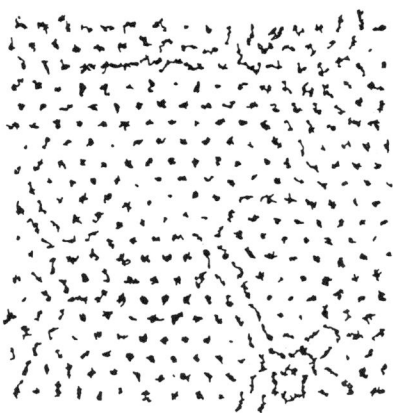

Figure 34: Plot of the trajectories of the vortices for the same system as in Fig. 1. The temperature is $t = 0.3$, diffusion has just started. The field and random potential is $b(0) = 0.1$, $\sigma = 0.1$.

In Fig. 33 we show the mean square displacement $R^2(\tau) = <[\mathbf{r}_i(\tau + \tau_0) - \mathbf{r}_i(\tau_0)]^2>$ as function of τ. The average is over the vortices and over τ_0. The presence of a random potential cause $R^2(\tau)$ to behave as τ^α with $\alpha \sim 0.2$ to 0.4 at times shorter than a cross-over time τ_{cr}. For times longer than τ_{cr} normal diffusive behavior is observed, i. e., $R^2(\tau) \propto \tau$. The cross-over time τ_{cr} increases with increasing σ and decreases with increasing temperature. The anomalous slow diffusion at short times is probably due to the restricting effect of the local minima in the random potential. [56] At longer times the vortices have experienced many thermal fluctuations among which many are large compared to the fluctuations in the random potential. Hence the effect of the random potential is lost at long times.

The diffusion is exponentially activated as is seen from Fig. 35. Here we plot the logarithm of the diffusion constant, D, as a function of the inverse temperature. The slope of the dashed line indicates that the temperature dependance of D follows a form $D = D_0 \exp(-\Delta E/k_B T)$ with an activation energy $\Delta E \simeq 2 \cdot 10^{-3} d\epsilon_0(0)$. The activation energy measured this way decreases for small σ ($0 \leq \sigma \leq 0.1$). As σ becomes larger than 0.1, ΔE starts to increase slowly. The cross-over from decreasing to slowly increasing $\Delta E(\sigma)$ presumably takes place when the fluctuations $\Delta E_p = [<U_{vp}^2> - <U_{vp}>^2]^{1/2}$ in the pinning potential become equal to the vortex-vortex energy barrier, ΔE_{bar}, for a string of vortices to move past the surrounding vortices. This barrier is connected with the shear properties of the lattice and hence is expected to decrease for small increasing σ due to the softening of the lattice.[24, 16]. An upper bound of ΔE_{bar} can be estimated from the potential barrier (per vortex) for a string of vortices to move in, say, the $<10>$ direction in the ideal lattice configuration. Using Eq. 52 we have (for $b = 0.1$) $\Delta E_{bar} \simeq 5 \cdot 10^{-3} d\epsilon_0(0)$. The fluctuations in the pinning potential are easily estimated if one neglects the distortion of the lattice. The result is

$$\Delta E_p = [\frac{\pi n_p}{2}]^{\frac{1}{2}} \xi \frac{(1-b)\sigma}{8} d\epsilon_0(0). \qquad (58)$$

For $b = 0.1$ we have $\Delta E_p = \Delta E_{bar}$ at $\sigma \sim 0.4$ this value compares reasonably with the

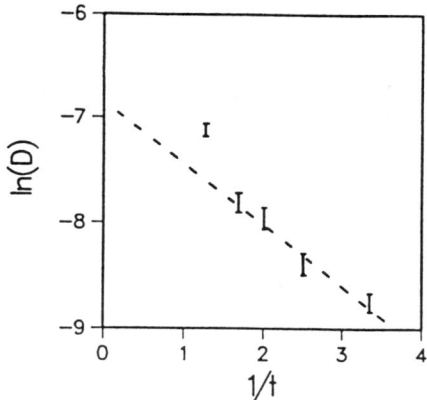

Figure 35: The diffusion constant as function of temperature. The system is the same as in Fig. 1. The strength of the random potential is $\sigma = 0.1$.

value in Fig. 32 at which the crossover occurs.

That the onset of diffusion takes place in the form of correlated string motion, as anticipated by the energy argument given above, can be seen from Fig. 35. Here we show the trajectories of the vortices for $\sigma = 0.1$ at temperature $t = 0.3$, i.e., just as diffusion has started. One notes that the fluctuations in the commensurability between the random potential and the flux line lattice produce two different types of regions. The pinning centers act more efficiently in places where the random potential matches the vortex lattice. The mobility of the vortices is therefore low in these regions. The less mobile areas are separated by channels in which the vortices diffuse more easily.

We do not observe any qualitative magnetic field dependence in the range accessible to us, $b(0) = 0.05$ to 0.2. For $b(0) = 0.1$ and 0.2, the diffusion temperatures for the pure system ($\sigma = 0$) are $t_D \equiv T_D/T_c = 0.625 \pm 0.025$ and 0.675 ± 0.025, respectively. for $\sigma = 0.1$, we have $t_D = 0.25 \pm 0.025$ and 0.525 ± 0.025, respectively. This behavior is to be expected since the flux line lattice becomes more rigid as the induction is increased for small inductions (see e.g. [6]).

Some comments about finite size effects are appropriate. Since the change from no observable diffusion to observable diffusion is difficult to pinpoint numerically, it is difficult to test for finite size effects. For strong random potentials the correlation length of the vortex lattice is of a few lattice spacings at any temperature. In this range we do not expect our results to change with system size.

The situation is different for weak randomness. It was argued above that the activation energy for diffusion in the weak random potential, $\Delta E(\sigma \ll 1)$, is connected with the barrier for plastic shear flow in a random potential background. The latter we expect to be related to the energy needed to produce topological defects in the vortex lattice. We found that the amplitude of the random potential, A_{plas}, at which the disorder is able to produce plastic deformations, at zero temperature, decreases logarithmically with the size of the system. Hence, we also expect $\Delta E(\sigma \ll 1)$ to be logarithmically decreasing with the system size. We estimate that the zero temperature value of A_{plas} for the system described in Fig. 32 corresponds to $\sigma \simeq 0.1$.

Markiewicz (see Ref. [47]) has considered an analytical model of two–dimensional flux lattice melting and the influence of pinning potentials on the melting temperature. For the Kosterlitz-Thouless [17] mechanism, the melting temperature is given by

$$k_B T_M = \frac{a_0^2}{4\pi} C_{66} d, \qquad (59)$$

where $a_0 = (\frac{4}{3})^{\frac{1}{4}} \sqrt{\Phi_0/B}$ is the lattice spacing, and C_{66} is the shear modulus of the flux line lattice. Using Brandt's [6] expression for C_{66} the melting temperature predicted by Eq. 59 for the system described in Fig. 32 at $\sigma = 0$ is found to be $t_M = 0.92$, while the diffusion temperature determined in the simulation is much lower, $t_D = 0.65$. The reason for this discrepancy could be that the onset of diffusion is connected with correlated string motion as shown in Fig. 35, rather than with unbinding of dislocations. As mentioned in Ref.[48], it is not possible from the simulation to distinguish unambiguously between Kosterlitz-Thouless–like behavior or an alternative mechanism. Markiewicz also points out that — as observed in the present simulation — the effect of a pinning potential can be to lower the diffusion temperature.

Finally, we mention that the onset of diffusion in the form of chain–like shear motion might explain why Tinkham's shear–limited phase slip model of the resistive transition [40] is so successful in spite of the fact that the model ignores the pinning potential.

7 Critical Dynamics

The time correlations of the motion of vortices just above the depinning transition are observed to be very long ranged. See for instance the paper by Yeh and Kao (Ref. [57]) and references there in. Yeh and Kao measured the density fluctuations in the vortex system. They found that for driving forces only slightly above the threshold for onset of vortex motion the power spectrum of the density fluctuations were of the form $1/f^\beta$ with $\beta \sim 1$.

When frequency analysis is made of temporal fluctuating signals, $N(t)$, one finds for many different systems that the power spectra, $S_N(f)$, exhibit power law behavior like $S_N(f) \sim 1/f^\beta$ over a broad frequency domain and especially in the low frequency limit. This is interesting because such a power spectrum do not contain a specific characteristic frequency, i.e., the phenomena under consideration lacks a time scale.[58, 59, 60] The power spectrum is the square modulus of the Fourier transform of the time signal. Hence, one can also express the power spectrum in terms of the autocorrelation function of the time signal. One have[59]

$$S_N(f) = 4 \int_0^\infty <N(\tau)N(0) - N^2(0)> \cos(2\pi f \tau) d\tau. \qquad (60)$$

From this expression it follows that if $S_N(f) \sim 1/f^\beta$ with $\beta \approx 1$ the autocorrelation function will decay slowly. The value $\beta = 1$ corresponds to a constant – or rather a logarithmically decaying – autocorrelation function. This is the reason why the value $\beta = 1$ is of particular interest. Moreover, one find $\beta \approx 1$ in a great diversity of systems.[58].

In principle Molecular Dynamics simulations should be able to throw some light on the observed spectra, but since the issue is the low frequencies, i.e., the long time behavior it seems not to be possible to overcome the numerical demands for the large system size of interest. The next best is to study discrete dynamics. For this purpose we have studied simple lattice gas models of flux motion.[61, 62]

7.1 Model.

Our lattice gas model was inspired by the experiment on flux flow in thin film type II superconductors performed by Yeh and Kao[57]. We want to model particles which follow diffusive dynamics. The equation of motion we have in mind is of the form $\eta \mathbf{v} = \mathbf{F}$, where η is a friction coefficient, $\mathbf{v}$ the velocity and $\mathbf{F}$ the total force on the particle. The model is defined as follows. Consider a lattice of $N_x \times N_y$ sites. Each site can contain one or zero particles. Particles on neighbor sites repel each other with a central force of unit strength. Let $\mathbf{f}_{par}$ denote the total force on a particle due to its neighbor particles. An additional driving force, $\mathbf{f}_{dr}$, can be applied to all the particles. Let $\mathbf{F}$ be the total force on a particle, i. e., $\mathbf{F} = \mathbf{f}_{par} + \mathbf{f}_{dr}$ and define the vector $\mathbf{n}$ as

$$n_x = [F_x/F] \text{ and } n_y = [F_y/F] \qquad (61)$$

where $[t]$ is the integer nearest to t. A particle on site $\mathbf{r}_0$ is moved to site $\mathbf{r}_1$ given by

$$\mathbf{r}_1 = \mathbf{r}_0 + \mathbf{n} \qquad (62)$$

if the new site $\mathbf{r}_1$ is unoccupied otherwise the particle is left on site $\mathbf{r}_0$. The whole lattice is simultaneously updated. In case two or more particles want to move into the same new site, the one with the largest force wins. The sites $(0, y)$ at the left edge of the lattice are occupied with fixed particles. This rim of particles tend to push particles on the sites $(1, y)$ into the lattice. In each time step particles in the column $(1, y)$ are first removed and then new particles are introduce on the sites $(1, y)$ with a probability p per site. Particles can freely leave the system over the right rim. The two dimensional systems are made periodic in the y-direction. We used: $N_x = 10, 20, 50, 100$, and 250, and $N_y = 6$ or 18, for the study of temporal features of the system, as well as systems of sizes 30×30, 60×60, and 128×128 for the spacial characterization discussed below.

Further more, we can choose a subset of sites which we denounce pinning centers. If a particle sits on such a site it will only be allowed to move off the site when $A_p F > 1$, where A_p is a number characterizing the strength of the pinning.

The particle density and the mean velocity depend on the values of p, A_p, and the density of pinning sites n_p. However, the qualitative features and the critical exponents do not depend on the specific values of these parameters.

The simulations are started out with some configuration of particles on the lattice (empty lattice, every second site occupied, random distribution of particles, etc.). After a certain transient time the system enters a statistically stationary state which is independent of the initial configuration. The total number of particles $N(t)$ is then recorded as function of time. Fig. 36 shows a typical measurement of $N(t)$ in the stationary state. In Fig. 36b we show a magnification of a section of Fig. 36a in order to exhibit the self-similar structure of the $N(t)$ curve.

The power spectrum is obtained by direct Fourier transformation of $N(t)$. In order to achieve sufficient statistics many power spectra of successive time sequences are averaged. Fig. 37 show typical measured spectra. Results for different system sizes are shown. Details

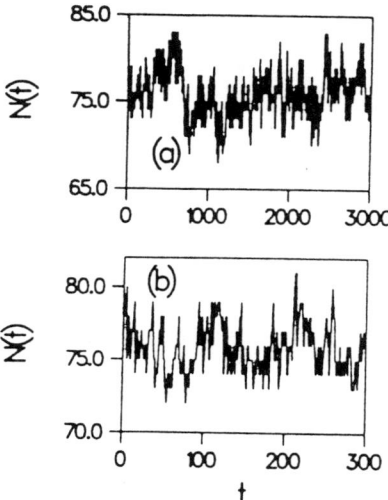

Figure 36: Time sequence of the number of particles on the lattice for a system of size 50×6, with nearest neighbor interaction, the density of pinning sites equal to $1/3$ and no driving force applied. b) shows a blow up of a section of a).

of the model can be found in Ref. [61, 62]. Let us here just remark that the models seems to be able to reproduce the $1/f^\beta$ spectra observed in the experiments.[57] As in the experiments we find $\beta \approx 1$ in the onset region for motion. Whereas for large driving force the exponent changes to $\beta = 2$. Below we discuss these findings analytically.

Let us look at energy dissipation in the model. The updating algorithm models an over damped equation of motion. We can think of energy being dissipated every time a particle is moved form one site to another. In order to study the spacial distribution of this energy dissipation we mark all the sites from which particles leave at time step t and mark in addition all the sites these particles move on to at time $t+1$. These are the sites on which energy has been dissipated during the update from t to $t+1$. The sites marked in this way form clusters containing different numbers, S, of sites, see Fig. 38. The distribution of cluster sizes, $D(S)$, follows a power law, $D(S) \sim 1/S^\gamma$. The exponent γ is approximately equal to 2.

7.2 Monte Carlo Simulations

The $1/f$ spectra observed in the lattice models with deterministic dynamics are to be compared with the fluctuation spectrum of the particle number observed in Monte Carlo simulations performed by Andersen, Jensen, and Mouritsen.[63] Lattice gas models were in Ref. [63] investigated using stochastic Monte Carlo dynamics instead of deterministic update rules. The fluctuation spectrum of the particle number were in the Monte Carlo simulations found to behave as $1/f^{1.5}$ for all dimension (one, two, and three) and irrespective of whether the interaction between the particles was taken to be repulsive, attractive or no interaction (except for the excluded double occupancy).

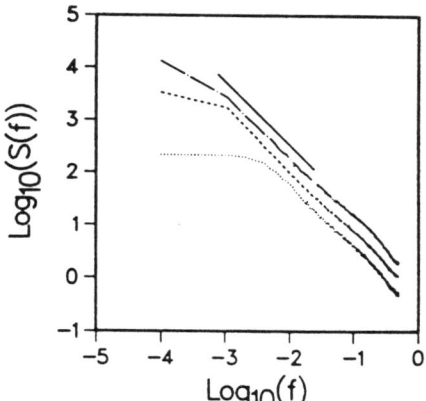

Figure 37: Power spectra of systems with next nearest neighbor interaction, the density of pinning sites equal to 1/3 and no driving force applied. Three different system sizes are shown: dotted line corresponds to 20 × 6, dashed line to 50 × 6, and dashed-dotted to 100 × 6. The data has been multiplied by factor 1, 4, and 8 - going from bottom to top - in order to keep the curves apart. The straight line is $1/f^{1.2}$.

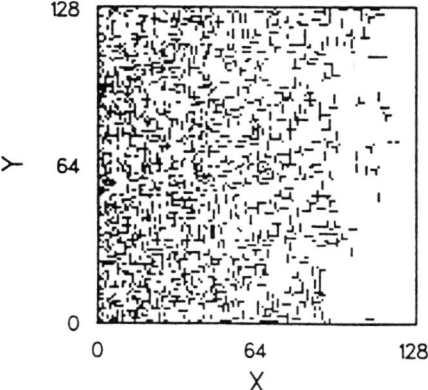

Figure 38: Clusters of energy dissipating sites in a system with nearest neighbor interaction without pinning sites and no applied driving force.

The deterministic models discussed above and the models considered by Andersen et al. are very similar and the only significant difference in their definitions is in the stochastic versus deterministic dynamics. Hence, it was concluded by Andersen et al. that microscopic stochastic dynamics leads to a $1/f^{1.5}$ spectrum in contrast to the $1/f$ spectrum produced by microscopic deterministic dynamics. This conclusion were further supported by the observation that an inclusion of a stochastic element in the update algorithm for the deterministic model changes the power spectrum of this model from $1/f$ to $1/f^{1.5}$.

7.3 $1/f$ from the Driven Diffusion Equation

The density fluctuations of the driven lattice gasses have recently been investigated in terms of diffusion equations and random walkers by several authors.[64, 65, 63, 66] From this work the following picture has emerged. The diffusion equation without any bulk noise

$$\frac{dn}{dt} = \gamma \vec{\nabla}^2 n(\mathbf{r}, t) \tag{63}$$

when driven by white noise at the boundary leads to a $1/f$ power spectrum for spatial integral of the diffusing quantity, $N(t) = \int n(\vec{x}, t) d\vec{x}$.

Let us briefly sketch here how this result comes about.[65, 67] Consider a system in D dimensions with a boundary S. We want to drive the system by fixing the value at the boundary to $n(\mathbf{r}, t) = B(\mathbf{r}, t)$ for $\mathbf{r} \in S$ for all t. The appropriate Green's function is a solution to

$$\frac{\partial G}{\partial t} - \gamma \vec{\nabla}^2 G = 4\pi \delta(\mathbf{r} - \mathbf{r}_0) \delta(t - t_0) \tag{64}$$

with the homogeneous Dirichlet condition[68] $G(\mathbf{r}, \mathbf{r}_0|tt_0) = 0$ for $\mathbf{r}_0 \in S$. The solution of Eq. 63 is given by (see Ref. [68])

$$n(\mathbf{r}, t) = -\frac{1}{4\pi} \int dt_o \int d\mathbf{S}_0 \cdot \vec{\nabla}_0 G(\mathbf{r}, \mathbf{r}_0|tt_0) B(\mathbf{r}_0, t_0) \tag{65}$$

To obtain the power spectrum of $N(t)$ we Fourier transform $n(\mathbf{r}, t)$. For a large system we will have approximately

$$n(\mathbf{q}, f) = \int dt_0 \int d\mathbf{S}_0 B(\mathbf{r}_0, t_0) \exp[i(\mathbf{q} \cdot \mathbf{r}_0 - 2\pi f t_0)] \tag{66}$$
$$\times \int dt \int d^D \mathbf{r} \vec{\nabla} G(\mathbf{r}, t) \exp[i(\mathbf{q} \cdot \mathbf{r} - 2\pi f t)]$$

The Green's function solution to Eq. 64 pertinent to an infinite domain in D dimensions is

$$G(\mathbf{r}, t) = [4\pi\gamma]^{1-D/2} t^{-D/2} \exp\{-r^2/4\gamma t\} \theta(t). \tag{67}$$

Here $\theta(t)$ is the Heaviside step function. The detailed form of the Green's function will depend on the boundary. For a large system, however, we can use Eq. 67 to conclude that

$$\int d^D \mathbf{r} \vec{\nabla} G(\mathbf{r}, t) \sim t^{-1/2}. \tag{68}$$

Assuming a white noise spectrum for the boundary factor in Eq. 66 it follows that the power spectrum $S_N(f)$ of $N(t)$ is given by

$$S_N(f) = |n(\mathbf{q} = 0, f)|^2 \sim |\int dt t^{-1/2} e^{i 2\pi f t}|^2 \sim 1/f. \tag{69}$$

Inclusion of a conserving bulk noise term on the right hand side of Eq. 63 produces a $1/f^{1.5}$ spectrum for the total particle number. An applied driving force will change the spectrum to $1/f^2$.[64] These result are independent of dimension.

That a driving current makes the spectrum change to a $1/f^2$ form can also be seen in the following way.[61] When f_{dr} is the dominating force the particles are pulled over the lattice moving one step to the right in each time step, i.e., speed equal to 1. The distribution of lifetimes becomes a delta function $D(T) = \delta(T - N_x)$ and the power spectrum becomes proportional to[61] $sin^2(\pi N_x f)/f^2$. This behavior is consistent with the experiment by Yeh and Kao[57], who found a cross over to a Lorentzian-like form well above threshold.

Nonlinear terms will always have to be include in the description of any real system, so also for the lattice gas models considered here. It was shown in the paper by Grinstein, Hwa, and Jensen that inclusion of nonlinear terms in Eq. 63 do not change the scaling behavior of the power spectrum of $N(t)$.[64]

From this discussion we note that the deterministic lattice gas behave according to the diffusion equation 63 with out any (conserving) bulk noise term. Where as the lattice gas with Monte Carlo dynamics behave according to 63 with a noise term included. The density fluctuations in a gas of non-interacting random walkers behave in the same way as the Monte Carlo driven lattice gas.[66]

8 Conclusions

We have studied the effect of different types of non-linear elastic instabilities in various models of elastic systems embedded in a quenched random potential.

First we considered a one dimensional system. We found that the response of an elastic medium to a set of sharp and dense pinning centers is controlled by non-linear instabilities. We studied the statistics of the instabilities in the case where the system is moved through the random potential. We found that the energy discontinuities obey a power law distribution. The power spectrum is also characterized by the presence of a non-trivial scaling regime. We should stress that these results do not seem a priori obvious. The *existence* of instabilities can be rather simply established analytically because of the one dimensional nature of the system. However, it is much more difficult to deal analytically with the hierarchy of different metastable states which are generated as the system size or the strength of the pinning potential are increased.

Next we investigated two dimensional models. We found that the response of an elastic medium to a set of sharp and dense pinning centers is controlled by non-linear instabilities. In the static adiabatic relaxation of the lattice the long range translational order only decays algebraically with a small exponent.

The random pinning potential was found to soften the elastic lattice. The elastic constants decreases with increasing strength of the pinning potential. We found that for a finite system regimes of different types of deformations of the elastic lattice exist. In the limit of infinite system size plastic deformations always occur as the elastic lattice is moved through the back ground potential.

For a two dimensional system a non-linear sector in the dependence of the center of mass velocity on the applied driving force always signals the occurence of plastic flow.

At finite temperature we found that a weak pinning potential increases the mobility of the vortices: the diffusion temperature decreases with increaseing pinning strength for weak pinning. This is consistent with the observed softening of the lattice.

The long ranged time correlations observed experimentally just above the threshold for depinning were simulated by use of *deterministic* lattice gas models. The qualitative behavior of the power spectrum was reproduced by the models. The power spectra of the lattice gases can be discussed by use of diffusion equations without any bulk noise term.

9 Acknowledgements

The work described in Chap 1 to 6 is the result of a collaboration with A. Brass, Y. Brechet, B. Doucot, An-Chang Shi, and A. J. Berlinsky. The content has been published previuosly together with changing combinations of groups of these collaborators. See Ref. [2, 4, 5, 8, 11, 16, 19, 24, 38, 48]

The work in Chap. 7 is partly a product of collaborations with J.V. Andersen, T. Bohr, T. Fiig, H.C. Fogedby, G. Grinstein, T. Hwa, M.H. Jensen, O.G. Mouritsen, H.H. Rugh, and Y.-C. Zhang. See Ref. [63, 64, 66]

References

[1] L.D. Landau and E.M. Liftshits, Chap. 21 in *Theory of Elasticity*, Pergamon Press 1970.

[2] H.J. Jensen, Y. Brechet, A. Brass, J. Low Temp. Phys. **74**, 293 (1989).

[3] W. F. Vinen in *Superconductivity*, Vol. II, ed. R. D. Parks (Marcel Dekker, New York, 1969). See also Y. B. Kim and M. J. Stephen, *ibidem*.

[4] H. J. Jensen, A. Brass and A. J. Berlinsky, Phys. Rev. Lett. **60**, 1676 (1988).

[5] A. Brass, H. J. Jensen, and A. J. Berlinsky, Phys. Rev. B **39**,102 (1989).

[6] E. H. Brandt, J. Low Temp. Phys. **53**, 41, 71(1983), and Phys. Rev. Lett. **50**, 1599 (1983).

[7] A. I. Larkin and Yu. N. Ovchinnikov in *Noneqµuilibrium Superconductivity*, ed. by D. N. Langenberg and A. I. Larkin (North Holland, 1986).

[8] H.J. Jensen, Y. Brechet, and B. Doucot, J. Phys. I *France* **3**, 611 (1993).

[9] O. Pla and F. Nori, Phys. Rev. Lett. **67**, 919 (1991).

[10] H. J. S. Feder and J. Feder, Phys. Rev. Lett. **66**, 2669 (1991) and **67**, 283 (1991) (erratum).

[11] H.J. Jensen, Y. Brechet, B. Doucot, A. Brass, Submitted to Europhys. Lett.

[12] C. Tang, K. Wiesenfeld, P. Bak, S. Coppersmith, and P. Littlewood, Phys. Rev. Lett. **58**, 1161 (1987).

[13] T. Nattermann, Phys. Rev. Lett. **64**, 2454 (1990). S.N. Coppersmith, *ibid.* **65**, 1044 (1990). E.M. Chudnovsky, *ibid.* **65**, 3060 (1990). J. Toner, *ibid.* **66**, 2523 (1991). A. Houghton, R. A. Pelcovits, and A. Sudbø, J. Phys.:Condensed Matter, **3**, 7527 (1991). M.P.A. Fisher, T.A. Tohuyasu, Phys. Rev. Lett. **66**, 2931 (1991). M.P.A. Fisher, Phys. Rev. Lett. **62**, 1415 (1989).

[14] A. I. Larkin and Yu. N. Ovchinnikov, J. Low Temp. Phys. **34**, 409 (1979).

[15] J.P. McTague, D. Frenkel, and M.P. Allen, in *Ordering in Two Dimensions*, ed. S. K. Sinha, (North Holland, 1980).

[16] Y. Brechet, B. Doucot, H.J. Jensen, and A.-Ch. Shi, Phys. Rev. B **42**, 2116 (1990).

[17] D.S. Fisher, Phys. Rev. B **31**, 1396 (1985).

[18] C.A. Murray, P.L. Gammel, D.J. Bishop, D.B. Mitzi, and A. Kapitulnik, Phys. Rev. Lett. **64**, 2312 (1990). D.G. Grier, C.A. Murray, C.A. Bolle, P.L. Gammel, D.J. Bishop, D.B. Mitzi, and A. Kapitulnik, *ibid.* **66**, 2270 (1991).

[19] H.J. Jensen, A. Brass, Y. Brechet, and A.J. Berlinsky, Phys. Rev. B **38**, 9235 (1988)

[20] This means that a rigid translation of the particle system requires a finite energy (see Ref. [16]) due to the fact that the random potential breaks the translational invariance.

[21] E. Akkermans and R. Maynard, Phys. Rev. B **32**, 7850 (1985), C.C. Yu and A.J. Leggett, Comments Condens. Matter Phys. **14**, 231 (1988).

[22] J.-P. Bouchaud, M. Mézard, and J.S. Yedidia, Phys. Rev. Lett. **67**, 3840 (1991).

[23] It is interesting to note that recent high precision neutron scattering experiments on the flux line lattice in a clean Niobium sample by R.N. Kleiman, T.E. Mason and D.J. Bishop found that the best orientationally ordered flux line lattice was established by entering the superconducting state by slowly decreasing the magnetic field through H_{c2} rather than field cooling or zero field cooling followed by an increase of the magnetic field. T.E. Mason, private communication.

[24] H.J. Jensen, A. Brass, Y. Brechet, and A.J. Berlinsky, Cryogenics **29**, 367 (1989).

[25] L.E. Rehn, P.R. Okamoto, J. Pearson, R. Bhadra, R. and M. Grimsditch, Phys. Rev. Lett. **59**, 2987, (1987).

[26] To be consistent one has to expand the vortex-vortex energy to the same order in **s** as the random potential. However, it turns out that the anharmonic elasticity terms are irrelevant for the point we consider here, see Ref. [16]

[27] R. Labusch, Cryst. Lattice Defects **1**, 1 (1969).

[28] J.P. Hirth and J. Lothe, *Theory of Dislocations* (McGraww-Hill, New York, 1968) p. 399; R. Hill, Proc. Phys. Soc. **65A**, 349 (1952).

[29] See *Theory of Elasticity*, L.D. Landau and E.M. Lifshitz (Pergamon Press, 1970) Sec. 21 for a discussion of elastic instabilities. This regime is "elastic" in the sense that the association of each particle with its neighbors in the flux lattice is unchanged by their motion through the random potential. By contrast, in the plastic regime, moving particles flow past neighbors which are pinned by the impurity potential.

[30] M. Tinkam, *Introduction to Superconductivity*, (McGraw-Hill, New York, 1975) Chap. 5.

[31] P. Thorel, R. Kahn, Y. Simon and D. Cribier, J. Physique, **34**, 447 (1973).

[32] R.P. Huebener, Phys. Rep. C **13**, 143 (1974).

[33] E.V. Thuneberg, J. Low Temp. Phys. **57**, 415,(1984).

[34] E.H. Brandt, Phys. Rev. B **34**, 6514, (1986).

[35] D.S. Fisher, Phys. Rev. Lett. **50**, 1486 (1983).

[36] F. de la Cruz, J. Luzuriaga, E.N. Martinez, and E.J. Osquiguil, Phys. Rev. B **36**, 6850 (1987).

[37] D.S. Fisher, in *Nonlinearity in Condensed Matter*, edited by Bishop, A. R., Campbell, D. K., Kumar, P. and Trullinger, S. E. (Springer-Verlag, Berlin, 1987).

[38] H.J. Jensen, A. Brass, A.-C. Shi, and A.J. Berlinsky, Phys. Rev. B **41**, 6394 (1990).

[39] T. T. Palstra, B. Batlogg, L. F. Schneemeyer, and J. .V. Waszczak, Phys. Rev. Lett. **61**, 1662 (1988).

[40] M. Tinkham, Phys. Rev. Lett. **61**, 1658 (1988).

[41] K. A. Müller, M. Takashige, and J. G. Bednorz, Phys. Rev. Lett. **58**, 1143 (1987).

[42] Y. Yeshurun and A. P. Malozemoff, Phys. Rev. Lett. **60**, 2202 (1988).

[43] P. H. Kes, J. Arts, J. van den Berg, C. J. van der Beek, and J. A. Mydosh, Supercond. Sci. Technol. **1**, 242 (1989).

[44] C. Rossel, Y. Maeno, and I. Morgenstern, Phys. Rev. Lett. **62**, 681 (1989).

[45] P. L. Gammel, L. F. Schneemeyer, J. V. Waszczak, and D. J. Bishop, Phys. Rev. Lett. **61**, 1666 (1988) and **62**, 2331 (1989).

[46] E. H. Brandt, P. Esquinazi, and G. Weiss, Phys. Rev. Lett. **62**, 2330 (1989).

[47] D. S. Fisher, Phys. Rev. B **22**, 1190 (1980); D. R. Nelson Phys. Rev. Lett. **60**, 1973 (1988); D. R. Nelson and H. S. Seung, Phys. Rev. B**39**, 9153 (1989); R. S. Markiewicz, J. Phys. C Solid State Phys. **21**, L1173 (1988); M. Moore, Phys. Rev. B. **39**, 136 (1989); H. Houghton, R. A. Pelcovits, and A. Sudbø, Phys. Rev. B, in press; E. H. Brandt, Phys. Rev. Lett. **63**, 1106 (1989).

[48] A. Brass and H. J. Jensen, Phys. Rev. B **39**, 9587 (1989).

[49] F. Reif, *Fundamentals of statistical and thermal physics*, Chap. 15.11 (McGraw-Hill, 1965).

[50] T. Schneider and E. Stoll, Phys. Rev B **17**, 1302 (1978).

[51] E. H. Brandt, Phys. Stat. Sol. **77**, 105 (1976).

[52] M. Tinkham, *Introduction to Superconductivity*, (McGraw-Hill, New York, 1975), Chap. 4.

[53] This is the same system as the one used in Ref. [48]. The value for $\lambda^2 T_c/d$ given in Ref. [48] contains a misprint. The correct value is $\lambda^2 T_c/d = 10^{-2} cm$.

[54] The lattice constant of the triangular Abrikosov lattice is given by $a_0 = (4/3)^{1/4}\sqrt{2\pi/b}\lambda/\kappa$.

[55] It should be noted that since B_{c_2} is temperature dependent the actual *reduced* field varies with temperature like $b(t) = b(0)(1+t^2)/(1-t^2)$. In the temperature range considered in Fig. 32 $b(t)$ varies from 0.11 at $t = 0.2$ to 0.45 at $t = 0.8$.

[56] J.P. Bouchaud, A. Comtet, A. Georges, and P. Le Doussal, J. Phys. *France* **48**, 1445 (1987).

[57] W.J. Yeh and Y.H. Kao, Phys. Rev. Lett. **53**,1590 (1984), and Phys. Rev. B **44**, 360 (1991).

[58] W.H. Press, Comments Astrophys. Space Phys. **7**, 103 (1978).

[59] P. Duta and P.M. Horn, Rev. Mod. Phys. **53**, 497 (1981).

[60] M.B. Weissman, Rev. Mod. Phys. **60**, 537 (1988).

[61] H.J. Jensen, Phys. Rev. Lett. **64**, 3103 (1990).

[62] T. Fiig and H.J. Jensen, J. Stat. Phys., To be published.

[63] J.V. Andersen, H.J. Jensen, and O.G. Mouritsen, Phys. Rev. B **44**, 439 (1991).

[64] G. Grinstein, T. Hwa, and H.J. Jensen Phys. Rev A **45**, R559 (1992).

[65] H.J. Jensen, Physica Scripta **39**, 593 (1991).

[66] H.C. Fogedby, M.H. Jensen, Y.-C. Zhang, T. Bohr, H.J. Jensen, and H.H. Rugh, Mod. Phys. Rev. Lett. B **5**, 1837 (1991).

[67] S.H. Liu, Phys. Rev. B **16**, 4218 (1977).

[68] P.M. Morse and H. Feshback *Methods of Theoretical Physics*, Chap 7. McGraw-Hill Book Company, New York (1953).

I-V CHARACTERISTICS OF HIGH TEMPERATURE SUPER-CONDUCTORS WITH CORRELATED DEFECTS

MATS WALLIN
Department of Theoretical Physics
Royal Institute of Technology
S-100 44 Stockholm
Sweden

and

S. M. GIRVIN
Department of Physics
Indiana University
Bloomington IN 47405
USA

ABSTRACT. Pinning of vortex lines in high temperature superconductors can be enhanced dramatically by correlations in the distribution of the pins. We study the nonlinear I-V characteristic of a model high temperature superconductor with columnar defects by nonequilibrium Monte Carlo simulation of the dynamics of a gas of interacting vortex lines in a random medium. Unlike previous work our simulation directly probes for the first time the vortex degrees of freedom, and takes screening of the vortex lines into account. From finite-size scaling we determine the previously unknown dynamical exponent z, and the exponent y in the I-V characteristic $V \sim I^y$ at the glass transition.

The electromagnetic properties of a high-temperature superconductor (HTS) in a magnetic field are largely determined by pinning of the vortex lines. For many applications it may be important to have large critical fields and currents, and this might possibly be accomplished by deliberate introduction of suitable pins into the superconducting material. There are different types of disorder possible in a bulk superconductor: (1) pointlike impurities (e.g. oxygen vacancies), (2) linearly correlated defects (columnar defects), and (3) planar defects (twin planes). The original vortex glass model [1] assumes point disorder. One signature of a vortex glass transition is a universal nonlinear current-voltage (I-V) characteristic, which has been observed in several experiments and in Monte Carlo (MC) simulations; for a review see Ref. [2]. However, point pinning tends to not be very effective in the highly anisotropic CuO materials and thus often gives small critical fields and currents. Furthermore, from MC simulations [3] it is not yet completely established that the vortex glass can exist in 3D at finite temperature. Columnar defects, on the other hand, give much stronger pinning than point pins when the field is aligned with the pins. Columnar defects can be produced in the lab as damage tracks from high-energy heavy ion bombardment of the sample, and a resulting dramatic shift upwards in the irreversibility line has been reported by several groups [4]. This columnar-defect-induced glass transition has been studied by Nelson and Vinokur [5] and Lyuksyutov [6] using an analogy to dirty boson localization, which shows that the columnar-defect glass phase is a bose glass whose existence is well established [7,8]. Planar defects in the form of naturally occurring twin planes can also be induced in the samples in a controlled way. Studies of planar defects and combinations of point and planar defects are currently underway.

We will now discuss our MC calculation of the nonlinear I-V characteristic for the case of columnar defects. A detailed account has been published elsewhere [9], and we here only discuss the main points. We consider the problem of a gas of interacting vortex lines in a random medium, which we model by the dirty boson action [8]

$$\beta H = \beta \sum_{\mathbf{r}} \left\{ \frac{1}{2} \mathbf{J}^2(\mathbf{r}) - v(\mathbf{r}_\perp) J_z(\mathbf{r}) \right\}, \tag{1}$$

where $\beta \equiv 1/k_B T$, $\mathbf{J} = (J_x, J_y, J_z)$ are integer variables on the lattice links representing the vortex degrees of freedom: $J = 0$ means no vortex line, $J = 1$ means a vortex line with positive orientation, and $J = -1$ means a vortex line with negative orientation. The field $\mathbf{J}(\mathbf{r})$ is constrained to be everywhere divergenceless, since flux lines cannot terminate in the material. The first term in H acts as a string tension and as a short-range repulsion, and the second term describes the columnar defects. $v(\mathbf{r}_\perp)$ is a random site energy with uniform probability distribution on the interval $[0, 1]$. All previous simulations have used either uniformly or randomly frustrated XY models in the phase representation where the vortices are implicit, but not explicit. This is, as far as we know, the first simulation ever done for the glass problem directly using the vortex degrees of freedom. The applied magnetic field is implemented as a fixed nonzero filling of penetrating vortex lines. We do not use any long-range interaction between the vortex lines. This assumes that the magnetic penetration length (measured with the vortices held *fixed*) is finite, and can be thought of as setting the lattice spacing. We use simple cubic lattices with periodic boundary conditions in all directions, in order to allow finite-size scaling to approach the thermodynamic limit.

In order to calculate the I-V characteristic we add a term $\Delta \epsilon = (\nabla \times \mathbf{J}) \cdot \mathbf{I}$ to H, representing a Lorentz force on the vortex lines from the applied finite current density I. We here assume that the current penetrates the sample. This is not right in the superconducting state, where an infinitesimal current would reside on the surface, but should happen at finite current density when the resistance becomes finite. Further, we assume uniform penetration. It is at present unclear if this is in the same universality class as a more realistic model where the current is modeled self-consistently to account for a finite penetration length and disorder. Our MC updates are attempts to add fluctuations consisting of closed vortex loops on the plaquettes of the lattice. The nonlinear response voltage can now be measured in our simulation as the rate of phase slip across the system in the plane normal to the applied current: $V \sim [\langle \dot{N}_+ - \dot{N}_- \rangle]$, where $\dot{N}_\pm$ is the rate of accepted loops of orientation $\pm$. To converge the simulation requires up to 10^7 sweeps through the lattice with updates of the vortex variables, denoted by $\langle\ \rangle$, which then has to be averaged over up to 20 quenched realizations of the columnar disorder, denoted by $[\ ]$.

At the glass transition the correlation length of the vortex line system diverges as $\xi \sim |T - T_g|^{-\nu}$, and the characteristic time scale is assumed to diverge as $\tau \sim \xi^z$, which defines the dynamical exponent z. The glass transition temperature $T_g = 0.248 \pm 0.002$ for the dirty boson model (1) was determined independently [8] from finite-size scaling of MC data for the dirty boson superfluid stiffness, which corresponds to the inverse tilt modulus of the vortex lines. MC results for the nonlinear I-V characteristic at the glass transition are shown in Fig. 1. The current is here applied perpendicular to the columns, and the applied magnetic field is aligned with the columns. The strength of the magnetic field here corresponds to a filling of $f = 1/2$ of penetrating vortex lines. Plotted in the figure are MC data for the measured voltage vs. the applied current, both scaled with appropriate powers of the system size to become dimensionless. A single parameter, the dynamical exponent z, was adjusted in the plot until all data for 6 different lattice sizes collapse onto the same curve. This curve thus represents the thermodynamic limit. This determines the value $z = 6 \pm 0.5$ for the dynamical exponent. The error estimate on z is from the rate at which the data collapse breaks down for z away from 6. We see that the I-V curve is an essentially perfect power law over many decades. The crossover to ohmic response at small current is a finite-size effect (the dashed line has slope 1). Scaling arguments fix the exponent of the power law $V \sim I^y$ to $y = (1+z)/3 = 7/3$ [5,9]. A straight solid line with this slope is drawn on top of the data, and the agreement with MC points is again essentially perfect. Note that this power law exponent was not fitted, the only fitting is the data collapse.

In summary, we have calculated the nonlinear I-V characteristic of a model HTS with columnar defects by Monte Carlo simulation. At the glass transition we obtain $z = 6 \pm 0.5$, and $V \sim I^y$ with $y = (1+z)/3 \approx 2.3$. for current perpendicular to the columns. Our number $y \approx 2.3$ is roughly consistent with that found in experiments: $y \approx 2.9$ [2]. In these samples it

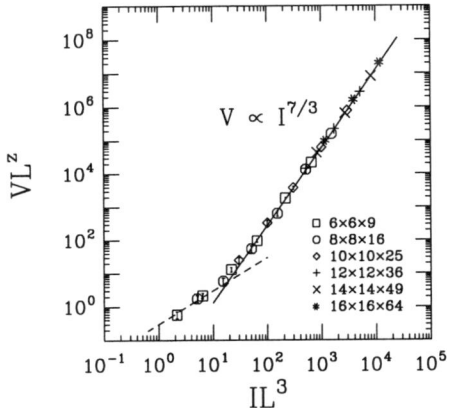

Fig. 1. MC data for the nonlinear I-V characteristic at the glass transition.

is not clear what types of pinning are actually present, i.e. some type of unintentional correlated pinning (dislocations, twins, etc.) might actually be important at the glass transition. In Ref. [9] we show that the response is qualitatively different when the current is parallel to the columns: $z_\| = 4 \pm 0.5$ and $y_\| = (2+z)/2 \approx 3$. Experimental results for the I-V characteristic of systems with deliberately introduced columnar defects, which have not yet been reported, are necessary for a direct comparison of our predictions with experiment.

MW is supported by the Swedish Natural Science Research Council and the Swedish National Board for Industrial and Technical Development. SMG is supported by DOE grant DE-FG02-90ER45427 and NCSA DMR-910014N.

References

[1] M.P.A. Fisher, Phys. Rev. Lett. **62**, 1415 (1989); D.S. Fisher, M.P.A. Fisher, and D.A. Huse, Phys. Rev. B **43**, 130 (1991).
[2] D.A. Huse, M.P.A. Fisher, and D.S. Fisher, Nature Vol. 358, 553 (1992).
[3] D.A. Huse and H.S. Seung, Phys. Rev. B **42**, 1059 (1990); J.D. Reger, T.A. Tokuyasu, A.P. Young, and M.P.A. Fisher, Phys. Rev. B **44**, 7147 (1991); M.J.P. Gingras, Phys. Rev. B **45**, 7547 (1992).
[4] L. Civale, A.D. Marwick, T.K. Worthington, M.A. Kirk, J.R. Thompson, L. Krusin-Elbaum, Y. Sun, J.R. Clem, and F. Holtzberg, Phys. Rev. Lett. **67**, 648 (1991); R. C. Budhani, M. Suenaga, and S.H. Liou, Phys. Rev. Lett. **69**, 3816 (1992). See also: Gerhäuser et al., Phys. Rev. Lett. **68** 879 (1992) and references therein.
[5] D.R. Nelson and V.M. Vinokur, Phys. Rev. Lett. **68**, 2398 (1992).
[6] I.F. Lyuksyutov, Europhys. Lett. **20**, 273 (1992).
[7] M.P. A. Fisher, P.B. Weichman, G. Grinstein, and D.S. Fisher, Phys. Rev. B **40**, 546 (1989).
[8] E.S. Sørensen, M. Wallin, S.M. Girvin, and A.P. Young, Phys. Rev. Lett. **69**, 828 (1992).
[9] M. Wallin and S.M. Girvin, Phys. Rev. B, in press (1993).

RELAXATION NEAR GLASS TRANSITION SINGULARITIES

W. GÖTZE
Institut für Theoretische Physik, Physik–Department
Technische Universität München, D–8046 Garching
Max–Planck–Institut für Physik (Werner–Heisenberg–Institut)
D–8000 München
Germany

ABSTRACT. Cooled liquids exhibit temperature sensitive slow structural relaxation phenomena. They are the reason for the production of amorphous solid non–equilibrium states of matter, called glasses. An outstanding feature of structural relaxation is the stretching of the dynamics over time or frequency intervals of several orders of magnitude.

The evolution of the structural relaxation patterns within the mesoscopic dynamical window upon cooling glass forming liquids has been observed by neutron scattering, light scattering and molecular dynamics studies. The appearance of the patterns is associated with the formation of two time fractals for localized excitations. There is a characteristic temperature T_c which separates regimes of qualitatively different spectra. The slowing down of the dynamics upon cooling towards T_c is governed by two time scales, which increase and separate upon decreasing $T - T_c$. The mesoscopic dynamics for $T < T_c$ is ruled by a time scale, which increases upon heating towards T_c, signalizing the instability of some frozen structure. Hard sphere colloidal suspensions exhibit analogues structural relaxation and cross over phenomena if the packing fraction φ is shifted through some critical value φ_c. This suggests that structural relaxation is a paradigm for the dynamics of disordered closely packed classical many particle systems.

A mathematical model can be formulated for a statistical treatment of many particle dynamics, which exhibits a transition from ergodic to non–ergodic motion if control parameters shift through certain critical values. There occur bifurcations whose novel features are caused by the interplay of non–linearities with diverging retardation intervals. The results for the dynamics are similar to those observed near T_c or φ_c thereby suggesting, that the experimentally known cross over phenomena are caused by a bifurcation singularity. According to Li, Du, Chen, Cummins and Tao the light scattering spectra for the structural relaxation of $CaKNO_3$ can be explained within the mentioned theory.

The model can be derived within the microscopic theory for simple liquid dynamics by treating the cage effect with Kawasaki's mode coupling approximation. All structural relaxation phenomena turn out to be implications of the equilibrium geometry of the canonical potential landscape. Preliminary analysis of photon correlation data by van Megen and Underwood indicate that the theory provides an ab initio treatment of structural relaxation and cross over phenomena for certain colloids.

The mesoscopic dynamics near the mentioned singularity is described within the extended mode coupling theory for the supercooled liquid dynamics by a two parameter scaling law. It deals with the interplay of the cage effect and phonon assisted hopping events. The results provide a consistent quantitative description of the neutron scattering and light scattering experiments done for $CaKNO_3$ near T_c.

1. Some Features of Structural Relaxation

1.1. α-RELAXATION PROCESSES

Already J.C. Maxwell anticipated the existence of low lying excitations in liquids and studied within his visco-elastic theory the implications of such excitations for the shear response [1]. Later these excitations have been detected explicitly by a variety of probing techniques.

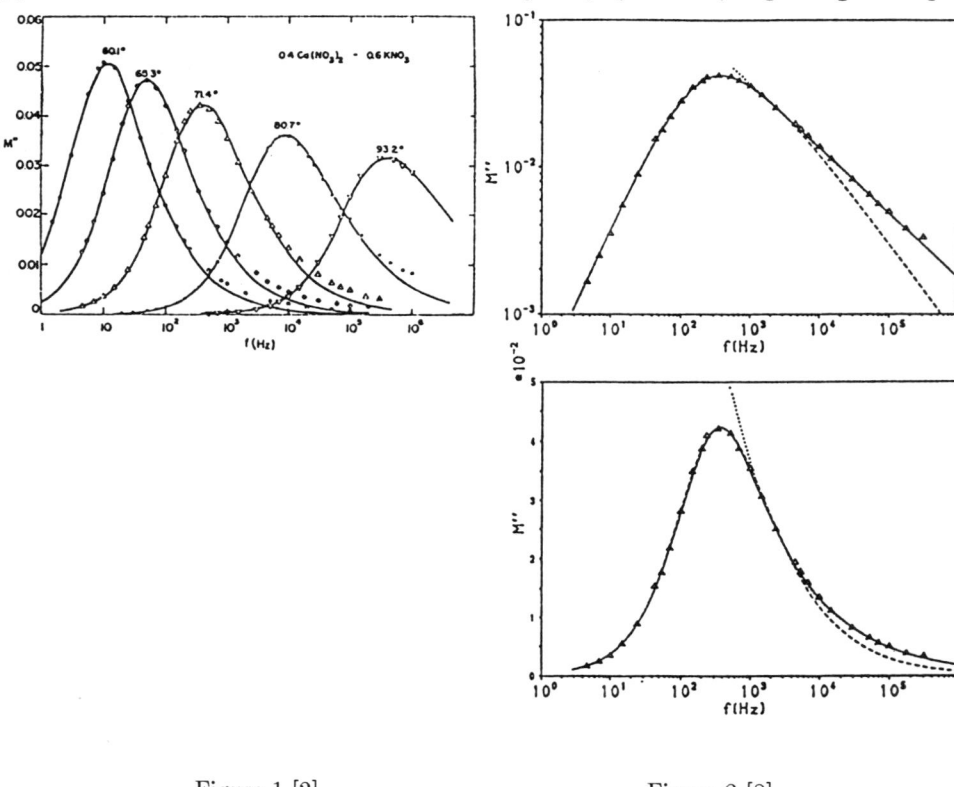

Figure 1 [2]   Figure 2 [8]

Fig. 1 [2] exhibits as example the spectrum M'' of the dielectric modulus for the mixed salt $0.4Ca(NO_3)_2 0.6K(NO_3)$ (CKN). The melting temperature of this system is $T_m \approx 165°C$, so that the shown data refer to the supercooled state. The spectra show bumps at certain frequencies $\omega_\alpha = 2\pi/\tau_\alpha$, called α-peaks. Indeed the shown excitations are *low lying*; ω_α is 6 to 11 orders of magnitude smaller than the frequencies characteristic for molecular motion in condensed matter, which are located in the THz band. The excitations describe processes whose relaxation time τ_α is macroscopic rather than the fs-scale, which specifies microscopic motion. Furthermore, ω_α is extremely *temperature sensitive*. A change of T by 10 degrees shifts ω_α by more than a factor 10. One assumes cooperative motion of complexes of molecules to be responsible for the specified excitations, and refers to them as *structural relaxation*. Structural relaxation processes occur in all cooled classical liquids and also in dense polymers. They show up in all structure sensitive probing: in experiments which test the dynamics over macroscopic or mesoscopic length scales, like shear response measurements, as well as in experiments testing motion on microscopic distances. They are not observed in ordered condensed matter like quantum liquids or crystalline solids. They should be considered as endemic for the dynamics of *strongly interacting disordered matter*.

The α-processes cause the glass transition and an understanding of the former is a prerequisite for the microscopic understanding of the latter. Let us extrapolate the times τ_α, shown in Fig. 1 to, say $0°C$ or $-30°C$; one finds them longer than 1 day or 3 years, respectively. Thus, for sufficiently low temperature, usual experiments cannot detect anymore the approach of the system towards its equilibrium state. The sample behaves as a solid, albeit an amorphous one. This non-equilibrium state of matter, which depends on the history of its preparation, is called a glass. Usually it is produced by cooling with speeds of some degrees per minute. The experiment can be characterized by some time scale t_{exp}, say the time needed to shift the α-peak position by the α-peak half width. Then one can define a temperature T_g by $\omega_\alpha(T = T_g)t_{exp} = 1$. This temperature T_g is *called calorimetric glass transition temperature*. For $T \sim T_g$ one shifts from probing the system below "resonance", $\omega_\alpha(T > T_g)t_{exp} > 1$, to above "resonance", $\omega_\alpha(T < T_g)t_{exp} < 1$. In the latter case the system responds stiffer than in the former. Thus T_g shows up in scanning curves, exhibiting an effective susceptibility versus T, by jumps. Usually one defines T_g via the specific heat jump. For CKN one gets $T_g \simeq 60°C$. The jumps and the typical hysteresis effects observed near T_g can be described by a proper extension of Maxwell's approach to non-linear phenomena [3]. Temperature T_g is an important number, characterizing the material from a technical point of view. From a scientific point of view the definition of T_g is arbitrary. Suppose, for example, one would study the system in a molecular dynamics experiment. In this case 10^{12} degrees per minute are the typical cooling rates. Then the time scale t'_{exp} is 10 or more orders of magnitude shorter than in the laboratory experiment defining T_g. Hence the cross over temperature T'_g, defined by $\omega_\alpha(T = T'_g)t'_{exp} = 1$, is considerably larger than T_g. Moreover, the cross over interval from soft liquid to stiff solid response is now expanded to 50 or more degrees [4].

An interesting facet of the structural relaxation phenomenon was discussed by R.D. Mountain [5]. Consider the compressibility $\kappa_q(\omega)$ for wave vector q and frequency ω. Evaluating this response function from Euler's equations for liquid dynamics, one gets

$$\kappa_q(\omega) \propto 1/\left[\omega^2 - q^2 M/\rho\right]. \tag{1.1}$$

This is an oscillator susceptibility with resonance frequencies $\Omega_q = cq$ for sound excitations. The sound speed c is given by the longitudinal acoustic modulus M and the mass density ρ as $c^2 = M/\rho$. There are standard equations relating susceptibilities like $\kappa_q(\omega)$ to correlation functions of time t. In our example the correlator $\phi_q(t)$ for fluctuation of densities with wavevector $\vec{q}, \rho_{\vec{q}}$, are relevant: $\phi_q(t) = <\rho^*_{\vec{q}}(t)\rho_{\vec{q}}>/S_q$. Here $S_q = <|\rho_{\vec{q}}|^2>$ denotes the liquid structure factor and $q = |\vec{q}|$. The correlation spectrum $\phi''_q(\omega)$ is the Fourier cosine transform of $\phi_q(t)$; it is proportional to the dynamic structure factor $S_q(\omega) \propto S_q \cdot \phi''_q(\omega)$ and $\kappa''_q(\omega) \propto \omega\phi''_q(\omega)$ [6]. From (1.1) one gets sharp sound wave resonances,

$$\phi_q^{B''}(\omega) = (\pi/2)[\delta(\omega - cq) + \delta(\omega + cq)], \tag{1.2}$$

which can be measured by inelastic light scattering as Brillouin peaks. If one uses Navier-Stokes equations for the liquid dynamics, the viscosities yield an imaginary part for the modulus, introducing a width Γ^B_q for the resonances. Structural relaxation induces a frequency dependence for the modulus: it increases with increasing ω from the below "resonance" value $M(\omega\tau_\alpha \ll 1) = M_0$ to the above "resonance" value $M(\omega\tau_\alpha \gg 1) = M_\infty$. This variation is connected via Kramers-Kronig relations to some spectral bump as discussed in connection with Fig. 1: $M(\omega) = M'(\omega) + iM''(\omega)$. As a result one gets sound speeds which vary with q and additional contributions to the width of the Brillouin peaks. In addition, there appears a resonance centered at zero frequency, a quasi elastic contribution, called Mountain's peak. Let us denote its normalized spectrum by $\phi_q^{M\prime\prime}(\omega)$ and its strength by f_q:

$$\phi_q''(\omega) = f_q \phi_q^{M\prime\prime}(\omega) + (1 - f_q)\phi_q^{B\prime\prime}(\omega). \tag{1.3}$$

This simple superposition holds only, if the width of the Mountain resonance $\Gamma_q^M \propto 1/\tau_\alpha$ is small compared to Ω_q, so that the resonances are well separated. One finds for the long wave length limit $f_{q\to 0} = f_0 = 1 - M_\infty/M_0$. The Mountain resonance implies the structural relaxation bump for the compressibility spectrum $\kappa_q''(\omega) \propto \omega\phi_q''(\omega)$. The structural relaxation contribution to one quantity, here the modulus $M(\omega)$, can be closely related to the one for another quantity, here the density spectrum $\phi_q''(\omega)$.

1.2. α-RELAXATION STRETCHING

Originally, exponential decay was anticipated for the structural relaxation: $\phi_D(t) = f \exp{-(t/\tau_\alpha)}$ [1]. In the present context such decay laws are named after P. Debye. Fourier transformation yields the Debye spectra for the correlations $\phi_D''(\omega) = f\tau_\alpha/[1 + (\omega\tau_\alpha)^2]$, or for the susceptibility: $\chi_D''(\omega) = f(\omega\tau_\alpha)/[1 + (\omega\tau_\alpha)^2]$. The latter describes an α-bump of half width of 1.14 decade. The χ_D'' versus $\log\omega$ plot is symmetric. However, typically one finds α-bumps being asymmetric on a $\log\omega$ abscissa and having a much larger width than χ_D''. The half width for the 71.4°C curve in Fig. 1 is about 1.7 decade, so that at half height the measured peak is four times broader than expected for Debye relaxation. The specified discrepancy between Debye formulae and experiment is even more pronounced if one compares the spectra on the level of 5% of the peak maximum. Here the mentioned bump is 100 times broader than the Debye bump. The spread of the spectra over many decades is called *stretching*. In order to scan 95% of the specified relaxation spectrum, one has to explore a dynamical window of more than 5 orders of magnitude. One has to follow a time increase over several decades, if one wants to detect the decay from 95% to 5% of the initial value $\phi(t = 0) = f$. To get quantitative information on structural relaxation the accessible dynamical window has to be larger than 2 decades, and if one wants to follow the evolution of the process with changes of temperature, even larger windows are needed. Data have to be presented on logarithmic time or frequency abscissas, if they should express some essential information on structural relaxation. The necessity to measure over huge windows is the major challenge for the experimentator studying supercooled liquid dynamics.

Stretching was observed already by R. Kohlrausch [7]. He reported, that his decay curves for dielectric relaxation follow closely the expression

$$\phi_K(t) = f \exp{-(t/\tau_\alpha)^\beta}, \tag{1.4}$$

a formula which modifies the exponential law $\phi_D(t)$ by the introduction of a *stretching exponent* $\beta < 1$. Many, but not all, later experiments confirmed, that the Kohlrausch law (1.4) is a good description of a major part of the α-processes. This is exemplified by Fig. 1, where the full lines are Kohlrausch fits with f, τ_α and β optimized for the data. The results for 71.4°C is shown for $\beta = 0.64$. Different structural relaxation processes, even if they refer to the same system, may exhibit different values for β. Obviously, the Kohlrausch law cannot be an exact description valid for all structural relaxation processes if $\beta < 1$. If one assumes (1.4) for the acoustic modulus in (1.1), one does not get a Kohlrausch shape for the Mountain bump in (1.3) and vica versa. This holds also if one considers the natural limit, that the Mountain bump is arbitrarily well separated from the Brillouin peaks. R. Kohlrausch reported indeed, that those of his data referring the largest dynamical window showed systematic deviations from (1.4). Fig. 1 exemplifies the phenomenon, the full fit curve for the 71.4°C experiment falls below the data for $\omega > 10^4$ Hz. The discrepancy between data and fit extends over more than a decade window.

Fig. 2 [8] reproduces the 71.4°C data from Fig. 1. The double logarithmic presentation follows a straight line of slope unity for $\omega\tau_\alpha \ll 1$. This means that the low frequency wing of the α-bump follows a white noise spectrum $\phi''_{\text{white}}(\omega\tau_\alpha \ll 1) = \text{const}$, $\chi''_{\text{white}}(\omega\tau_\alpha \ll 1) \propto \omega$, as does the Debye spectrum. There is no particular stretching phenomenon in simple systems, as opposed to in many polymers, for the low frequency part of the α- spectrum. The stretching phenomenon is a feature for the structural relaxation referring to frequencies large on the α-scale, $(\omega\tau_\alpha) \gg 1$, or to times short on the α-relaxation scale, $(t/\tau_\alpha) \ll 1$. The high frequency data in the upper part of Fig. 2 follow also a straight line over almost three decades with a slope to be denoted by $(-b)$. This implies power law variations:

$$\chi''(\omega\tau_\alpha \gg 1) \propto 1/(\omega\tau_\alpha)^b \quad \text{or} \quad \phi''(t\tau_\alpha \ll 1) = f - B(t/\tau_\alpha)^b. \tag{1.5}$$

The anomalous exponent $b = 0.44$ is considerably smaller than the Debye value unity for the slope. This observation was discussed by E. von Schweidler [9] for a variety of disordered systems. The *von Schweidler exponent* b is not universal, different systems yield different values for b. The von Schweidler law (1.5) should not be considered as limit of (1.4), since one finds regularly $\beta > b$, 0.64 versus 0.44 in the case shown in Fig. 2. Since double logarithmic plots have some trend to favor power law fits, the results are reproduced in Fig. 2 also semilogarithmically. The Kohlrausch fit from Fig. 1 is now shown in dashed and (1.5) in dotted. For the description of the high frequency α- peak wing, the von Schweidler fit is superior to the Kohlrausch fit.

Let us assume, that the measured variable A, like the dipole moment, can be represented as properly normalized sum of various contributions A_i, $i = 1,\ldots, N : A = (A_1 + A_2 + \ldots + A_N)/\xi_N$. The variables A_i, leading to some relaxation function ϕ_i with relaxation time τ_i, refer to different complexes of molecules. The total function $\phi = \Sigma \phi_i$ is thus a superposition implying a distribution of rates $1/\tau_i$. The central limit theorem would suggest, that for $N \to \infty$ one gets a Gaussian correlator $\phi(t) = f \exp -(t/\tau)^2$. The corresponding resonance is narrower than a Debye peak and cannot explain stretching. The limit theorem assumes quadratic spectral moments to exist, and this is not the case, if every complex relaxes according to $\phi_i(t/\tau_i \ll 1) - f_i \propto -(t/\tau_i)^b$. In this case the limit theorem can be modified. A proof by P. Lévy [10] yields as limit the Kohlrausch function. Therefore ϕ_K is also called *Lévy's characteristic function* of stable symmetric distributions. It is the expected decay law for sums of infinitely many infinitely small independent contributions, relaxing according to (1.5). If one could understand, that α-relaxation follows the von Schweidler law so, that for a given system all variables exhibit the same exponent b, independent sums would relax according to (1.4) with $b = \beta$. Since complexes of molecules overlap, the assumption of independence is only approximately valid. Thus the $N \to \infty$ result (1.4) should only be an idealization, and in reality the Kohlrausch formulae should exhibit $\beta \neq b$, resulting in systematic discrepancies from experiment. Understanding of (1.5) would imply a qualitative understanding of α-relaxation stretching.

1.3. THE CROSS OVER PROBLEM

Structural relaxation bumps in simple systems are fully developed as soon as their scale has moved into the macroscopic region: $\omega_\alpha \leq$ MHz. Shifting ω_α within this window by five or more decades is not connected with any obvious anomaly, as is exemplified by Fig. 1. On the other hand, there are no structural relaxation processes at high temperatures, say for T near the triple point. There the normal liquid state dynamics is specified by the fs scale, which is the relevant one also for ordered condensed matter. Exploring this normal liquid state e.g. by Raman or neutron scattering spectroscopy one finds the excitations to be located within the THz band, and for lower frequencies one merely finds white noise. Thus, there appears the question: how do structural relaxation patterns evolve within the mesoscopic MHz-GHz window upon cooling the liquid? How do dynamical processes with a ps-ns scale appear if disordered matter is condensed by lowering the temperature T or increasing the density n? Obviously, one has to look on the dynamics through the specified broad mesoscopic window, if one wants to see the key for the explanation of the supercooled liquid dynamics in general, and for the glass transition in particular.

The specified issue was formulated 1969 by M. Goldstein [11], who extended the reasoning as follows. At low enough temperatures the molecules are arrested close to their equilibrium positions. A particle experiences the interaction with its neighbors as a potential landscape. Thermally activated jumps over saddles are the elementary step, causing structural relaxation. The exponential sensitivity on temperature of the jump probabilities leads to the strong T dependence of ω_α for $T \sim T_g$. But this picture for the strongly supercooled state looses its justification for T increasing towards some *characteristic temperature* $T_c^{(1)}$. For high enough temperatures it is impossible to separate the particles into those producing a quasistatic landscape and those percolating via phonon assisted hopping in the frozen random potential. The picture for activated motion is irrelevant if the time of a jump is of the same length as the life time of the barrier for the jump. In normal liquids it is more adequate to view particle motion as orbits like in dense gases. The interaction can be treated near the triple point with kinetic equations modified by molecular fields and accounting for excluded volume effects. Below the triple point a particle is surrounded by a polarization cloud and it can move only via the creation of a streaming pattern. The motion in the normal and moderately supercooled state requires the understanding of rattling motion of particles, trapped in cages formed by their neighbors, of the statistical averaging over all the various cage formations, and of the building of flow patterns. This view was confirmed by neutron scattering work, molecular dynamics studies, and theoretical analysis during the past 20 years [12]. The traditional way of fitting transport coefficients like viscosity by Arrhenius formulae within the normal or moderately supercooled state [1], has no rational basis for simple liquids. But this view will loose its relevance if the temperature is lowered towards some $T_c^{(2)}$. Orbits and flow patterns are inadequate concepts if the blocking effects force the particles to correlated jumps, which are separated by long waiting periods.

Thus one expects, that at some temperature $T_c \sim T_c^{(1)} \sim T_c^{(2)}$ the dynamics crosses over from one endemic for the normal and moderately supercooled liquid state to one endemic for the strongly supercooled and glassy state. In simple liquids one expects that for $T \geq T_c$ the structural relaxation patterns appear within the window just below the microscopic excitation band.

2. Some Recent Discoveries

2.1. THE CRITICAL DECAY

The dynamical structure factor $S_q(\omega)$ of CKN for $q \approx 1.9 Å^{-1}$, measured by W. Knaak, F. Mezei and B. Farago with a time of flight neutron scattering spectrometer, is shown in Fig. 3 [13]; here only the measured points over the spectrometer window TOF shall be considered. The data c) refer to $70°C$, a temperature close to the one discussed in Figs. 1,2. The expected α-peak is below the small frequency edge of the used instrument. For frequencies ω below $1THz \approx 4meV$ one does not observe a white noise spectrum, but an increase of $S_q(\omega)$ by about a factor 5. This *mesoscopic spectrum* is *temperature insensitive*. The data change somewhat if T is raised from $-73°C$ (data a)) to $+90°C$ (data d)); but

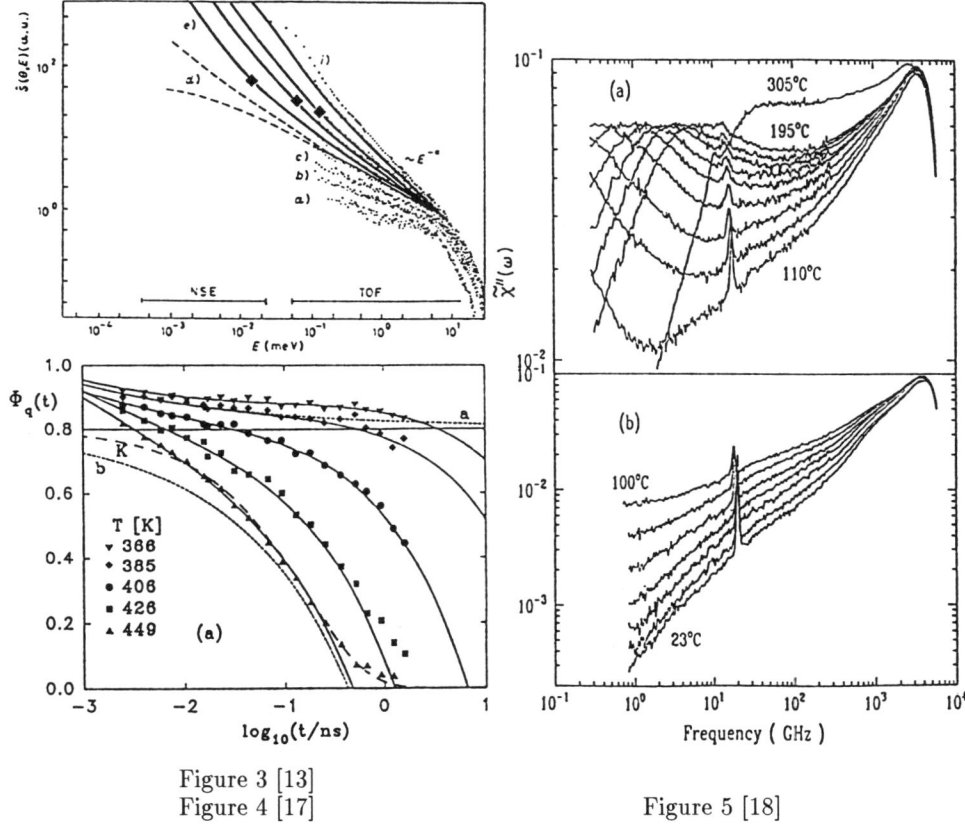

Figure 3 [13]
Figure 4 [17]
Figure 5 [18]

this is caused mainly by the Bose factor relating the dynamical structure factor $S_q(\omega)$ and density spectrum $\phi_q''(\omega)$.

The data for the double logarithmic presentation in Fig. 3 follow a straight line, i.e.

$$\omega \phi_q''(\omega) \propto \kappa_q''(\omega) \propto \omega^a \quad \text{or} \quad \phi_q(t) - f_q \propto 1/t^a . \tag{2.1}$$

The same spectrum has been found also with light scattering [14], and both experiments yield for the anomalous exponent for $T \sim 100°C : a = 0.30 \pm 0.05$. The process, leading to the power laws (2.1) shall be called critical decay. The critical susceptibility spectrum $\kappa_q''(\omega)$ increases sublinearly with increasing frequency. The high frequency α-peak wing decreases with increasing ω and it is very T-sensitive. Therefore, the critical decay is not considered as a part of the α-process.

2.2. THE TWO STEPS RELAXATION SCENARIO

The $\phi_q(t)$ versus $\log t$ diagram for the critical decay exhibits positive curvature: $d^2\phi/d^2 \log t > 0$. The curvature of the corresponding diagram for the initial part of the α-process (1.5) is negative. Thus both mentioned parts are separated by some inflexion point at a certain time t_i^β. The complete relaxation pattern outside the transient, say $t > t_0$, appears as a *two step process*. First the decay curve follows (2.1) towards some *plateau value* f_q.

But at a time t_i^β, large compared to t_0 but small compared to τ_α, the α-process takes over and describes the decay towards the equilibrium value zero. The α-process does not describe the complete structural relaxation process outside the transient motion, but rather the decay from some plateau value f_q, which was reached at the mesoscopic time t_i^β. The α-part of the $\phi(t)$ versus $\log t$ diagram exhibits also an inflexion point at some time t_i^α, due to cross over from the initial part to the long time part with positive curvature. This time can also be used to define a scale for the α-process: $\tau_\alpha \propto t_i^\alpha$. Similarly, one can use $\tau_\beta \propto t_i^\beta$ as characteristic time scale for the identified mesoscopic process.

The process, where $\phi_q(t)$ is close to the plateau value, is occasionally referred to as β-*relaxation*. The center of the β-relaxation window is τ_β. Notice, that this process describes the approach towards f_q according to (2.1) for $t_0 \ll t < t_i^\beta$ and also the initial part of the α-process, for $t_i^\beta < t \ll \tau^\alpha$. In this terminology the link between α- and β-process is the interval between τ_β and τ_α, which is shared by both processes. This interval describes times, which are long on the β-scale τ_β and short on the α-scale τ_α. An ad hoc interpolation between (1.5) and (2.1) reads:

$$\phi_{\rm int}(t) - f = C\left[b(t_i^\beta/t)^a - a(t/t_i^\beta)^b\right]. \tag{2.2}$$

Normalized correlation functions $\phi_q(t)$ for density fluctuations of CKN with $q \approx 1.8 \text{Å}^{-1}$ have been obtained by neutron spin echo spectroscopy. Results within the window $0.04\text{ns} < t < 2\text{ns}$ could be fitted by (1.4) with T-independent $\beta = 0.58$ and $f = 0.84$ [15]. This plateau value was below the value reached within the transient motion, exemplifying the two step scenario. The experiments have been extended by F. Mezei [16] to cover a 2.7 decade window; and a part of his data is reproduced as Fig. 4 [17]. The dashed line with label a is $f + (t^*/t)^a$ for $f = 0.80$, $a = 0.25$ and t^* chosen as to match the data for $385K$. The data follow the specified critical decay law for an interval of almost 2 decades, as expected from Fig. 3. The data do not allow to read off inflexion points compellingly, leave aside the 5 parameters for the fit formula (2.2). One has to impose much more severe limits on data quality for $\phi(t)$ versus $\log t$ results than on ones for $\log \chi''(\omega)$ or $\log \phi''(\omega)$ versus $\log \omega$ graphs, if one wants to arrive at conclusions of comparable certainty.

The evolution of the $\alpha - \beta$-susceptibility spectrum of CKN, as measured by G. Li, W.M. Du, X.K. Chen, H.Z. Cummins and N.J. Tao with depolarized light scattering, is shown as Fig. 5 [18]. The microscopic excitations produce a bump at 4 THz; it was measured by Raman scattering. The data within the GHz band were obtained with a tandem Fabry Perot interferometer. Upon heating, the α-peak position ω_α moves to higher frequencies and enters the spectrometer window at about $130°C$. Heating further, ω_α increases further in qualitative agreement with Fig. 1. The peak can be fitted with a Kohlrausch spectrum using a T-independent exponent $\beta = 0.55 \pm 0.05$ [18]; the fit by (1.4) yields the α-scale τ_α. The α-peak moves over the $\log \omega$ abscissa, without changing its half width. Even at $195°C$, i.e. 30 degrees above T_m, it can still be identified. *No trend of the α-process towards stochastic dynamics upon heating* is observed [18]; such trend would manifest itself as shrinking of the peak width or an increase of β. For still higher temperatures the α-peak merges with the microscopic excitations, but even at $T = 305°C = T_m + 140°C$ it causes a strong anomalous spectrum within a one decade window. The α-peak is not placed on top of a white noise background $\chi''_{\rm white}(\omega) = C_w(\omega t_0))$. The latter would be a straight line of slope unity in the double logarithmic presentation of Fig. 5, which matches the microscopic excitation band for $\omega = \omega_{mic}$. Such spectrum would predict $\chi''_{\rm white}(\omega = 10\text{GHz}) = 10^{-3.5}$ (compare the dashed line in Fig. 38 below). The measured spectrum for $T = 23°C = T_g - 37°C$ is enhanced by about a factor 5 above the specified $\chi''_{\rm white}$ for $\omega = 10\text{GHz}$, and the spectrum for $110°C$ exhibits enhancement by about a factor 100. As discussed above in connection with Fig. 3, this comes about because the critical spectrum $\chi''_{cr}(\omega) = C_c(\omega t_0)^a$ is arbitrarily much larger than $\chi''_{white}(\omega)$ if ω tends to zero. Notice, that indeed the spectrum for $100°C$ follows a straight line with slope $a \sim 0.3$ closely [14].

The high frequency α–peak tail (1.5) and the critical spectrum (2.1) produce a susceptibility minimum at $\omega = \omega_{min}$, where $\chi_{min} = \chi''(\omega = \omega_{min})$. Fig. 5 exhibits these minima for $T \geq 110°C$. Such minima would exist also, if the α–peak would be placed on top of a white noise spectrum. However, in the latter case χ_{min} would be much smaller. The two step relaxation scenario manifests itself by a *susceptibility enhancement near the minimum*. This enhancement is recognized directly, if one knows the microscopic scale $\chi''_{white}(\omega = \omega_{min})$ from measurements within the microscopic excitation band. The minimum position can be used to define a scale for the β-process, to be denoted by $\tau_\beta^- = 1/\omega_{min} \propto t_i^\beta$. Fourier transformation of (2.2) yields an ad hoc interpolation between von Schweidler decay and critical decay:

$$\chi''_{int}(\omega) = \chi_{min}\left[b(\omega/\omega_{min})^a + a(\omega_{min}/\omega)^b\right]/(a+b). \qquad (2.3)$$

2.3. CROSS OVER PHENOMENA

Within the GHz band the spectra of Fig. 5 exhibit *three regimes*. For *high temperatures*, say $T \geq 120°C$, the $\log \chi''$ versus $\log \omega$ curves are concave. The corresponding curves for *low temperatures*, say $T \leq 80°C$, are convex; they have a knee at some frequency ω_e, where the spectral intensity shall be denoted by $\chi_e = \chi''(\omega = \omega_e)$. There is a change of slope of the low temperature $\log \chi''$ versus $\log \omega$ curves from a value close to a for $\omega > \omega_e$ to some larger one for $\omega < \omega_e$. There is no such knee for the high temperature curves, nor is there one for the data in the intermediate regime for $80°C < T < 120°C$. The shape of the minimum within this *intermediate regime*, unfortunately only measured for $T = 110°C$, seems to be different from the one observed for $T \geq 120°C$ in the sense, that the slope of the $110°C$ curve for $\omega < \omega_{min}$ is larger than the one for high temperatures. These observations suggest, that T_c for CKN is located between $80°C$ and $120°C$.

Some elementary data analysis corroborates the preceding conclusion. Fitting the minimum with (2.3) is possible for $T \geq 120°C$, if one chooses [18]:

$$CKN : \qquad a = 0.27 \pm 0.03 , \quad b = 0.46 \pm 0.08 . \qquad (2.4)$$

In Fig. 6 [18] fit curves for T–independent $a = 0.27$, $b = 0.46$ are shown as full lines in comparison with the high temperature data from Fig. 5. Such Fitting procedure does not work anymore for $T = 110°C$; the corresponding fit curve in Fig. 6 accounts for the spectrum for $\omega > \omega_{min}$ only on the expense, that its minimum is shifted below the experimental one. Since the exponents a and b are kept fixed for the fit, the various functions (2.3) differ by the two scale factors χ_{min} and ω_{min} only; the shape of the various $\log \chi''$ versus $\log \omega$ curves is the same for all T. The different full lines in Fig. 6 can be made to coincide, if shifted parallel to the vertical axis by $\log \chi_{min}$ and parallel to the horizontal axis by $\log \omega_{min}$. In order to exclude, that this result is an artifact of choosing an oversimplified interpolation formula (2.3), Fig. 7 [17] is reproduced. Here the data from Fig. 5 for $T = 195°C, 180°C, 170°C$ and $160°C$ are parallel shifted, so that the minima coincide. Indeed, the spectra follow a common curve over a two decade (ω/ω_{min}) interval. *The interval for (ω/ω_{min}), where the data collapse, expands upon cooling.* If one carries out similar shifts for the low temperature data, one gets Fig. 8 [18]. The necessary shifts are used to quantitatively define the knee parameters ω_e, χ_e. Hence also for low temperatures *the data can be reduced to a common curve by parallel shifts, but that curve differs drastically from the one found for high temperatures.* In order to exclude the possibility, that the found temperature variations for ω_e, χ_e are artifacts of data manipulations, Fig. 9 [17] is reproduced. It exhibits a $\log(\chi''(\omega)/\sqrt{\omega})$ versus $\log \omega$ plot for the temperatures $28°C, 45°C, 60°C, 70°C$ and $80°C$. Since the curves in Fig. 5 for $T < 80°C$ exhibit a slope larger than $(1/2)$ for $\omega < \omega_e$ and smaller than $1/2$ for $\omega > \omega_e$, the $\chi''/\sqrt{\omega}$ versus $\log \omega$ graphs exhibit maxima, which are marked by dots in Fig. 9. The position of the maximum ω_m can equally well be used to define the knee; $\omega_m \propto \omega_e$. The $\chi''/\sqrt{\omega}$ curves do not exhibit a maximum in the intermediate nor in the high temperature region.

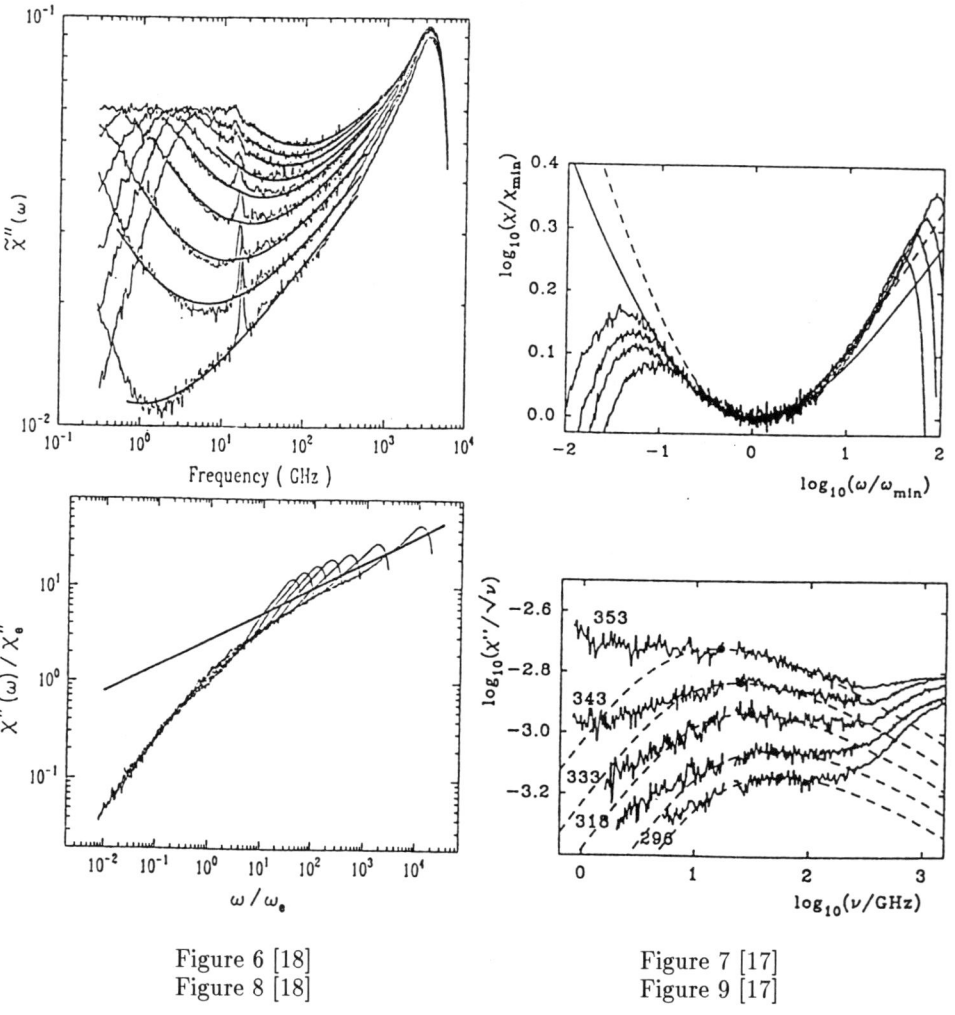

Figure 6 [18]   Figure 7 [17]
Figure 8 [18]   Figure 9 [17]

Fig. 10 [18] shows the measured times $\tau_\beta^-, \tau_\beta^+ = 1/\omega_e$ and τ_α as function of temperature. The times τ_β^- and τ_α increase with decreasing temperature. This reflects the *slowing down of the dynamics upon cooling* as a freezing precursor. The time τ_β^+ increases with increasing temperature. There is *slowing down of the dynamics upon heating*, apparently a precursor phenomenon of the instability of some frozen structure. Fig. 11 [18] exhibits the squares of the spectral intensities χ^2_{min} and χ^2_e as function of temperature. Figures 6–9 demonstrate the relevance of the cross over concept and Figs. 10, 11 suggest [18]:

$$CKN: \quad T_c = (105 \pm 5)°C. \quad (2.5)$$

Temperature T_c is just in the middle between calorimetric glass transition temperature T_g and melting temperature T_m. The ratio τ_α/τ_β^- increases by about one order of magnitude, if the temperature is lowered from $195°C$ to $120°C$. This *separation of scales* upon cooling is another facet of the two step relaxation scenario.

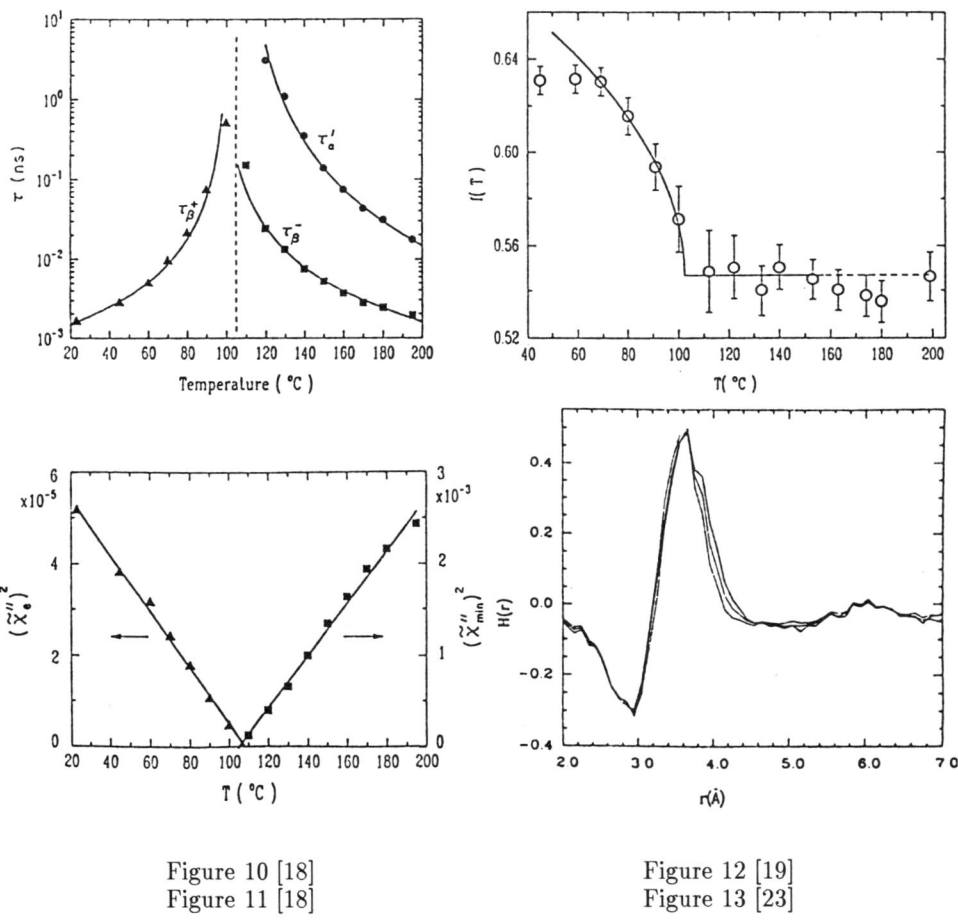

Figure 10 [18]
Figure 11 [18]

Figure 12 [19]
Figure 13 [23]

An instability upon heating the strongly supercooled CKN towards T_c is also indicated by the temperature variation of the Mountain peak area f in (1.3), as is shown in Fig. 12 [19]. The result was obtained in Ref. [19] by extracting the α–contribution for the elastic modulus and then determining f_0 from M_0 and M_∞. The complete modulus, needed to explain the measured Brillouin spectrum, contained a β–contribution, which was assumed to be proportional to the spectrum shown in Fig. 5. The characteristic temperature manifests itself as a cusp of the f_0 versus T graph.

The cross over phenomena, discussed in Figs. 5–12 for CKN, have also been found with light scattering for the van der Waals liquid salol [20, 21]. A similar system is orthoterphenyl, where extensive studies by neutron scattering and other techniques have established $T_c = (290 \pm 10)K$, as can be inferred from Ref. [22] and the papers quoted there.

2.4. THE FACTORIZATION PROPERTY

Typical microscopic excitations in crystalline solids are phonons. In the simplest case they are described by a wave equation, which formulates the connection of the evolution of perturbations in space with that in time. The spectra exhibit the strong correlation between wave vector and frequency changes, given by some dispersion relation $\omega = \Omega_q$; compare equ. (1.2). Also in simple liquids such dispersion relations can be identified, e.g. as peak positions of current relaxation spectra [12]. For the α-process appreciable correlations between space and time variations manifest themselves as q-variations of the relaxation time $\tau_\alpha(q)$ defined e.g. via the decay scale of the density correlator $\phi_q(t)$. Increasing q from 0.1Å^{-1} to 0.9Å^{-1} in CKN, the time $\tau_\alpha(q)$ increases by about a factor 8 [15]. However, the spectrum within the β-window does not change at all if q is varied, except for a change in scale. The spectra factorize in one term h_q and another term to be denoted by $G''(\omega)$. The former does not depend on frequency and it varies only weakly with T. The latter is q-independent and it carries the strong T-dependence of $\omega_{\min}$:

$$\phi_q''(\omega) = h_q G''(\omega), \quad \chi_q''(\omega) = h_q \chi''(\omega), \quad \phi_q(t) = f_q^c + h_q G(t). \qquad (2.6)$$

In the following $G(t)$ shall be referred to as β-correlator, its Fourier cosine transform $G''(\omega)$ as β-spectrum, and $\omega G''(\omega) = \chi''(\omega)$ as β-susceptibility spectrum. Fourier-backtransformation of $S_q \phi_q(t)$ with respect to q yields the space–time density correlation function $G(r,t) = \langle \delta\rho(\vec{r}t)\delta\rho(\vec{0},0)\rangle$ and (2.6) is then equivalent to $G(r,t) = F(r) + H(r)G(t)$. Function $G(r,t)$ has been obtained for CKN via molecular dynamics experiments by G.F. Signorini, J. L. Barrat and M.L. Klein. Fig. 13 [23] reproduces their results for $[G(rt) - F(r)]$ divided by a properly chosen $G(t)$. The curves refer to $177°C$ and are calculated for t = 20 ps, 30 ps and 40 ps. For all three times one gets the same $H(r)$, verifying (2.6). Function $H(r)$ exhibits variations with r, which reflect strong short ranged spacial order.

The specified absence of correlations between space and time variations distinguishes the mesoscopic β-regime on the one hand from the high frequency microscopic as well as the low frequency α-regime on the other hand. The result (2.6) can also be phrased by saying, that β-relaxation deals with *localized modes*. Equations (2.6) can be used as precise definition for the concept "localized excitations".

2.5. COLLOIDAL SUSPENSIONS

The simplest model for matter is the hard sphere system (HSS): particles, whose interaction is given by infinite repulsive cores of radius R, are placed in a box with density n. There is only one control parameter, viz. the packing fraction $\varphi = 4\pi n R^3/3$. The HSS freezes at $\varphi_f = 0.495$ and it melts at $\varphi_m = 0.543$. The system models instantaneous collisions and excluded volume effects and nothing else. The HSS is a good model for certain colloidal suspensions. In this case all thermodynamic functions, like the structure factor S_q, are the same as for the HSS. The dynamics is different, however. The one particle dynamics in suspensions is Brownian rather than Newtonian and hydrodynamic interactions lead to velocity dependent forces. The sphere diameter 2R is of the order of 4000Å and thus comparable to the wave length of light. Therefore photon correlation spectroscopy can be used to measure density correlators $\phi_q(t)$ for qR of order unity, thereby probing the dynamics on the interparticle distance scale. The properties of the dilute system are well explored [24].

It was discovered by P.N. Pusey and W. van Megen [25] that the mentioned suspensions exhibit some critical point φ_c in the following sense. For $\varphi_f < \varphi < \varphi_c$ homogeneous nucleation leads to crystallization within some minutes. For $\varphi > \varphi_c$ no normal crystallites were formed even if one waited for hours. This indicates absence of diffusion for the high density phase. For $\varphi < \varphi_c$ correlations decay to zero as expected for an ergodic liquid, but spontaneous arrest of density fluctuations was observed for large packing:

$$\phi_q(t \to \infty) = 0 \quad \text{if} \quad \varphi < \varphi_c, \quad \text{and} \quad \phi_q(t \to \infty) = f_q > 0 \quad \text{if} \quad \varphi > \varphi_c. \tag{2.7}$$

This result means, that the system is non-ergodic for $\varphi > \varphi_c$, more precisely non-mixing, a property, which is shared with conventional glasses for $T < T_g$. Therefore the limit f_q is referred to as *non-ergodicity parameter*. For $\varphi > \varphi_c$ the Mountain resonance in (1.3) has zero width: $\phi_q^{M''}(\omega) = \pi f_q \delta(\omega)$. Thus f_q is the probability for γ-scattering without transfer of recoil. Such possibility is endemic for crystals, where f_q is called *Debye-Waller factor*. The periodic structure of crystals implies, that f_q is non-zero there only for the discrete sets of wave vectors on the reciprocal lattice. For colloids one finds $f_q \neq 0$ for all q. Hence, the phase for $\varphi > \varphi_c$ is an *amorphous solid*. For the listed reasons the high density phase was called a glass and φ_c a glass transition point. Notice that φ_c, as opposed to the T_g, is an equilibrium concept. No anomaly for S_q was observed for φ near φ_c.

Figure 14 [26] reproduces decay curves measured by W. van Megen and S.M. Underwood for wave vectors q below, near and above the structure factor peak position $q_0 \approx 3.4/R$ [26]. The results for $q \sim q_0$ look similar to those shown in Fig. 4 for CKN. The slowing down of the dynamics for $\varphi > 0.5$ and the φ-sensitivity of the decay curves is evident. A Debye curve is added as dashed line; its decay time $t_0 \approx 10^4 \mu s$ is adjusted so, that the initial transient motion is matched. The stretching of relaxation for $t > t_0$ manifests itself by the $\phi_q(t)$ versus $\log t$ curves to be flatter than the Debye curve. The cross over from decay specific for $\varphi < \varphi_c$ to that specific for $\varphi > \varphi_c$ is quite drastic and yields [26]:

$$\text{HSS}: \quad \varphi_c = 0.578 \pm 0.004. \tag{2.8}$$

The plateau value f_q^c for $\varphi \sim \varphi_c$ is about 0.85 for $q \sim q_0$, but it is considerably smaller for the two wave vectors off the peak position. In the latter cases the positive curvature of the $\phi(t)$ versus $\log t$ graphs, which is characteristic for the critical decay, is obvious and inflexion point times τ_i^β can be read off. The α-relaxation time $\tau_\alpha(q)$ are longer for $q \sim q_0$ than for q off the peak position. The increase of τ_α/τ_β^- with increasing φ is also obvious from the data shown in Fig. 14. Thus, the two step relaxation scenario as discussed in section 2.3 is demonstrated. It would be quite impossible to force the shown data for $q \neq q_0$ and $t \gg t_0$ into a fit by equ. (1.4). This conclusion was arrived at also for decay curves obtained for a somewhat different colloidal suspension in Ref. [28], where it was tested explicitly, that a 6 parameter fit by a sum of two Kohlrausch functions (1.4) could not account for the observed decay process. The Debye-Waller factor for the HSS was measured and a found anomaly could be interpolated by [27]: $f_q = f_q^c + \tilde{h}_q \sqrt{\varphi - \varphi_c}$. This formula is in agreement with what was found in Ref. [19] for CKN. The analogues of Figs. 8 and 11 have been found for colloids also [27] and the graph for the time scales $\tau_\beta^\pm, \tau_\alpha$ versus φ [26] is quite similar to Fig. 10.

One can find φ-independent *functions* of wave vector f_q^c and h_q so, that $[\phi_q(t) - f_q^c]/h_q$ is a q-independent function $G(t)$. This is demonstrated in Fig. 15 [26], where the upper part refers to $\varphi > \varphi_c$ and the lower one to $\varphi < \varphi_c$. The q-independence does not hold for the transient dynamics $t \leq t_0$, nor for the final stage of the α-decay. But it holds for the mesoscopic window $10^{4.5} \mu s \leq t \leq 10^{7.5} \mu s$. The parameters f_q^c and h_q, shown as Fig. 16 [26] exhibit appreciable q-dependence due to the short ranged order in the dense colloids.

Fig. 15 verifies the factorization property (2.6) for a 3 decade dynamical window. Function $G(t)$ depends sensitively on φ and it is qualitatively different for $\varphi > \varphi_c$ and $\varphi < \varphi_c$.

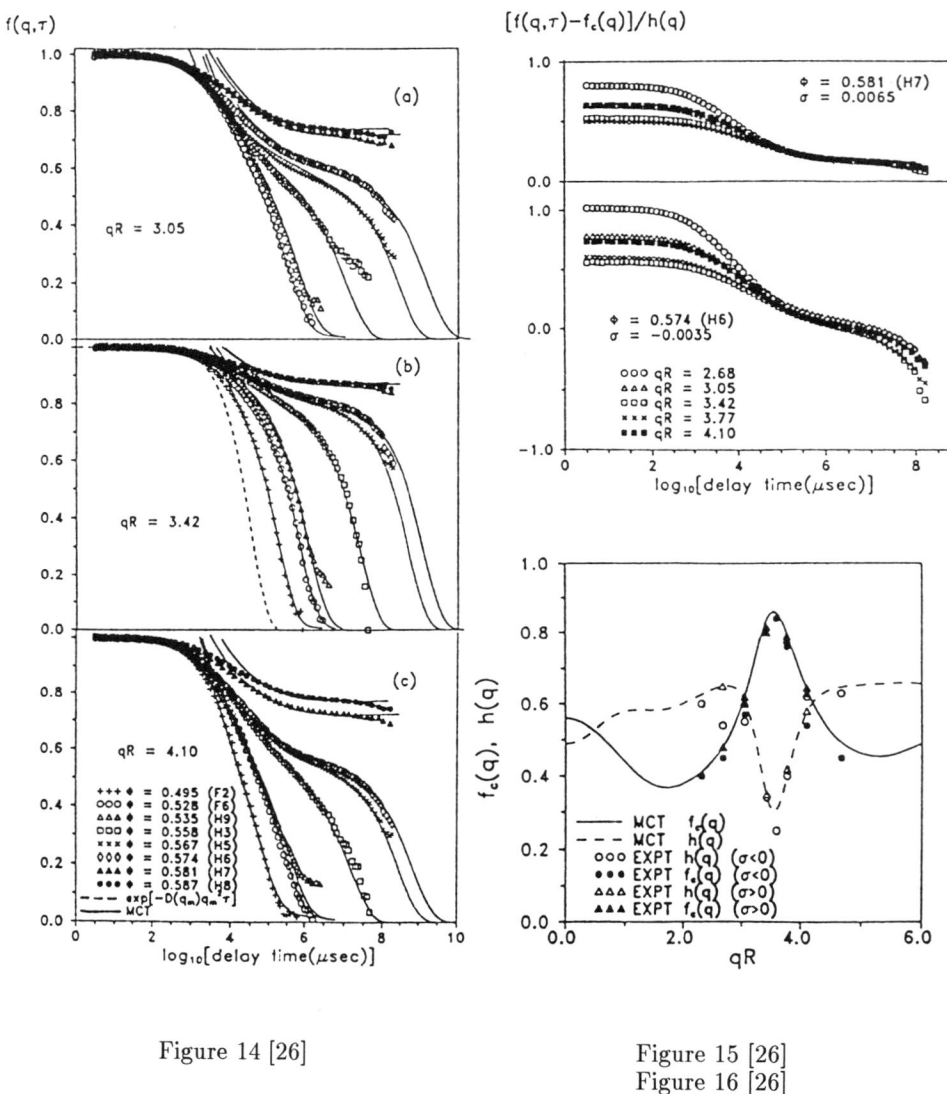

Figure 14 [26]

Figure 15 [26]
Figure 16 [26]

One concludes that the critical packing ratio φ_c of the HSS is the analogue of the cross over temperature T_c identified for CKN and in some other systems. The same cross over phenomena are found for the HSS as for simple conventional systems and thus one infers, that *dense random packing is the reason for structural relaxation phenomena* in general and for the T_c or φ_c peculiarities in particular. The T_c or φ_c cross over does not depend on complexities of intermolecular interactions nor on subtleties of the structure, rather it seems to be a *paradigm for the dynamics of strongly interacting disordered matter*.

3. Ideal Glass Transitions

During the past years the mode coupling theory (MCT) for the supercooled liquid dynamics has been developed. First proposed in 1984 [29], that approach has evolved to a complex machinery whose output has repeatedly been used to analyze experiments. In this section the basic version of that theory, considered as a model for a dynamics, shall be presented. A comprehensive discussion of the MCT and its relation to experiment can be found in a recent review [30], to which the reader is also referred for a list of original publications. Mathematical derivations will be omitted in the following; they are reviewed in Ref. [31].

3.1. BASIC EQUATIONS

The MCT deals with a set of M functions of time t, denoted by $\phi_q(t)$, $q = 1, 2, \ldots, M$. They have continuous derivatives up to second order and the initial conditions are: $\phi_q(t=0) = 1, \partial_t \phi_q(t=0) = 0$. They obey the M equations:

$$\partial_t^2 \phi_q(t) + \Omega_q^2 \phi_q(t) + \Omega_q^2 \zeta_q \partial_t \phi_q(t) + \Omega_q^2 \int_0^t m_q(t-t') \partial_{t'} \phi_q(t') = 0. \tag{3.1}$$

These are generalized oscillator equations with Ω_q^2 and ζ_q specifying Hooke's restoring forces and Newton's friction constants, respectively. Function $m_q(t)$ is a friction kernel, also called retardation or *memory kernel*; it is the constant of proportionality between a force at time t and a velocity at time $t' \leq t$. The frequency $\Omega_q > 0$ and the time $\zeta_q \geq 0$ are scales for the short time evolution. Kernel $m_q(t)$ is coupled to the functions $\phi_k(t)$:

$$m_q(t) = \mathcal{F}_q\left(\vec{V}, \phi_k(t)\right), \quad \mathcal{F}_q(\vec{V}, f_k) = \sum_e \sum_{k_1 \cdots k_e} V^{(e)}(q, k_1 \ldots k_e) f_{k_1} \cdots f_{k_e}. \tag{3.2}$$

The polynomial $\mathcal{F}_q$ is referred to as *mode coupling functional*. The coefficients $V^{(e)}$, which are requested to be non-negative, are the coupling constants of the theory; they are called mathematical *control parameters*. There are N of them, where N depends on the number M of components and on the degree ℓ_0 of the polynomial $\mathcal{F}$. They shall be combined to a vector $\vec{V} = (V_1, \ldots, V_N)$, specifying a point in an abstract N--dimensional control parameter space $\mathbf{K}$. The formulated equations define the basic version of the MCT, also referred to as the simplified or *idealized MCT*.

The specified equations have a unique solution. If the dependence on control parameters shall be emphasized, notations like $\phi_q(\vec{V}, t)$ shall be used. For finite time intervals, $\phi_q(\vec{V}, t)$ depends smoothly on the $(N+1)$ variables $\vec{V}$ and t. Moreover there is the Fourier-representation

$$\phi_q(t) = \int e^{i\omega t} \phi_q''(\omega) d\omega / \pi, \quad \phi_q''(\omega) \geq 0. \tag{3.3}$$

Thus the $\phi_q(t)$ are positive definite functions, called correlation functions or *correlators*. The $\phi_q''(\omega)$ are the *correlation spectra* and $\chi_q''(\omega) = \omega \phi_q''(\omega)$ are the *susceptibility spectra*. The formulated equations of motion exhibit two kinds of non-linearities. The first one is given by the product under the integral in (3.1). The second one occurs in (3.2) if $\mathcal{F}_q$ is of higher than first order. Kernel m_q governs the dynamics of mode q as formulated by (3.1). But the kernel is not given, rather it is determined by the M modes ϕ_q. Kernels and modes have to be determined self-consistently. There is a feed back mechanism between the kernel governing the mode dynamics and the modes governing the kernel evolution.

The simplest specializations of the MCT are one component models dealing with a single correlator $\phi(t)$ only. Then (3.1) holds with index q dropped, and (3.2) reduces to an ordinary polynomial of some degree ℓ_0. For example, with $\ell_0 = 3$ one gets

$$m(t) = \mathcal{F}(\vec{V}, \phi(t)); \; \mathcal{F}(\vec{V}, f) = v_1 f + v_2 f^2 + v_3 f^3 \,. \tag{3.4}$$

Here **K** is three dimensional and $\vec{V} = (v_1, v_2, v_3)$.

Two comments shall be added in order to indicate, that the formulated equations are not quite unmotivated. The first is based on the Fourier–Laplace transform of (3.1), rewritten for the dynamical susceptibility $\kappa_q(\omega)$, and on the Fourier transform $m_q''(\omega)$ of (3.2):

$$\kappa_q(\omega) = -\Omega_q^2/\left\{\omega^2 - \Omega_q^2 + \Omega_q^2 \omega \left[i\zeta_q + m_q(\omega)\right]\right\}, \tag{3.5}$$

$$m_q''(\omega) = \sum_e \sum_{k_1 \ldots k_e} V^{(e)}(q, k_1 \ldots k_e)$$
$$\int d\omega_1 \cdots \int d\omega_e \delta\left[\omega - (\omega_1 + \cdots \omega_e)\right] \phi_{k_1}''(\omega_1) \cdots \phi_{k_e}''(\omega_e)/(2\pi)^{\ell-1} \,. \tag{3.6}$$

Equation (3.5) generalizes a typical oscillator susceptibility by replacing a friction term $i\zeta_q$ by a frequency dependent function $M_q(\omega) = i\zeta_q + m_q(\omega)$. Normalized spectra can be viewed as probability densities, so that (3.3) identifies $\phi_q(t)$ as characteristic function of the probability density $\phi_q''(\omega)/\pi$. The latter is the probability density for finding a contribution to the mode q, fluctuating with frequency ω. Thus, equation (3.6) formulates the idea, that the probability density $m_q''(\omega)$ is composed of that for sums of independently contributing fluctuating complexes. The second comment points out that T.D. Kirkpatrick and D. Thirumalai [32] and, with some reservation, V.N. Prigodin [33] have discovered certain models for many particle systems, for which the MCT specialization (3.4) yields the exact description for the dynamics. Thus, (3.1, 3.2) can be considered as generalizations of well controlled examples for problems in statistical physics.

3.2. GLASS TRANSITION SINGULARITIES

The MCT equations of motion are regular. No ad hoc assumptions on singularities, transitions, fractal behaviour and the like are made. The interesting properties concern the long time behaviour of the correlators and the low frequency properties of the spectra. Let us consider first the dependence on $\vec{V}$ of the long time asymptotes $\phi_q(\vec{V}, t \to \infty) = f_q(\vec{V})$. These M functions are solutions of the coupled implicit algebraic equations

$$f_q/(1 - f_q) = \mathcal{F}_q(\vec{V}, f_k); \; 0 \leq f_k < 1; \; q, k = 1, \ldots, M \,. \tag{3.7}$$

There is the trivial solution: $f_q^{(0)} = 0$. For a given $\vec{V}$ there may be Z non–trivial solutions: $f_q^{(\alpha)}(\vec{V}), \alpha = 1, \ldots, Z$. One finds the *maximum property*:

$$f_q(\vec{V}) \geq f_q^{(\alpha)}(V); \quad q = 1, \ldots, M; \quad \alpha = 0, 1, \ldots, Z \,. \tag{3.8}$$

The $f_q(\vec{V})$ are not smaller than the corresponding $f_q^{(\alpha)}(\vec{V})$ for any solution of (3.7).

The regularity properties of the $f_q^{(\alpha)}(\vec{V})$ are governed by the M by M stability matrix, constructed from the derivatives of the mode coupling polynomial: $C_{qk}(\vec{V}) = [\partial \mathcal{F}_q(\vec{V}, f_k)/\partial f_k](1 - f_k)^2$. The $f_q^{(\alpha)}(\vec{V})$ depend smoothly on $\vec{V}$, unless matrix C has eigenvalue unity. There is a *maximum eigenvalue* $E(\vec{V})$; it is real, non–degenerate and exceeds the moduli of all other eigenvalues of the stability matrix. There appear several generic possibilities. There is an open domain $\mathcal{D}_L$ in $\mathbf{K}$, which contains the small coupling region $|\vec{V}| \sim 0$, where the long time limit is trivial

$$\vec{V} \epsilon \mathcal{D}_L : f_q(\vec{V}) = 0. \qquad (3.9a)$$

There is an open domain $\mathcal{D}_G$, which contains the strong coupling region $|\vec{V}| \sim \infty$, where there is a non-trivial long time limit depending smoothly on $\vec{V}$:

$$\vec{V} \epsilon \mathcal{D}_G : f_q(\vec{V}) \geq 0, \quad 0 < E(\vec{V}) < 1. \qquad (3.9b)$$

The boundary points $\vec{V}_c$ of $\mathcal{D}_G$ form a hypersurface, to be denoted by $\mathcal{D}_C$. Here $f_q(\vec{V})$ exhibits singularities:

$$\vec{V} = \vec{V}_c \epsilon \mathcal{D}_C : f_q(\vec{V}_c) = f_q^c \geq 0; \quad E(\vec{V}_c) = 1. \qquad (3.9c)$$

Anticipating that every point $\vec{V}_c \epsilon \mathcal{D}_c$ can be connected with the large coupling region by a path avoiding $\mathcal{D}_c$ one finds: $\mathbf{K} = \mathcal{D}_L \cup \mathcal{D}_G \cup \mathcal{D}_C$. For reasons, explained in section 2.5, the points in $\mathcal{D}_G$ are called *ideal glass states*. S.F. Edwards and P.W. Anderson [34] pointed out explicitly, that glass states shall be characterized by some correlation function having a non-vanishing long time limit. Therefore $f_q(\vec{V})$ is also called *Edwards-Anderson parameter*. The points in $\mathcal{D}_L$ are called *liquid states*. The truly important parameters of the MCT are the $\vec{V}_c \epsilon \mathcal{D}_C$, called *critical states* or glass transition singularities.

Let us compile a series of concepts, introduced to describe the MCT dynamics near critical states. The Edwards-Anderson parameters at the singularities are called *critical non-ergodicity parameters* $f_q^c \geq 0$. From the dangerous eigenvector of the stability matrix a *critical amplitude* $h_q \geq 0$ is constructed. On the hypersurface $\mathcal{D}_C$ a continuous function $0 \leq \lambda(\vec{V}_c) \leq 1$ is defined, called *exponent parameter*: $\lambda < 1$ specifies inner points of $\mathcal{D}_C$ and for $\lambda \to 1$ endpoints are found. For a neighbourhood of inner points of $\mathcal{D}_C$ a smooth function of $\vec{V}$ is defined, called *separation parameter* $\sigma(\vec{V})$; it is used to characterize the critical hypersurface by the implicit equation: $\sigma(\vec{V}_c) = 0$. The convention is chosen such, that $f_q(\vec{V})$ increases with σ. One gets

$$f_q(\vec{V}) - f_q^c = h_q \sqrt{\sigma/(1-\lambda)} + 0(\sigma), \qquad \sigma \geq 0, \qquad (3.10)$$

and finds for the long time decay at the singularity

$$\phi_q(\vec{V}_c, t) = f_q^c + h_q(t_0/t)^a + 0\left((t_0/t)^{2a}\right). \qquad (3.11)$$

The exponent a is the solution of $\lambda = \Gamma(1-a)^2/\Gamma(1-2a)$, $0 < a \leq 1/2$. The time scale t_0, depending smoothly on $\vec{V}_c$, is called *matching time*. Let us follow the practice in singularity theory and consider smooth curves $C : x \to \vec{V}(x)$, which intersect $\mathcal{D}_C$ transversally. The curve parameter x plays the role of an *external control parameter*. The value for the singularity, $\vec{V}(x = x_c) = \vec{V}_c$, is called *critical parameter* x_c. The convention shall be adopted, that $\sigma > 0$ for $x > x_c$. Then one can connect the separation parameter with x by

$$\sigma = C_c \cdot \varepsilon + O(\varepsilon^2); \qquad \varepsilon = (x - x_c)/x_c. \qquad (3.12)$$

Here $C_c > 0$ is a gauge constant relating σ to the dimensionless distance ε. For simple cases as e.g. specified by (3.4) one can evaluate all the mentioned quantities elementarily. But the general expressions for λ, σ, h_q are rather involved.

The hypersurface $\mathcal{D}_C$ exhibits a strange shape. There are sharp edges and self crossing pieces. Therefore a digression might be adequate. Singularities of the solutions of equations like (3.7) are called bifurcations. The special cases of interest here, where the singularity is caused by a non-degenerate dangerous eigenvalue $E(\vec{V})$, are denoted as A_ℓ bifurcations, $\ell = 2, 3, \ldots$. For A_ℓ there are ℓ solutions of (3.7), which coalesce if $\vec{V} \to \vec{V}_c$. The inner points of the $(N-1)$-dimensional manifold $\mathcal{D}_C$ are A_2 bifurcations, or *Whitney fold singularities*. If two smooth pieces of $\mathcal{D}_C$ are joint in a $(N-2)$-dimensional edge, one gets an A_3 etc. There are two generic possibilities for fold bifurcations. Either $f_q^c = 0$ for all q or $f_q^c > 0$. The first case is referred to in MCT as *type A* glass transition singularity, and the second case as *type B* singularity. Type A transitions, where the non-ergodicity parameter changes continuously, can occur only, if the mode coupling polynomial $\mathcal{F}_q$ has linear terms, and then (3.12) is to be modified to $\sigma \propto \varepsilon^2$. In the following only type B folds shall be discussed.

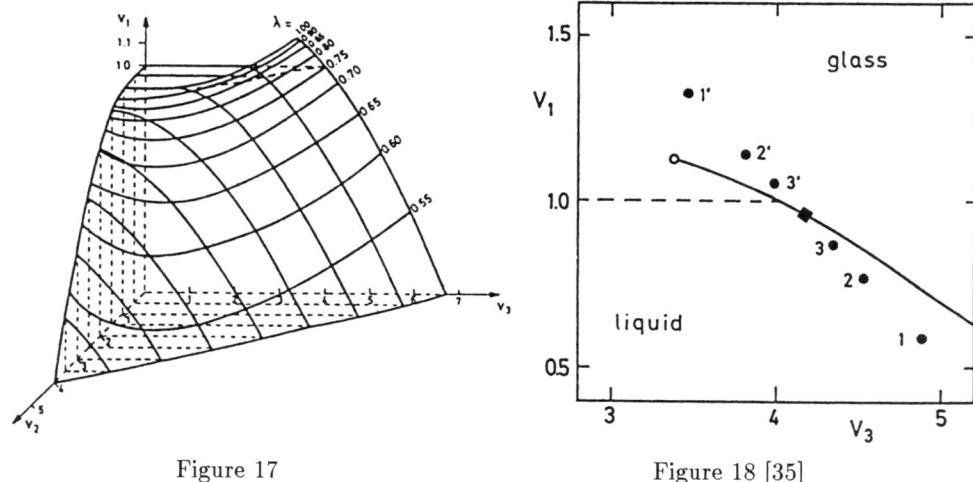

Figure 17   Figure 18 [35]

Figure 17 exhibits the surface $\mathcal{D}_C$ for the model (3.4). Cuts with planes $v_2 = 0.0, 0.5, 1.0, \ldots, 3.5$ are marked and also the lines of constant λ are drawn. The liquid region $\mathcal{D}_L$ is separated from the glass region $\mathcal{D}_G$ by a smooth surface of type B folds for $v_1 \le 1$. The other boundary piece is the plane $v_1 = 1$, where type A transitions occur. For $v_2 < 1$, a type B surface piece cuts into the glass region for $v_1 > 1$. This piece terminates in a line of A_3-singularities, characterized by $\lambda = 1$. Figure 18 shows a cut $v_2 = 0$ of the parameter space. This is a part of the parameter plane $\vec{V} = (v_1, v_3)$ for the one component model with $\mathcal{F}(\vec{V}, f) = v_1 f + v_3 f^3$. The dashed line is the type A part of $\mathcal{D}_C$ and the full one is the type B transition line. The A_3 endpoint singularity is marked by a circle.

Figures 19 - 21 [35] exhibit a set of correlators and susceptibilities evaluated for model (3.4) with $\zeta = 5/\Omega$. Parameter points with label n, n' are chosen on a straight line: $v_1 = v_1^c \mp 0.730/2^n$, $v_2 = 0$, $v_3 = v_3^c \pm 1.424/2^n$. The critical point has coordinates $v_1^c = 0.9549$, $v_2^c = 0$, $v_3^c = 4.1733$. The exponent parameter is $\lambda = 0.72$ and the critical non-ergodicity parameter reads $f^c = 0.52$; it is indicated by a horizontal line in Fig. 19. Some of the parameter points are marked as dots in Fig. 18. A Debye curve $\exp -(t/t_m)$ is shown as dashed line in Fig. 19 with label w. The time t_m is chosen such, that the short time transient dynamics is matched. For the model (3.4) this curve represents what one would consider as the normal liquid state dynamics. Similarly, the dashed lines with label w in Figs. 20, 21 represent estimation for white noise spectra, matched to the high

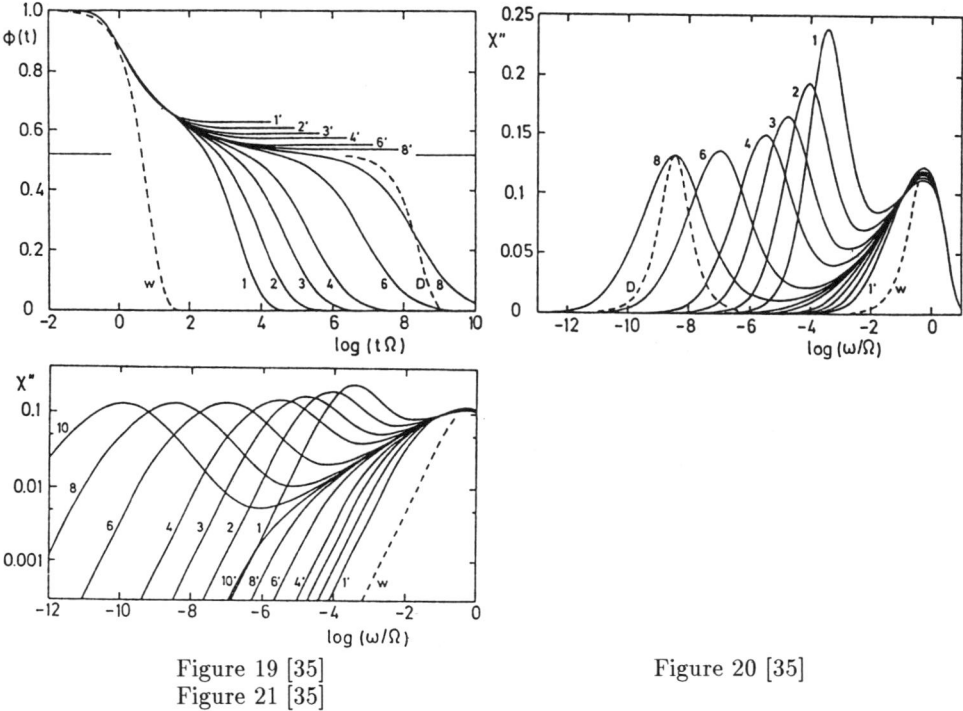

Figure 19 [35]
Figure 20 [35]
Figure 21 [35]

frequency excitation band $\omega \sim \Omega$. A Debye curve $f^c \exp -(t/\tau)$ is also added as dashed curve with label D in Fig. 19. Similarly, the dashed spectrum with label D in Fig. 20 represents Debye relaxation. Obviously, the shown results exhibit a two step relaxation scenario as discussed in section 2.2. The Debye curves are added to those figures, in order to make it plain, that the results show the most important phenomenon for cooled liquid dynamics: stretching of the relaxation.

3.3. THE FIRST SCALING LAW REGIME

Let us introduce the deviations of the correlator from the critical plateau value: $\eta = \phi_q(\vec{V}, t) - f_q^c$. The equations of motion (3.1), (3.2) can then be simplified by asymptotic expansions using η as small quantity. In leading order one gets $\eta = h_q G(t)$, i.e.

$$\phi_q(\vec{V}, t) = f_q^c + h_q G(t) + O_q(\eta^2). \tag{3.13a}$$

Here $G(t)$ is the solution of

$$\sigma + \lambda G(t)^2 = (d/dt) \int_0^t G(t - t') G(t' dt') \tag{3.13b}$$

obeying the initial condition

$$\lim_{t \to 0} G(t)(t/t_0)^a = 1. \tag{3.13c}$$

The quantities f_q^c, h_q, t_0, λ and σ are the same as used in section 3.2. The found result is referred to as the *reduction theorem*. The problem of solving M coupled equations for M correlators is reduced to the problem of solving one equation (3.13b) for the function $G(t)$. The problem of studying the dependence of correlators on N coupling constants $V_1, \ldots, N_N$ is reduced to studying the dependence of G on the separation parameter σ, which appears as the only *relevant control parameter*. The determination of the influence of the short time motion on the long time dynamics is reduced to fixing the time t_0:

$$G(t) = g(t/t_0, \sigma). \qquad (3.13d)$$

The dependence on details of the model is reduced to the dependence of g on the parameter λ. Within the implicitly specified region of $(\vec{V}, t)$, the solutions of the MCT equations exhibit *robustness*. They reflect, up to obvious scales, only the topology of the bifurcation hypersurface via the number λ. For example, all models exhibit the same susceptibility spectra $\chi_q''(\omega) = h_q \omega G''(\omega)$ up to the scales h_q, t_0, provided they refer to the same value for λ. This holds not only for the dependence on frequency ω but also for the dependence on all the coupling constants, which enter only as the combination $\sigma(\vec{V})$.

Function g exhibits a homogeneity property, called *scaling law*:

$$g(t/t_0 y, \sigma \cdot y^{2a}) = y^a g(t/t_0, \sigma), \qquad y > 0. \qquad (3.14)$$

The function G changes by a factor y^a if the time is rescaled by $1/y$ and σ is rescaled by y^{2a}. Let us choose $y = |\sigma|^{-1/2a}$ and introduce the σ–independent *master functions* $g_\pm(\hat{t}) = g(t/t_0, \pm 1)$, determined by λ. One obtains

$$G(t) = c_\sigma g_\pm(t/t_\sigma), \quad \sigma \gtrless 0; \qquad c_\sigma = |\sigma|^{\frac{1}{2}}, \quad t_\sigma/t_0 = |\sigma|^{-\frac{1}{2a}}. \qquad (3.15)$$

Thus, the control parameter σ enters $G(t)$ via the *correlation scale* c_σ and *time scale* t_σ only. If one considers $\log \chi''$ versus $\log \omega$ diagrams, curves for different σ have the same shape. Different curves coincide if shifted by $\log c_\sigma$ parallel to the vertical axis and by $\log t_\sigma$ parallel to the horizontal axis. There is slowing down of the dynamics for $\sigma \to 0$ in the sense, that the scale $t_\sigma \to \infty$. The scale diverges if x in (3.12) is lowered within the glass region and also if x is raised within the liquid region towards x_c. The preceding equations (3.13 - 3.15) formulate a limit theorem for the MCT equations of motion (3.1, 3.2):

$$\lim_{\vec{V} \to \vec{V}_c} \left[\phi_q(\vec{V}, \hat{t} t_\sigma) - f_q^c \right] / [c_\sigma h_q] = g_\pm(\hat{t}), \quad \sigma \gtrless 0. \qquad (3.16)$$

The result (3.13a, 3.15) describes the dynamics in the limit $\vec{V} \to \vec{V}_c$, $t \to \infty$ so that the rescaled time $\hat{t} = t/t_\sigma$ is fixed. It describes the solution for parameters $\vec{V}$ near the glass transition singularity $\vec{V}_c$ on scale t_σ.

The master functions interpolate smoothly between large and small time asymptotes and quantitative details can be inferred from published tables. For short rescaled times $\hat{t}$ one gets the power series expansion

$$g_\pm(\hat{t}) = \left[1/\hat{t}^a \right] \pm A_1 \hat{t}^a + \cdots. \qquad (3.17)$$

This series is dominated by the *critical decay law*, $g_\pm(\hat{t} \to 0) \sim 1/\hat{t}^a$, specified by the *critical exponent* a. This result together with (3.15, 3.13a) reproduces (3.11). For large rescaled times the glass solution approaches arrest

$$g_+(\hat{t} \to \infty) = 1/\sqrt{1-\lambda}. \qquad (3.18)$$

Together with (3.15, 3.13a) this result reproduces (3.10). The liquid solution exhibits a power law divergency for large rescaled times:

$$g_-(\hat{t}) = -[B\hat{t}^b] + B_1/[B\hat{t}^b] + \cdots . \tag{3.19a}$$

This asymptotic expansion is dominated by the *von Schweidler law*, $g_-(\hat{t} \to \infty) \sim -B\hat{t}^b$, specified by the *von Schweidler exponent* b. All the numbers like A_1, B, B_1 are given by λ. In particular b has to be determined from $\Gamma(1+b)^2/\Gamma(1+2b) = \lambda$, $0 < b \le 1$. Substituting the leading term from (3.19a) into (3.15) and this into (3.13a) one gets the von Schweidler law in the form

$$\phi_q(t) = f_q^c - h_q B(t/t'_\sigma)^b . \tag{3.19b}$$

Here a *second time scale* t'_σ is introduced by

$$t'_\sigma = t_0/|\sigma|^\gamma, \qquad \gamma = (1/2a) + (1/2b) . \tag{3.20}$$

Figures 22, 23 exhibit the master functions $g_\pm(\hat{t})$ and the master spectra $\hat{\chi}''_\pm(\hat{\omega})$ for $\lambda = 0.72$, where $a = 0.318$, $b = 0.608$. The dashed lines show the critical decay laws.

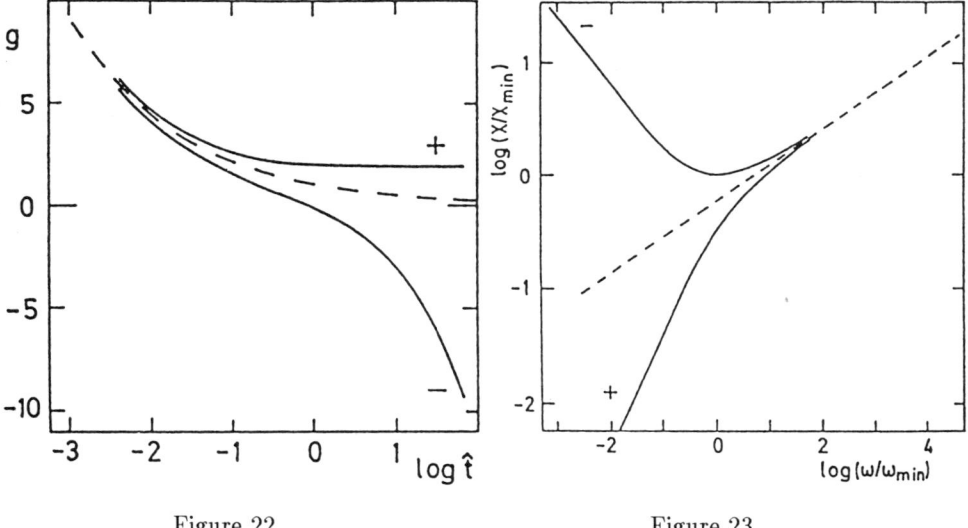

Figure 22   Figure 23

For $\hat{t} \ll 1$, i.e. for times t short compared to the scale t_σ, the $\log \phi$ versus $\log t$ diagram exhibits positive curvature as signature for the critical decay. In leading order, the $\log \chi''$ versus $\log \omega$ diagram is a straight line with slope a for large rescaled frequencies $\hat{\omega} = \omega t_\sigma$. The next to leading term in (3.17) implies, that the $\log \chi''$ versus $\log \omega$ graph has positive curvature for $\sigma < 0$ but negative curvature for $\sigma > 0$. For $t > t_\sigma$, the glass correlator $G(t)$ levels off at the plateau value (3.18) and this implies, that $\chi''(\omega)$ crosses over to white noise behaviour: $\chi''_+(\omega \ll 1/t_\sigma) \propto c_\sigma(\omega t_\sigma)$. Thus the $\log \chi''$ versus $\log \omega$ curve exhibits a knee at some frequency ω_e, where $\chi''(\omega_e) = \chi_e$. Since $a < 1/2$, the $\chi''(\omega)/\sqrt{\omega}$ versus ω graph for the glass exhibits a maximum at some frequency ω_m, where $\chi''(\omega_m) = \chi_m$. The identified spectral features vary sensitively with control parameter x, and from (3.15) one obtains

$$\omega_e = c_e^1/t_\sigma \,,\, \omega_m = c_m^1/t_\sigma \,,\, \chi_e = c_e^2/c_\sigma \,,\, \chi_m = c_m^2/c_\sigma \,. \tag{3.21}$$

For $t > t_\sigma$ the liquid correlator crosses over to (3.19a), so that the curvature of the ϕ versus $\log t$ diagram becomes negative. There is an inflexion point at some time t_i^β, where the correlator has some value $G_i = G(t_i^\beta)$. Similarly, with $\hat{\omega} = \omega t_\sigma$ decreasing, the susceptibility spectrum increases for $\hat{\omega} \ll 1$. Thus the χ'' versus ω graph exhibits a minimum at some frequency $\omega_{\min}$, where $\chi''(\omega_{\min}) = \chi_{\min}$. The identified signatures are σ-sensitive as follows from the scaling law (3.15):

$$\omega_{\min} = c_{\min}^1/t_\sigma \,,\, t_i^\beta = c_i^1 t_\sigma \,,\, \chi_{\min} = c_{\min}^2 c_\sigma \,,\, G_i = c_i^2 c_\sigma \,. \tag{3.22}$$

Let us reiterate, that the 8 coefficients c of proportionality in (3.21, 3.22) are given by λ.

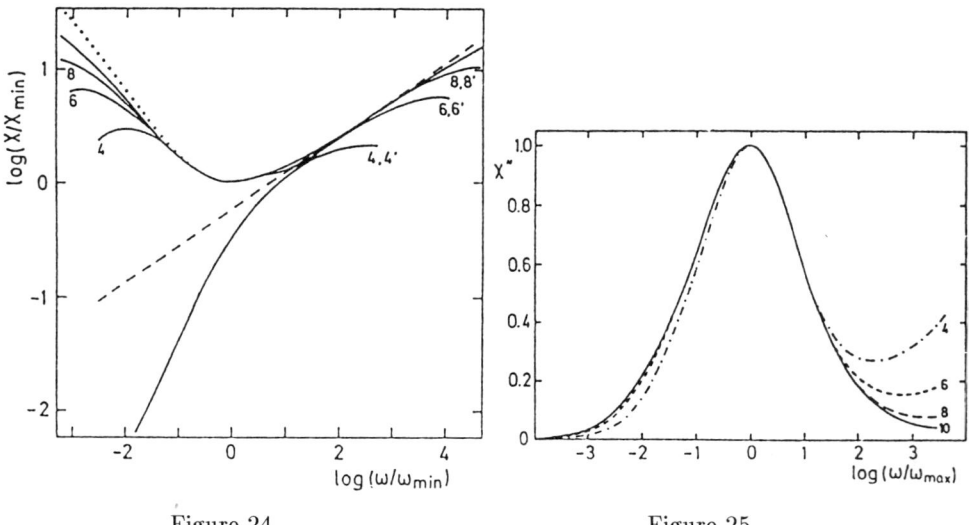

Figure 24   Figure 25

The divergency of (3.19b) for $t \to \infty$ implies, that the correction terms $O(\eta^2)$ in (3.13a) become of order unity for $t \approx t'_\sigma$. It is not allowed to let $\hat{t} \to \infty$ for fixed $\sigma < 0$. The derived formulae for $\sigma < 0$ describe only the first part of the two step relaxation process. There is, however, separation of the scales $t'_\sigma/t_\sigma \to \infty$ for $\sigma < 0, \sigma \to 0$. Figure 24 shows the spectra from Fig. 21 shifted so that for every $|\sigma|$ the minima of the liquid results coincide. The dotted curves reproduce the master spectra from Fig. 23. The various spectra coincide near the minimum and near the knee, respectively. The interval of rescaled frequencies $\hat{\omega} = \omega/\omega_{\min}$, where the spectra follow the master curves, expands upon lowering $|\sigma|$, and this is the contents of (3.16).

The obtained results are similar to those discussed in section 2 for the mesoscopic dynamics as was anticipated already by the choice of the notations. Therefore G is called β-correlator, and the first scaling law regime is called β-regime of the MCT.

3.4. THE SECOND SCALING LAW REGIME

The dynamics for parameter $\vec{V}$ near a glass transition singularity $\vec{V}_c$ and times exceeding t_σ is governed by the asymptotic law

$$\lim_{\vec{V}\to\vec{V}_c}\phi_q(\vec{V},\tilde{t}t'_\sigma) = F_q(\tilde{t}). \tag{3.23}$$

Here the σ-independent function F_q is obtained as solution of the equation

$$F_q(\tilde{t}) = m_q^c(\tilde{t}) - (d/d\tilde{t})\int_0^{\tilde{t}} m_q^c(\tilde{t}-\tilde{t}')F_q(\tilde{t}')d\tilde{t}'. \tag{3.24a}$$

Kernel $m_q^c(\tilde{t}) = \mathcal{F}_q(\vec{V}_c, F_k(\tilde{t}))$ is the mode coupling functional evaluated at the singularity. The equation (3.24a) is scale invariant: if $F_q(\tilde{t})$ is a solution, the same is true for $F_q(\Omega\tilde{t})$ for all $\Omega > 0$. The scale is to be fixed by an initial condition. For the limit from the glass side, i.e. $\vec{V} \to \vec{V}_c$ but $\sigma(\vec{V}) > 0$, one gets $F_q(\tilde{t}) = f_q^c$. This case shall not be considered any further, since it is already discussed completely in section 3.3. The following discussion can be restricted to $\sigma < 0$, and there the initial condition reads:

$$F_q(\tilde{t}) = f_q^c - Bh_q\tilde{t}^b + O_q(\tilde{t}^{2b}). \tag{3.24b}$$

The various symbols have the same meaning as in the preceding sections: f_q^c, h_q are critical Debye Waller factor and critical amplitude respectively, b is the von Schweidler exponent, t'_σ is the second time scale and B is the coefficient entering (3.19a). The result (3.23) describes again a *scaling law*:

$$\phi_q(\vec{V},t) = F_q(t/t'_\sigma). \tag{3.25}$$

Near the transition, the correlators depend on control parameters sensitively only via a time scale t'_σ. The *master functions* $F_q(\tilde{t})$ do not exhibit a sensitive dependence on the control parameter x. Equivalent scaling laws hold for all other quantities. For example, the susceptibility spectrum obeys $\chi''_q(\vec{V},\omega) = \tilde{\chi}''_q(\omega t'_\sigma)$. It is given by a σ-insensitive master spectrum, $F''_q(\tilde{\omega})\tilde{\omega} = \tilde{\chi}''_q(\tilde{\omega})$. The sensitive dependence on x enters only via the formation of a rescaled frequency $\tilde{\omega} = \omega t'_\sigma$. The scaling laws are equivalent to the statement that the ϕ versus $\log t$ diagrams or the χ'' versus $\log\omega$ graphs have a σ-independent shape. Curves referring to different x can be superimposed by shifts parallel to the $\log t$ or $\log\omega$ abscissa by $\log t'_\sigma$. Therefore the scaling law (3.25) is referred to as the *superposition principle*. In Fig. 25 the normalized susceptibility peaks of Fig. 20 are reproduced but shifted so, that the maxima coalesce. The interval of rescaled frequencies $\tilde{\omega}$, where the various spectra coincide, expands upon lowering the separation $|\sigma|$; and this is the contents of (3.23).

Equation (3.24b) is the von Schweidler law, appearing here in the same form as discussed in section 1.2. This power law is obtained as the double limit for the solutions of (3.1, 3.2):

$$\lim_{\tilde{t}\to 0}\lim_{\vec{V}\to\vec{V}_c}\left[\phi_q(\vec{V},\tilde{t}t'_\sigma) - f_q^c\right]/\left[h_q\tilde{t}^b\right] = -B. \tag{3.26a}$$

It describes the dynamics near the transition for long times t, where however the rescaled time $\tilde{t} = t/t'_\sigma$ becomes short. It deals with the short on scale t'_σ dynamics and describes the high frequency part of the α-process. The same law was found in section 3.3, albeit for the quite different double limit:

$$\lim_{\hat{t}\to\infty}\lim_{\vec{V}\to\vec{V}_c}\left[\phi_q(\vec{V},\hat{t}t_\sigma) - f_q^c\right]/\left[c_\sigma h_q\hat{t}^b\right] = -B. \tag{3.26b}$$

It describes the dynamics near the transition for long times, where the rescaled time $\hat{t} = t/t_\sigma$ also becomes large. It deals with the long on scale t_σ dynamics and describes the low frequency part of the β–minimum. The von Schweidler law is valid on the time interval $t_\sigma \ll t \ll t'_\sigma$, where the first scaling law (3.15) and the second scaling law (3.25) yield identical results. Similarly, the α–spectrum and the β–spectrum coincide for $1/t'_\sigma \ll \omega \ll 1/t_\sigma$, as is exemplified by Figs. 24, 25.

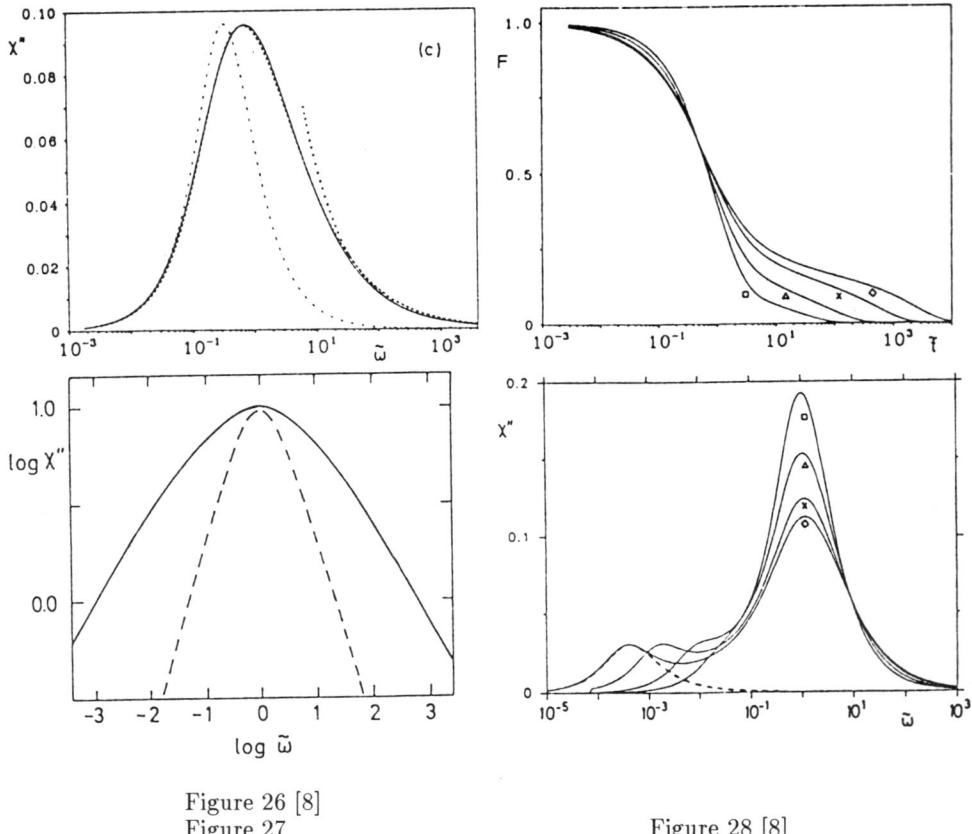

Figure 26 [8]
Figure 27
Figure 28 [8]

The solutions $F_q(\tilde{t})$ of (3.24) decay monotonically. For rescaled times $\tilde{t} \approx 1$ the master functions are fallen so much below the plateau $f_q(\vec{V}_c)$, that the glass transition point $\vec{V}_c$ ceases to be relevant for the evolution. One expects as generic possibility, that $F_q(\tilde{t})$ relaxes quickly to zero. Hence one expects a simple α–bump for the susceptibility spectrum with white noise behaviour for $\omega \ll \omega_{\max}$ and $\chi''(\omega) \propto 1/\omega^b$ for $\omega \gg \omega_{\max}$. The von Schweidler decay is identified as the generic reason for the stretching of the α–process and the structural relaxation peak would have the asymmetric form discussed in section 1.2. Let us consider two examples. Fig. 26 [8] shows the spectrum obtained by solving (3.24) for the one component model (3.4). Parameters were chosen such, that $v_3 = 0$ and $\lambda = 0.7$. The dotted line exhibits the von Schweidler tail with $b = 0.64$. The chain curve is a Debye curve. A Kohlrausch law (1.4) with a stretching exponent $\beta = 0.59$ is shown as dashed line. It fits the spectrum almost perfectly over a window larger than 6 decades. In this example one would have to expand the window to even larger $\tilde{\omega}$, in order to notice, that the Kohlrausch law is not a correct form for the MCT spectrum. Usually, the deficiencies of (1.4) are more obvious, as is demonstrated in Fig. 2. Remember, that there the dashed line was the best Kohlrausch law fit for the shown CKN spectra measured for 71.4°C. The full line in Fig. 2, which matches the data over a 5 decade window, was calculated from (3.24) for an $M = 3$ model.

The solutions of (3.24) may be more subtle than discussed in the preceding paragraph. For the long time limit $f'_q(\vec{V}_c) = F_q(\tilde{t} \to \infty)$ one derives the same equation (3.7) as before. The whole discussion of section (3.2) can start anew. The N-dimensional control parameter space $\mathbf{K}$ is to be replaced by the $(N-1)$-dimensional manifold $\mathcal{D}_C$ near $\vec{V}_c$. The initial condition $\phi_q(t \to 0) = 1$ is to replaced by $F_q(\tilde{t} \to 0)/f_q^c = 1$. In the maximum property (3.8) the largest solution has to be ignored. An interesting new long time dynamics appears, if $\vec{V}_c$ moves near another glass transition singularity $\vec{V}'_c$, connected with a bifurcation of the second largest solution of (3.7). For this dynamics t'_σ plays the role of the matching time t_0 and the $\alpha - \beta$-relaxation pattern, discussed above, plays the role of the short time dynamics. The singularity $\vec{V}'_c$ is specified by a second exponent parameter λ' and a second separation parameter $\sigma'(\vec{V}_c)$, so that $\sigma'(\vec{V}_c \to \vec{V}'_c) \to 0$. Whenever stretching is present for $t \gg t'_\sigma$ or $\omega \ll \omega_{\max}$, it can be attributed to another glass transition singularity $\vec{V}'_c$. Fig. 27 exhibits an example, where the α-peak looks rather symmetric because the low frequency wing follows the law $\chi''(\omega \ll \omega_{\max}) \propto \omega^{1/2}$. The large frequency wing exhibits a von Schweidler decay with $b = 0.56$. The dashed line is a Debye spectrum. The result was calculated for the model (3.4) for $v_1 = 1$, $v_2 = 0$, $v_3 = 4$. At this point the part of $\mathcal{D}_C$ containing $\vec{V}'_c$, shown in Fig. 18 in dashed, crosses the part with $\vec{V}_c$, shown as full line. Fig. 28 [8] exhibits the master functions $F(\tilde{t})$ and susceptibility spectra $\chi''(\tilde{\omega})$ for a succession of 4 points $\vec{V}_c$, moving on $\mathcal{D}_C$ towards some $\vec{V}'_c$. The curves have been calculated for some two component model, $M = 2$.

The α-peak shapes depend on the details of the mode coupling functional $\mathcal{F}_q$ in (3.2). There is no few parameter fit formula of general validity.

3.5. MODE COUPLING THEORY BIFURCATION SCENARIOS

The standard bifurcation dynamics deals - as far as the time dependencies are concerned - with differential equations. The equations of motion relate the time derivatives $\dot{\phi}_q(t)$ at time t with these functions at the same time. The conventional bifurcation dynamics provides a paradigm for two phenomena. First, there is slowing down of the motion. It manifests itself by a divergence of some relevant time scale t_σ. This phenomenon is shared by the MCT scenarios. Second, conventional bifurcation theory yields power law decay near the critical points: $\phi(t) - f \propto 1/t^x$ or $\phi(t) - f \propto -t^y$. The exponents x and y are universal. For the case of a fold bifurcation one gets $x = y = 1$. These results render conventional bifurcation theory irrelevant for a description of structural relaxation stretching. However, these results are not shared by the MCT scenarios. Here $0.4 > x = a \neq b = y$ and both exponents a and b can be arbitrarily small.

The novelties of the MCT scenarios are caused by the retardation effects. Such retardations appear in all dynamical theories, where closed equations of motion are derived for a subset of the degrees of freedom. The necessity to eliminate some of the variables leads to memory effects as expressed by the integral term in (3.1). The mode q disturbs at some time t' a hidden mode $\bar{q}$. This eliminated mode $\bar{q}$ acts back on the mode q at some later time t. In conventional theories one anticipates, that the memory effects exhibit a characteristic time scale t_c: one assumes, that there is no serious variation of the spectrum $m''(\omega)$ of the memory kernel for $\omega t_c < 1$. Then one approximates $m''(\omega) \approx m''(\omega = 0)$. Thereby the memory effects disappear for the dynamics on scale $\omega \ll 1/t_c$. The approximation amounts to a coarse graining: the evolutions $\phi(t)$ are smoothened on intervals of length t_c. The hidden modes enter via transport coefficients, friction constants or rates [6]. But the cited assumption is questionable if there are bifurcation points. The above mentioned power laws for the decay may imply that t_c diverges. Then there is no separation between the time scale for the correlator dynamics and the time scale t_c for the kernel m. The memory kernel may not have any scale at all but rather exhibit similar power laws as the functions $\phi_q(t)$. Thus there appears the problem to understand the interplay of non-linearities and divergencies for the retardation times. The equations (3.1, 3.2) define a mathematical theory, exemplifying the identified problem. The found robustness suggest the conjecture, that those results of the MCT, which merely reflect the topology of the bifurcation hypersurface, will be present also in more complicated theories.

The results reported in sections 3.2–3.4 together with those derived for the other singularities imply a mathematical understanding of the solutions of the equations (3.1, 3.2). Relevant small parameters have been identified, viz. properly defined separations from the glass transition singularities. These can be employed to derive asymptotic solutions. Thereby the MCT introduces a series of new concepts and also a set of formulae relating these concepts. As example let us reconsider the power laws for the critical decay and for the von Schweidler decay. The oldest example for power law dynamics in liquids, governed by an anomalous exponent, is Einstein's law for the evolution of the mean squared displacement of a tagged particle. The anomalous exponent mirrors the fractal dimensionality of the so-called local time, a concept introduced by P. Lévy to quantify the Cantor set of visiting time distributions of the underlying standard Brownian process. For general Brownian motion one expects more subtle Hausdorff–Besicovitch dimensionalities to specify the local time and this will lead to less obvious time fractals than used in standard theories [36]. The MCT is a mathematical model showing, how two time fractals can appear from regular equations of motion and how a relation between the two dimensionalities a and b can be established.

Strong deviations from the reported leading order results for states near a simple fold singularity $\vec{V}_c$ may occur, if there is a second singularity $\vec{V}'_c$ nearby. These deviations may completely mask the simple pattern, described in section 3.3, 3.4, if $\vec{V}'_c$ is of higher order, say an A_3-singularity. In case one has to employ formulae, which handle $\vec{V}_c$ and $\vec{V}'_c$ simultaneously. Examples have been found, which follow the more complex MCT scenarios as can be inferred from Ref. [37]. Many examples are known for polymer dielectric loss spectra, where the α–peak exhibits the stretching behaviour discussed above in connection with Fig. 27. There are also examples known, where the α–process exhibits a splitting into two subprocesses as discussed above in connection with Fig. 28.

The applicability of the MCT for an explanation of structural relaxation can be judged best from quantitative data analysis. Let us consider as examples the findings for CKN, reported in section 2. In this case the temperature T was used as experimental control parameter, so that (3.12) is to be applied with $x = 1/T$, i.e. $\sigma \propto (T_c - T)$. The scaling laws are tested for the mesoscopic regime in Figs. 6, 7 for $\sigma < 0$ and in Figs. 8, 9 for $\sigma > 0$. The figures show that the frequency interval, where the data follow a common master curve, expands properly upon lowering T for the high temperature data $T > 120°C$ and also upon increasing T for the low temperature data $T < 80°C$. The master spectrum χ''_- of the MCT is reasonably well interpolated by the expression (2.3). Li et al. [18] have used this interpolation formula with both exponents a and b as free fit variables. They arrived at the result (2.4), which verified the MCT relation $\Gamma(1-a)^2/\Gamma(1-2a) = \Gamma(1+b)^2/\Gamma(1+2b)$ and suggested $\lambda = 0.81 \pm 0.05$. The full lines in Fig. 6 exhibit the master function χ_{int}. Full and dashed lines in Fig. 7 exhibit the MCT functions $\chi''_-(\hat{\omega})$ for $\lambda = 0.85$ and $\lambda = 0.81$ respectively. The comparison of the high temperature data for $T = 195, 180, 170$ and $160°C$ restrict the uncertainty interval for the exponent parameter to $\lambda = 0.85 \pm 0.02$. The parameter λ fixes the theoretical exponents for the scales $\tau_\beta^+ \propto \tau_\beta^- \propto t_\sigma$ and $\tau_\alpha \propto t'_\sigma$ and Fig. 10 shows the MCT result as full lines in comparison with the measurements. The spectral intensities for the minimum and the knee follow a square root laws $\chi_{\min} \propto \chi_e \propto \sqrt{|T - T_c|}$ according to (3.21, 3.22), and so do the measurements as shown in Fig. 11. The scaling laws imply that the graphs for $\omega_{\min}^{2a}$, ω_e^{2a}, $\omega_{\max}^{1/\gamma}$, $\chi_{\min}^2$ and χ_e^2 versus temperature should be linear and intersect the abscissa at T_c. This is the case for the CKN measurements [18] and the graphs yield 5 estimations for the critical value T_c with the result (2.5). The cusp behaviour for the non-ergodicity parameter shown in Fig. 12 confirms the MCT result (3.10) and yields a sixth measurement for T_c, compatible with (2.5) [19]. The factorization property, discussed in connection with Fig. 13 is identical with the equation (3.13a) of the reduction theorem.

The described basic version of the MCT cannot be a complete description of structural relaxation in supercooled liquids like CKN, since it ignores those dynamical processes which restore ergodicity for $T \leq T_c$. There is no ideal structural arrest for $T < T_c$, as is demonstrated by Fig. 1. The α-peak position ω_α is three to four orders of magnitude smaller for $T \simeq T_c$ than the natural microscopic frequency scale $\omega_{\mathrm{mic}} \approx 4$ THz, but it is not zero. For $T = T_c$ the relaxation in spectrum $\phi''(\omega = 0)$ is 10^3 to 10^4 times larger than the microscopic white noise background, but it is not infinity. The explained theory was proposed as an idealization in the sense, that small numbers like $\omega_\alpha(T \leq T_c)/\omega_{\mathrm{mic}}$ or $\omega_{\min}(T \leq T_c)/\omega_{\mathrm{mic}}$ are replaced by zero and large numbers like $\phi''(\omega \sim 0)/\phi''_{\mathrm{white}}$ by infinity. Some discrepancies between idealized description and experimental facts are obvious. The spectrum for $110°C$, which is the lowest curve in Fig. 6, exceeds the master spectrum χ''_- for $\omega < 1$ GHz. The master curves χ''_+ fall below the data for $\omega < 5$ GHz, as shown in Fig. 9.

4. Mode Coupling Theory for Simple Liquids

4.1. THE CAGE EFFECT

In this section a system of structureless particles of mass m shall be considered. The statistical description is done via canonical averages $\langle \cdots \rangle$. Thermodynamic functions can be evaluated from the structure factor $S_q = \langle |\rho_{\vec{q}}|^2 \rangle$. Here $\rho_{\vec{q}}$ denotes density fluctuations of wave vector $\vec{q}$, and q abbreviates the modulus $|\vec{q}|$. The discussion shall be restricted to such stable or metastable states, where S_q is a smooth function of temperature T and density n. The time evolution of phase space functions A shall be characterized conventionally by a generating operator $\mathcal{L}$: $A(t) = \exp(i\mathcal{L}t)A$. The statistical description of the dynamics shall be done as usual via correlators $\phi_A(t) \propto \langle A(t)^* A \rangle$. The Fourier-Laplace transforms of the latter for complex frequency z can be represented as Stieltjes continued fraction: $\phi(z) = -c_0/z - c_1/z \ldots$. Here the c_ℓ are given by the moments of the spectrum $\phi''(\omega) = Im\phi(\omega + i0)$. Denoting by $\tilde{M}_R(z)/c_{R-1}$ the fraction from step R onward, one gets:

$$\phi(z) = -c_0/z - c_1/z - \ldots c_{R-1}/z - \tilde{M}_R(z)/c_{R-1}. \qquad (4.1)$$

Such representations of correlators in terms of coefficients c_ℓ and kernels $\tilde{M}_R(z)$ have been introduced in statistical physics by R. Zwanzig and H. Mori. They showed that $\tilde{M}_R(z)$ can be written as Fourier-Laplace transform of a fluctuating force correlator: $\tilde{M}_R(t) = \langle F_R(t)^* F_R \rangle$. Here the fluctuating force reads $F_R(t) = \exp(i\mathcal{L}_R t)F_R$, with $\mathcal{L}_R = Q_R \mathcal{L}_{R-1} Q_R$, $F_R = \mathcal{L}_R \mathcal{L}_{R-1} \ldots \mathcal{L}_1 A$ and $Q_1, Q_2, \ldots, Q_R$ denoting a set of recursively defined projectors. The dynamical susceptibility for variable A is related to the correlator by $\chi_A(\omega) = [z\phi_A(z) + c_0]/(k_B T)$, $z = \omega + i0$. Details can be inferred e.g. from Refs. [6, 12].

The structure of simple liquids is specified by the distribution of the particle positions. Therefore, the simplest function characterizing structural relaxation theoretically, is the density correlator $\phi_q(t) = \langle \rho_q^*(t)\rho_{\vec{q}}\rangle/S_q$. This quantity is also of relevance for discussion of experiments, since it is measured by neutron scattering, photon correlation and Brillouin scattering spectroscopy. It shall be represented as (4.1) with $R = 2$:

$$\phi_q(\omega) = -1/\left[\omega - \Omega_q^2/\left[\omega + \Omega_q^2 M_q(\omega)\right]\right]. \tag{4.2}$$

Here appears the characteristic frequency Ω_q, a blurred phonon dispersion: $\Omega_q^2 = (qv)^2/S_q$, with $v = \sqrt{(k_B T)/m}$ denoting the thermal velocity. The relaxation kernel $M_q(\omega)$ follows from $M_q(t) = \langle F_{\vec{q}}(t)^*\rangle S_q/[k_B T q^4]$, with $F_{\vec{q}} = mQ\ddot{\rho}_{\vec{q}}$, $F_{\vec{q}}(t) = \exp(iQ\mathcal{L}Qt)F_{\vec{q}}$ and Q projecting perpendicular to $\rho_{\vec{q}}$ and $\dot{\rho}_{\vec{q}}$. Formula (4.2) is the same as (3.1), but ζ_q is missing and $m_q(t)$ is to be replaced by $M_q(t)$. If one considers the $q \to 0$ limit, one gets (1.1). Thus $[v^2 nm/S_q][1 - \omega M_q(\omega)]$ is a generalization of the longitudinal elastic modulus M to a function of wave vector and frequency. The memory kernel $M_q(t)$ appears, because the density fluctuations are not a closed set of dynamical variables: $\dot{F}_{\vec{q}} \neq 0$.

The fluctuating force F for zero wave vector is a combination of pairs of density fluctuations $F = \sum_{\vec{k}} v(\vec{k})\rho_{-\vec{k}}\rho_{\vec{k}}$, because the fluctuation of forces are due to fluctuations of the distances between particle pairs. Thus $M(t)$ can be related to functions like $P(t) = \langle \rho_1(t)^*\rho_2(t)^*\rho_3\rho_4\rangle$. It is suggesting to approximate averages of products by products of averages: $P(t) \approx \langle \rho_1(t)^*\rho_3\rangle\langle \rho_2(t)^*\rho_4\rangle + (1 \leftrightarrow 2)$. The contents of this approximation was indicated above in connection with (3.6). The probability to find a pair $\rho_1\rho_2$ to fluctuate with frequency ω is related to the probability for ρ_1 to fluctuate with frequency ω_1 and ρ_2 to fluctuate with frequency ω_2 so, that $\omega = \omega_1 + \omega_2$. The simplifying assumption is that, in the evaluation of P, ρ_1 and ρ_2 are treated as statistically independent. The strong interactions in densely packed liquids require, that short ranged correlations between density fluctuations are properly accounted for. To do this one does not factorize directly but one first projects the fluctuating forces $F_{\vec{q}}$ onto the subspace spanned by density pairs $\rho_{\vec{q}-\vec{k}}\rho_{\vec{k}}$ and then one factorizes the correlations of the latter. The described procedure has been introduced in statistical physics by K. Kawasaki in some other context. His approximation, leading to the coupling of force correlations to density correlations in the case considered here, is the essence of the MCT equations of motion for supercooled liquid dynamics. One finds as contribution to the kernel:

$$m_q(t) = \sum_{\vec{k}+\vec{p}=\vec{q}} \left\{ nS_q S_k S_p \left[\vec{q}(\vec{k}C_k + \vec{p}C_p)\right]^2 /(2q^4) \right\} \phi_k(t)\phi_p(t). \tag{4.3a}$$

Here C_k denotes the direct correlations, which are connected with the structure factor via the Ornstein–Zernike equation $S_k = 1/[1 - nC_k]$. The other contributions to the kernel shall be denoted by m_q^0, so that

$$M_q(\omega) = m_q^0(\omega) + m_q(\omega). \tag{4.3b}$$

From the theory of liquids one knows, that kernel $m_q(t)$ describes the so–called cage effects. Assume, that one could discuss the motion of a tagged particle in the liquid so as if all the other particles are fixed in space. Because of the dense packing the tagged particle could not move very far; diffusion over long distances would be impossible. The particle would be trapped and merely rattle in a cage formed by its neighbours. The continuous spectra for its correlation functions would result from averaging its localized motion over the many shapes of the cages. But the cages are not given static entities. If a cage forming neighbour is kicked by the tagged particle it starts moving, thereby exploring its possibilities for opening the cage. A diffusion of the tagged particle becomes possible, if the neighbour moves out of the way because its neighbour moves out of the way etc. Ergodic liquid dynamics is established because the particles built up backflow patterns as one knows it for the motion of a sphere in an ideal incompressible liquid. Kernel $m_q(t)$ describes approximately the indicated cooperative motion in normal simple liquids [12].

The microscopic theory of normal state liquids is so complicated, because there kernel m_q^0 in (4.3b) is of equal importance as m_q. Moreover, kernel m_q^0 occasionally describes trends opposite to those described by m_q, so that subtle compensations have to be handled. For example, kernel m_q^0 describes vortex motion, accompanying the tagged particle. This phenomenon makes diffusion more coherent than for a random traveller, leading to the well known long time tail contribution to the tagged particle velocity autocorrelation function $\psi(t) = \langle \vec{v}(t)\vec{v} \rangle$. This contribution reads $\psi(t \to \infty) \sim 1/t^{3/2}$, and it dominates for large times. The cage effect, on the other hand, yields negative contributions to $\psi(t)$. To avoid the discussion of the indicated normal liquid state problems, the range of parameters n or T shall now be restricted to such values, that the cage effect contribution $m_q(\omega)$ dominates over $m_q^0(\omega)$. This implicit restriction will require that the dynamical window is chosen outside the microscopic transient regime and that the state is only metastable relative to the proper stable crystalline phase. It is evident, that such range of parameters ω and (T,n) exists, if one anticipates that there are states close to ideal glass states. Spontaneous arrest of density fluctuations, $\phi_q(t \to \infty) = f_q > 0$, implies via (4.3a) also spontaneous arrest of the fluctuating force correlations: $m_q(t \to \infty) = \mu_q > 0$. This in turn implies a pole singularity for the relaxation kernel: $-m_q(z)z \to \mu_q$ for $z \to 0$. Compared to this zero frequency divergency, $m_q(z) \sim -\mu_q/z$, any finite contribution from $m_q^0(z)$ can be neglected for small frequencies. So one carries out a coarse graining approximation, i.e. one restricts ω to such small values that $m_q^0(\omega)$ can be replaced by its zero frequency value:

$$m_q^0(\omega) = i\zeta_q, \qquad \zeta_q \geq 0. \tag{4.4}$$

It is this approximation, which eliminates most of those phenomena, studied by established liquid theories. Loosing the conventionally discussed effects is of no harm, however, since these well known results are quite irrelevant for the understanding of all those structural relaxation phenomena, which were mentioned in sections 1 and 2.

The equations (4.2–4.4) are equivalent to the equation of motion (3.1). Equation (4.3a) implies the second equation of motion (3.2) of the MCT with a specific form of the functional $\mathcal{F}_q$. The basic version of the MCT, discussed in section 3, is obtained within the microscopic theory of liquids as the simplest meaningful treatment of the cage effect.

4.2. THE IDEALIZED MODE COUPLING THEORY

The described approximate treatment of the cage effect leads to the existence of ideal glass states, and hence the theory is also referred to as the idealized MCT. It was recapitulated in section 4.1, that the equations (3.1, 3.2) have been known to describe the cooperative motion in dense liquids, which leads, for example, to the strong decrease of the diffusivity D upon cooling. In section 3.2 it was shown, that this known picture for the dynamics is obtained only if the external control parameters $1/T$ or n do not exceed a certain critical value $1/T_c$ or n_c, respectively. If the parameters are too large, the cage effect brings out an entirely different solution. A tagged particle is trapped in a cage forever, because its neighbours cannot move out of the way because their neighbours cannot move out of the way etc. Treating the cage effect in the simplest plausible manner one finds ideal liquid to glass transitions and one gets the subtle structural relaxation phenomena discussed in section 3. The treatment of the cage effect in section 4.1 missed the well known activated transport mechanisms completely. The idealization inherent in that approach is the study of the dynamics in regions, where the phonon assisted relaxation can be ignored. Within the MCT the limitations of the idealized version and improvements beyond those limitations are discussed by considering extended versions of the theory as will be described below.

Other correlation functions besides $\phi_q(t)$ are also of interest. Within the MCT they are treated by Kawasaki's approximation in quite a similar manner as described above. Let us consider two examples. First, let $I(t)$ denote the correlator, whose spectrum $I''(\omega)$ is the intensity for depolarized light scattering. The scattering is caused by fluctuations of the dielectric constant, which in turn are mainly due to fluctuations of the distances between the particles. Thus, the simplest contributions to $I(t)$ are composed of the above mentioned pair correlators $P(t) = \langle \rho_1(t)^* \rho_2(t)^* \rho_3 \rho_4 \rangle$; and one arrives at the analogue of (4.3):

$$I(t) = I_0(t) + \sum_{\vec{k}+\vec{p}=\vec{0}} V(\vec{k},\vec{p}) \phi_k(t) \phi_p(t) . \tag{4.5}$$

The first term I_0 leads to a regular background, which is ω–independent for frequencies below the typical band for Raman excitations. The coupling constants $V(\vec{k},\vec{p})$ are given by the molecules' polarization properties and liquid structure factors as discussed in lowest approximation in Ref. [38]. Second, let us consider the correlators for the longitudinal and transversal current densities $j_q^{L,T}$:

$$\phi_q^{L,T}(t) = \langle j_q^{L,T}(t)^* j_q^{L,T} \rangle . \tag{4.6a}$$

The longitudinal function can be expressed in terms of the density correlator with the aid of the continuity equation. With (4.2) one finds:

$$\phi_q^L(t) = nv^2\omega / \left[\omega^2 - \Omega_q^2 + \omega \Omega_q^2 M_q(\omega)\right] . \tag{4.6b}$$

The transversal current correlator $\phi_q^T(t)$ cannot be treated by factorization directly. Therefore one applies the Zwanzig–Mori formula (4.1) with $R = 1$ in order to express $\phi_q^T(t)$ in terms of a fluctuating force kernel M_q^T:

$$\phi_q^T(\omega) = (-nv^2) / \left[\omega + q^2 M_q^T(\omega)/(nm)\right] . \tag{4.6c}$$

Kernel M_q^T is treated as explained above and one arrives at the analogue of (4.3). Let us notice that the zero frequency spectrum of the kernel is the shear viscosity $\eta = M_0''(\omega = 0)$. The treatment of the cage effect within the ideal MCT leads to the closed equations (3.1, 3.2) for the density correlator $\phi_q(t)$. Other functions are related directly or indirectly to $\phi_q(t)$ as explained for $I''(\omega)$ and η. Thereby the structural relaxation phenomena for light scattering, for the shear viscosity, and the like are obtained as corollaries of the relaxation phenomena exhibited by the density fluctuations.

The discussion of the preceding paragraph is oversimplified and therefore a digression shall be added. It is necessary to handle the interplay of low lying long wave length hydrodynamic excitations with low lying structural relaxation excitations. This has been done by proper modifications of the Zwanzig–Mori formalism, as can be inferred from Ref. [39]. The same holds for the necessary extensions of (4.5), which are e.g. required by the coupling of shear fluctuations to the light scattering correlator [40]. There are the mentioned classical long time tails, which also can be incorporated into the MCT for the liquid to glass transition as was shown for a special case in Ref. [41]. All these subtleties are not of primary importance for the understanding of structural relaxation phenomena.

A further digression concerns the experimental test of MCT. So far all applications have been done with the leading order asymptotic formulae, derived in sections 3.3, 3.4 for parameters approaching the glass transition singularity. But in the above cited analysis of CKN data, temperature variations over more than $160K$ have been considered. It is unreasonable to expect, that such large intervals for control parameter variations can be treated by leading order Taylor expansions with respect of T for the mode coupling coefficients in e.g. (4.3a) or (4.5). Rather the microscopic derivation makes the following suggestion plausible. If control parameters are shifted as strongly as indicated e.g. in Figs. 4, 5, one should interpret data with asymptotic expansions only so, that parameters like h_q or t_0 are allowed to be smooth functions of the control parameters T or n.

According to (4.3a), the coefficients $V(q, k, p)$ of the functional $\mathcal{F}_q$ in (3.2) are given by S_q; i.e. they are determined by equilibrium correlations. The latter are fixed by the weight factor in configuration space $w = \exp[-U/(k_B T)]$, defining the so-called potential landscape. Here U denotes the potential energy of the many particle system. Inertia effects or any other feature of microscopic dynamics as given by the regular kernel $m_q^0(\omega)$ in (4.3b), do not enter the mode coupling functional $\mathcal{F}_q$. As a result, one identifies the non–ergodicity parameter f_q, the critical amplitude h_q and the separation parameter σ as equilibrium quantities. The same is true for the exponent parameter λ, which fixes the master functions for the β–relaxation and also the exponents governing the two time scales t_σ, t'_σ. Similarly, all the master functions for the density relaxation $F_q(\tilde{t})$ or for the longitudinal and transversal elastic moduli, for the velocity correlations or for the stress correlations are completely determined by w. In this sense the MCT deals with stochastic geometry. It provides approximate information on the statistics of possible orbits in configuration space rather than in phase space. The time fractals, which are the essential features of the discussions of section 3, are indeed expected subtleties for the projections of orbits onto a one dimensional time axis [36]. All details of the motion, which occur within the natural time or frequency window of the many particle system, are reduced to the parameter t_0. This is the time scale for the exploration of the potential landscape. It does not matter for the structural relaxation phenomena, whether the exploration is done according to Newton's equations of motion as in a conventional simple liquid or according to Brownian rules like in a colloidal suspension.

4.3. THE HARD SPHERE SYSTEM

The idealized MCT is definitive in the sense, that all quantities for the structural relaxation, except the time scale t_0, are determined by the structure factor S_q. The latter can be evaluated with reasonable accuracy from the pair potential [12]. A complete determination of all the quantities for the discussion of α- and β-relaxation, has been published so far only for the HSS. In this case the exponent parameter is $\lambda = 0.77$ implying the anomalous exponents: $a = 0.30, b = 0.53, \gamma = 2.6$. Remember, that λ fixes also the master functions $g_\pm$ for the dynamics within the first scaling region. Comparison of the ergodicity parameter f_q, as obtained from (3.7), with the asymptotic law (3.10), indicates, that the critical amplitude drifts by about 15% if $\varepsilon = (\varphi - \varphi_c)/\varphi_c$ is increased to 5% [42].

The various master functions of the α–process have been calculated by M. Fuchs, I. Hofacker and A. Latz [43]. Let us consider a few of their results. Figure 29 [43] reproduces as full line the spectrum of the longitudinal acoustic modulus. The α–peak is stretched to about 1.8 decade. Comparison with a Debye–peak, shown in dashed, demonstrates the peak asymmetry and confirms, that the low frequency part of the α–bump is close to white noise. The von Schweidler law, shown in long dashes, describes that part of the high frequency α–peak tail, where the spectrum is fallen below 20% of the maximum value. A Kohlrausch fit with stretching exponent $\beta = 0.61$, shown as dotted curve, accounts for a major part of the α–spectrum; but it falls below the master curve for large frequencies. Indeed, the theoretical HSS result in Fig. 29 exhibits all those structural relaxation features, which were discussed in connection with Fig. 2. Fitting the α–process for the density fluctuations by a Kohlrausch function, requires that the stretching exponent β and the relaxation time

τ are q-dependent: $F_q(\tilde{t}) = f_q^c \exp[-(\tilde{t}/\tau_q)^{\beta_q}]$. The upper part of Fig. 30 [43] shows, that there is only little stretching for wave vectors q near the structure factor peak position q_0. But for q off q_0 the stretching exponent β_q falls to conventional values. For large q the Kohlrausch exponent β_q tends towards the von Schweidler exponent b. The time τ_q is found to vary considerably with q, as is shown by the lower part of Fig. 30.

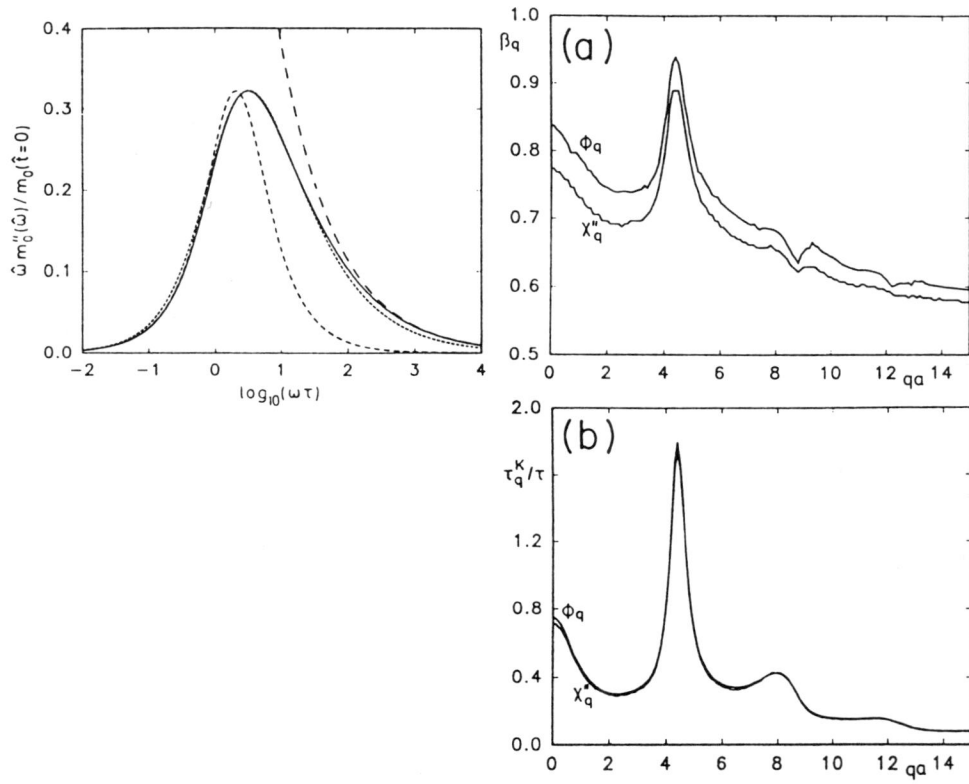

Figure 29 [43]   Figure 30 [43]

W. van Megen and S.M. Underwood compared MCT results for the ideal liquid to glass transition for the HSS with their photon correlation spectroscopy data for colloidal suspensions [26, 27]. Let us consider some of their findings. It was already discussed in section 2, that Fig. 15 tests the factorization of spacial and temporal correlations within the mesoscopic window. The shown data confirm the basic equation (3.13a) of the reduction theorem, and they indicate that the first scaling law regime of the MCT extends over a window of about three orders of magnitude. Figure 16 compares the experimental results for the critical Debye–Waller factor $f_c(q) = f_q^c$ and for the critical amplitude $h(q) = h_q$ with the theoretical results, shown as full and dashed line, respectively. The measurements for the glass states $\sigma > 0$ and the liquid states $\sigma < 0$ yield the same results for f_q^c and h_q within the experimental uncertainties. This finding demonstrates, for example, that the amplitude h_q' for the square root anomaly of the Edwards–Anderson parameter,

$f_q - f_q^c = h_q'\sqrt{\varepsilon/(1-\lambda)}$, is the same as the amplitude h_q'' for the von Schweidler decay, $\phi_q(t) - f_q^c = -h_q'' B \cdot (t/t_\sigma')^b : h_q' = h_q'' = h_q$. The relaxation curves for the glass states follow the theoretical first scaling law formulae. This is shown for three wave vectors and two values $\varphi > \varphi_c$ by the upper two curves in Figs. 14a, b, c.

In Fig. 14 also theoretical results for the liquid states $\varphi < \varphi_c$, shown as full lines, are compared with relaxation data. The curves combine the MCT results for the first and the second scaling regimes. The comparison was done in Ref. [26] as follows. First, the theoretical master curves for the α–process have been matched with the theoretical curves for the β–process and this with the data. Here enter only t_σ' and σ as fit parameters. Contrary to the discussion of the β–process done in connection with Fig. 15, neither f_q^c nor h_q have been used as fit parameter for the combined analysis. They are contained in the α–master function $F_q(\tilde{t})$ because of (3.24b). Contrary to the discussion of the relaxation for $\varphi > \varphi_c$, the β–relaxation scale t_σ cannot be used as fit parameter for the combined analysis either. Rather it is given as $t_\sigma = t_\sigma' \mid \sigma \mid^{1/2b}$ according to (3.15, 3.20). Notice, that Fig. 14 also implies a verification for the predicted q–variations of the α–decay scale τ_q. Second, it was tested that t_σ' follows the power law (3.20) and that the separation parameter varies according to (3.12): $\sigma = C(\varphi - \varphi_c)/\varphi_c$. Finally, the two constants C and φ_c are found to be close to the theoretical prediction. For example, the critical packing fraction was measured as $\varphi_c = 0.58$ while the MCT formulae lead to 0.52. The reported findings suggest, that the idealized MCT provides a rather realistic description of the structural relaxation and the glass transition of hard sphere colloidal suspensions.

4.4. THE EXTENDED MODE COUPLING THEORY

An extended frame for approximations of simple liquid dynamics has been formulated by L. Sjögren and A. Sjölander [44]. They start with an infinite set of variables $A_{\vec{q}}^\nu = \sum_{\vec{k}} H_\nu(\vec{k}) f_{\vec{q}}(\vec{k})$. Here the $f_{\vec{q}}(\vec{k})$ are the phase space densities and the $H_\nu(\vec{k})$ denote the three dimensional Hermite polynomials. The Zwanzig–Mori fraction representation can be generalized to matrices of correlators $\phi_{\vec{q}}^{\nu\mu}(t) = \langle A_{\vec{q}}^\nu(t)^* A_{\vec{q}}^\mu \rangle$. The matrix of memory kernels is written as combination of products $\phi_{\vec{k}}^{\alpha\beta}(t)\phi_{\vec{p}}^{\kappa\lambda}(t)$ by means of Kawasaki's mode coupling approximation. The resulting closed equations imply a separation of short time transient dynamics as caused by renormalized binary collision events from cooperative effects as described by mode coupling functionals. The equations can then be simplified further by restricting parameters and frequencies so, that structural relation phenomena dominate. The most important correlators are those for the densities since these and only these may exhibit zero frequency poles due to spontaneous arrest $\phi_q(z \to 0) \sim -f_q/z$. Keeping only these functions in the kernels, one reproduces the equations of the idealized MCT. As a leading improvement one keeps also the current correlators. The result is (4.2) with

$$M_q(\omega) = [i\zeta_q + m_q(\omega)] / \{1 - \delta_q(\omega)\Omega_q^2 [i\zeta_q + m_q(\omega)]\} . \quad (4.7)$$

Here ζ_q and Ω_q have the same meaning as in section 4.1. Also the kernel $m_q(t)$ is given by the same cage effect functional (4.3a) as above. The new feature of the extended theory is the additional kernel $\delta_q(t)$, given by a new mode coupling functional:

$$\delta_q(t) = \sum_{\vec{k}+\vec{p}=\vec{q}} [V_1'(\vec{q};\vec{k},\vec{p})\partial_t\phi_k(t)\partial_t\phi_p(t) \\ + V_2'(\vec{q};\vec{k},\vec{p})\phi_k(t)\phi_p^L(t) + V_3'(\vec{q};\vec{k},\vec{p})\phi_k(t)\phi_p^T(t)] . \quad (4.8)$$

For the longitudinal current correlator ϕ^L one obtains again (4.6b). For the transversal current correlator ϕ^T one derives (4.6c); and the kernel $M_q^T(\omega)$ is expressed by kernels

ζ_q^T, m_q^T and δ_q^T quite analogues to (4.7, 4.8). The resulting closed equations are the basis of the extended MCT [45]. The coefficients V_e' and $V_e^{T'}$ contain information on the short time dynamics; they are not equilibrium quantities.

The zero frequency limit of the new kernels is given by integrals over positive density and current spectra. Therefore one gets for $\delta_q(\omega = 0) = i\delta_q$ and $\delta_q^T(\omega = 0) = i\delta_q^T$ with

$$\delta_q > 0, \qquad \delta_q^T > 0, \tag{4.9}$$

unless all the coupling coefficients V_e' and $V_e^{T'}$ vanish. This result is not compatible with spontaneous arrest. Mode coupling to currents as described by (4.7, 4.8) with $V_e' \neq 0$, $V_e^{T'} \neq 0$ excludes the possibility for non-ergodic motion; the δ-terms restore ergodicity. Within the MCT for the Fermi-liquid to Fermi-glass transition it was shown, that coupling to currents as described by (4.8), reproduces conventional approximations for phonon assisted hopping transport. It was shown in particular, that inclusion of δ-terms replaces the ideal ergodic to non-ergodic transition by a smooth cross over, as can be inferred from Ref. [46] and the papers quoted there. Hence it is obvious, that the δ-kernels describe approximately hopping transport in the sense discussed in section 1.3. Numerical solution for a particular model has shown, that the interplay of the cage effect with hopping effects leads to a strong suppression of δ if T decreases below T_c. Thus the α-peak position is very temperature sensitive [45]. Coupling of long wave length phonons with short wave length density fluctuations can lead to an Arrhenius temperature dependence of the α-relaxation time, as can be inferred from Ref. [47].

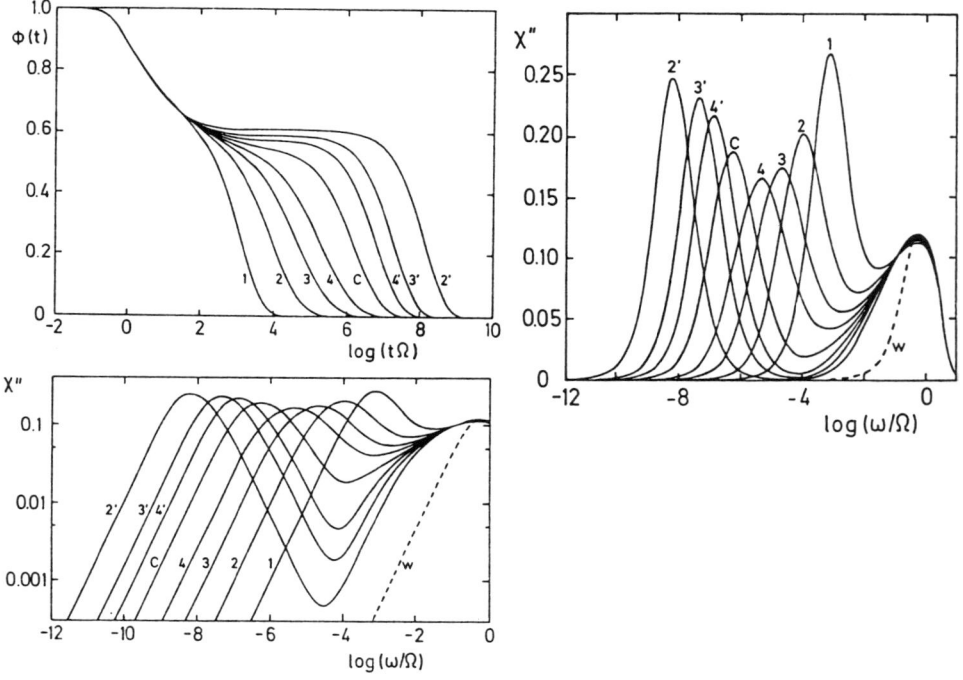

Figure 31 [35]

Figure 32 [35]

Figure 33 [35]

As before, let us combine all the coefficients like $V(\vec{q},\vec{k},\vec{p})$, entering the functionals for the cage effect kernels m_q and m_q^T, to some vector $\vec{V}$, denoting a point in some N–dimensional space $\mathbf{K}$. Similarly, let us combine all the coefficients like $V'_e(\vec{q};\vec{k},\vec{p})$, entering the hopping effect kernels δ_q and δ_q^T, to some vector $\vec{V}'$ in some N'–dimensional space $\mathbf{K}'$. The solution of the extended MCT equations of motion are then determined, if one specifies the control parameter vector $\vec{X} = (\vec{V},\vec{V}')$, denoting a point in an abstract $(N+N')$–dimensional parameter space $\mathbf{R} = \mathbf{K} \times \mathbf{K}'$. The extended theory contains the ideal MCT as special case for those states $\vec{X} = (\vec{V},\vec{O})$, for which current couplings vanish. There are ideal glass states in $\mathbf{R}$ given by the parameter points $(\vec{V},\vec{O})$ with $\vec{V}\epsilon\mathcal{D}_G$. In particular there are glass transition singularities $\vec{X}_c = (\vec{V}_c,\vec{O})$, $\vec{V}_c\epsilon\mathcal{D}_C$. The ideal glass states and the glass transition singularities are determined by the cage effect mode coupling functional as described in section 3.2. All states $\vec{X}$ with $\vec{V}\epsilon\mathcal{D}_L$ or $\vec{V}' \neq \vec{O}$ are ergodic states. The results on α- and β–relaxation, discussed in section 3 within the ideal MCT, are reproduced by the extended theory within certain regions of parameters $\vec{X}$ near the glass transition singularities $\vec{X}_c$ provided that times or frequencies are restricted to certain windows.

Figures 31–33 [35] illustrate the results of the extended MCT for model (3.4). Parameter points, shown as dots in Fig. 18, are the same as used in Figs. 19–21 for the illustration of the idealized theory. The hopping kernel was not calculated but used as $\delta(\omega) = i\delta$, with δ chosen to decrease exponentially upon moving from the liquid to the glass. The curves with label C refer to $\vec{V} = \vec{V}_c$. Obviously, the results exhibit structural relaxation phenomena with the proper stretching also for $T < T_c$. Even though δ is largest in the liquid, the cage effect dominates there, so that the results are close to those shown in Figs. 19–21 for $\delta = 0$. There is a pronounced mesoscopic β–spectrum for $T < T_c$ as well as for $T > T_c$. Upon decreasing T, the point $\vec{X}$ moves from liquid states to points close to ideal glass states. The latter exhibit other dynamics than the former. Pronounced cross over phenomena occur for T near T_c, since there $\vec{X}$ passes nearby the glass transition singularity. All functions vary smoothly with T throughout, but it is the glass transition singularity which causes the structural relaxation patterns and cross over features.

5. Mesoscopic Dynamics Near T_c

5.1. THE TWO PARAMETER SCALING LAW

Also within the extended MCT there exists a range of control parameters and a mesoscopic dynamical window, where the equations of motion can be simplified using the deviation $\eta = \phi_q(t) - f_q^c$ of the correlator from the critical plateau as small quantity. In leading order in η one reproduces the factorization theorem (3.13a). This theorem can be generalized to all correlators $\phi_A(t) = \langle A^*(t)A\rangle/\langle |A|^2\rangle$ of variables A, coupling to density fluctuations or to products of density fluctuations:

$$\phi_A(t) = f_A^c + h_A G(t). \qquad (5.1)$$

The critical non-ergodicity parameter f_A^c and the critical amplitude h_A are equilibrium parameters, specified by the liquid structure and characterizing the variable chosen. If one carries out Fourier transforms one finds the corresponding factorization for the correlation spectra or for the susceptibility spectra $\phi_A''(\omega)\omega = \chi_A''(\omega) = h_A\chi''(\omega)$, where $\chi''(\omega) = \omega G''(\omega)$. For example, with $A = \rho_{\vec{q}}$ denoting the density fluctuations one gets the result (2.6). If A denotes the function describing the light-matter coupling for depolarized scattering, one derives for the cross section $I''(\omega) = h_{es}G'''(\omega)$, where one finds from (4.5)

$h_{es} = 2\sum_{\vec{k}} V(\vec{k}, -\vec{k})h_k$. The frequency dependence and the sensitive temperature dependence of the spectra $\phi''_A(\omega)$ of all variables A is given by the same function $G''(\omega)$, except for a scale h_A. The result (5.1) mirrors the basic finding of the MCT: the structural relaxation dynamics is due to a singularity, which is caused by a single dangerous eigenvector of a matrix, characterizing the instability of a cage effect enforced frozen structure.

Also the other aspects of the reduction theorem, considered in section 3.3 for the idealized MCT, have a proper generalization within the extended theory. The dependence of G on the $N+N'$ mathematical coupling constants, entering the various mode coupling functionals, is reduced to the dependence on *two relevant control parameters*. These are complicated but smooth functions of the control parameters $\vec{X}$. One is the known *separation parameter* σ, which measures the strength of the cage effect. Its zero defines the critical external control parameters like the cross over temperature T_c via (3.12). The other is the *hopping parameter* $\delta \geq 0$. Parameter δ can vanish only if there are no mode couplings to the currents: $\vec{V}' = \vec{0}$. The β–correlator G is a function of the three variables t/t_0 of σ and of $\delta \cdot t_0$:

$$G(t) = g(t/t_0, \sigma, \delta \cdot t_0). \tag{5.2}$$

Here t_0 is the matching time from (3.11). The solution of the equations of motion of the extended MCT is reduced to solving the equation for G, which generalizes (3.13b) to

$$\sigma - \delta \cdot t + \lambda G(t)^2 = (d/dt)\int_0^t G(t-t')G(t')dt'. \tag{5.3}$$

It has to be solved with the critical decay law $G(t \to 0) \sim (t_0/t)^a$ as initial condition. The function g is determined by the exponent parameter λ; and hence it is an equilibrium quantity. All properties of g can be understood by asymptotic expansions and a routine to solve (5.3) numerically is available, as can be inferred from Ref. [48].

The construction of the functions σ and δ implies a mapping of the original control parameter space **R** on the *control parameter half plane* **E**, spanned by $(\sigma, \delta t_0 \geq 0)$:

$$\vec{X} \to (\sigma, \delta t_0). \tag{5.4}$$

The mesoscopic dynamics for $\vec{X}$ near $\vec{X}_c$ is completely specified by the points in **E**. The *glass transition singularity* $\vec{X}_c$ under study is the origin in **E**; and the liquid states $\mathcal{D}_L$ and the *ideal glass states* $\mathcal{D}_G$, which were studied in the basic version of the MCT in section 3, are the left and right part of the abscissa in **E**, respectively:

$$\vec{X}_c \to (0,0); \qquad \mathcal{D}_L \to (\sigma < 0, 0); \qquad \mathcal{D}_G \to (\sigma > 0, 0). \tag{5.5}$$

The solution of (5.3) at the singularity is the critical law: $g(t/t_0, 0, 0) = (t_0/t)^a$. The solution on the half abscissas it is given by two $\sigma - \delta$-independent master functions $g_\mp$, according to the scaling law (3.15): $g(t/t_0, \sigma, 0) = c_\sigma g_\pm(t/t_\sigma)$, $\sigma \gtrless 0$.

The solution G of (5.3) is a homogeneous function. There holds the extension of (3.14), called a two parameter scaling law:

$$g(t/t_0 y, \sigma \cdot y^{2a}, \delta t_0 \cdot y^{1+2a}) = y^a g(t/t_0, \sigma, \delta t_0), \qquad y > 0. \tag{5.6}$$

Let us define *scaling lines* as the curves in **E**:

$$\sigma = \hat{\sigma} y^{2a}, \qquad \delta \cdot t_0 = \hat{\delta} y^{1+2a}. \tag{5.7a}$$

Here $y > 0$ is the curve parameter. A scaling line is a generalized half parabola through some point $(\hat{\sigma}, \hat{\delta})$, which terminates in the glass transition singularity for $y \to 0$. Changing y, the correlator changes in a self similar manner. Expanding the time according to $t/t_0 = \hat{t}/y$ the correlator g merely decreases by a scale factor y^a: $g(\hat{t}/y, \sigma, \delta t_0) = y^a g(\hat{t}, \hat{\sigma}, \hat{\delta})$. Shifting parameters along a scaling line, the shape of the $\log[\phi_A(t) - f_A^c]$ versus $\log t$ curve is fixed. Similarly, the shape of the $\log \chi_A''(\omega)$ versus $\log \omega$ curve is fixed. These results have been discussed in all details in section 3 for the idealized MCT. In that case there were only two scaling lines of relevance: the two lines specified by (5.5), which are obtained from (5.7a) for $\hat{\delta} = 0$ and $\hat{\sigma} = \mp 1$ respectively. The extended theory needs a continuum of scaling lines, and hence a continuum of master functions, in order to cover the whole parameter plane **E**. Let us make this result more explicit for states with $\delta t_0 > 0$. One can introduce a natural scale σ_0 for the separation parameter, as it is induced by the hopping rate:

$$\sigma = s\sigma_0, \qquad \sigma_0 = (\delta t_0)^{\frac{2a}{1+2a}}. \tag{5.7b}$$

The number s specifies the scaling line and δt_0 replaces y as curve parameter. One gets

$$g(t/t_0, \sigma, \delta t_0) = c_\sigma g_s(t/t_\sigma). \tag{5.7c}$$

Here the correlation scale c_σ and the time scale t_σ have the same meaning as before in (3.15). The result (5.7c) has the standard form of a scaling law. The master function g_s is σ–independent: $g_s(\hat{t}) = g(\hat{t}/s^{\frac{1}{2a}}, s, 1)/s^{\frac{1}{2}}$. The results of the idealized MCT are obtained as the limit $g_s(\hat{t}) \to g_\pm(\hat{t})$ for $s \to \pm\infty$. The shape functions are pairwise qualitatively quite different depending on whether $s \ll -1, |s| \ll 1$ or $s \gg 1$. If one wanted to emphasize this feature one could introduce the following fuzzy concepts. States with $\sigma \ll -\sigma_0$, $|\sigma| \ll \sigma_0$ and $\sigma \gg \sigma_0$ could be referred to as liquid states, transition states and glass states respectively.

The preceding formulae (5.1-5.7) are implications of the extended MCT equations of motion in the sense of a limit theorem. Consider a path $y \to X(y)$ in parameter space **R**, which terminates for $y \to 0$ in the glass transition singularity $\vec{X}_c$. Assume that the mapping (5.4) yields a curve, which approaches some scaling line (5.7a). The correlator will be a function of y: $\phi_A^y(t)$. One gets:

$$\phi_A^y(t_0 \hat{t}/y) - f_A^c = y^a h_A g(\hat{t}, \hat{\sigma}, \hat{\delta}) + O(y^{2a}). \tag{5.8}$$

Corresponding results hold for the spectra. If the singularity is approached for $y \to 0$, and if the time tends to infinity like $t = t_0 \hat{t}/y$, the correlator approaches the plateau value f_A^c. The difference $\phi_A(t) - f_A$ vanishes in leading order like y^a; and this leading order term is described by the scaling law. The corrections to this result vanish proportional to y^{2a}. The interval of rescaled frequencies $\hat{\omega} = \omega t_0/y$, where the susceptibility spectrum is given by

$$\chi_A''(\omega)/y^a = h_A \hat{\omega} g''(\hat{\omega}, \hat{\sigma}, \hat{\delta}) + O(y^a), \tag{5.9}$$

expands for small as well as for large values, if y tends to zero. The corrections to the leading order results, i.e. the size of the O-terms in (5.8, 5.9) may depend on the variable A. The corrections O diverge for $\hat{\omega} \to \infty$ and, if $\delta t_0 \neq 0$, also for $\hat{\omega} \to 0$. The formulae deal with frequencies small enough, so that the transient dynamics can be ignored. But the frequencies have also to be chosen sufficiently above the α–peak position.

5.2. SOME CROSS OVER FEATURES

Let us consider some of the phenomena described by the two parameter scaling law. Figure 34 [48] exhibits from bottom to top the decay curves for fixed δt_0 and $s = -13.9, -7.2, -3.7, -1.9, -1.0, -0.07, 0, 0.07, 1.0, 1.9, 3.7, 7.2, 13.9$. Figure 35 [48] shows from top to bottom the corresponding susceptibility spectra. The diagrams refer to $\lambda = 0.91$, i.e. $a = 0.20$, $b = 0.28$. They exhibit the evolution of the mesoscopic dynamics upon cooling for fixed hopping rate. For $t \to 0$ all correlators approach the critical decay law $G(t) \propto 1/t^a$. Similarly, for large frequencies, the susceptibility spectra approach the sublinear variation $\chi''(\omega) \propto \omega^a$. This part of the dynamics does not exhibit a sensitive dependence on the separation parameter. If the formulae are applied to experiment, the mentioned part of the spectra may exhibit some smooth drift with changes of control parameters, caused by the drifts of the critical amplitude h_A in (5.1) or of the matching time t_0 in (3.13c). For $t \to \infty$ the correlators $G(t)$ diverge towards negative values according to a von Schweidler power law.

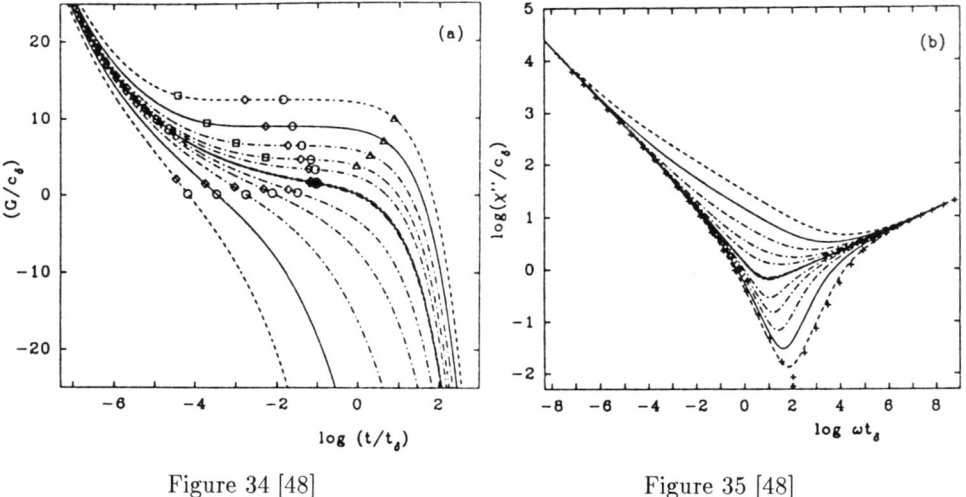

Figure 34 [48] Figure 35 [48]

This decay below the plateau value of $\phi_A(t)$ is the start of the α–process. Equivalently, χ'' increases with decreasing ω; the theory describes the fractal behaviour of the high frequency α–peak tail. The center of the dynamical window, described by the results of section 5.1, is the position $\omega_{\min}$ of the susceptibility minimum. The corresponding times $t_{\min} = 1/\omega_{\min}$ are marked by circles in Fig. 34. They are near the inflection points t_i of the $G(t)$ versus $\log t$ diagrams, marked in Fig. 34 by diamonds.

For $s \ll -1$ the master function g_s in (5.7c) becomes s–independent for $t \sim t_{\min}$. Therefore the susceptibility minimum for the liquid states is well described by the scaling laws of the idealized MCT. Figure 36 [48] reproduces the susceptibility spectrum for $s = -7.2$. The dashed line, denoted by χ_{int}, is the interpolation formula (2.3) between the critical fractal and the von Schweidler fractal. It fits the exact result very well. The slowing down of the dynamics upon cooling towards T_c manifests itself by the decrease of $\omega_{\min}$ and of $\chi''(\omega_{\min}) = \chi_{\min}$, as was discussed in detail in section 3.3. If the frequency becomes very small, or if the time becomes very large, the hopping effects render the idealized MCT unvalid. The $\delta \cdot t$–term in (5.3) becomes dominant for $t \to \infty$ no matter how small is δ. This yields to a cross over of the von Schweidler law $\chi'' \propto 1/\omega^b$ with $b < 0.5$ to $\chi'' \propto 1/\omega^{1/2}$, as is demonstrated in Fig. 35. The dynamics for the transition states, $\mid s \mid \ll 1$, is not very σ–sensitive. The minimum for $T \sim T_c$ has quite a different shape compared to the one for the liquid or glass states. The trend towards stochastic motion, caused by the hopping effects, suppresses the leading corrections to the two fractal laws. Therefore the change from ω^a- to ω^{-b}–behaviour is so abrupt on a $\log \omega$–abscissa, that a fit by the interpolation formula (2.3) is quite impossible. The cross over from the critical decay to almost arrest leads for the glass states to a knee at some frequency ω_e for the $\log \chi''$ versus $\log \omega$ curves. The corresponding times $t_e = 1/\omega_e$ are marked by squares in Fig. 34. For $s \gg 1$ the dynamics for $\omega \sim \omega_e$ is insensitive to variations of s and can be described by the scaling laws for the idealized MCT. This is shown by the crosses in Fig. 35, which fit the knee for $s = 13.9$. These crosses represent an empirical interpolation formula for the master spectrum: $\chi''_+ \approx 2\chi_e/[(\omega_e/\omega)^a + (\omega_e/\omega)]$. The softening of the almost arrested cage effect induced structure upon heating causes the decrease of the frequency ω_e and the decrease of χ_e with increase of T, as was discussed in section 3.

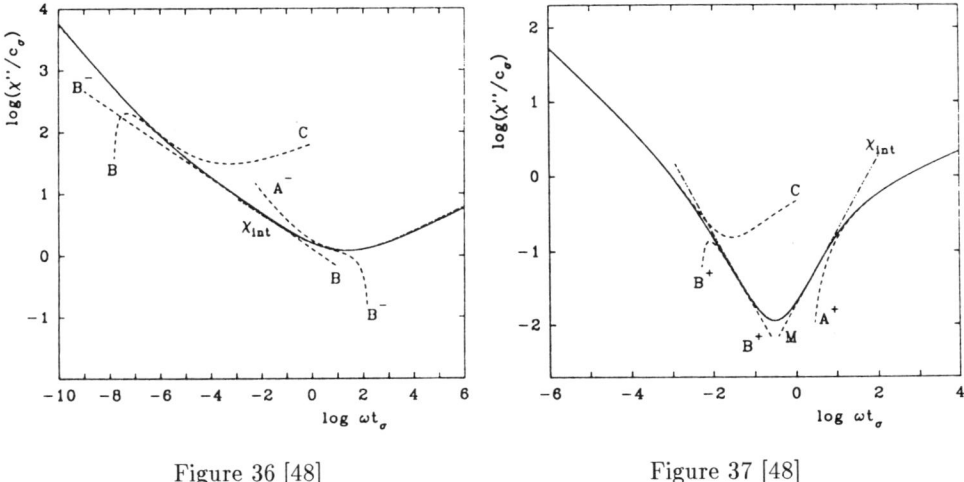

Figure 36 [48] Figure 37 [48]

The hopping effects produce a spectral minimum also for $T < T_c$. For fixed δt_0 the mentioned slowing down of the dynamics upon heating leads to a decrease of $\omega_{\min}$ with increasing temperature. The breaking up of the frozen structure leads to an increase of the spectral intensity $\chi_{\min}$ with increasing T. The minimum parameters depend on σ and on δt_0 sensitively. With constants $D_{1,2}$, given by λ, one gets:

$$\omega_{\min} t_0 = D_1 (\delta t_0)^{\frac{1}{2}} \, |\sigma|^{\frac{1-2a}{4a}}, \quad \chi_{\min} = D_2 (\delta t_0)^{\frac{1}{2}} / |\sigma|^{\frac{1}{4a}}; \quad \sigma \gg \sigma_0. \tag{5.10}$$

The hopping induced minimum for the glass states is described by a spectrum, varying regularly with ω. It can be fitted well by the interpolation $\chi''_{\text{int}} = \chi_{\min}[(\omega/\omega_{\min}) + (\omega_{\min}/\omega)]/2$, as is shown in Fig. 37 [48] by the dashed curve with label χ_{int} for the $s = 7.2$.

In order to identify the specified qualitative changes of the liquid dynamics upon crossing over from $T \gg T_c$ through $T \approx T_c$ to $T \ll T_c$, one has to study dynamical windows of several decades and variations of spectral intensities over several orders of magnitude. This prediction formulates also the great difficulties, which the experimentator has to master, if he wants to study the appearance of structural relaxation in liquids upon cooling.

The microscopic theory of section 4 provides the mathematical control parameter $\vec{X}$ as function of external variables like temperature T or density n. Let us consider the set up where only T is varied. Then the states are located on a path in parameter space $\mathbf{R} : T \to \vec{X}(T)$. Via (5.4) this path is mapped on a curve C in the plane $\mathbf{E}$.

$$C : T \to (\sigma(T), \delta(T) t_0). \tag{5.11}$$

The function $\sigma(T)$ will increase with decreasing temperature, so that according to (3.12): $\sigma \propto (T_c - T)$. The hopping parameter $\delta(T)$ will decrease strongly with decreasing T as described e.g. by an Arrhenius law. One cannot vary σ and δ independently. One cannot scan the plane **E** and therefore one cannot test all implications of the results in section 5.1. One cannot test the scaling law (5.7) since one cannot drive the states along a scaling line. In particular one cannot decrease $\delta(T_c)$ to arbitrarily chosen small values. The simple results of the ideal MCT can be expected only in those special regions, where the function g_s in (5.7) does not depend sensitively on s.

5.3. A MCT ANALYSIS OF SOME STRUCTURAL RELAXATION DATA

G. Li, W.M. Du, X.K. Chen, H.Z. Cummins and N.J. Tao have observed the appearance of the structural relaxation in CKN by depolarized light scattering, as reported in sections 2.2, 2.3. They have analyzed their data for the mesoscopic regime within the idealized MCT, as discussed above in section 3.5. Since some discrepancies between measurements and data have been identified, the experiments have been reanalyzed within the extended MCT by Cummins et al. [17]. Let us consider some of their findings. Figure 38 [17] reproduces the susceptibility measurements from Fig. 5; the parasitic spectral spikes at 20 GHz are suppressed. A dashed line is added in order to show the estimated white noise background spectrum due to the transient dynamics. The spectral enhancement above this line, which can be larger than two orders in magnitude, is caused by structural relaxation. The mesoscopic window under discussion extends from the high frequency wing of the α–bump to the conventional Raman band near 1 THz. The full lines in Fig. 38 represent the MCT results from equations (5.1-5.3) for exponent parameter $\lambda = 0.85$.

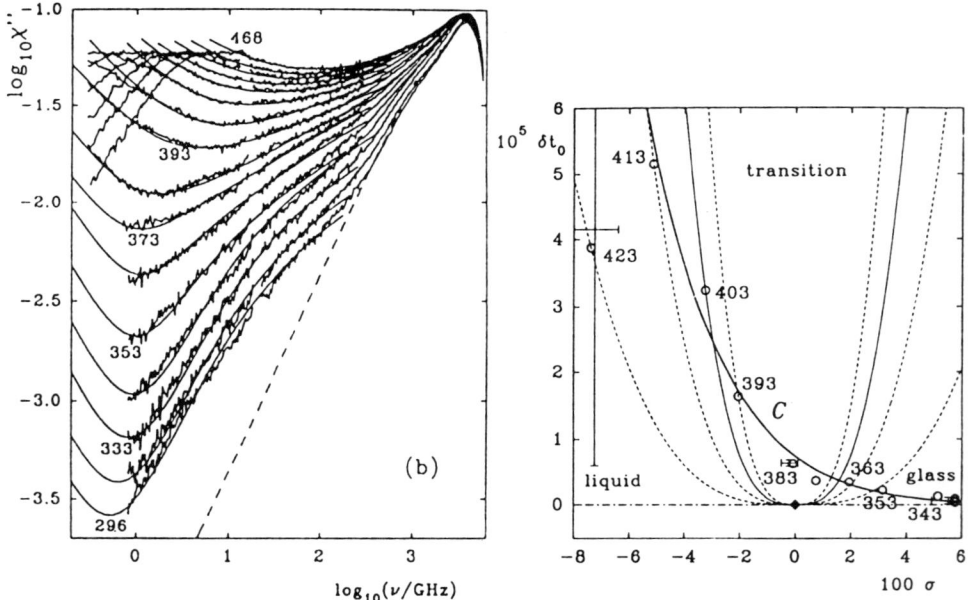

Figure 38 [17]   Figure 39 [17]

For every temperature T the two relevant control parameters σ and δ have been adjusted to get an optimal fit. The results are shown as circles in Fig. 39 [17] for $T \leq 423K$. The critical amplitude h_{es} in (5.1) was allowed to drift with T. Because of the scaling law (5.6), the parameters h_{es} and t_0 are not independent and in Ref. [17] the arbitrary choice $t_0 = 1\text{ps}/2\pi$ was made. The shown analysis confirmed (2.5) for T_c but decreased the uncertainty to $\pm 1K$. The found best λ is within the previously determined interval [18] and the uncertainty could be reduced to ± 0.02. Within the interval $T_c - 40K \leq T \leq T_c + 40K$, the separation parameter varies linearly with T according to (3.12) and δ could be interpolated by an Arrhenius law. The corresponding curve is shown as line C in Fig. 39. If T is lowered towards $T_c = 378K$, the frequency interval, where the scaling law results describe the data, expands properly. At $T = 383K$ the extended MCT describes the dynamics over a frequency interval of three orders of magnitude extension. If the system is heated towards T_c, the fit interval also expands on the high frequency end, as predicted. So far it was not possible to measure the spectral minimum for $T < T_c$.

Scaling lines (5.7a) are added to Fig. 39 as dashed in full generalized parabolas, where the latter refer to $\sigma = \pm \sigma_0$. For $T = 423K$ the largest hopping rate was used for the fit. But in this case $|\sigma|/|\sigma_0| = s$ is already so large, that the master function becomes s-insensitive. This results in a big uncertainty interval for the δ. For even larger T, $-s$ increases even further, since the scaling lines in Fig. 39 are steeper than the curve C. Therefore the spectra can be fitted well for $T \geq 433K$ within the ideal MCT, i.e. with $\delta t_0 = 0$, as was discussed in connection with Fig. 7. Even though the hopping rate is largest for the four highest measured temperatures, the fit cannot be improved by acknowledging $\delta \neq 0$.

According to (5.1) there should be a Debye–Waller factor f_q^c and a critical amplitude h_q so that F. Mezei's neutron spin echo data for CKN [16] are given by $\phi_q(t) = f_q^c + h_q G(t)$, with $G(t)$ determined by the light scattering analysis. This is indeed the case as shown in Fig. 4. The full curves are obtained for a fixed $f_q = 0.80$ and an amplitude h_q which drifts with T similar to the amplitude h_{es} for the light scattering spectra. The identified β-interval excludes only that part of the α-process, where $\phi(t) < 0.2$.

Acknowledgments

I thank cordially H.Z. Cummins, G. Li and their collaborators as well as W. van Megen and S.M. Underwood for the permission to present their unpublished data and M. Fuchs for preparing figures 25, 27. Helpful discussions with Herman Cummins, Matthias Fuchs, Siegfried Großmann, Bill van Megen and Lennart Sjögren are also gratefully acknowledged. I am indebted to Frau H. Sprzagala for the great efforts she invested in the preparation of the manuscript.

References

1. Frenkel J 1955 *Kinetic Theory of Liquids* (Dover Publ., New York)
2. Howell F S, Bose R A, Macedo P B and Moynihan C T 1974 J. Phys. Chem. 78 639
3. Scherer G W 1986 *Relaxation in Glass and Composites* (Wiley, New York)
4. Angell C A and Torell L M 1983 J. Chem. Phys. 78 937
5. Mountain R D 1966 J. Res. Nat. Bur. Standards 70 A 207
6. Forster D 1975 *Hydrodynamic Fluctuations, Broken Symmetry, and Correlation Functions* (Benjamin, Reading)
7. Kohlrausch R 1854 Pogg. Ann. Phys. 91 56, 179
8. Fuchs M, Götze W, Hofacker I and Latz A 1991 J. Phys.: Condens. Matter 3 5047
9. Von Schweidler E 1907 Ann. Phys. 24 711
10. Gnedenko B V and Kolmogorov A N 1954 *Limit Distributions for Sums of Independent Random Variables* (Addison–Wesley, Reading)

11. Goldstein M 1969 J. Chem. Phys. 51 3728
12. Boon J P and Yip S 1980 *Molecular Hydrodynamics* (McGraw-Hill, New York) – Hansen J P and McDonald I R 1986 *Theory of Simple Liquids* 2nd edn (Acad. Press, London)
13. Knaak W, Mezei F and Farago B 1988 Europhys. Lett. 7 529
14. Tao N J, Li G and Cummins H Z 1991 Phys. Rev. Lett. 66 1334
15. Mezei F, Knaak W and Farago B 1987 Phys. Rev. Lett. 58 571 and Phys. Scr.T 19 363
16. Mezei F 1991 Ber. Bunsenges. Phys. Chem. 95 1118 – J. Non–Cryst. Sol. 131–133 317
17. Cummins H Z, Du W M, Fuchs M, Götze W, Hildebrand S, Latz A, Li G and Tao N J 1993 Phys. Rev. E in print
18. Li G, Du W M, Chen X K, Cummins H Z and Tao N J 1992 Phys. Rev. A 45 3867
19. Li G, Du W M, Hernandez J and Cummins H Z 1993 Phys. Rev. E submitted
20. Li G, Du W M, Sakai A and Cummins H Z 1992 Phys. Rev. A 46 3343
21. Dreyfus C, Lebon M J, Cummins H Z, Toulouse J, Bonello B and Pick R M 1992 Phys. Rev. Lett. 69 3666
22. Kiebel M, Bartsch E, Debus O, Fujara F, Petry W and Sillescu H 1992 Phys. Rev. B 45 10301
23. Signorini G F, Barrat J L and Klein M L 1990 J. Chem. Phys. 92 1294
24. Pusey P N 1991 *Liquids, Freezing and the Glass Transition* J P Hansen et al ed. (North Holland, Amsterdam), p 763
25. Pusey P N and van Megen W 1987 Phys. Rev. Lett. 59 2083
26. van Megen W and Underwood S M 1993 preprint
27. van Megen W and Underwood S M 1993 Phys. Rev. E 47 248
28. Bartsch E, Antonietti M, Schupp W and Sillescu H 1992 J. Chem. Phys.97 3950
29. Leutheusser E 1984 Phys. Rev. A 29 2765 – Bengtzelius U, Götze W and Sjölander A 1984 J. Phys. C: Solid State Physics 17 5915
30. Götze W and Sjögren L 1992 Rep. Prog. Phys. 55 241
31. Götze W 1991 *Liquids, Freezing and the Glass Transition* ed. J P Hansen et al. (North Holland, Amsterdam) 287
32. Kirkpatrick T R and Thirumalai D 1988 Phys. Rev. B 37 5342
33. Prigodin V N 1992 J. Phys.: Condens. Matter 4 785
34. Edwards S F and Anderson P W 1975 J. Phys. F: Met. Phys. 5 965
35. Cummins H Z, Du W M, Fuchs M, Götze W, Latz A, Li G and Tao N J 1993 preprint
36. Sjögren L 1989 Z. Phys. B 74 353
37. Sjögren L 1991 J. Phys.: Condens. Matter 3 5023 – Flach S, Götze W and Sjögren L 1992 Z. Phys. B 87 29
38. Stephen M J 1969 Phys. Rev. 187 279 – Tao N J, Li G, Chen X, Du W M and Cummins H Z 1991 Phys. Rev. A 44 6665
39. Bengtzelius U and Sjögren L 1986 J. Chem. Phys. 84 1744 – Götze W and Latz A 1989 J. Phys.: Condens. Matter 1 4169
40. Fuchs M and Latz A 1991 J. Chem. Phys. 95 7074
41. Götze W, Leutheusser E and Yip S 1981 Phys. Rev. A 23 2634
42. Fuchs M, Götze W, Hildebrand S and Latz A 1992 Z. Phys. B 87 43
43. Fuchs M, Hofacker I and Latz A 1992 Phys. Rev. A 45 898
44. Sjögren L and Sjölander A 1979 J. Phys. C 12 4369 – Sjögren L 1980 Phys. Rev. A 22 2866
45. Götze W and Sjögren L 1987 Z. Phys. B 65 415 – 1988 J. Phys. C 21 3407
46. Belitz D and Schirmacher W 1984 J. Non–Cryst. Solids 61 1073
47. Sjögren L 1990 Z. Phys. B 79 5
48. Fuchs M, Götze W, Hildebrand S and Latz A 1992 J. Phys.: Condens. Matter 4 7709

NEUTRON SCATTERING AT THE GLASS TRANSITION

U. BUCHENAU
Institut für Festkörperforschung
Forschungszentrum Jülich
Postfach 1913
D-5170 Jülich
Federal Republic of Germany

ABSTRACT. In the close neighbourhood of the glass transition temperature, neutron scattering shows fast relaxations with a practically temperature-independent time constant of the order of a picosecond. Recent neutron data indicate a close connection between these fast relaxations and the soft vibrations which coexist and interact with the sound waves in the glassy state, giving rise to the maximum in C_p/T^3 (C_p specific heat, T temperature) and to the boson peak in neutron and Raman scattering.

The slow α-relaxation of the flow process can only be studied by neutrons at higher temperatures, where the relaxation times enter the nanosecond range. Spin-echo measurements have shown a stretched exponential Kohlrausch time dependence and the validity of the time-temperature Vogel-Fulcher-Tamman scaling even at these very short relaxation times. Furthermore, the separation of the Johari-Goldstein relaxation from the α-process could be observed. Several predictions of the mode coupling theory have been verified in four different substances.

1 Introduction

There are many different concepts trying to explain the puzzling universal features of undercooled liquids [1]. One of them, the mode coupling theory [2], makes detailed predictions for neutron scattering measurements and is introduced by W. Götze in this volume. The mode coupling theory is based on concepts from the theory of simple liquids [3]. Inspite of its generality, it has been found to describe remarkably well the essential features of neutron scattering data for many different substances. In its simplest version ("idealized mode coupling theory"), it is limited to elevated temperatures and does not reach down to the glass transition temperature T_g (defined either by a viscosity of 10^{12} Pa s or by the step in the specific heat). However, as explained by W. Götze in his article, the theory can be extended to include hopping processes at lower temperatures.

Alternatively, for lower temperatures one can try a solid state description in terms of normal modes. However, the concept of normal modes, so successful in explaining the dynamics of crystals and molecules, has met with only limited success in disordered condensed matter. The concept requires a stable atomic structure at a local minimum of the potential energy with respect to the atomic displacements. At first sight, one would think the concept applies in glasses, where the atoms have fixed equilibrium positions, but does not apply in liquids, where the atoms diffuse away. As it turns out, this is not the main difficulty. The diffusive motion is strongly temperature-dependent. Lowering the temperature, one can always achieve a situation where the diffusive motion is negligible within many vibrational periods, while the temperature is still well above the glass transition. In fact, the main hindrance for the straightforward application of the normal mode concept to disordered matter is found in the glass itself. There is convincing experimental evidence [4] for tunneling modes in glasses. Though their number is small, of the order of 10^{-5} tunneling states per atom, their existence implies that one has to deal with a multiminimum situation in the potential energy, with small barriers between the different minima. If one insists on an expansion of the potential energy in powers of atomic displacements, such a situation is only stable taking at least fourth order terms into account.

Within the last decade, this main difficulty has been addressed with some success by the phenomenological soft potential model [5, 6], an extension of the well-known tunneling model to include soft vibrations into the picture. The model starts from the coexistence of sound waves and localized modes. The localized modes are assumed to have small force constants which can be positive or negative, but are stabilized by a positive fourth order term. The model is supported by findings in glasses at low temperatures. It will be discussed by Yu. Galperin in this volume.

The present article discusses these and other concepts in the context of a review of recent neutron data.

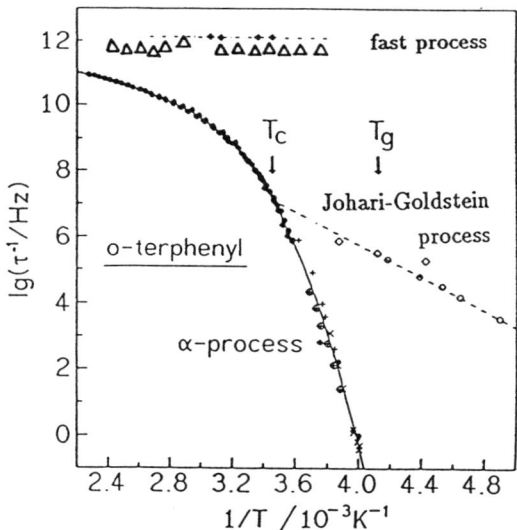

Fig. 1. Relaxation rates of ortho-terphenyl as given by different methods; the lines are guides for the eye. α-process: + dielectric relaxation [7]; × dynamic Kerr effect [7]; ⊕ light scattering [8]; • NMR [9]. Johari-Goldstein process: o dielectric relaxation [10]; ◇ time resolved optical spectroscopy [11]. Fast process: ♦ neutron scattering [12]; △ light scattering [13].

To illustrate the many competing relaxations on different time scales, Fig. 1 summarizes the results of several different experimental techniques in a single Arrhenius plot of the inverse relaxation time τ on a logarithmic scale against the inverse of the temperature T. In such a plot, a thermally activated process shows up as a straight line, the slope of which gives the activation energy. The specific example shown here is ortho-terphenyl, a molecular glassformer, taken from a compilation in reference [14] together with recent light scattering data for the fast process [13]. It is typical for glassformers in general in the following features:

(i) The α-process, the relaxation connected with the viscous flow, shows a non-Arrhenius curved behaviour and becomes extremely slow at the calorimetric glass transition temperature T_g.

(ii) At a temperature of $\sim 1.2\ T_g$, a process which we denote here as Johari-Goldstein process [10] separates from the α-process. From thereon towards lower temperatures, the Johari-Goldstein process follows an Arrhenius behaviour with barrier heights of 10 to 20 $k_B T_g$.

(iii) There is a fast relaxation with a practically temperature-independent time constant of the order of 10^{-12} s, not understandable in terms of a thermally activated process.

Apart from these three universal features, one often observes additional slow relaxations between the Johari-Goldstein relaxation and the fast relaxation. However, while the Johari-Goldstein relaxation seems to be a universal phenomenon, the additional ones are often attributable to specific molecular configuration changes or side group rotations [10, 15].

In the specific example of Fig. 1, the fast process was measured both by neutron and light scattering. The fast relaxation and its developement from the boson peak observed in the glass phase was indeed seen in light scattering many years before neutron scatterers became aware of it [16].

Neutrons are best suited to scan atomic motions in the picosecond range. Therefore, we will first focus on the fast relaxations. Section II presents neutron evidence to show that these relaxations are related to soft vibrations in the glass phase. Section III discusses the findings in glasses in that frequency range, giving a common description of the neutron data and the low temperature anomalies of glasses in terms of the soft potential model. Section IV contains a simplified version of a recent treatment of the quasielastic scattering from anharmonic modes by Condat and Jäckle [17]. Neutron measurements of the slow relaxations at higher temperatures and tests of the mode coupling predictions are reviewed in Section V. Section VI discusses some of the current concepts of the undercooled liquid in the light of the neutron data. The results are summarized in section VII.

2 Fast relaxations

The first neutron measurements of the fast relaxation in undercooled liquids were done by Fujara and Petry [18] on α-TNB, a molecular liquid, and by Frick et al [19] on amorphous polybutadiene. Fig. 2 shows the α-TNB data. The time-of-flight data of Fig. 2 (a) show a very broad quasielastic feature, consistent with a relaxation time of 3.6 picoseconds. As the temperature increases, that broad quasielastic scattering does not change its width within experimental accuracy, but merely increases its height. Fig. 2 (b) shows a backscattering measurement with better resolution. There, the elastic line remains sharp, again within the experimental accuracy (naturally, one would in principle see the broadening of the elastic line by the slow flow processes at much higher resolution). The broad quasielastic feature shows up as a constant background, increasing essentially with Q^2 (Q scattering vector). The proportionality to Q^2 shows that we deal with motional amplitudes which are small compared to interatomic distances. No dependence of the width on the scattering vector could be detected, indicating that the atoms remain localized and do not diffuse away in this fast motion.

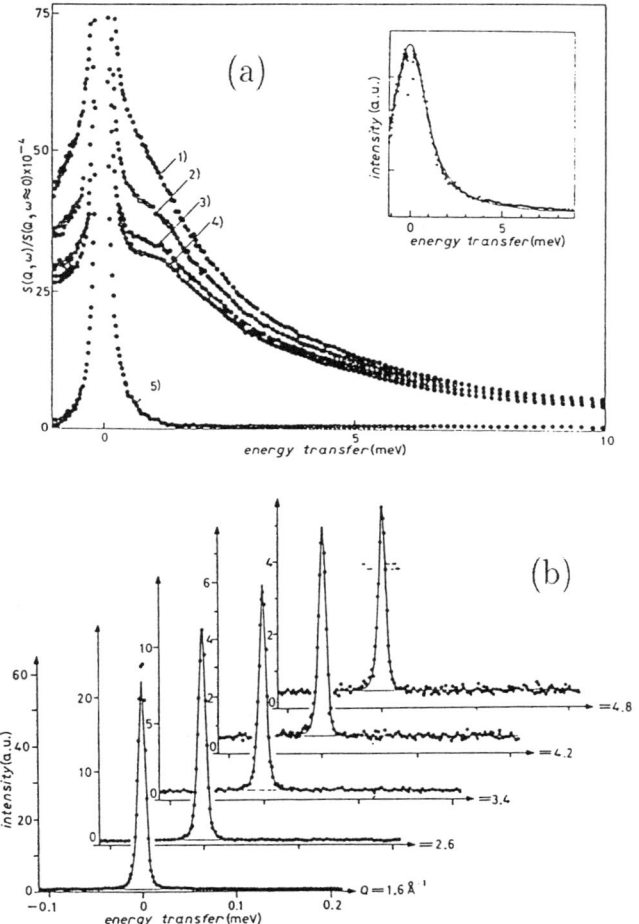

Fig. 2. (a) Time of flight measurements (IN6, ILL Grenoble) of tri-α-TNB at different temperatures, together with a vanadium measurement to show the resolution. Curve 1) at T_g+27K; curve 2) at T_g+7K; curve 3) at T_g-19K; curve 4) at T_g-42K; curve 5) vanadium. The glass transition is at T_g=342 K.

(b) High energy resolution spectra measured on the IN13 backscattering spectrometer at T_g+27K [18].

In this example, as well as in the very similar results on polybutadiene [19, 20], ortho-terphenyl [21], $Ca_{0.4}K_{0.6}(NO_3)_{1.4}$ [22] and polystyrene [23], the fast relaxations appear as a broad lorentzian, with no indication of any vibrational character of this fast motion. However, such a vibrational character is suggested by measurements at lower temperatures in the glass, where one finds a broad peak, clearly apart from the elastic line, in the same frequency range. That peak is absent in polycrystalline samples of the same material. Here we will follow the custom of Raman scatterers and denote it as boson peak.

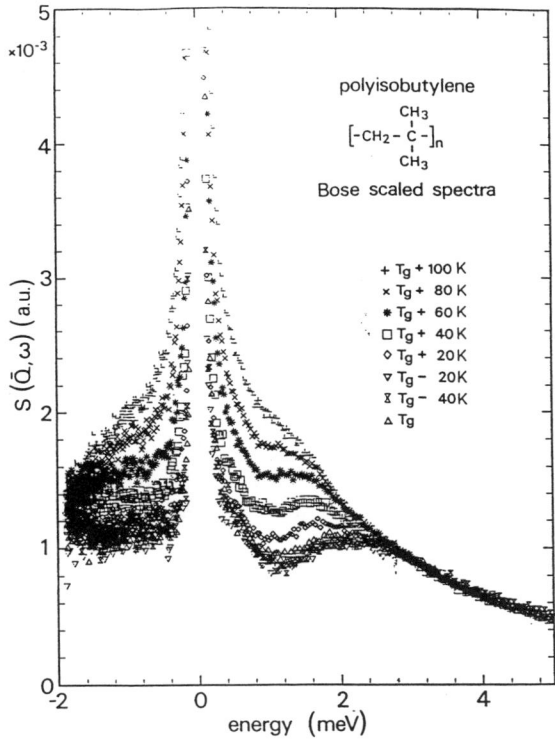

Fig. 3. Time-of-flight spectrum of polyisobutylene corrected for Debye-Waller factor and Bose factor as measured on the IN6, ILL. The measured curves are scaled to $T=200$ K, the glass transition temperature of polyisobutylene [24].

The boson peak is not merely a vibrational feature of the glass, to be subtracted from the liquid data after proper Bose scaling. This is demonstrated by a very recent neutron time-of-flight experiment [24] on polyisobutylene, displayed in Fig. 3. In this substance, the boson peak is still unchanged at T_g. As the temperature increases, the minimum between elastic line and boson peak is not filled by quasielastic scattering emerging from the elastic line. Rather, one observes a shift of the boson peak towards lower frequencies, which remains a clearly pronounced shoulder even 80 K above T_g (in this substance, the glass transition occurs at about 200 K). The experiment shows convincingly that the fast motion cannot be considered as purely relaxational. On the other hand, it is obvious from Fig.

3 that it is neither purely vibrational; something dramatic happens to these modes in the undercooled liquid.

Polyisobutylene is not an isolated case. Neutron data for selenium (see section IV) show the same behaviour. Similar data have been obtained earlier by light scattering from glycerole [16] and from GeSBr$_2$ [25]. It is important, however, to confirm this information by neutron scattering, because light scattering data suffer from uncertainties concerning the frequency dependence of the coupling between the light and the vibrational modes [26, 27], a disadvantage which neutrons do not share.

Another very recent neutron experiment [23] on polystyrene (Fig. 4) shows that the softening of the boson peak does not necessarily occur in the undercooled liquid. This finding is already suggested by the quasielastic character of the fast motion in many substances at T_g (as seen, for instance, in Fig. 2 (a)). However, the softening in the earlier examples seems still to occur very near to T_g. Polysterene is an example where it occurs deep in the glass phase, between 100 and 200 K (see Fig. 4). The glass transition temperature T_g of polystyrene is at 373 K, more than a factor of two higher.

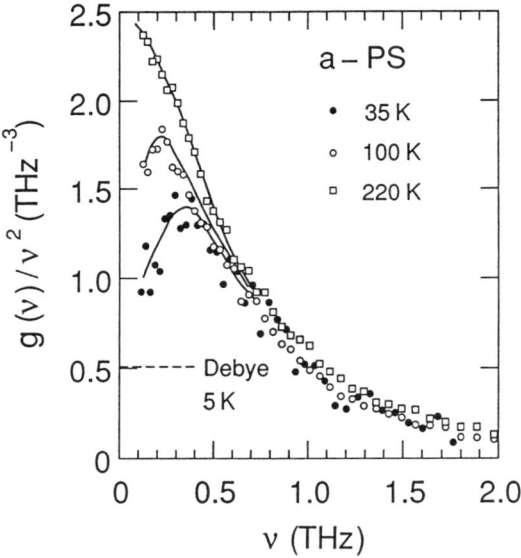

Fig. 4. Quasiharmonic vibrational density of states fitted to time-of-flight data (IN6, ILL) of polystyrene [23] at different temperatures. The density of states $g(\nu)$ is plotted as $g(\nu)/\nu^2$ versus the frequency ν. The plot corresponds essentially to the Bose and Debye-Waller factor scaling in Fig. 3.

From Raman data [28], PMMA, another polymer, may be a similar case.

Additional neutron evidence for a close relation between the soft vibrations of the boson peak and the fast relaxation in the glass phase has been recently observed [29] in BAF4, a technical silicate glass with a relatively low glass temperature. In BAF4, $T_g = 760$ K, while pure a-SiO$_2$ has $T_g = 1473$ K, nearly twice as high. Earlier experiments [30] in a-SiO$_2$ showed a very characteristic Q-dependence of the inelastic scattering of the boson peak, not compatible with sound waves, but compatible with coupled SiO_4 librations in the corner-connected statistical network of SiO_4 tetrahedra. The same dependence is seen for the boson peak modes of BAF4. Above the glass transition of BAF4, the additional fast relaxation shows again the same Q-dependence, indicating that the fast relaxations have the same motional eigenvector as the modes at the boson peak.

We conclude that the vibrational modes of the boson peak soften at a temperature which can be either below or above the glass transition temperature. In most cases, this softening occurs in the neighbourhood of the glass transition, but there are clear exceptions from that rule. Therefore we have to add still another characteristic temperature to the many which are already defined for the undercooled liquid [1], namely the temperature of the softening of the boson peak.

3 Findings in glasses

As shown in the preceding section, the fast relaxations of the undercooled liquid are closely related to the soft vibrational modes of the boson peak in glasses. These modes have been studied in detail within the last decade [31]. It has become gradually clear that they in turn are closely related to the anomalous properties of glasses at low temperatures. These cannot be understood in terms of the standard Debye model. Nonetheless, they seem to be universal [32, 4, 33, 34]. In the following, we give a short introduction to these glassy anomalies and their connection to the boson peak.

Below 1 K, one finds a linear specific heat and a thermal conductivity increase with T^2 (T temperature). These features as well as the unusual low temperature behaviour of the sound velocity and absorption [34, 35] can be explained within the tunneling model [36].

Above 1 K, one finds glassy anomalies which can no longer be explained in terms of the tunneling model. The specific heat C_p rises stronger than the Debye T^3-term, the thermal conductivity shows a plateau and the sound absorption rises. At still higher temperatures, one finds a peak in C_p/T^3 and a second rise of the thermal conductivity. In terms of the vibrational density of states $g(\nu)$ and the frequency ν, the peak in C_p/T^3 is due to a maximum in $g(\nu)/\nu^2$, observed universally in Raman and neutron scattering [27]. This is the boson peak introduced in the preceding section. The boson peak appears at a frequency at which the corresponding crystals still have only sound waves with a wavelength of the order of ten to twenty interatomic spacings. In glasses, the soft vibrational modes of the boson peak coexist with sound waves of that wavelength [30]. Their number is by two to three orders of magnitude higher than the number of tunneling states which dominate the properties below 1 Kelvin. Consequently, these vibrational modes do not only appear in the specific heat, but become accessible to neutron and Raman scattering [30, 27] as well as to numerical simulations [37].

It has recently been shown that all these glassy anomalies can be described consistently by the soft potential model [5, 6]. The soft potential model postulates soft localized modes with an effective mass M and a stabilizing fourth order term in the potential

$$V(x) = W\left[D_1\left(\frac{x}{d}\right) + D_2\left(\frac{x}{d}\right)^2 + \left(\frac{x}{d}\right)^4\right], \quad (1)$$

The origin of the configurational coordinate x is chosen such that the third order term of the potential vanishes. The coefficients D_1 and D_2 are supposed to be random with a gaussian distribution in D_1. The distance d is fixed by the condition $W = \hbar^2/2Md^2$. With this condition, the quantum mechanical balance between potential energy and kinetic confinement energy leads to level splittings which are greater than W for any single-well potential. Smaller level splittings are only achieved in the double-well tunneling case (see Fig. 5). Consequently, the energy W marks the crossover between tunneling and vibrational states. In order to explain the universal anomalies in the acoustic properties and the thermal conductivity of glasses [4], both vibrational and tunneling states are assumed to interact with the sound waves.

Fig. 5 shows the division of the $D_1 - D_2$ plane into double-well and single-well regions. The inserts show the potentials and levels of a typical tunneling state (left) and a typical vibrational state (right). The levels were calculated using a numerical search for stationary solutions of the Schrödinger equation.

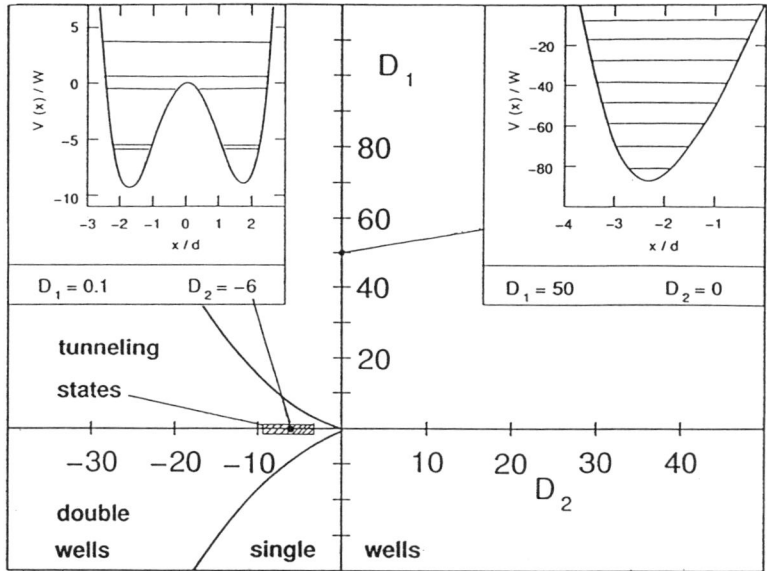

Fig. 5. Single- and double-well regions in the $D_1 - D_2$ plane of the soft potential model. Inserts: Potentials and levels of a typical tunneling state (left) and a typical vibrational state (right). (taken from reference [38]).

The soft potential model allows to describe the whole low temperature and low frequency behaviour of the glass. As an illustration, Fig. 6 shows a comparison of numerical calculations to measured specific heat, thermal conductivity and vibrational density of states data of vitreous silica [38]. The fit assumed a broad gaussian distribution of D_1 values, independent of D_2 and centered around zero. The position of the boson peak depends on the width of that gaussian. In [38], this width was attributed to thermal strains freezing in at the glass temperature. This assumption yields the relations

$$T_{max} = 1.07\, T_{min}^{3/4} T_g^{1/4} \quad \text{and} \quad h\nu_{max} = 4.5\, k_B T_{max}, \tag{2}$$

which relate the position ν_{max} of the boson peak to the glass transition temperature and to the position T_{min} of the minimum in C_p/T^3 (see Fig. 6). That minimum is in turn related to the energy W of the soft potential model via $W = 1.8\, k_B T_{min}$. T_{max} is the position of the peak in C_p/T^3 corresponding to the boson peak in $g(\nu)/\nu^2$.

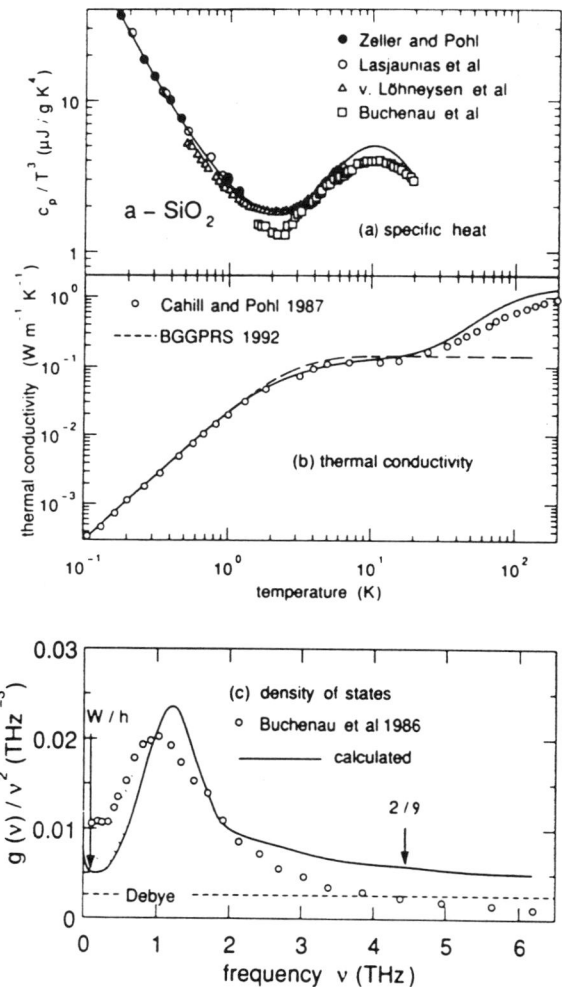

Fig. 6. Comparison of the soft potential model calculations [38] (continuous lines) to experimental data in vitreous silica. (a) Specific heat [32, 30, 39, 40], plotted as C_p/T^3 versus temperature. (b) Thermal conductivity [32, 42] versus temperature (dashed line: earlier calculation [41]). (c) Density of states [30], plotted as $g(\nu)/\nu^2$ versus frequency ν. The shaded area shows an estimate of the relaxational contribution to the experimental data.

For the purpose of the next section, we are particularly interested in the upper barrier cutoff of the double well part of the soft mode distribution. This cutoff is indeed seen in low temperature relaxation data. The best studied example is vitreous silica [34, 43, 44], where one infers a barrier cutoff at heights corresponding to 600 to 800 K in temperature, about half the glass temperature. Data for the temperature dependence of the sound absorption in selenium [45] are shown in Fig. 7. The dashed line in that figure was calculated from the soft potential model [41]. As seen from Fig. 7, the measured data fall short of the theoretical expectation by a factor of two already at 10 Kelvin. Thus there must be a barrier cutoff already at fairly low barriers. If the attempt frequency is of the order of 1 Thz, the peak in the absorption at 30 Kelvin should be due to barriers of 300 Kelvin, the order of the glass temperature ($T_g = 303$ K). For the $D_1 - D_2$ - distribution of the soft potential model, the results imply a cutoff towards negative D_2 values around $D_2 = -30$. This cutoff is of small influence for the results in Fig. 6, but becomes clearly visible in the temperature dependence of the acoustic absorption in Fig. 7. An early deviation from a $T^{3/4}$ law towards lower values is indeed seen universally in acoustic and dielectric absorption data in glasses [35], often followed by a low temperature peak. We conclude that the double-well part of the soft mode distribution in glasses has an upper barrier cutoff around or even below the thermal energy at the glass transition temperature.

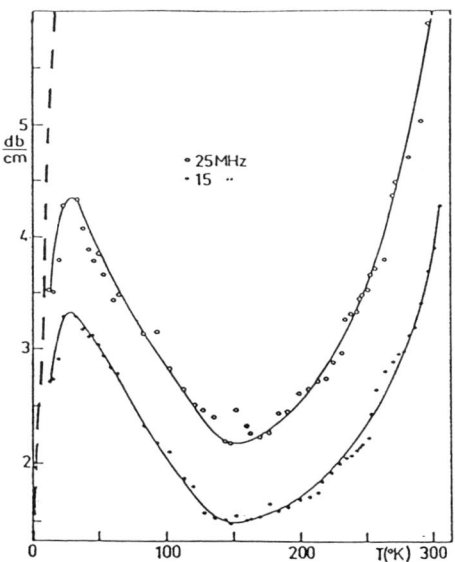

Fig. 7. Sound absorption at 15 and 25 MHz in amorphous selenium as a function of temperature [45]. The dashed line shows the prediction of the soft potential model.

Towards higher temperatures, as one approaches the glass transition, Fig. 7 shows again a rise of the acoustic absorption, which looks like a precursor of the strong α-relaxation setting in at the glass temperature and connected with the flow processes. This is the Johari-Goldstein relaxation introduced in section I. It is often denoted as β-relaxation, but here we denote it as Johari-Goldstein relaxation in order to distinguish it from the fast picosecond relaxation in the undercooled liquid. This fast relaxation has also been denoted as β-relaxation in the context of mode coupling theories [2], but is clearly separated by orders of magnitude in time from the Johari-Goldstein relaxation at temperatures slightly above the glass transition [14]. In the glass phase, the Johari-Goldstein relaxation shows an Arrhenius behaviour [10, 15] with barrier heights of the order of 20 $k_B T_g$, nearly two orders of magnitude higher than the upper cutoff of the barriers of the soft potential model. Thus, while a soft potential model relaxation occurs within a given glass configuration, with jump eigenvectors which resemble the vibrational eigenvector of the vibration within one of the wells (this picture has been confirmed very recently by a numerical study of low temperature relaxations in a model glass [46]), the Johari-Goldstein relaxation must be a jump to an entirely new local configuration. In that case, the vibrations in the two wells should be different and the jump eigenvector should be composed of many modes of either the old or the new well.

4 The soft modes in the undercooled liquid

In the preceding section, we saw that there seems to be a large number of soft anharmonic localized modes in glasses, coexisting with the sound waves of wavelengths larger than ten interatomic spacings and giving rise to the boson peak in neutron and Raman scattering. These modes do not exist in textbook crystals and must be an intrinsic feature of the undercooled liquid, freezing in at the glass transition.

The soft potential model which describes these modes is entirely phenomenological. Consequently, it does not tell anything about the origin of these modes. Nevertheless, it allows some conclusions on their behaviour at higher temperatures. Since the potential barriers in the double-well part of these modes are lower than the thermal energy in the undercooled liquid, we may take the pure quartic potential as representative for the whole distribution, in order to get a feeling for orders of magnitude. The mean square displacement of the pure quartic potential, calculated classically [47], is given by

$$< x^2 > = 0.338 d^2 (k_B T/W)^{1/2} \qquad (3)$$

With the soft potential model fits [48] of the low temperature specific heat of vitreous silica, amorphous selenium and vitreous boron trioxyde, this relation gives atomic amplitudes in a single soft mode of the order of 0.3 to 0.4 Å, about one tenth of the relevant interatomic spacings, at the glass transition temperatures of these three glasses. Atomic amplitudes of that size would still be consistent with the observed Q^2 dependence of the intensity of the scattering from the fast process [18, 19, 20, 23].

In view of the large atomic amplitudes and the strong anharmonicity of these modes, the question arises whether they should not be overdamped in the undercooled liquid. However, from the experimental evidence of section II, these modes need not be overdamped even above T_g. If these modes were overdamped, one would only see a broad quasielastic scattering centered around the frequency zero. This is what one does indeed observe at higher temperatures. Fig. 8 shows this behaviour in neutron data from glassy and liquid selenium. The figure displays the results [49] in terms of a quasiharmonic density of states $g(\nu)$, plotted as $g(\nu)/\nu^2$ (apart from minor corrections, this is the scattering function divided by the temperature). In a perfect harmonic system, this quantity should not depend on the temperature. In fact, there is not a big difference between the results at 100 and 300 K. Both show the boson peak in reasonable agreement with a soft potential model fit. At 360 K, 57 Kelvin above the glass transition, the boson peak is still seen as a separate peak, though at a slightly lower frequency. Apart from the softening, there seems to be an increase in the number of soft modes with increasing temperature. At 440 K, there is no longer a separate boson peak and the scattering looks quasielastic as expected for overdamped modes.

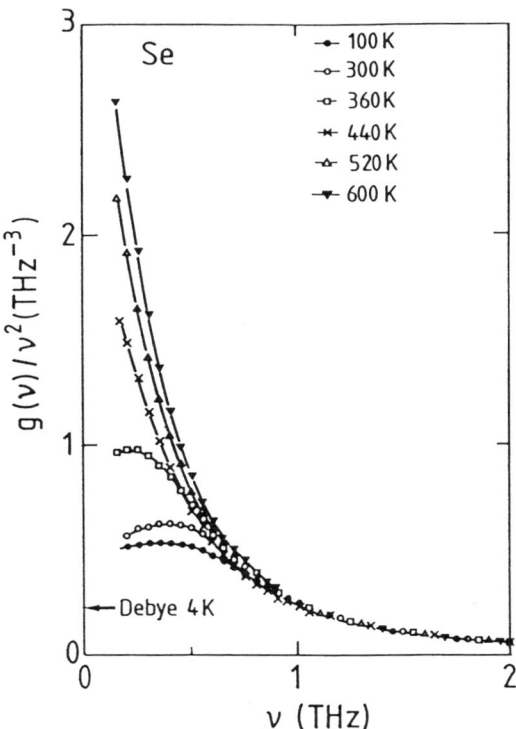

Fig. 8. Quasiharmonic vibrational density of states fitted to neutron data [49] in glassy and liquid selenium.

If one assumes the soft potential model to be a valid description, there should be an important difference in the scattering from these soft modes and crystalline vibrations. The difference concerns the quasielastic scattering connected with the damping of these modes. For the crystalline modes, vibrating in a reasonably harmonic potential, the quasielastic scattering [50] rises only with a weak Q^4 term with increasing momentum transfer Q. This is different for strongly anharmonic modes. The difference has been quantified in a recent theoretical treatment of the scattering from onedimensional potentials by Condat and Jäckle [17]. They showed that the mean square displacement of an anharmonic mode may be subdivided into two parts. The first comes from a properly weighted average over the possible vibrational levels of the potential. It leads to a reduction of the elastic intensity.

The missing intensity appears in the inelastic scattering, at the vibrational frequencies and their higher harmonics, well separated from the elastic line. The second part of the mean square displacement comes from changes between different vibrational levels. It appears only if the level change is accompanied by a displacement. Therefore it is absent in symmetric single well potentials as, for instance, the harmonic potential. It is, however, familiar for the case of classical Arrhenius relaxation over the potential barrier in a double-well potential, though one usually does not think about that case in terms of a mean square displacement connected with the relaxational jumps. In Condat and Jäckle's generalization, the anharmonic single-well potential of the right insert of Fig. 5 has also such a relaxational component in its mean square displacement. Numerical calculations [51] show that this component is not negligible at T_g. In fact, Fig. 5 shows that the center of gravity of these energy levels shifts to the right with increasing level energy. Thus one has a kind of localized diffusional motion as the system samples the different energy levels. Again, this second part of the mean square displacement reduces the elastic intensity, but now the missing intensity does not appear well-separated from the elastic line, but as quasielastic scattering centered around the elastic line. The width of that quasielastic line depends on the lifetime of the levels. If this is ten vibrational periods, the line width will be of the order of one tenth of the vibrational frequency.

We conclude that the validity of the soft potential model implies the existence of a fast quasielastic scattering component, increasing with the square of the momentum transfer and comparable in intensity with the boson peak. As the temperature increases, the width of that quasielastic scattering should increase until it merges with the boson peak. If one interprets the neutron and Raman measurements in different liquids in terms of this picture, the soft modes get overdamped above or below T_g, depending on the system (in most cases near to T_g). The concept is able to explain the temperature independence of the fast process.

5 Neutron scattering from slow relaxations

For neutrons, the investigation of the slow relaxations is restricted to high resolution techniques like spin-echo and backscattering. Even so, one needs relaxation times below 10^{-8} s in order to study these relaxations with neutrons. In terms of temperature, this implies temperatures at and above the separation of the Johari-Goldstein relaxation from the α-process (see Fig. 1).

Inspite of this restriction, neutrons have been able to make significant contributions to this field. The first of these concerned the two most striking features known from relaxations at lower frequencies and temperatures, namely the stretched exponential time dependence and the time-temperature superposition principle. As could be shown, these continue to hold at the very short relaxation times of the neutron experiments [52, 53]. Fig. 9 (a) shows spin-echo data [52], taken on the ionic glass $Ca_{0.4}K_{0.6}(NO_3)_{1.4}$ at several temperatures above T_g. In Fig. 9 (b), the same data are scaled together using the stretched exponential Kohlrausch form

$$\Phi(t) = \Phi_0 exp(-t/\tau)^\beta. \qquad (4)$$

Here Φ denotes the neutron spin polarization, providing immediately the intermediate scattering function $S(Q,t)$, and τ is the temperature-dependent relaxation time of the sample. The scaling was done using the Vogel-Fulcher-Tamman law for the viscosity of the substance. The stretching parameter β was chosen at 0.58, at a slightly smaller value than that measured in mechanical relaxation.

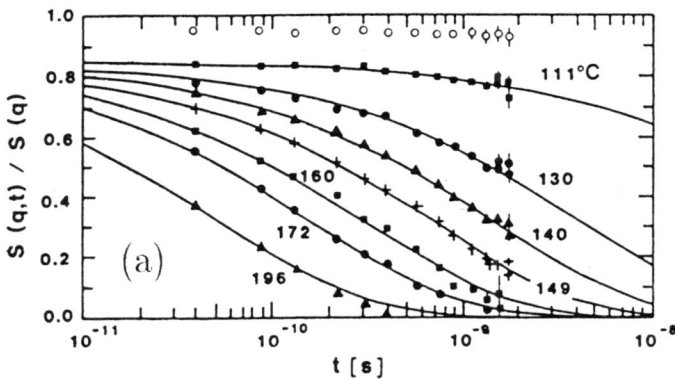

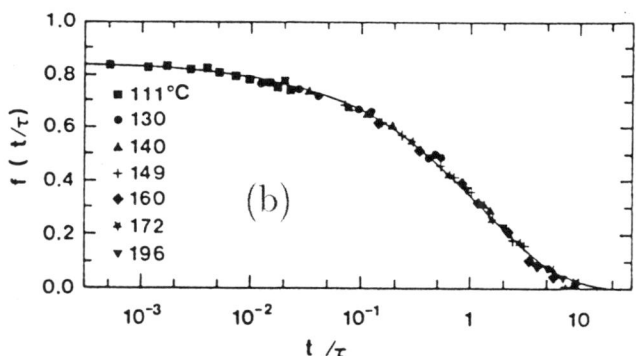

Fig. 9. (a) Spin echo data (IN11, ILL) [52] in $Ca_{0.4}K_{0.6}(NO_3)_{1.4}$ at different temperatures. (b) The same data as in (a) scaled together by the temperature scaling of the viscosity.

Very recently, it has turned out to be possible to see the separation of the Johari-Goldstein relaxation from the viscosity curve in polybutadiene with neutrons, again with the help of the spin-echo technique [54]. The main result is shown in Fig. 10, where the measured spin-echo relaxation time is plotted on a logarithmic scale against $1/T$ and compared to the relaxation time of the viscosity (the continuous curve).

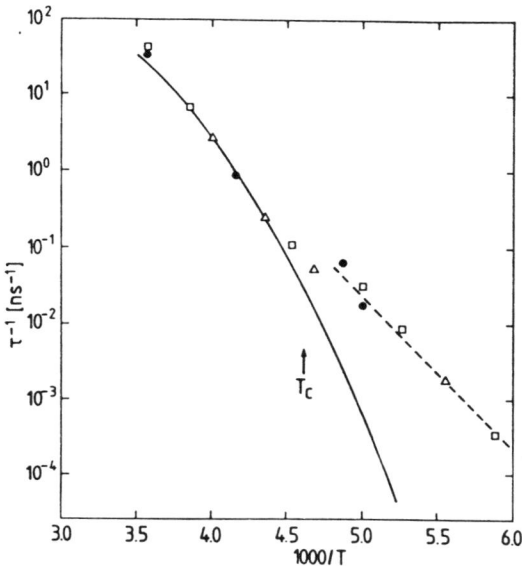

Fig. 10. Arrhenius plot of inverse relaxation times of spin-echo data in polybutadiene [54].

As seen from Fig. 10, it was possible to pursue the Johari-Goldstein relaxation down to rather low temperatures, corresponding to rather long relaxation times. This was done exploiting the strong Q-dependence of the effective relaxation time observed for the α-relaxation, measuring at the highest accessible Q value, at the first minimum of $S(Q)$ of polybutadiene.

Another interesting question addressed by neutron experiments [55] is the spatial distribution of the probability of an atom initially at zero after a given time t. If this is a gaussian, and if the probability decay follows a stretched exponential, one can argue that the Q-dependent width of the quasielastic scattering from the flow process should increase with $Q^{2/\beta}$, where β is again the exponent of the stretched exponential. The presented neutron backscattering data [55] seem to support this conjecture.

Much of the neutron work on undercooled liquids has been motivated by a fascinating recent theoretical developement, the mode coupling theory [2]. The mode coupling theory is based on a successful treatment of the fast picosecond motion in simple liquids. It explains the rapid rise of the viscosity with decreasing temperature in terms of a simplified equation of motion for the density correlation function of the liquid. This equation leads to a divergence of the viscosity at a critical temperature T_c *above* the glass temperature. This implies that the fast picosecond relaxations should suffice to describe the flow process and that the high-barrier processes need not be invoked, at least above T_c. The power law predicted for the viscosity places T_c in the investigated cases [52, 53, 59, 56, 12, 21] at $T_c \approx 1.2T_g$. From thereon to lower temperatures the fast processes, though still existent, should not contribute any longer to the flow. The flow is supposed to be maintained by thermally activated jumps over configurational barriers.

The mode coupling theory makes detailed predictions for the scattering, in particular for the region above T_c. According to the theory, which is given in more detail by W. Götze in this volume, the density-density correlation function $\Phi_Q(t)$ (t time) should display a two-step decay, the second step corresponding to the α-process of the flow. In terms of frequencies, the main feature of the predicted density fluctuations is the existence of two scaling regions governed by two scaling frequencies

$$\omega_{min} = \Omega \cdot |\epsilon|^{1/2a} \quad \text{and} \quad \omega_\alpha = \omega_{min} \cdot |\epsilon|^{1/2b} \tag{5}$$

where ϵ is the control parameter of the transition and $\Omega \sim 1$ THz is a characteristic microscopic frequency. To first order ϵ is proportional to the distance from the critical temperature T_c. The exponent parameters a and b ($0 < a < 0.5$, $0 < b < 1$) are not universal, but related to each other. They are the two solutions of $\lambda = \Gamma^2(1-x)/\Gamma(1-2x)$ for a given λ. ω_α governs the α-process in the liquid and leads to a power law for the viscosity: $\eta \sim \epsilon^{-1/2a - 1/2b}$. Another scaling region is expected at higher frequencies around ω_{min}. The scattering should be describable as

$$S(Q,\omega) = f \cdot \delta(\omega) + \sqrt{\epsilon} \cdot h_Q \cdot f(\hat{\omega})/\omega_{min}, \tag{6}$$

where the nonergodicity parameter $f = 0$ for $T > T_c$, $\hat{\omega} = \omega/\omega_{min}$ and the amplitude factor h_Q depends on Q but not on ω or ϵ. At ω_{min}, the susceptibility $(\omega/k_B T) \cdot S(Q,\omega)$ should have a minimum.

For $\hat{\omega} \ll 1$ and $\hat{\omega} \gg 1$ the master function $f(\hat{\omega})$ approaches two power laws:

$$f(\hat{\omega}) = \hat{\omega}^{-(1-a)} \quad \text{for} \quad 1 \ll \hat{\omega} \ll \frac{\Omega}{\omega_{min}} \tag{7}$$

$$f(\hat{\omega}) = \hat{\omega}^{-(1+b)} \quad \text{for} \quad \frac{\omega_\alpha}{\omega_{min}} \ll \hat{\omega} \ll 1, \tag{8}$$

respectively, where the second one, the von Schweidler power law, exists only in the liquid and the first one appears in both phases.

The predictions of the mode coupling theory have been checked quantitatively by neutron scattering in a number of substances: $Ca_{0.4}K_{0.6}(NO_3)_{1.4}$ [22], polybutadiene [53, 59, 58], orthoterphenyl [57, 21] and even in a biological substance, myoglobin [56]. Such a quantitative check is by no means an easy task. The energy $\hbar\omega_{min}$ ranges from ~ 10 μeV up to about 0.3 meV for different temperatures. This implies that one has to combine a high resolution measurement (spin-echo or backscattering) with a medium resolution measurement, mostly on a time-of-flight spectrometer. The two measurements have to be calibrated to each other. The resolution function of the time of flight has to be evaluated very carefully, including possible tails of the elastic line. In the case of a combination with spin-echo data, a reliable Fourier transform of at least one set of data is required.

In all four cases, the predictions of the mode coupling theory were found to be fulfilled within reasonable expectation, taking into account the technical difficulties of the task. In all cases, a minimum of the susceptibility was found. With the exception of polybutadiene, the minimum showed the correct scaling behaviour with temperature. In $Ca_{0.4}K_{0.6}(NO_3)_{1.4}$, the spectral behaviour turned out to be well described by the pair of exponents $a=0.28$ and $b=0.23$ in reasonable compatibility to the relation between the exponents and an exponent $\gamma = 4 \pm 0.2$ fitted to the temperature dependence of the viscosity. Similar values were found in the other three cases. Doster et al [56] report $\lambda = 0.7 \pm 0.05$, $a = 0.33 \pm 0.05$ and $b = 0.64 \pm 0.1$ for the myoglobin case.

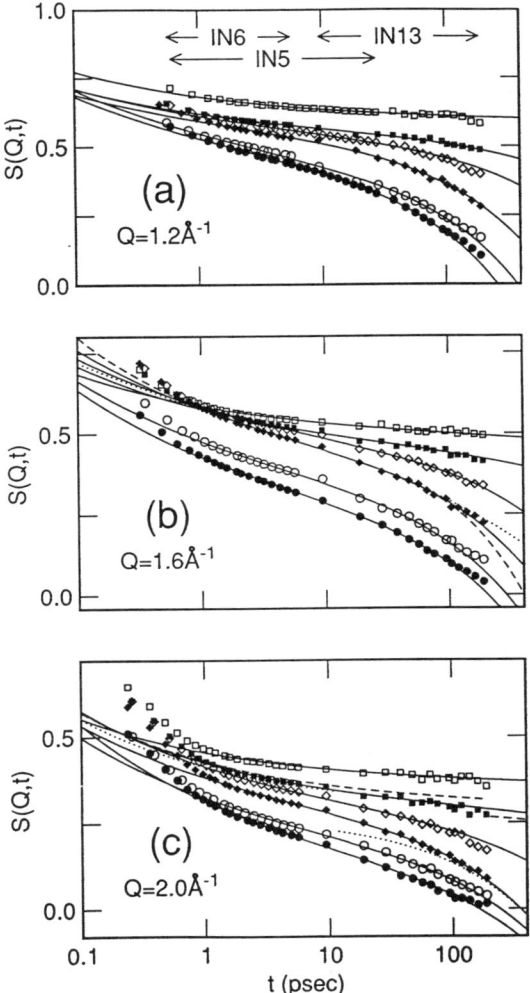

Fig. 11. Intermediate scattering law $S(Q,t)$ of ortho-terphenyl [21] for three values of Q as determined from measurements on the IN5, IN6 and IN11 at the ILL. Temperatures 293 K (□), 298 K (■), 306 K (◇), 312 K (◆), 320 K (○) and 327 K (●). The solid lines are mode coupling fits with a fixed $\lambda = 0.77$.

An exceptionally careful evaluation, including corrections for multiple and multiphonon scattering and numerical corrections to the mode coupling laws, has been done for ortho-terphenyl [21]. All exponents were taken from a power law fit of the viscosity, which gave $\lambda = 0.765$ and $T_c = 290$ K. Fig. 11 shows data from two time-of-flight spectrometers and a backscattering instrument, all three Fourier-transformed into the time domain.

The determination of T_c need not be done from a fit to the viscosity. Alternatively, one can use the square root singularity of nonergodicity parameter, as one approaches T_c from below. Fig. 12 shows such a determination for polybutadiene [59]. The open circles are intensities, integrated over a range of ± 200 μeV around the elastic line of a time-of-flight spectrum at $Q = 1.76$ Å^{-1}, plotted as a function of temperature. These raw data do not show any trace of the singularity. Taking into account, however, that above T_c one should include the tails of the α-process which begin to show up outside the elastic line and using the spectral form and the time constants determined on the spin-echo, one calculates the crosses which show the singularity. Note, however, that this procedure requires that the spectral form of the α-relaxation is maintained up to frequencies in the range of the boson peak discussed in the first sections.

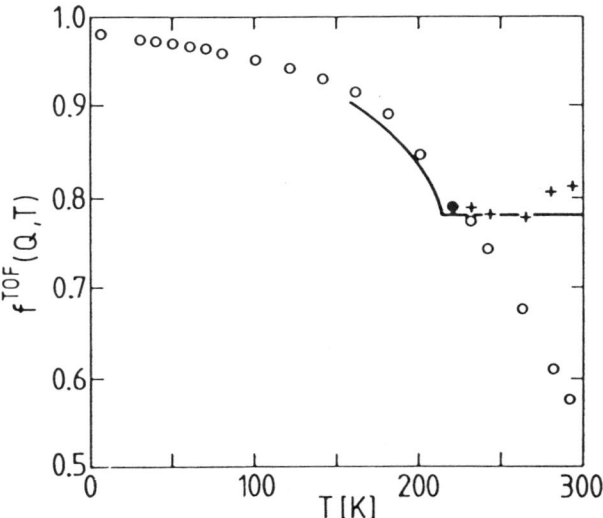

Fig. 12. Normalized elastic intensities from a time-of-flight experiment on polybutadiene [59] (IN6, ILL). ∘ uncorrected data, + data plus calculated tails of the α-relaxation. The solid line is a fit result from spin-echo data.

We conclude that neutron measurements in several different systems do indeed find a dynamic anomaly at the critical temperature obtained from a power law fit to the measured viscosity. The temperature and frequency scaling of the scattering in the neighbourhood of the minimum of the susceptibility is remarkably close to the predictions of the mode coupling theory. The theory does not predict the separation of the Johari-Goldstein process from the α-relaxation, at least not in its present form. Similarly, it does not predict the formation of the boson peak, though it is based on a description of the fast relaxation. We will return to that point in the discussion.

6 Discussion

In the interpretation of the preceding sections, the fast picosecond relaxation observed in undercooled liquids is due to soft localized anharmonic modes which become overdamped as one raises the temperature. In the following, this interpretation is related to current ideas on the undercooled liquid.

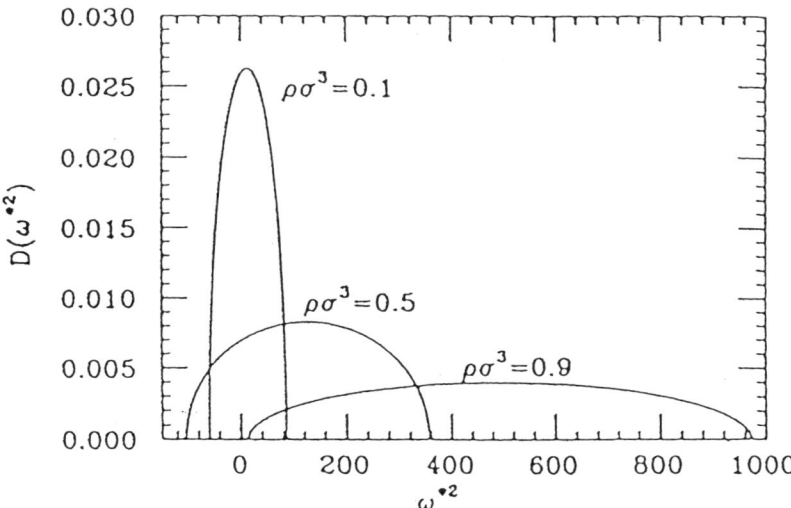

Fig. 13. The averaged distribution of eigenvalues of the dynamical matrix for a simple Lennard-Jones liquid treated in mean-field theory. ω^* is the frequency in reduced units. The three different curves show three different densities (in reduced units) ranging from a near ideal gas to a liquid close to its triple point (taken from reference [62]).

We begin with the connection to a nearly related idea, the instantaneous normal mode theory [60, 61, 62, 63] used in the interpretation of molecular dynamics calculations of liquids. It is based on the evaluation of the dynamical matrix for the instantaneous atomic positions. In such a calculation, one finds not only positive eigenvalues, i.e. stable modes, but also negative eigenvalues corresponding to unstable modes. The higher the temperature and the lower the density, the higher becomes this fraction of unstable modes. This is illustrated in Fig. 13 (taken from reference [62]), which shows the eigenvalue distribution vs. frequency squared for three different densities of a Lennard-Jones liquid. The modes with small positive or with negative eigenvalues should be identical with the soft modes of the soft potential model. In the instantaneous normal mode interpretation of the molecular dynamics calculations, one does indeed find an increase of the number of soft modes with increasing temperature, as postulated in the preceding section to explain the rise in the quasiharmonic density of states at the boson peak between 300 and 360 K in Fig. 8. These investigations are still at the beginning; in principle, they should be able to provide a much more detailed picture of these soft modes and their damping than the one that we have now.

Next we turn to the connection between the soft mode picture and the traditional free volume approach [64]. This connection has been discussed in detail recently [65], again for the case of selenium. The free volume was identified with the part $<u^2>_{loc}$ of the mean square displacement due to the localized soft modes. The reason for that identification is seen from Fig. 14, taken from that paper, which shows the mean square displacements of glassy, liquid and crystalline selenium as a function of temperature. In fact, if one postulates a relation between the viscosity and $<u^2>_{loc}$, in the spirit of the derivation of the Vogel-Fulcher law from the free volume idea, one gets a much better fit of the viscosity of selenium than that of the Vogel-Fulcher law [65]. This is seen in Fig. 15, which shows that this relation holds from below the glass transition temperature up to temperatures high above the melting point.

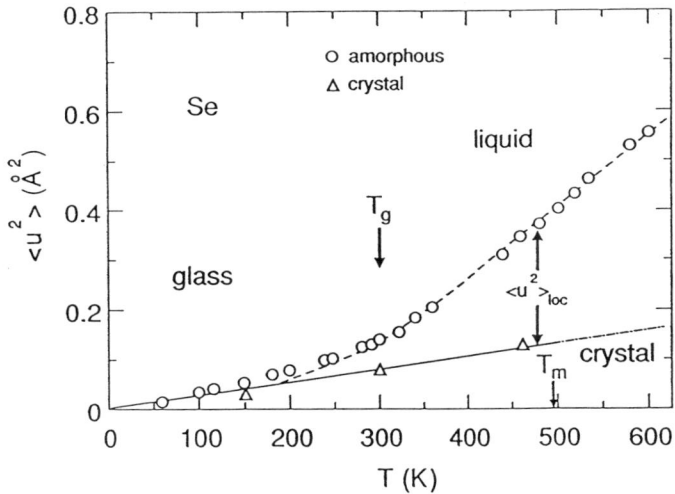

Fig. 14. Mean-square displacements in glassy, liquid and crystalline selenium determined from neutron data for motions with frequencies above 10^{11} Hz (from reference [65]).

The relation to Angell's classification [66] of undercooled liquids in strong and fragile ones is supplied by the connection between that classification scheme and the free volume approach. In the latter, fragile glasses are able to get very near to the Vogel-Fulcher temperature, where the free volume vanishes. So, they have only little free volume left in the glass. In the soft potential model, that means that the number of soft modes gets comparatively low. This can be indeed seen at low temperatures. In the strong glass vitreous silica, the boson peak amplitude in Fig. 4 (c) is six to seven times higher than the signal from the sound waves, while the one in the more fragile selenium is approximately of the same size as the sound wave signal. The same information can be taken from the height of the peak in C_p/T^3 in Fig. 4 (a).

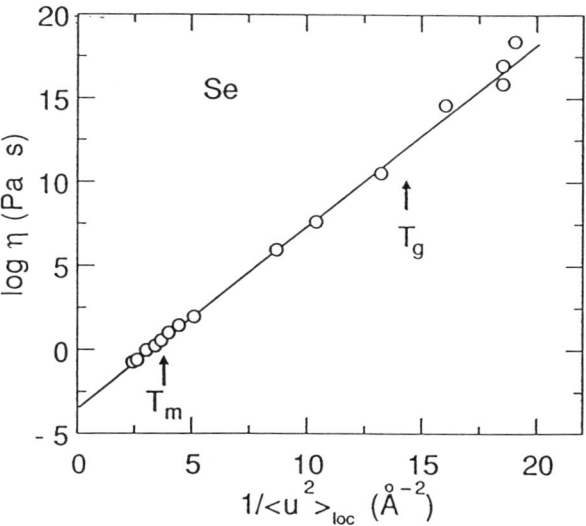

Fig. 15. Linear relation between the logarithm of the viscosity and the inverse of the soft mode mean square displacement in selenium (taken from reference [65]).

Finally, we come to the consequences of the soft potential model for the mode coupling theory [2]. This point is difficult, because it requires a combination of ideas from the solid and from the liquid. At the present stage, the mode coupling theory enables a successful description of the dynamic anomaly in the neighbourhood of its critical temperature, observed both in neutron (see section V) and in light scattering [67]. However, it does not describe the formation of the boson peak from the fast relaxations in that frequency region, nor does it explain the branching off of the Johari-Goldstein process from the flow process, though this occurs at the critical temperature of the theory. On the other hand, as discussed in section IV, the slow relaxations are outside of the scope of the soft potential model. Possibly, a successful combination of these ideas holds the promise of a deeper microscopic understanding of the liquid.

7 Summary

The successsful description of glassy low temperature anomalies in terms of the empirical soft potential model implies the existence of soft anharmonic localized modes in undercooled liquids. In agreement with the results of numerical work, the number of these modes seems to increase with increasing temperature above the glass transition. The damping of these modes gives rise to a fast quasielastic component in neutron and light scattering data.

The experimental finding of a nearly temperature-independent width of the quasielastic scattering at higher temperatures can be explained assuming that these modes become overdamped at a temperature above the glass transition, in the close neighbourhood of the critical temperature of the mode coupling theory.

Acknowledgements. Helpful discussions with G. Meier and D. Richter are gratefully acknowledged. Thanks are due to B. Frick, F. Fujara, K. Linder and R. Zorn for communicating their results prior to publication.

References

[1] Jäckle J. (1986), Rep. Prog. Phys. **49**, 171

[2] Bengtzelius U., Götze W. and Sjölander A. (1984) 'Dynamics of supercooled liquids and the glass transition', J. Phys. **C 17**, 5915-5934; Leutheusser E. (1984), Phys. Rev. **A 29**, 2765; Götze W. (1991), in *Liquids, Freezing and the Glass Transition*, edited by Hansen J. P., Levesque D. and Zinn-Justin J., (North-Holland, Amsterdam) p. 287

[3] Hansen, J. P. and MacDonald I. R. (1986) The Theory of Simple Liquids, Academic Press, New York

[4] Phillips W. A. (ed.)(1981), *Amorphous Solids: Low temperature properties*, Springer, Berlin

[5] Karpov V. G., Klinger M. I. and Ignat'ev F. N. (1983), Zh. Eksp. Teor. Fiz. **84**, 760 [Sov. Phys– JETP **57**, 439]

[6] Il'in M. A., Karpov V. G. and Parshin D. A. (1987), Zh. Eksp. Teor. Fiz. **92**, 291 [Sov. Phys.– JETP **65**, 165]

[7] Beevers M. S., Crossley J., Garrington D. C. and Williams G. (1977) J. Chem. Soc. Faraday Trans. II 73, 458

[8] Fytas G., Wang C. H., Lilge D. and Dorfmüller Th. (1981) J. Chem. Phys. 75, 4247

[9] Dries Th., Fujara F., Kiebel M., Rössler E. and Sillescu H. (1988) J. Chem. Phys. 88, 2139

[10] Johari G. P. and Goldstein M. (1970), J. Chem. Phys. **53**, 2372 Johari G. P. (1973), J. Chem. Phys. **58**, 1766

[11] Hyde P. D., Evert T. E., Cicerone M. T. and Ediger M. D. (1991) J. Non-Cryst. Solids 131-133, 42

[12] Bartsch E., Debus O., Fujara F., Kiebel M., Sillescu H. and Petry W. (1991) Ber. Bunsenges. Phys. Chem. 95, 1146

[13] Steffen W., Patkowski A., Meier G. and Fischer E. W., (1992) 'Depolarized light scattering studies of ortho-terphenyl dynamics above T_g', J. Chem. Phys. 96, 4171-4179.

[14] Rössler E. (1992) 'Comment on "Decoupling of time scales of motion in polybutadiene close to the glass transition"; Phys. Rev. Lett. 69, 1620-1620.

[15] Heijboer J. (1978), in *Molecular Basis of Transitions and Relaxations* ed. by Meier D. J., Gordon and Breach, New York, p. 75

[16] Wang C. H. and Wright R. B. (1970), J. Chem. Phys. **55**, 3300

[17] Condat C. A. and Jäckle J. (1992), Phys. Rev. B **46**, 8154

[18] Fujara, F. and Petry W. (1987) Europhys. Lett. 4, 921-927

[19] Frick B., Richter, D., Petry W. and Buchenau U. (1987) Z. Phys. B 70, 73

[20] Kanaya T., Kaji K. and Inoue K. (1991) 'Local motions of cis-1,4-polybutadiene in the melt. A quasielastic neutron-scattering study', Macromolecules 24, 1826-1832

[21] Wuttke J., Kiebel M., Bartsch E., Fujara F., Petry W. and Sillescu H. (1993) 'Relaxations and phonons in viscous and glassy orthoterphenyl by neutron scattering', Z. Phys. B (submitted)

[22] Knaak W., Mezei F. and Farago B. (1988) 'Observation of scaling behaviour of dynamic correlations near liquid-glass transition', Europhys. Lett. 7, 529-536 (see also the discussion of these data by W. Götze in this volume)

[23] Linder K. (1993) Diplomarbeit RWTH Aachen

[24] Frick B. and Richter D. (1993) unpublished

[25] Krüger M., Soltwisch M., Petscherizin I. and Quitmann D. (1992), J. Chem. Phys. **96**, 7352

[26] Shuker R. and Gammon R. W. (1970) 'Raman-scattering selection-rule breaking and the density of states in amorphous materials', Phys. Rev. Lett. 25, 222-225.

[27] Malinovsky V. K., Novikov V. N., Parshin P. P., Sokolov A. P. and Zemlyanov M. G. (1990), Europhys. Lett. **11**, 43

[28] Malinovsky V. K., Novikov V. N., Sokolov A. P. and Bagryansky V. A. (1988) 'Light scattering by fractons in polymers', Chem. Phys. Lett. 143, 111-114.

[29] Buchenau U., Nücker N., Dianoux A. J. and Krause D. (1993) Conference on Dynamics of Disordered Materials II, Grenoble, March 1993

[30] Buchenau U., Prager M., Nücker N., Dianoux A. J., Ahmad N. and Phillips W. A. (1986), Phys. Rev. B **34**, 5665; Buchenau U., Zhou H. M., Nücker N., Gilroy K. S., and Phillips W. A. (1988), Phys. Rev. Lett. **60**, 1318

[31] Buchenau U. (1990) , in *Basic Features of the Glassy State*, ed. by Colmenero J. and Alegria A., World Scientific, Singapore, p. 297

[32] Zeller R. C. and Pohl R. O. (1971), Phys. Rev. B **4**, 2029

[33] Freeman J. J. and Anderson A. C. (1986), Phys. Rev. B **34**, 5684

[34] Hunklinger S. and Arnold W. (1976), in *Physical Acoustics* Vol. XII, ed. by Mason W.P. and Thurston R.N., Academic Press, New York, p. 155.

[35] Hunklinger S. and von Schickfus M. (1981), in Ref. 2, p. 81

[36] Phillips W. A. (1972), J. Low. Temp. Phys. **7**, 351; Anderson P. W., Halperin B. I., Varma C. M. (1972), Phil. Mag. , **25**, 1

[37] Laird B. B. and Schober H. R. (1991), Phys. Rev. Lett. **66**, 636; Schober H. R. and Laird B. B. (1991), Phys. Rev. B **44**, 6746

[38] Gil L., Ramos M. A., Bringer A. and Buchenau U. (1993) Phys. Rev. Lett. 70, 182-185

[39] Lasjaunias J. C., Ravex A., Vandorpe M. and Hunklinger S. (1975), Solid State Commun. **17**, 1045

[40] Löhneysen H. v., Rüsing H. and Sander W. (1985), Z. Physik B **60**, 323

[41] Buchenau U., Galperin Yu. M., Gurevich V. L., Parshin D. A., Ramos M. A. and Schober H. R. (1992), Phys. Rev. B **46**, 2798

[42] Cahill D. G. and Pohl R. O. (1987), Phys. Rev. B **35**, 4067; D. G. Cahill (1989), Ph. D. Thesis, Cornell University

[43] Gilroy K. S. and Phillips W. A. (1981), Phil. Mag. B **43**, 735

[44] Tielbürger D., Merz R., Ehrenfels R. and Hunklinger S. (1992), Phys. Rev. B **45**, 2750

[45] Carini G., Cutroni M., Galli G. and Wanderlingh F. (1978), J. Non-Crystalline Solids **30**, 61

[46] Schober H. R., Oligschleger C. and Laird B. B. (1993), J. Non-Crystalline Solids (accepted)

[47] Buchenau U. (1992), Phil. Mag. B **65**, 303

[48] Buchenau U., Galperin Yu. M., Gurevich V. L. and Schober H. R. (1991), Phys. Rev. B **43**, 5039

[49] Phillips W. A., Buchenau U., Nücker N., Dianoux A. J. and Petry W. (1989), Phys. Rev. Lett. **63**, 2381

[50] Vineyard C. M. (1959), Phys. Rev. **110**, 999

[51] Bhattacharya K., Kehr K. and Buchenau U. (1993) (unpublished)

[52] Mezei F., Knaak W. and Farago B. (1987) 'Neutron spin-echo study of dynamic correlations near the liquid-glass transition', Phys. Rev. Lett. 58, 571-574.

[53] Richter D., Frick B. and Farago B. (1988) Phys. Rev. Lett. 61, 2465

[54] Richter D., Zorn R., Farago B., Frick B. and Fetters L. J. (1992) 'Decoupling of time scales of motion in polybutadiene close to the glass transition', Phys. Rev. Lett. 68, 71-74.

[55] Colmenero J., Alegria, A., Arbe A. and Frick B. (1992) Phys. Rev. Lett. 69, 478-481

[56] Doster W., Cusack S. and Petry W. (1990) Phys. Rev. Lett. 65, 1080-1083

[57] Kiebel M., Bartsch E., Debus O., Fujara F., Petry W. and Sillescu H. (1992) 'Secondary relaxation in the glass transition regime of ortho-terphenyl observed by incoherent neutron scattering', Phys. Rev. B 45, 10301-10305

[58] Zorn R. (1993) Dynamics of Disordered Materials II, Conference Grenoble March 1993

[59] Frick B., Farago B. and Richter D. (1990) 'Temperature dependence of the nonergodicity parameter in polybutadiene in the neighbourhood of the glass transition', Phys. Rev. Lett. 64, 2921-2924.

[60] Hahn H. and Matzke M. (1984), PTB-Bericht FMRB-105, ISSN 0341-6666, Physikalisch-Technische Bundesanstalt, Braunschweig; (1987) Zeitschrift für Physikalische Chemie 156, 365; Matzke M. and Hahn H. (1988), PTB-Bericht PTB-FMRB-116, ISSN 0341-6666 and ISBN 3-88314-763-X, Physikalisch-Technische Bundesanstalt, Braunschweig

[61] Seeley G. and Keyes T. (1989), J. Chem. Phys. 91, 5581

[62] Xu B. C. and Stratt R. M. (1990), J. Chem. Phys. 92, 1923

[63] Buchner M., Ladanyi B. and Stratt R. M. (1992), J. Chem. Phys. 97, 8522

[64] Cohen M. H. and Turnbull D. (1959), J. Chem Phys. 31, 1164

[65] Buchenau U. and Zorn R. (1992) 'A relation between fast and slow motions in glassy and liquid selenium', Europhys. Lett. 18, 523-528.

[66] Angell C. A. (1984), in *Proceedings of the Workshop on Relaxation Effects in Disordered Systems*, ed. by Ngai K. and Lee T. K., McGregor and Werner Inc, New York, p. 3.

[67] Tao N. J., Li G., Chen X., Du W. M. and Cummins H. Z. (1991), Phys. Rev. B 44, 6665; Tao N. J., Li G. and Cummins H. Z. (1991), Phys. Rev. Lett. 66, 1334

SUSCEPTIBILITY STUDIES OF SUPERCOOLED LIQUIDS AND GLASSES

SIDNEY R. NAGEL
The James Franck Institute and The Department of Physics
The University of Chicago
5640 South Ellis Avenue
Chicago, Illinois 60637
U.S.A.

ABSTRACT. These lectures review the results of a number of susceptibility studies of supercooled liquids and glasses. Dielectric response and specific-heat spectroscopy can investigate the motions that occur at the glass transition, T_g, as the liquid slows down and approaches an amorphous solid. In contrast to predictions of mode-coupling theory, these experiments give no evidence of a critical slowing down occurring at high temperature but rather indicate a divergence of the relaxation-time scales at a much lower value close to the Kauzmann temperature where the extrapolation of the entropy of the liquid state crosses that of the crystal. In addition, the dielectric relaxation of the liquid (for all temperatures and samples measured) can be scaled onto a single master curve. In addition to this primary relaxation, dielectric susceptibility can give detailed information about the secondary (Johari-Goldstein) relaxation occurring in the glass phase below T_g. For several glasses, the dielectric studies indicate that the secondary relaxation is due to the activation of single, uncoupled, entities over barriers which have a Gaussian distribution of energies.

1. Introduction

1.1 NUCLEATION AND GLASS FORMATION

As a liquid is slowly supercooled below its equilibrium freezing temperature, it normally undergoes a first-order phase transition into a crystal. Crystallization happens via a nucleation and growth process: a spontaneous fluctuation of the liquid produces a crystalline nucleus which, if larger than a critical size, continues to grow until it envelopes the entire system. However, since a critical nucleus can take a long time to form, one can prevent nucleation by quenching the liquid sufficiently rapidly.[1] Upon sufficient supercooling, the liquid forms an amorphous solid called a glass. This is a ubiquitous phenomenon and happens in many different types of liquids including ones with metallic, ionic, covalent or Van der Waals bonds.[2]

As a liquid is supercooled, its viscosity increases dramatically so that it becomes less likely for a spontaneous nucleation event to occur. This makes it possible to slow down the cooling rate so that experiments can be performed close to the region of the glass transition. The temperature, T_g, at which the liquid turns into a glass is, to some extent, a matter of convention. An often used criterion is that T_g is where the viscosity, η, reaches 10^{15} poise [1 poise = 0.1 Ns m^{-2}]. This is the value of the viscosity where the the liquid takes approximately one day to respond to a small external stress and should be contrasted with the value $\eta \approx 10^{-2}$ poise for water at room temperature.[1] Clearly there is nothing inherently special about a relaxation time of one day except that it is considered to be the limit of an experimentalist's patience in doing a measurement. A less arbitrary and more satisfactory definition for the glass-transition temperature would be to use the value at which the viscosity, or equivalently some relaxation time, actually diverges. Unfortunately, it is impossible to measure arbitrarily long times or large viscosities so that the use of such a temperature most rely on an extrapolation of the data taken above the true glass transition. Such extrapolations are dangerous since different fitting forms will lead to different temperatures for the divergence. For example two common forms used to fit the temperature dependence of the viscosity and the relaxation times are the Vogel-Fulcher form:

$$\tau = \tau_0 \exp\left(\frac{A}{T-T_0}\right) \tag{1a}$$

(which for $T_0 = 0$ is just Arrhenius, or activated, behavior: $\tau = \tau_0 \exp[E/k_B T]$) and a scaling law:

$$\tau = \tau_0 \left(\frac{T-T_0}{T_0}\right)^{-\alpha}. \tag{1b}$$

Extrapolations from these forms lead to very different values for the temperature T_0.[3] It has even been questioned whether there is any divergence at all. It is one of the purposes of these lectures to review the available evidence for a diverging relaxation time at a finite temperature.[4] My view is that the data support such a divergence of the relaxation times at a finite temperature T_0. This would indicate a true freezing of the liquid state. There is however considerable debate as to whether the viscosity diverges at a finite value of T_0 or whether it decouples from other relaxation times in the liquid.[5]

Neither the Vogel-Fulcher nor the scaling-law forms fit the viscosity or relaxation-time data over the entire range of temperature for most materials. Different liquids obey these functional forms over differing portions of their temperature range. Angell[6] has classified glass-forming liquids into two categories: fragile and strong. He plots log η versus $\frac{T_g}{T}$ where T_g is the temperature where the viscosity reaches 10^{13} poise. Thus the data for all the liquids coincide at $\frac{T_g}{T} = 1$. At high temperatures ($\frac{T_g}{T} = 0$), the data again approaches a common value[7] since in the non-viscous limit, all simple liquids have approximately the same value for the viscosity $\eta \approx 10^{-2}$ poise. Different supercooled liquids have different amounts of curvature on such a plot. Those glasses that Angell calls "strong", such as the covalently-bonded network glass SiO_2, show no curvature and therefore have an Arrhenius dependence of the viscosity on temperature. Glasses which show the most curvature, such as the associated liquids salol and o-terphenyl, are deemed to be "fragile". The viscosity of these liquids tends to decrease rapidly above T_g so that they flow easily at slightly elevated temperatures. This is the rationale for the nomenclature "fragile" and "strong". Most of the data that I will discuss in this review will be on the fragile liquids.

Along with the dramatic increase in the viscosity with decreasing temperature, there is an anomaly near T_g in the temperature dependence of the specific heat, c_p, of the liquid. Figure 1a shows a schematic plot of the specific heat in a crystal, a liquid, a supercooled liquid and a glass. As the liquid is supercooled below the freezing temperature, c_p at first maintains the same large value as it had at higher temperatures in the equilibrium liquid state. However, at a temperature close to T_g, the specific heat drops rapidly (by approximately a factor of two) to the value associated with the solid, crystalline, state and remains close to the crystalline value to lower temperature. The temperature at which the specific heat drops depends on the *rate* at which the liquid is cooled. The slower the rate of cooling, the lower will be the value at which c_p drops as shown by the two curves in the figure. The dashed curve represents the faster cooling rate. This can naturally be interpreted in terms of an experimental time, determined by the cooling rate, becoming shorter than an inherent relaxation time in the liquid which grows as the temperature is decreased. Thus the value of "T_g" determined from the drop in the specific-heat data is again a matter of convention since it depends on the experimental cooling rate. Again, the less arbitrary value for the glass transition would be to extrapolate to the case of infinitely slow cooling. As we shall see below this extrapolated value taken from the specific-heat measurements is close to the value T_0 obtained from extrapolating the relaxation times as discussed above.

1.2 THE KAUZMANN ENTROPY ARGUMENT

The specific heat is important for another reason related to the thermodynamics of the liquid. We have seen above that there is the possibility that a liquid can be supercooled all the way to $T = 0$ and still not form a glass. This could happen for example if the viscosity and the relaxation times increased Arrheniusly. However, an argument due to Kauzmann[8] argues that a glass transition must occur at a finite temperature in a wide class of materials. Even though the supercooled liquid is not in equilibrium, the specific entropy, s, of this state can be calculated in the usual way by integrating over the measured specific heat. Using the thermodynamic relation

$$s(T_f) - s(T_0) = \int_{T_0}^{T_f} \frac{c_p(T)}{T} dT \qquad (2)$$

we use the specific heat, measured from $T = 0$ to the melting temperature, T_m, in the crystal phase, to determine the entropy of the crystal at the melting point. We can obtain the entropy of the liquid by measuring the latent heat of melting. Upon recooling the liquid below T_m, we again measure c_p and use Eq. 2 to determine the entropy of the supercooled liquid. This is shown schematically in Fig. 1b. In the liquid and supercooled liquid, where c_p has its largest values, the slope of s versus T must also be largest. As temperature is decreased, the supercooled liquid looses its entropy faster than does the crystal at the same temperature. At T_m the entropy of the liquid is higher than that of the crystal because of the release of latent heat. However, as the temperature drops, the entropies of the two phases quickly approach one another.

A continuation of this behavior to arbitrarily low temperature would imply that the entropy of the liquid would eventually cross that of the crystal at the "Kauzmann temperature", T_K. It is certainly counter-intuitive to expect the liquid to have an entropy smaller than that of the ordered crystal. However, it does not violate thermodynamics. A violation of the third law would occur if the liquid entropy continued to decrease much below T_K without any change in slope since then we would find that s becomes negative well above $T = 0$. (Since glasses and crystals have comparable amounts of entropy carried by their vibrations, the glass, once it has been formed, cannot have an entropy much less than that of the corresponding crystal. The only alternative is to postulate an "ideal glass" that has very different properties, i.e., vibrations, than any solid we have yet encountered.) The slower one cools the liquid the lower will be the value of T_g where c_p drops. The Kauzmann argument effectively puts a limit on how low a value of T_g one can achieve since the glass transition must intervene at a temperature above T_K so that the entropy of the glass always remains positive. Thus we expect that T_K, although not a rigorous bound, must be a good estimate for how low a liquid can be supercooled before the glass transition must intervene:

$T_g \gtrsim T_K$. Thus, no matter how slowly one cools the liquid, there should be a specific-heat anomaly (i.e., a drop in c_p) at a temperature above the Kauzmann temperature.

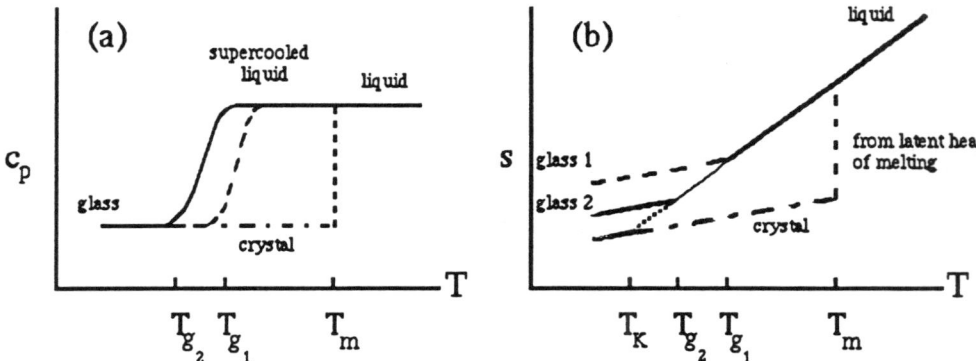

Figure 1. A schematic diagram of the temperature dependence of (a) the specific heat, c_p, and (b) the specific entropy, s, of a crystal, liquid, supercooled liquid and glass. Glasses 1 and 2 are obtained with different cooling rates and have different apparent glass transition temperatures. Glass 1, shown by the dashed curve, represents the result of a faster cooling rate than that used to produce glass 2, the solid curve.

The Kauzmann "paradox" is more important for fragile than for strong glass formers.[6] In these liquids the difference between the liquid and crystalline specific heat is relatively large so that T_K is often not far below the experimentally accessible temperature range. In a strong liquid like SiO_2 the liquid and solid specific heats are nearly the same so that the extrapolated value of T_K is close to $T = 0$ K. Since, the Kauzmann temperature puts a limit on how far a liquid can be supercooled without undergoing a transition, the fragile liquids are the systems of choice for experimentalists seeking to investigate the nature of the glass transition.

2. Specific-Heat Spectroscopy

Clearly, the specific heat of a supercooled liquid is of importance if one wants to investigate the glass transition. Not only does c_p have a dependence on cooling rate, which indicates an intrinsic temperature-dependent relaxation time in the sample, but c_p is also a thermodynamic parameter which, via the Kauzmann argument, cannot remain large down to indefinitely low temperature. The Kauzmann argument therefore sets a lower bound on how low the glass-transition temperature can be even for an infinitely slow cooling rate of the liquid. However, as we see in Fig. 1a, a conventional measurement of the specific heat is unsatisfactory. As one cools the liquid, the specific heat remains smooth and featureless until it starts to drop near T_g. This interesting feature however only indicates that the sample is no longer in equilibrium. If the experimentalist had only been more patient and cooled the liquid at a slower rate, then the liquid behavior would have persisted to lower temperature. It is very discouraging to get interesting behavior only when the experiment is out of equilibrium and therefore, to some extent, out of the experimentalist's complete control. The reason for this problem, of course, is that the relaxation times in the liquid are getting arbitrarily long as the temperature is lowered so that, no matter how slowly the sample is cooled, it will always fall out of equilibrium at some temperature well above T_0.

In order to overcome these problems, one should confront the basic dilemma which is that the specific heat has an intrinsic time scale which increases as the temperature is lowered. Thus one should measure the specific heat dynamically rather than statically. The idea is to heat the equilibrium liquid with a sinusoidal time dependence and then measure the temperature variation at the same frequency. By measuring the frequency dependence of the response one can obtain a frequency-dependent specific heat:

$$c_p(\nu) = \frac{\delta Q(\nu)}{\delta T(\nu)}. \tag{3}$$

In the limit $\nu \rightarrow 0$, this is the conventional specific heat. In order to insure that the sample remains in equilibrium, the liquid must be cooled extremely slowly so that the time scale for cooling is much longer than the lowest frequency used in the measurement of c_p. This decouples the time scale of interest, ν, from the time scale determined by the cooling rate so that there will no longer be any hysteresis in the measurement. By using a very small perturbation $\delta Q(\nu)$ one can stay in the linear-response regime. This frequency-dependent specific heat will in general have a real and an imaginary part indicating an in-phase and out-of-phase component of the heat input and temperature rise. By varying the frequency over a wide range, one can measure the "specific-heat spectrum" of the supercooled liquid and find the complete distribution of relaxation times for the enthalpy, that is, for how the heat relaxes into the different modes of the sample.

In the traditional adiabatic technique[9] for measuring specific heat, the heat must diffuse across the entire sample. This requires that the measurement time (i.e., the inverse of the frequency) must be long enough for the heat diffusion to take place. In liquids, the low thermal diffusivity ($\sim 10^{-3}$ cm^2/sec) makes this diffusion time long, even for extremely thin samples, and places a severe constraint on the highest useful frequency for this technique. (Even for a thickness as small as d=0.1mm, $\nu <$ 1 Hz.) To circumvent this obstacle, a non-adiabatic technique was developed in order to measure c_p which allows the use of much higher frequencies.[10] This was done by measuring the temperature at the same point at which the heat was generated: the same electrical resistor was used for the thermometer as well as for the heater. In addition, a simple heater geometry was used so that the temperature profile created by the heat diffusion could be solved for explicitly. This eliminated the necessity of using a very thin sample and allowed an extension of the measurements to much higher frequencies.

The heater, of resistance R, is a thin metal film evaporated onto a glass substrate. This is immersed in a bath of the liquid to be measured. A sinusoidal current, i, at frequency $\nu/2$ through the heater, produces a power dissipation, i^2R, with both a D.C. component (which produces a constant temperature gradient in the sample cell) and an A.C. component at frequency ν. This latter contribution sends diffusive thermal waves into the surrounding medium at frequency ν. Because the geometry of the heater is a wide and long flat ribbon, the heat diffuses into the surrounding liquid according to a one-dimensional heat diffusion equation:

$$\frac{\partial Q}{\partial t} = \kappa \frac{\partial^2 T}{\partial z^2} \tag{4}$$

where z is the direction perpendicular to the plane of the heater. The solution to this one-dimensional diffusion equation shows that the temperature oscillations at the heater have a magnitude proportional to $(\nu c_p \kappa)^{-1/2}$ and a phase lag of 45° with respect to the heat oscillations, where c_p and κ are respectively the specific heat and thermal conductivity of the medium surrounding the heater.

Because the resistor is made out of a metal with a large temperature coefficient of resistance, the resistance varies with the temperature of the heater. Thus the resistance has a small component oscillating at frequency ν proportional to the temperature oscillations of the film. The voltage across the heater oscillates with a large component at $\nu/2$ and a small component at 3ν/2 due to the mixing of the current at frequency ν/2 with the resistance oscillations at ν. By subtracting the large ν/2 component (by using a Wheatstone bridge), both the magnitude and phase of the 3ν/2 component can be measured with a lock-in amplifier. This technique covers the frequency range 0.1Hz < ν < 6 kHz.

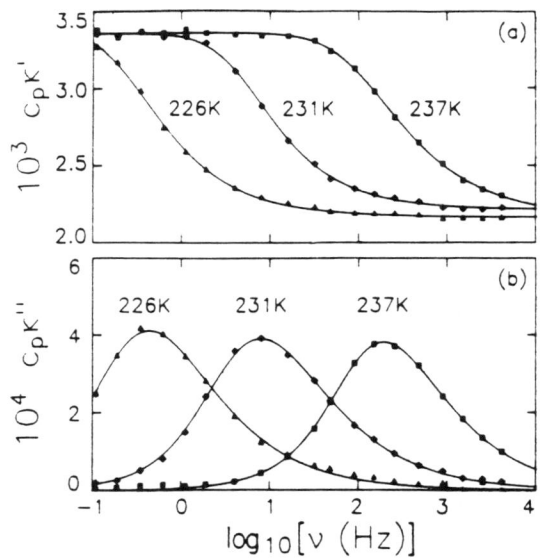

Figure 2. The (a) real $c_p\kappa'$ and (b) imaginary $c_p\kappa''$ parts of $c_p\kappa$ versus the log of the frequency for the supercooled liquid salol. Figure is taken from Dixon (Ref. 11). Measurements on other liquids indicate that the frequency dependence is contained almost entirely in the specific heat and not in the thermal conductivity. The solid lines through the data points are the best fits to the stretched-exponential form given by Eq. 15.

We have studied the frequency-dependent specific heat near the glass transition in several liquids which are good glass formers. Figure 2a shows $c_p\kappa'$ versus frequency for the case of salol for three different temperatures.[11] For each temperature there is a frequency where $c_p\kappa'$ drops by roughly a third. The frequency at which $c_p\kappa'$ drops depends strongly on the measurement temperature, indicating that the characteristic relaxation times in the liquid increase as T is lowered. As mentioned above, the frequency dependent specific heat is a complex quantity. Because there is dispersion in the real part, $c_p\kappa'$, there must also be an imaginary part, $c_p\kappa''$, as required by the Kramers-Kronig relations. Figure 2b shows this quantity which peaks at the frequency where the real part is decreasing most rapidly.

We do not usually think of the specific heat as a complex, frequency-dependent, quantity. However, just as for other susceptibilities, it can be related to the Fourier transform of an equilibrium correlation function. In the case of the specific heat, the appropriate correlation function is that between the energy fluctuations in the liquid at two separated times.[12] Normally the imaginary part of a linear susceptibility signifies a net absorption of energy by the sample from the applied field. But during a complete cycle of the specific heat experiment there is no net exchange of energy between the sample and the surrounding heat bath. However the *entropy* of the bath does change during a complete cycle. If the experiment is carried out at a nominal temperature T with small oscillations of magnitude ΔT, then the net increase of entropy of the heat bath during one cycle is:

$$\Delta s = \pi \, c_p'' \left(\frac{\Delta T}{T}\right)^2. \tag{5}$$

The second law of thermodynamics insures that $c_p'' > 0$.

Note that this experimental method measures the product $(c_p\kappa)$ rather than just c_p. However the changes in the product $(c_p\kappa)$ are dominated by the behavior of c_p. This was demonstrated by varying the width of the heater and using the corrections to the simple one-dimensional heat diffusion equation to evaluate the contributions of κ and c_p separately.[13] This is shown in Fig. 3 for another fragile glass former, o-terphenyl mixed with 9% o-phenylphenol. Computer simulations on simple model liquids undergoing the glass transition can also investigate the properties of the specific heat[12] and the thermal conductivity.[14] These simulations likewise show that the thermal conductivity does not have any appreciable frequency dependence at low frequencies whereas the specific heat does. In a measurement of c_p in glycerol, Christensen[15] used an adiabatic technique (which as discussed above limited the frequency range to $\nu < 1$ Hz) which showed that the specific heat varied in the same manner as did the product $c_p\kappa$ measured in the non-adiabatic experiment.

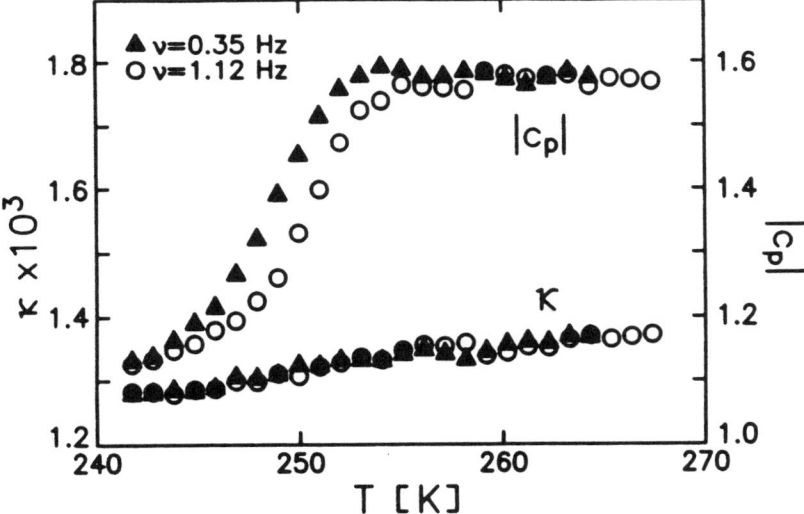

Figure 3. The thermal conductivity, κ [J/sKcm], and the magnitude of the specific heat, $|c_p|$ [J/Kcm^3], vs T for the fragile glass former, o-terphenyl mixed with 9% o-phenylphenol, at two frequencies: $\nu = 0.35$ Hz and $\nu = 1.12$ Hz. Note that the thermal conductivity does not show any frequency dependence and is smooth over the temperature range studied whereas the drop in the specific heat shifts to lower temperature as the frequency is lowered. This is what was shown schematically in Fig. 1a. This figure is taken from Ref. 13.

3. Susceptibility as a Probe of Relaxation Phenomena

Many experiments have probed the relaxation phenomena in glasses.[16] One important class of experiments measures the frequency-dependent susceptibility of the liquid to a small perturbation. As discussed above, specific-heat spectroscopy[10] measures the enthalpy relaxation due to a small perturbation of the temperature. Likewise, dielectric susceptibility[17] detects the reorientation of polar molecules due to a small oscillating electric field, and ultrasonic attenuation[18] measures the density response to a pressure perturbation. I will present here an elementary discussion of how to interpret the data produced by this class of experiments in terms of the relaxation processes in which we are interested.

I will focus on dielectric susceptibility although the same results will be applicable to any other linear susceptibility simply by changing the names of the variables. In the case of dielectric susceptibility the perturbation is a small electric field E and the response is the electric displacement vector D. The displacement vector $D(t)$ at time t is the sum of the system's response to the electric field, $E(t')$, at all earlier times t':

$$D(t) = \int_{-\infty}^{\infty} \varepsilon(t-t') \, E(t') \, dt' \quad (6)$$

where $\varepsilon(t-t')$ is the dielectric constant. Causality demands that there cannot be a response until the field is present which means: $\varepsilon(t-t') = 0$ for $(t-t') < 0$. We Fourier transform both sides of this equation in order to find the response to a perturbation varying sinusoidally with frequency $\omega = 2\pi \nu$. We find:

$$D(\omega) \equiv \int_{-\infty}^{\infty} D(t) \, e^{i\omega t} \, dt = \int_{-\infty}^{\infty} \varepsilon(t-t') \, e^{i\omega(t-t')} \, d(t-t') \int_{-\infty}^{\infty} E(t') \, e^{i\omega t'} \, dt'$$

$$= \varepsilon(\omega) \, E(\omega). \quad (7)$$

The response to a step function in $E(t')$ will generally be a complicated function $D(t) = D_0 \phi(t)$. However it is instructive to examine the response of a system with only a single relaxation time, τ. This is the Debye model. For such a system, the response to a step function will be an exponential decay from the initial to the final state: if

$$E(t) = \begin{matrix} E_0 & & t < 0 \\ 0 & & t \geq 0, \end{matrix}$$

then $\quad (8)$

$$\begin{aligned} D(t) &= D_0 = \varepsilon_0 \, E_0 & t < 0 \\ &= (\varepsilon_0 - \varepsilon_\infty) E_0 \, \phi(t) = \Delta\varepsilon \, E_0 \, e^{-t/\tau} \, . & t \geq 0. \end{aligned}$$

Here $\Delta\varepsilon \equiv (\varepsilon_0 - \varepsilon_\infty)$ where ε_0 and ε_∞ are, respectively, the dielectric constant evaluated at $\omega = 0$ and evaluated at the high frequency limit $\omega \to \infty$. A nonzero value for ε_∞ implies an instantaneous jump in $D(t)$ at $t = 0$ when the field is turned off. In Eq. 7, in order to calculate $D(\omega)$ we need to know the response to a delta-function field at time t'. This can be immediately obtained from Eq. 8 where $\phi(t)$ is the response to a step function:

$\varepsilon(t) = \varepsilon_\infty \, \delta(t) - \Delta\varepsilon \, \dfrac{d\phi(t)}{dt}$. Thus we arrive at the response:

$$D(\omega) = E(\omega)\left(\varepsilon_\infty + \Delta\varepsilon \int_0^\infty e^{i\omega t}\frac{d}{dt}\{-e^{-t/\tau}\}\,dt\right)$$

$$= E(\omega)\left(\varepsilon_\infty + \Delta\varepsilon \frac{1}{1-i\omega\tau}\right) \quad (9)$$

$$= E(\omega)\left(\varepsilon_\infty + \Delta\varepsilon\left(\frac{1}{1+\omega^2\tau^2} + i\frac{\omega\tau}{1+\omega^2\tau^2}\right)\right).$$

The dielectric constant for the Debye model is therefore:

$$\varepsilon(\omega) = \varepsilon_\infty + \Delta\varepsilon\left(\frac{1}{1+\omega^2\tau^2} + i\frac{\omega\tau}{1+\omega^2\tau^2}\right). \quad (10)$$

The imaginary part of the dielectric response, ε'', is symmetric when plotted versus $\log_{10}\omega$ and has a full width at half maximum of $W_D = 1.14$ decades. The frequency, $\omega_p \equiv 2\pi\nu_p$, at which ε'' is a maximum occurs when $\omega = \frac{1}{\tau}$. Thus by measuring the susceptibility over a range of frequencies at different temperatures one can determine the temperature dependence of τ.

Susceptibility data, such as that shown in Fig. 2, has certain qualitative features which are similar to what is found in the Debye model. The real part of the spectrum is large and constant at low frequencies. This corresponds to the high temperature limit where all the relaxations have a chance to take place during each cycle of the perturbation. At higher frequencies, the response falls off to another constant value. This high-frequency value corresponds to what happens at low temperatures, in the solid, where no relaxation can occur during one cycle. Meanwhile, the imaginary part of the spectrum has a peak at the frequency where the real part of the spectrum is decreasing most rapidly. Since τ increases in a supercooled liquid as the glass transition is approached, ν_p and the susceptibility curves move to lower frequency, with decreasing temperature. This is seen in the data from specific-heat spectroscopy for salol shown in Fig. 2.

However, it is rare to find quantitative agreement with the simple Debye model. For example, not only are the peaks in the imaginary part broader than that given by a Debye spectrum, Eq. 10 ($W_D = 1.14$ decades), but, as in the data shown in Figure 2, they generally tend to be asymmetric with a tail to high frequencies. This is true not only of data from specific-heat spectroscopy, but is also true for dielectric susceptibility, ultrasonic attenuation[19], shear modulus[20] and other probes. A natural explanation for why the width is broader than that given by the Debye model is that there is more than one relaxation mechanism each with its own relaxation time in the material. The total response is therefore a sum of the responses from each of the different relaxation mechanisms. We therefore take

$$\phi(t) = \int_0^\infty \rho(\tau) e^{-t/\tau}\,d\tau$$ with $\rho(\tau)$ being the distribution of relaxation times. In the frequency domain we therefore have:

$$\varepsilon(\omega) = \varepsilon_\infty + \Delta\varepsilon \int_0^\infty \rho(\tau) \frac{1}{1-i\omega\tau} d\tau \qquad (11)$$

$$= \varepsilon_\infty + \Delta\varepsilon \int_0^\infty \rho(\tau) \frac{1}{1+\omega^2\tau^2} d\tau + i \Delta\varepsilon \int_0^\infty \rho(\tau) \frac{\omega\tau}{1+\omega^2\tau^2} d\tau.$$

A number of phenomenological forms have been suggested to fit the data for relaxation phenomena. These forms lack a solid theoretical justification, but they have been moderately successful at fitting the data over limited ranges of frequency. One of these forms was suggested by Davidson and Cole[17]:

$$\varepsilon(\omega) = \varepsilon_\infty + \Delta\varepsilon \frac{1}{(1-i\omega\tau)^\beta} \qquad \text{for } 0 \leq \beta \leq 1. \qquad (12)$$

This form is the same as the Debye form when $\beta = 1$. As β decreases, the peak in ε'' becomes broader and develops an asymmetric high frequency tail. A second form proposed by Cole and Cole[21] is also broader than Debye but produces a symmetric peak:

$$\varepsilon(\omega) = \varepsilon_\infty + \Delta\varepsilon \frac{1}{1+(i\omega\tau)^{1-\alpha}} \qquad \text{for } 0 \leq \alpha \leq 1. \qquad (13)$$

Again when $\alpha = 0$ this reduces to the Debye case. A form, which is essentially a combination of the two forms given by Eqs. 12 and 13, was proposed by Havriliak and Negami:[22]

$$\varepsilon(\omega) = \varepsilon_\infty + \Delta\varepsilon \frac{1}{[1+(i\omega\tau)^{1-\alpha}]^\beta}. \qquad (14)$$

This has one more parameter than the other forms and allows an asymmetric shape with different power laws governing the low-frequency increase and the high-frequency decrease of the peak. The stretched-exponential form (also known as the Kohlrausch[23]-Williams-Watts[24] form) also gives a wide and asymmetric peak. This form must be numerically integrated from its form in the time domain since it cannot be expressed analytically as a function of frequency. In the time domain it is:

$$\phi(t) = \phi_0 e^{-(t/\tau)^\beta} \qquad \text{for } 0 \leq \beta \leq 1. \qquad (15)$$

There have been a number of theoretical derivations of this form for the relaxation.[25]

With this background we can now start to analyze data of the type shown in Fig. 2. Our first goal is to obtain the relaxation time as a function of temperature. Since the susceptibility curves are all wider than the simple Debye form, the frequency $v_p = \frac{\omega_p}{2\pi}$, where $c_p\kappa''$ has a peak at any given temperature, does not measure the unique relaxation frequency of the liquid but rather determines an average value of a distribution of relaxation frequencies.

If we plot $\log v_p$ versus the inverse temperature, T^{-1}, we can obtain how the average relaxation times varies with temperature. A straight line on such a graph would indicate

Arrhenius, or activated, behavior. Fig. 4a shows such a plot for specific-heat data obtained[13] on a mixture of o-terphenyl with 9% o-phenylphenol. It is clear that the data do not follow a straight line. The solid line is a Vogel-Fulcher fit to the data as given by Eq. 1a and that the relaxation times are tending to diverge at a finite temperature, T_0. When extrapolated (from a series of different mixtures) to the case of pure o-terphenyl the value T_0 = 184 ± 13 K is rather close to the Kauzmann temperature, T_K = 200 ± 10 K. This is not unique to this particular glass. We have found that the Kauzmann temperature, T_K, is reasonably close to the divergence temperature, T_0, in the Vogel-Fulcher equation for a number of different glass-forming liquids.

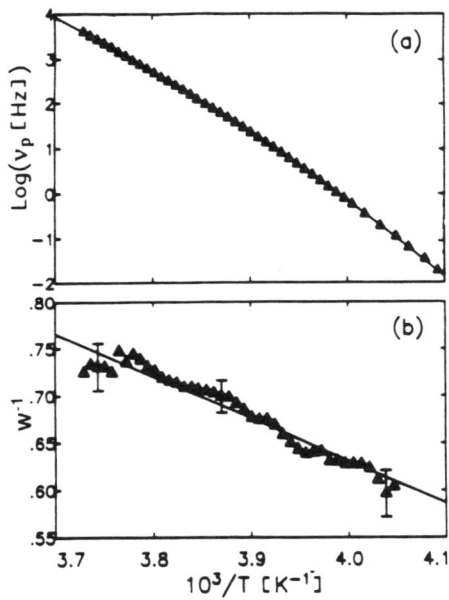

Figure 4. Parameters from the specific-heat data for the fragile glass former o-terphenyl mixed with 9% o-phenylphenol. (a) The logarithm of the mean relaxational frequency, v_p, versus the T^{-1}. The solid line is a best fit to the data using a Vogel-Fulcher form, Eq. 1a. (b) The inverse of the normalized width of the peak in the imaginary part of the specific heat versus T^{-1}. The figure is from reference 13.

As mentioned above, the width of the peak in the imaginary part of the susceptibility is larger than that given by the Debye model, W_D = 1.14 decades. The width therefore gives information about the distribution of relaxation times in the liquid and allows us to measure how the distribution broadens as a function of temperature. It is convenient to normalize the width to that of the Debye model, $w = \frac{W}{W_D}$, where W is the full width at half maximum of the peak. When w approaches one, then the relaxation is approaching a Debye model with a single relaxation time. In Fig. 4b, the inverse normalized width, w^{-1} is plotted against T^{-1}. Not only is w significantly different from one over the entire temperature range measured, but it also has a strong temperature dependence. As the temperature decreases towards T_0 the width increases. Because w is temperature dependent it is not correct to use the time-temperature superposition analysis that has often been employed to treat glass-forming liquids.[26] Such a procedure can only be used if the shape of the relaxation curve (on a logarithmic time or frequency axis) does not change with temperature.

We have found a strong temperature dependence for w in many different glass forming liquids.[27]

Data from other susceptibility probes can be treated in the same manner as was done here for the specific-heat spectroscopy. In the case of glycerol, ultrasonic[19] and shear modulus[20] data was also measured. In both cases the average relaxation times, given by the peak frequency in the imaginary part of the susceptibility, had approximately the same temperature dependence as that found in the specific heat.[28] However, the widths of the peaks, w, were not the same. This again appears to be a common feature of glass-forming liquids: the temperature dependence of the average relaxation times measured by several different probes are almost identical whereas the widths measured by these probes are very different. This indicates that the different probes are all coupling to the same slow dynamical process but weight the different relaxation times differently. This is most clearly seen in comparing the dielectric susceptibility to the specific heat, where the peak positions are very close to each other but the widths can be much larger in some cases and much smaller in others. We have not been able to identify a trend in the comparative widths given by different susceptibilities.

4. Dielectric Susceptibility Data

As just mentioned different susceptibility probes give approximately the same values for the average relaxation times. Therefore, it would be convenient to use a probe which has the widest frequency range in order investigate the behavior of v_p over as wide a range of temperature as possible. Dielectric susceptibility is an excellent candidate since not only can it be measured over a very wide frequency range but it is extremely sensitive as well. It does not couple to all relaxation times as does the specific heat but measures the reorientation of the permanent dipoles in the presence of an electric field. Figure 5 shows the dielectric susceptibility data for the fragile glass-forming liquid salol[11,29] which is typical of many other materials. The frequency range covered is 13 decades: 10^{-3} Hz $< v < 2 \times 10^{10}$ Hz.

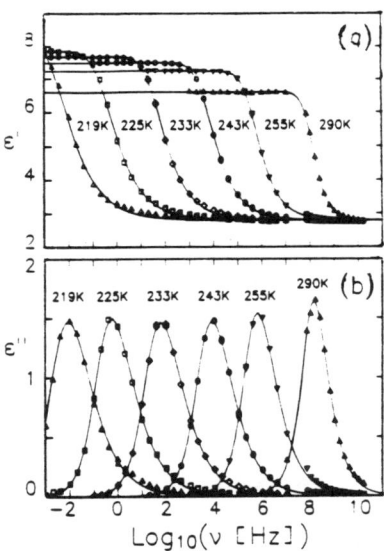

Figure 5. The real and imaginary parts of the dielectric susceptibility of salol as a function of $\log_{10} v$. As marked in the figure each curve represents a different temperature. The solid lines are the best fit curves using the stretched-exponential form. The data is taken from ref. 29.

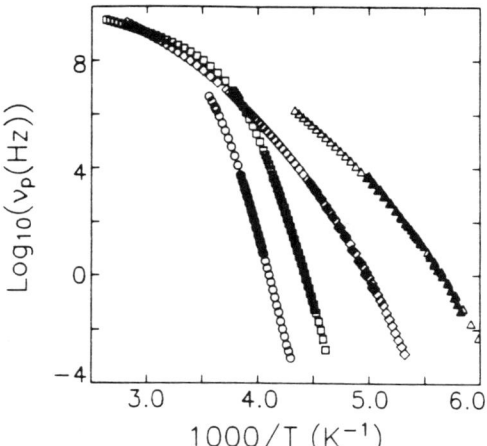

Figure 6. The peak frequency in the susceptibility curves versus the inverse temperature for four liquids. The circles are for o-terphenyl, the squares are for salol, the diamonds are for glycerol and the triangles are for propylene glycol. The open symbols are taken from dielectric susceptibility measurements and the closed symbols are taken from specific heat spectroscopy data. Except for propylene glycol for which T_K is not available, $T_o \approx T_K$. Figure is taken from reference 30.

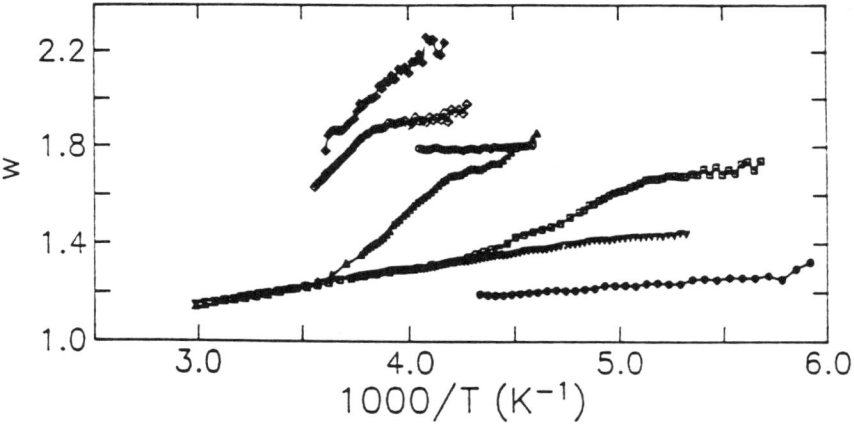

Figure 7. The normalized width of $\varepsilon''(\nu)$, $w = W/W_D$, plotted versus temperature for several glass-forming liquids: glycerol ($\triangledown$), propylene glycol (solid circle), salol (solid triangle), dibutyl-phthalate (open square), α-phenyl-o-cresol with 13% o-terphenyl impurity (open circle), and o-terphenyl with 9% (solid diamond) and 33% (open diamond) o-phenylphenol impurity. Data is taken from reference 29.

The first thing to check is whether the relaxation times obtained from dielectric susceptibility data are indeed the same as those determined by other probes. Figure 6 shows a plot for four supercooled liquids (glycerol, salol, propylene glycol and o-terphenyl) of $\log_{10}(\nu_p)$ versus T^{-1} obtained by dielectric susceptibility and by specific-heat spectroscopy.[30] The open symbols are taken from dielectric susceptibility and the solid ones are from specific-heat spectroscopy measurements. The data for $\varepsilon(\nu)$ were taken over a very wide range of frequency (in the case of glycerol 14 decades) whereas the specific-heat data could only be obtained over a range of roughly 5 decades. Nevertheless, where the frequency regimes overlap, the two probes give nearly identical values for v_p leading us to conclude that the two probes measure essentially the same average relaxation time.

As we saw in the specific-heat spectroscopy data shown in Figure 4, the width, W, of the peak in the imaginary part of the susceptibility varies as a function of temperature. Figure 7 shows the width of the $\varepsilon''(v)$ normalized to the value given by the Debye model: $w = W/W_D$. As the temperature is increased[29] (which also means that the peak frequency is getting large) w approaches 1. Thus, when v_p approaches a typical phonon frequency, a single relaxation-time model appears to fit the data. As mentioned above, although the different probes measure the same average the relaxation times, they find very different widths of the distributions.

5. Universal Curves for Dielectric Response

The susceptibility behavior of these samples have many features in common. They are all broader than predicted by a Debye model, they are all slightly asymmetric with a tail to high frequency, they all have widths that approach the Debye value at high temperatures (and frequencies), and they all have average relaxation frequencies that appear to diverge at a finite temperature (often correlating with the Kauzmann temperature, T_K). From what has been said so far, it is not clear whether these curves demonstrate any truly *universal* features. To demonstrate that such generic behavior exists, we would have to be able to place the data, appropriately scaled, from one sample directly on top of that from another.

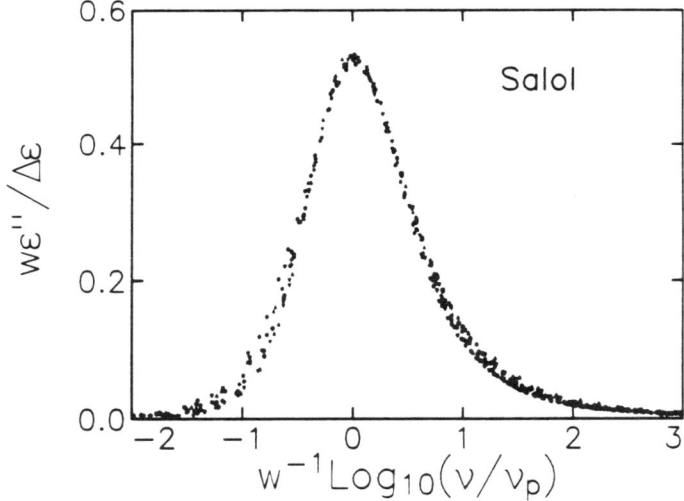

Figure 8. An attempt, using the most natural scaling variables, at making a master curve for the salol dielectric susceptibility data. The figure is taken from reference 30.

We might expect that all that needs to be done to make such a master curve is to align the peaks positions and then take account of the variation in width. This can be achieved by plotting the data versus $\frac{1}{w} \log_{10}(v/v_p)$. The division by v_p brings the peaks into alignment and the division by w makes all the half widths the same. In order to get the height to be uniform the values of ε'' should be divided by $\Delta \varepsilon \equiv \varepsilon_0 - \varepsilon_\infty$ and multiplied by w. (From Kramers-Kronig analysis we know that the area under $\varepsilon''(v)$ plotted logarithmically in frequency is just $\Delta \varepsilon$. Therefore $\Delta \varepsilon / w$ should be proportional to $\varepsilon''(v_p)$.) Figure 8 shows such an attempt to make a master curve for salol.[30] This attempt must be regarded as unsuccessful. Data taken at different temperatures do not lie on top of one another in the tails of the curve. A a log-log plot would emphasize these discrepancies. A more unconventional scaling procedure must be used to collapse all the data onto a single curve.

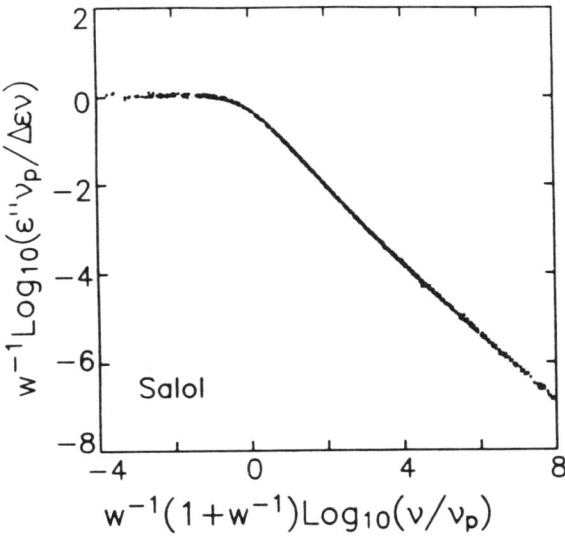

Figure 9. A successful attempt at making a master curve for the salol dielectric susceptibility data using the scaling axes: $\frac{1}{w}(1+\frac{1}{w}) \log_{10}(\nu/\nu_p)$ and $\frac{1}{w} \log_{10}\frac{\varepsilon''\nu_p}{\nu\Delta\varepsilon}$. The figure is taken from reference 29.

Figure 9 shows a much more successful procedure which does collapse all the salol data onto a single curve[29]. The ordinate is $\frac{1}{w} \log_{10}\frac{\varepsilon''\nu_p}{\nu\Delta\varepsilon}$. The division of $\varepsilon''/\Delta\varepsilon$ by the normalized frequency, ν/ν_p, in the logarithm tilts the curve and decreases its slope on a logarithmic plot so that it monotonically decreases. The abscissa is $\frac{1}{w}(1+\frac{1}{w}) \log_{10}(\nu/\nu_p)$. Multiplication by $(1+\frac{1}{w})$ produces the same high frequency slopes for all temperatures. Dividing by w on both axes shifts the curves so that they lie on top of each another at all temperatures. This division by w on both axes is reminiscent of the multifractal fitting procedure.[31] If we plot the susceptibility data for different samples on the same graph we find that the data do collapse: all the data for all the temperatures on these simple glass-forming samples lie on a single "universal" curve as is shown in Figure 10.

We can conclude several features about the data. At low frequencies, $\nu < \nu_p$, the fact that the data in Fig. 10 lie on a horizontal line for negative values of the abscissa, indicates that $\varepsilon''(\nu)$ is proportional to ν. For positive values of the abscissa (i.e., $\nu > \nu_p$) there are two distinct regions which are separated roughly at a value of 2.5 on that axis. Both the low and high frequency regions are well fit by straight lines with slopes of -1 and of approximately $-\frac{3}{4}$ respectively. This indicates that $\varepsilon''(\nu)$ has two separate power-law regimes with exponents that are related to each other.

There does appear to be a master curve for the susceptibility data on glass-forming liquids albeit not of a very simple kind. It appears only when the data is scaled in an unconventional way. It is clear from this scaling procedure that none of the forms that have been suggested in the literature can fit the data over its entire range. This dielectric data covers an exceptionally large range of frequency - in some cases over 14 decades. Thus this behavior is found over almost the entire range that is experimentally accessible. It will be difficult to extend the range by many more decades.

There is, as yet, no adequate explanation for the existence of this master curve nor for the unconventional scaling procedure that was used to bring the data onto a single curve. The most notable attempt was that due to Chamberlain[32] who proposed a percolation model for the relaxing entities. It was not clear, however, in that model what was percolating and why the scaling form should have the form that it did. In particular one might expect that the correct explanation would indicate why the form $\frac{1}{w}\left(1+\frac{1}{w}\right)\log\left(\frac{v}{v_p}\right)$ should emerge as the natural variable for the relaxation.

It is natural to ask how universal this scaling is. Does it extend to different probes and to different kinds of samples? Schönhals, Kremer and Schlosser[33] have claimed that this type of data analysis does not yield a master curve when applied to several polymer systems: poly (p-chlorostyrene), poly (propylene glycol) and poly (vinyl acetate). In particular they claimed that the low frequency behavior of $\varepsilon''(v)$ did not vary linearly with frequency. These same polymer liquids have recently been remeasured[34] and, in contradiction to the Schönhals et al. data, they were found to fit onto the same master curve as the other simple liquids shown in Fig. 10. The source of the discrepancy between these two measurements may have been the way the conductivity background, which is particularly important for these samples, was subtracted from the data.

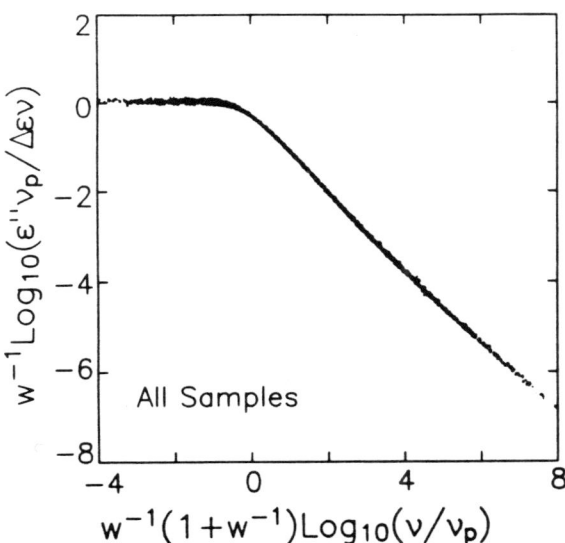

Figure 10. A master curve of all the dielectric data at all temperatures on several different samples. The data is taken from reference 29.

6. Relation of Glasses to Spin Glasses

Except for the similarity in their names there has been scant evidence to connect the physics of spin-glasses to that of the structural glasses studied here. However, the experimental results presented above do imply an unexpected connection between these two systems[35]. From the form of the master curve of the susceptibility data it was shown that the normalized width, w, of the susceptibility cannot become greater than a critical value which is approximately 3. This is reminiscent of what has been found[36] in spin-glasses.

The argument rests on two assumptions. The most controversial one is that the master curve presented in Fig. 10 is indeed *universal*. This implies that it can be used even in the region close to the glass transition which is not experimentally accessible. Although it is impossible to reach T_0, this scaling form appears to be the best way of extrapolating into that regime. The second assumption is that above v_p, $\varepsilon''(v)$ should be a monotonically decreasing function of v. This simply says that the structure of the relaxation should have a single peak. If $\varepsilon''(v)$ were to start to increase again above v_p then we would expect this to be part of a second relaxation process. To see what this implies for the data, we take the master curve and calculate, in terms of the high-frequency slope, what is the asymptotic behavior of $\varepsilon''(v)$.

For large values of $\frac{v}{v_p}$, the slope of the master curve, which plots $\frac{1}{w}\log\left(\frac{\varepsilon'' v_p}{v \Delta \varepsilon}\right)$ versus $\frac{1}{w}\left(1+\frac{1}{w}\right)\log\left(\frac{v}{v_p}\right)$, is close to $(-\frac{3}{4})$. This implies that $\varepsilon''(v)$ varies as v^s with the exponent $s = \left(1-\frac{3}{4}\left(1+\frac{1}{w}\right)\right)$. The assumption that ε'' decreases monotonically in this region implies that s must be negative so that $w < 3$.

In simulations of spin-glass dynamics in the high-temperature paramagnetic phase, when the relaxation is fit by a modified stretched exponential, $\phi(t) = \phi_0 t^{-\alpha} \exp(-(t/\tau)^\beta)$, the exponent β approached the value $\frac{1}{3}$ at the spin-glass transition temperature. For a stretched exponential curve[11] $\beta \cong w^{-1}$ Thus the result in glasses that $w < 3$ is equivalent to the result in spin-glasses that $\beta > \frac{1}{3}$. One can speculate that similar physics may give rise to the same phenomenon, with the same numerical values for the limiting widths, in glasses as in spin-glasses[37].

7. Relevance to Mode-Coupling Theory

Mode-coupling theory has made a number of predictions about the shape and temperature dependence of the relaxation in a supercooled liquid. Experimental evidence for many of these predictions has been given in the lecture by Götze in these proceedings[38]. One prediction that has been discussed at length in that lecture is the existence of a critical temperature, T_c, well above the observed values of T_g where the thermodynamic anomalies take place at a finite cooling rate (and therefore even farther above the temperature T_0 where a divergence in the extrapolated time scales appears to exist). Evidence for such a temperature, where critical slowing down is predicted to take place, is taken from the light-scattering studies on the ionic glass former CKN [39] and on salol[40]. Here, I will make a few remarks about the relevance of the dielectric data on salol to the interpretation of a critical T_c made in the analysis of the light scattering data on the same liquid.

As seen in Figs. 5, 6, 7 and 9, the data on salol varies smoothly as a function of both frequency and temperature throughout the entire range of measurement. The data from the light scattering claim to see a critical temperature at $T = 253$ K. In Fig. 6, the frequency of the peak position of $\varepsilon''(v)$ varies smoothly through this temperature. This has already been shown by Dixon[11] in his analysis of the specific-heat and dielectric-susceptibility data. In Fig. 9, it was shown that the data from all temperatures both above and below 253K could be fit onto a single master curve. There is no break in the behavior as this value of temperature is crossed. Even at T = 253 K the frequency window in the dielectric measurements is sufficiently wide so that there are approximately 4 decades of frequency that have been measured above the peak. Thus the data from this temperature span much of the interesting range found on the master curve including the change of slope on the high frequency side of the peak. When the data is plotted as a function of temperature (for various fixed frequencies) it is again apparent that there is no indication of any critical temperature whatsoever in the vicinity of the purported value given by the light scattering data analysis. It is important to note that the highest frequencies measured in the dielectric data overlaps with the lower frequencies measured by the light-scattering measurements. Thus the lack of any discernible feature in the dielectric data cannot be explained away as due to a low frequency cut-off for the critical slowing down since the two measurements cover the same frequencies.

In addition to the prediction of a critical temperature, the mode-coupling theory predicts a sequence of frequency scales that should occur in the relaxation process. Kim and Mazenko[41] have reviewed these predictions in light of the dielectric measurements and suggest that the two distinct power-law regimes found in in the data for the region $v > v_p$ are manifestations of the frequency scales predicted by the theory. The power law that occurs at lower frequency is the stretched-exponential relaxation of the α peak and the power law occurring at the highest frequencies in the data is the von Schweidler relaxation. Because the master curve shown in Fig. 10 appears to work for all temperatures and samples, they conclude that there is a relationship between the von Schweidler exponent b and the exponent β governing the stretched exponential: $(1 + b) = 3(1 + \beta)/4$. Although the mode coupling theory does predict the sequence of times scales it does not yet predict this relationship between the two exponents.[41]

There is a final important point to be made in regard to the extended mode-coupling theories. We have seen that the best evidence from the dielectric and specific-heat data indicate that the peak frequencies, v_p, vary according to a Vogel-Fulcher form at low temperatures. They do not, as has been sometimes asserted, vary Arrheniusly. Thus much care must be taken when including any hopping mechanism into the mode-coupling theories to insure that the hopping does not simply follow an activated form down to zero temperature. Such a form is not consistent with the data.

8. Secondary Relaxation

So far we have only considered the primary relaxations that occur as the liquid is cooled down towards the glass transition temperature. However there are also secondary relaxations that occur in the glass phase, below the temperature where the primary relaxation is so slow as to be unmeasurable in our frequency range. Studies of these secondary (or β) relaxation in glasses were initiated by Johari and Goldstein[42,43] who interpreted these relaxations as intrinsic to the glass-transition process. Their measurements showed that the frequency of the peak in the imaginary part of the susceptibility followed an Arrhenius dependence as distinct from the primary relaxation discussed above. At temperatures slightly above T_0, the α- and β- relaxation processes often have comparable frequencies so that it is difficult to distinguish the shapes of the two contributions separately.

Mode-coupling theory[44] has attempted to calculate the form of the relaxation processes that occur as a liquid is supercooled. This theory predicts that there would be a second relaxation in addition to the primary ones but at higher frequencies.[45] This process was also names "β relaxation". It now seems clear that this is a very different process than that identified by Johari and Goldstein. The mode-coupling theory has attempted[45,46] to fit both types of β peaks including those originally identified by Johari and Goldstein.

Susceptibility data can investigate the origin of these relaxations. As shown below, the available experimental evidence suggests that the Johari-Goldstein peaks are due to simple activation of independent entities over an approximately fixed distribution of barriers. The distribution of barrier heights is Gaussian, as one might naturally expect for a disordered system. This explanation of the relaxation mechanism is very simple and does not rely on any cooperative motions. In particular, it does not necessitate the use of mode-coupling theory.

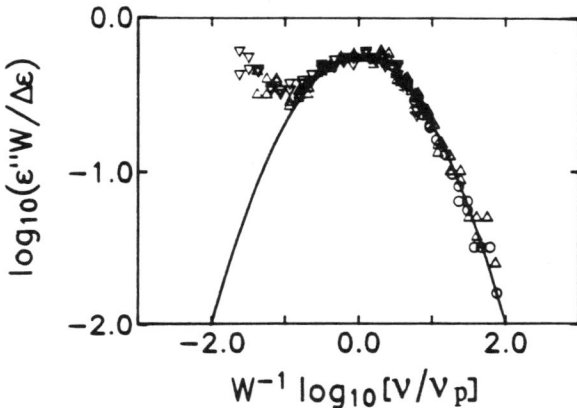

Figure 11. A collapse of the dielectric data at different temperatures for o-terphenyl showing $\varepsilon''W_e/\Delta\varepsilon$ versus $W_e^{-1}\log_{10}[\nu/\nu_p]$. The solid curve is a parabola representing a log-normal fit of the form given by Equation 16. Each different symbol represents the data taken at a different temperature. Figure is from Ref. 43.

One of the first glasses studied by Johari and Goldstein was o-terphenyl. An overview of the frequency/temperature behavior of the different relaxation processes for this sample is given in Fig. 1 of the lecture by Buchenau in these proceedings[47]. A more recent study[48] has measured the real and imaginary part of the dielectric response in the range 10^{-1} Hz $< \nu < 4 \times 10^4$ Hz. As shown in Figure 11, the relaxation peak in the imaginary part can be fit with a log normal form:

$$\varepsilon''(\nu) = \frac{\Delta\varepsilon}{\sqrt{\pi}W_e} \exp[-(\log \nu - \log \nu_p)^2/W_e^2] \qquad (16)$$

where W_e is the width where the peak reaches $(1/e)$ of its peak height [i.e., where $\varepsilon''(\nu_p 10^{\pm W_e})$ is $1/e$ of $\varepsilon''(\nu_p)$]. The solid line is a parabola which indicates a log-normal frequency dependence of the form of Eq. 16. The data fall on the parabola except on the low frequency side where the primary relaxation processes overlap with the secondary peak and begin to dominate the spectrum. From the high frequency portion of the data we can determine the parameters ν_p and W_e as a function of temperature. These are shown in Figures 12a and b respectively. As was observed by Johari and Goldstein[37], ν_p varies Arrheniusly. As is shown in the bottom curve, the width W_e also depends strongly on temperature. A fit to the data gives $W_e = -0.15 + 493/T$.

This behavior can be understood by expressing $\varepsilon''(\nu)$ as due to activation over energy barriers with a distribution of barrier heights. A convolution of Debye relaxation over a single energy barrier with a distribution of activation energies $P(E)$ produces:

$$\varepsilon(\nu,T) = \varepsilon_\infty + \Delta\varepsilon \int_0^\infty dE\, P(E)\, \frac{1}{1-i(\nu/\nu_t)e^{E/k_BT}}. \tag{17}$$

Assuming a Gaussian distribution of activation energies

$$P(E) = \frac{1}{\sqrt{\pi}\sigma}\, e^{-(E-\overline{E})^2/\sigma^2} \tag{18}$$

gives

$$\varepsilon(\nu,T) = \varepsilon_\infty + \Delta\varepsilon \int_0^\infty dE\, \frac{1}{\sqrt{\pi}\sigma}\, e^{-(E-\overline{E})^2/\sigma^2}\, \frac{1}{1-i(\nu/\nu_t)e^{E/k_BT}}. \tag{19}$$

If the width of the distribution, σ, is broad, this produces an approximately log-normal distribution for $\varepsilon''(\nu)$ with the width in energy, σ, related to the frequency width, W_e, by:

$$\sigma/k_B \approx W_e T \ln(10) \tag{20}$$

For o-terphenyl, the data for W_e indicates that $\sigma/k_B = (\sigma_0 - BT)/k_B = 1135\text{ K} - 0.35\text{ T}$. It remains unclear, however, whether the primary and secondary (Johari-Goldstein) relaxations merge smoothly with one another as the temperature is increased or whether they intersect abruptly at some finite temperature.[49]

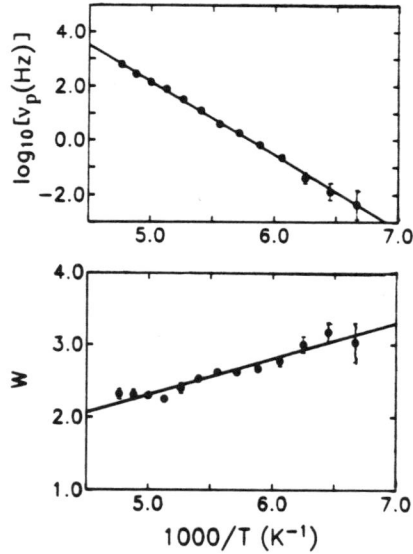

Fig. 12 Upper curve: The temperature dependence of ν_p, the peak position of the imaginary part of $\varepsilon''(\nu,T)$. The straight line on the $\log_{10} \nu_p$ versus $1000/T$ plot indicates Arrhenius behavior. Lower curve: The width of the dielectric response peak W_e as a function of T^{-1}. Figure is from Ref. 43.

Another glass where the secondary relaxation has been studied by dielectric techniques is a mixture of benzyl chloride and toluene[50]. In that system, not only were the same features seen as were just described for o-terphenyl, but it was even possible to assign a tentative configuration for the atoms which move during the hopping process. A third example is $(KBr)_x(KCN)_{1-x}$ which is crystalline in that its molecules are arranged in a periodic structure, but is an *orientational* glass in that the CN molecules point in random directions. Dielectric studies of this system at low temperatures in the glassy phase, where the axes of the CN molecules were frozen so that the molecules could only undergo head-to-tail flips, showed a relaxation process similar to the secondary relaxation in glasses.[51] Thus there are three systems for which the shape of the secondary relaxation has been determined: one (crystalline) orientational glass, $(KBr)_x(KCN)_{1-x}$ and two structural glasses, i) a mixture of benzyl chloride and toluene and ii) o-terphenyl. All three cases show Arrhenius temperature dependence of the peak frequency and a wide log-normal susceptibility curve which can be interpreted naturally in terms of a Gaussian distribution of activation barriers with only a slight temperature dependence to its width.

In the case of (KBr)(KCN), Sethna and collaborators[52] have suggested that the same motions that contribute to the dielectric response at high temperatures (where the Johari-Goldstein β relaxation takes place) is also responsible for the tunnelling levels that produce the linear temperature dependence of the specific heat at very low temperatures below 1 K. The same my be true for the real structural glasses as well. In addition, for (KBr)(KCN) the dielectric response measured the quadrupolar order parameter that was responsible for the freezing of the axes CN molecules. If a similar interpretation could be applied to structural glasses, this would imply that the dielectric susceptibility could measure the order parameter of the glass transition itself. A tentative analysis of this sort was given of the o-terphenyl data.[48]

9. Conclusions

In these lectures, I have given an overview of the results from two types of susceptibility measurements on supercooled liquids and glasses. Specific-heat spectroscopy is a generalization of the ordinary static specific-heat measurements that allows a dynamic measurement of enthalpy relaxation. Dielectric susceptibility is a probe which couples to the permanent dipoles in the liquid and can measure their reorientation times. It is a particularly useful probe since it can cover an enormous range of frequency with high precision. It appears that the average relaxation times measured by these different probes as a function of temperature are the same and that only the widths of the response functions differ between the two measurements.

It should be emphasized again that these results present some substantial puzzles for understanding supercooled liquids. The average relaxation times do behave as if they are about to diverge at a finite temperature which, in the cases where we have the data, is not far from the Kauzmann temperature where the entropy of the liquids extrapolates to be below that of the crystalline solid. No adequate theory yet exists for such behavior. Mode-coupling theory mainly addresses the behavior at much higher temperature. The question that naturally comes up is whether this temperature, T_o, marks an underlying phase transition. However, if that were the case one would expect to see some correlation length increasing as the temperature is lowered. Searches for such a length scale have been attempted (via non-linear susceptibility[50], computer simulations[14] and viscosity measurements[53]) but without much success.

In discussing the mode-coupling theory, a problem came up in analyzing the salol data in the context of recent light-scattering data. It appears that either one of the analyses is wrong or else the two probes couple in a very different way to the relaxation at the glass transition. This would be very surprising since both probes clearly see the onset of the slow relaxation.

A further series of questions is generated by the dielectric data. The scaling plot which allows all the dielectric data to be scaled onto one curve is not understood. In particular, no motivation is given for why the scaling should take such an unconventional from. It is surprising that the scaling form indicates that there is a limiting width for the dielectric-susceptibility curves which is the same as the one discovered for spin glasses. Spin glasses have quenched disorder (i.e., the spins have a random placement in the lattice which is fixed) whereas the disorder in liquids is annealed. One would expect that the physics of these two situations would be quite different. However, given this similarity in the limiting width, one might suspect that there is some underlying similarity. On the experimental side, one would certainly want to know how universal is the master curve: Is it only good for the dielectric susceptibility or does it work for other probes as well? Does it work for all supercooled liquids or only a subset of them? Can similar scaling be found in other systems such as spin glasses? Although many questions remain, it is encouraging that so much progress has been made in recent years both on the theoretical and experimental fronts.

Acknowledgements

I would like to thank all my collaborators in this research on glasses. These include Shobo Bhattacharya, Norman Birge, John Carini, Paul Dixon, Richard Ernst, Gary Grest, Yoon Jeong, Narayan Menon, Kevin O'Brien, Bruce Williams and Lei Wu. I am also indebted to many scientists who have helped me understand various aspects of glass formation. In particular, Austen Angell, Daniel Kivelson, Gene Mazenko and James Sethna have guided me many times into productive paths. This work was supported by NSF DMR 91-11733.

---

1 D. Turnbull, Contemp. Phys. **10**, 473 (1969).

2 Although there is no proof, many people believe that all liquids can be quenched into a glassy state if the cooling rate is great enough. [For a discussion see: F. Spaepen and D. Turnbull, in *Rapidly Quenched Metals; Second International Conference, Section I*, edited by N. J. Grant and B. C. Giessen page 205 (MIT Press, 1976).] For example, the Lennard-Jones computer "liquid", which is a good approximation to liquid argon, can be quenched into a glass using cooling rates which are much greater than can be currently attained in the laboratory. See e.g.: A. Rahman, M.J. Mandell and J.P. McTague, J. Chem. Phys. **64**, 1564 (1976); J. Fox and H.C. Andersen, J. Phys. Chem. **88**, 4019 (1984).

3 To see the relationship between the values of T_0 obtained from these two fits see: J. Souletie, J. Phys. (Paris) **49**, 1211 (1988).

4 If the time scale, or the viscosity, diverged only at $T = 0$ there would be no need to postulate the existence of a glass transition.

5 See: E. Rössler, Phys. Rev. Lett. **65**, 1595 (1990); D. Richter, R. Zorn, B. Farago, B. Frick and L. J. Fetters, Phys. Rev. Lett. **68**, 71, (1992); E. W. Fisher, E. Donth and W. Steffen, Phys. Rev. Lett. **68**, 2344 (1992).

6   C. A. Angell, in *Proceedings of the Workshop on Relaxations in Complex Systems,* edited by K. L. Ngai and G. B. Wright (National Technical Information Service, U.S. Dept. of Commerce, 5285 Port Royal Rd., Springfield, VA, 22161, 1984) p 3.

7   T. A. Witten, Phys. Today, **43** #7 (July) 21 (1990).

8   W. Kauzmann, Chem. Rev. **43**, 219 (1948).

9   P.F. Sullivan and G. Seidel, Phys. Rev. **173**, 679 (1968).

10   N. O. Birge and S. R. Nagel, Phys. Rev. Lett. **54**, 2674 (1985); N. O. Birge and S. R. Nagel, Rev. Sci. Instr. **58**, 1464 (1987); N. O. Birge, Phys. Rev. B **34**, 1631 (1986).

11   P. K. Dixon, Phys. Rev. B **42**, 8179 (1990).

12   G. S. Grest and S. R. Nagel, J. Phys. Chem. **91**, 4916 (1987).

13   P. K. Dixon and S. R. Nagel, Phys. Rev. Lett. **61**, 341 (1988).

14   R. M. Ernst, S. R. Nagel and G. S. Grest, Phys. Rev. B **43**, 8070 (1991).

15   T. Christensen, J. Phys. (Paris) Colloq. **46**, C8-635 (1985).

16   See J. Wong and C. A. Angell, *Glass: Structure by Spectroscopy* (Dekker, New York, 1976) for an excellent review of much of the literature on the glass transition up until 1976.

17   D. W. Davidson and R. H. Cole, J. Chem. Phys. **19**, 1484 (1951).

18   T. D. Davis and T. A. Litovitz, *Physical Acoustics* Vol. 2B, Academic Press, Inc., N. Y. (1965).

19   Y. H. Jeong, S. R. Nagel and S. Bhattacharya, Phys. Rev. A **34**, 602 (1986).

20   Y. H. Jeong, Phys. Rev. A **36**, 766 (1987).

21   K. S. Cole and R. H. Cole, J. Chem. Phys. **9**, 341 (1941).

22   S. Havriliak and S. Negami, J. Polym. Sci. Polym. Symp. **14**, 89 (1966).

23   R. Kohlrausch, Pogg. Ann. Phys. **91**, 198 (1854).

24   G. Williams and D. C. Watts, Trans. Faraday Soc. **66**, 80 (1970)

25   M. H. Cohen and G. S. Grest, Phys. Rev. B **24**, 4091 (1981); R. G. Palmer, D. Stein, E. Abrahams and P.W. Anderson Phys. Rev. Lett. **53**, 958 (1984); J. T. Bendler and M. F. Shlesinger, J. Molecular Liquids **36**, 37 (1987); I. A. Campbell, J. -M.

Flesselles, R. Jullien and R. Botet, Phys. Rev. B **37**, 3825 (1988); J. Kakalios, R. A. Street and W. B. Jackson, Phys. Rev. Lett. **59**, 1037 (1987); V. Degiorgio, T. Bellini, R. Piazza, F. Mantegazza and R. E. Goldstein, Phys. Rev. Lett. **64**, 1043 (1990).

26   F. Mezei, W. Knaak and B. Farago, Phys. Rev. Lett. **58**, 571 (1987).

27   In the case of glycerol, the temperature dependence was less than we could distinguish by specific-heat spectroscopy (see Ref. 10). However, using dielectric spectroscopy the temperature dependence of w was cearly discernable.

28   N. O. Birge, Y. H. Jeong and S. R. Nagel, in *Dynamic Aspects of Structural Change in Liquids and Glasses*, Annals of the New York Acad. of Sciences vol 484. Edited by C. Austen Angell and M. Goldstein.(New York Academy of Sciences, New York, 1986), page 101.

29   P. K. Dixon, L. Wu, S. R. Nagel, B. D. Williams and J. P. Carini, Phys. Rev. Lett. **65**, 1108 (1990).

30   L. Wu, P. K. Dixon, S. R. Nagel, B. D. Williams and J. P. Carini, J. Non-Cryst. Solids **131**, 32 (1991).

31   T. C. Halsey, P. Meakin and I. Procaccia, Phys. Rev. Lett. **56**, 854 (1986); T. C. Halsey, M. H. Jensen, L. P. Kadanoff, I. Procaccia and B. I. Shraiman, Phys. Rev. A **33**, 1141 (1986); L. P. Kadanoff, S. R. Nagel, L. Wu, and S.-m. Zhou, Phys. Rev. A **39**, 6524, (1989).

32   R. V. Chamberlin, Phys. Rev. Lett. **66**, 959 (1991); R. V. Chamberlin and D. N. Haines, Phys. Rev. Lett. **65**, 2197 (1990).

33   A. Schönhals, F. Kremer and E. Schlosser, Phys. Rev. Lett. **67**, 999 (1991).

34   N. Menon and S. R. Nagel (to be published).

35   P. K. Dixon, L. Wu, S. R. Nagel, B. D. Williams, and J. P. Carini, Phys. Rev. Lett. **66**, 960 (1991).

36   A. T. Ogielski, Phys. Rev. B **32**, 7384 (1985).

37   I. A. Campbell, J. M. Flesselles, R. Julien and R. Botet J. Phys. C**20**, L47 (1987). For a review see also D. L. Stein and R. G. Palmer in "Complex Systems, SFI Studies in Sciences of Complexity" Ed. D. L. Stein (Addison-Wesley Longman Publishing Group Ltd.,1989) p. 1.

38   See: W. Götze, these proceedings.

39   G. Li, W. M. Du, X. K. Chen, H. Z. Cummins and N. J. Tao, Phys. Rev. A **45**, 3867 (1992).

40   G. Li, W. M. Du, A. Sakai and H. Z. Cummins, Phys. Rev. A **46**, 3343 (1992).

41   B. Kim and G. F. Mazenko, Phys. Rev. A **45**, 2393 (1992).

42   G. P. Johari and M. Goldstein, J. Phys. Chem. **74**, 2034 (1970); J. Chem. Phys. **53**, 2372 (1970); J. Chem. Phys. **55**, 4245 (1971).

43   G. P. Johari, J. Chem. Phys. **58**, 1755 (1973); Ann. N.Y. Acad. Sci. **279**, 117 (1976).

44   E. Leutheusser, Phys. Rev. A **29**, 2765 (1984); U. Bengtzelius, W. Götze and A. Sjölander, J. Phys. C **17**, 5915 (1984); S. P. Das and G. F. Mazenko, Phys Rev. A **34**, 2265 (1986).

45   For a comprehensive review see: W. Götze and L. Sjögren, Rep. Prog. Phys. **55**, 241 (1992), and references therein.

46   W. Götze and L. Sjögren, J. Phys.: Condens. Matter **1**, 4183 (1989).

47   See: U. Buchenau, these proceedings.

48   L. Wu and S. R. Nagel, Phys. Rev. B **46**, 11198 (1992).

49   C. A. Angell, Chemical Review **90**, 523 (1990).

50   L. Wu, Phys. Rev. B **43**, 9906 (1991).

51   N. O. Birge, Y. H. Jeong, S. R. Nagel, S. Bhattacharya and S. Susman, Phys. Rev. B **30**, 2306 (1984); L. Wu, R. M. Ernst, Y. H. Jeong, S. R. Nagel and S. Susman, Phys. Rev. B. **37**, 10444 (1988); R. M. Ernst, L. Wu, S. R. Nagel and S. Susman, Phys. Rev. B **38**, 6246 (1988).

52   J. P. Sethna and K. S. Chow, Phase Transitions **5**, 317 (1985); M. Meissner, W. Knaak, J. P. Sethna, K. S. Chow, J. J. De Yoreo and R. O. Pohl, Phys. Rev. **B32**, 6091 (1985); J. P. Sethna, N. Y. Acad. Sci. **484**, 130 (1986).

53   P. K. Dixon, S. R. Nagel and D. A. Weitz, J. Chem. Phys. **94**, 6924 (1991).

HIERARCHICAL MELTING OF ONE–DIMENSIONAL INCOMMENSURATE STRUCTURES

R.Schilling
Institute of Physics
Johannes Gutenberg–Universität
Postfach 3980
6500 Mainz, F.R.G.

Abstract. We study the low–temperature properties of quasi one–dimensional, incommensurate structures which are described by a Frenkel–Kontorova–like model. A new type of renormalization method will be presented, which is determined by the continued fraction expansion of the incommensurability ratio ζ. This method yields a hierarchy of renormalized Hamiltonians $\mathcal{H}^{(n,p)}$ describing the thermal behavior for temperatures $T = O(T^{(n,p)})$, where $T^{(n,p)}$ follows from the continued fraction expansion of ζ. By means of this method the low–temperature specific heat $c(T)$ and the static structure factor $S(q)$ are calculated for fixed ζ. $c(T)$ possesses a hierarchy of Schottky anomalies related to the rational approximates of ζ and $S(q)$ exhibits more and more satellites when the temperature is decreased. Our theoretical approach predicts a high sensitivity on a small change of ζ. For instance $c(T)$ and $S(q)$ may change by several orders of magnitude if ζ is changed by, e.g one per cent only. Finally our results are compared with experimental data.

1. Introduction

Many physical properties of solids sensitively depend on their structure. We can distinguish between three classes of different structures: periodic, quasi–periodic and aperiodic. Examples are presented in figure 1.
The *periodic* or *crystalline* structures are well studied due to their simplicity. With the knowledge of the unit cell, the extended structure is uniquely defined. The position $\vec{R}_n^{(0)}$ of the n–th atom in a d–dimensional, primitive lattice is given by

$$\vec{R}_n^{(0)} = \sum_{\alpha=1}^{d} n_\alpha \vec{a}_\alpha, \quad n_\alpha \in \mathbb{Z} \qquad (1)$$

where $\vec{a}_\alpha$ are the elementary lattice vectors (see figure 1a).

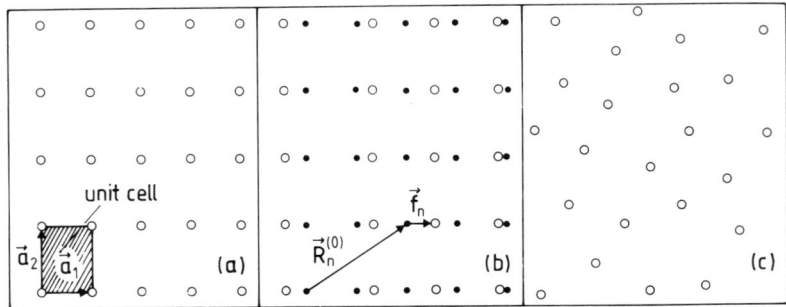

Figure 1: Schematic representation of, e.g. two–dimensional (a) periodic, (b) quasi–periodic and (c) aperiodic structures.

The *quasi–periodic* structures can be decomposed into two subclasses, the conventional *incommensurate* (IC) structures and the recently discovered *quasicrystals*. Both have in common that there exists infinite many subsets of the atomic positions which are *almost* periodic. The conventional IC–phases are modulated structures, where $\vec{R}_n$ can be represented as follows:

$$\vec{R}_n = \vec{R}_n^{(0)} + \vec{f}_n, \quad \vec{f}_n = \vec{f}(\vec{q} \, \vec{R}_n^{(0)} + \varphi) \quad . \tag{2}$$

The *hull function* $f_\alpha(x) = f_\alpha(x+1)$, $\alpha = 1, 2, ..., d$ is periodic with period one and it describes the modulation of the underlying periodic lattice defined by $\{\vec{R}_n^{(0)}\}$. The modulation can be characterized by the modulation vector $\vec{q}$ and a phase $\varphi \in [0, 2\pi)$. If $(\vec{q} \cdot \vec{a}_\alpha)/(2\pi) = \zeta_\alpha$ is an *irrational* number at least for one of the α–values, the corresponding configuration is incommensurate. Figure 1b presents an example where ζ_1 is irrational, only. The atomic positions in a quasicrystal, obtained from a high–dimensional lattice, when projected to the d–dimensional space, can in general not be represented by (2), since no underlying lattice exists such that the hull function is *bounded*. For more details on incommensurate structures the reader is referred to the review by Janssen and Janner (1987).

All the structures which are neither periodic nor quasiperiodic are *aperiodic*. Typical aperiodic structures are amorphous ones. But also regular and irregular fractals belong to the class of aperiodic structures.

A *direct* characterization of structures follows from the static structure factor $S(\vec{q})$ which can be measured by x–ray or neutron scattering. Figure 2 shows the qualitative features of $S(\vec{q})$ for the three different classes of structures for dimension d=3. Because for d=3 the periodic and quasi–periodic configurations exhibit long range order even at finite temperatures T, δ–peaks occur in $S(\vec{q})$. Besides the Bragg–reflections (figure 2a) related to the periodic lattice $\{\vec{R}_n^{(0)}\}$, *satellites* (figure 2b) exist for an IC–structure. Their positions in reciprocal space are given by

$$\vec{q}_{\nu,\{\nu_\alpha\}} = \sum_{\alpha=1}^{d} \nu_\alpha \vec{q}_\alpha + \nu \vec{q}, \quad \nu_\alpha \in \mathbb{Z}, \nu \in \mathbb{Z} \tag{3}$$

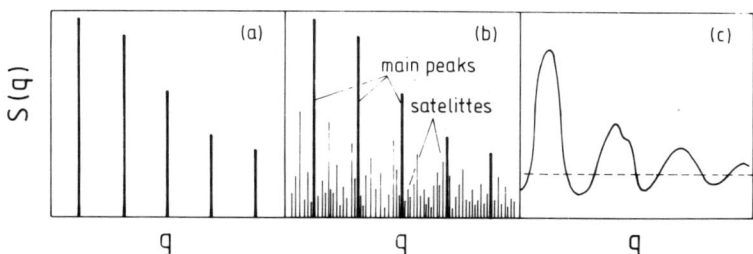

Figure 2: Static structure factor for a (a) periodic, (b) quasi–periodic and (c) aperiodic (amorphous) structure as function of $q \equiv q_x (q_y, q_z$ fixed).

where $\{\vec{q}_\alpha\}$ are the reciprocal lattice vectors related to $\{\vec{a}_\alpha\}$. The imcommensurability implies that the modulation vector $\vec{q}$ can *not* be represented as a linear combination of the $\vec{q}_\alpha$ with coefficients which are *rational numbers*. Therefore the satellites densely fill the reciprocal space or a subspace of it. An aperiodic (amorphous) structure does not posses long range order. Therefore $S(\vec{q})$ has broad reflection peaks and it converges to a constant for $|\vec{q}| \to \infty$ (figure 2c).

As already stated above, these different types of structures also reveal different behavior, e.g. for the electrical, magnetic, thermal etc. properties. Let us concentrate on the low–temperature specific heat (figure 3).

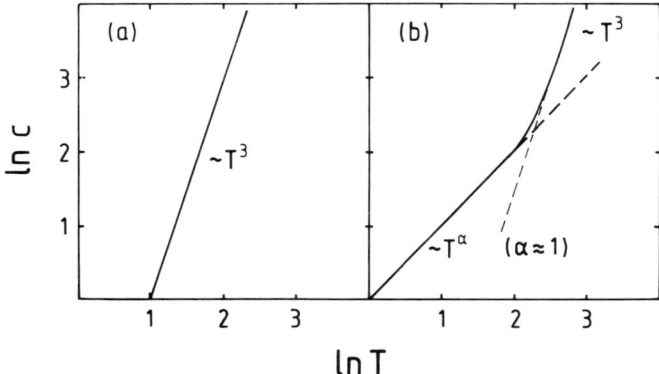

Figure 3: Low–temperature specific heat for a three–dimensional (a) periodic and (b) aperiodic (amorphous) structure

The specific heat c(T) of a three–dimensional crystalline (periodic) and amorphous (aperiodic) solid at low temperatures, i.e. $T < \approx 1K$, obeys a power law

$$c(T) \sim T^\alpha$$

with $\alpha = d$ and $0.4 \approx \alpha \approx 1.4$, respectively. The difference in α is not just of quantitative nature, but is a manifestation of the qualitatively different low–energy exitations. These are phonons in case of a crystal and localized configurational excitations, called two–level–systems, for amorphous materials. For incommensurate structures no similar result exists for c(T). Therefore we will investigate this question in this paper. Since there exists quasi one–dimensional IC–structures in nature, we will restrict ourselves to d=1.

The outline of our paper is as follows. In the next section we discuss a class of materials which form quasi one–dimensional, incommensurate structures and we show how these systems can be modelled. Section 3 contains basic properties like ground state configuration and elementary excitations of the model which will be used. A new type of renormalization method which yields the thermodynamic properties of our model, will be described in section 4 and applied in section 5 to calculate the static structure factor and the low–temperature specific heat. A summary and some conclusions are given in the final section 6.

2. Model

The super–ionic conductors hollandite are examples for solids with quasi one–dimensional properties. One example, K–hollandite $K_{2\rho}Mg_\rho Ti_{8-\rho}O_{16}$ is schematically shown in figure 4.

K–hollandite consists of a skeleton of edge – and corner–sharing TiO_6 and MgO_6 octahedra forming parallel channels of nearly square cross section along the c–axis (figure 4a). The substochiometric K–ions can move within each channel (figure 4b). Figure 4b already demonstrates the existence of *two competing length scales* which are the lattice constant $c \simeq 2.98$ Å in c–direction and $\ell = \rho^{-1}$, the mean distance between neighboring K–ions. Their ratio:

$$\zeta = \ell/c$$

is called the *incommensurability ratio*, which corresponds to a modulation vector $\vec{q} = (0,0,q_1 = \frac{2\pi}{c})$ pointing into the c–direction.

Due to screening effects and the rather large distance between adjacent channels, the interactions of K–atoms in different channels will be neglected. In addition we assume the skeleton to be rigid. Therefore, two *competing interactions* exist only: the interaction between the K–ions in one channel (which is also screened) and the interaction of the K–ions with the periodic skeleton. The ion–ion interaction (within a channel) is approximated by a *harmonic* nearest neighbor coupling and the ion–skeleton interaction is taken to be piecewise harmonic as shown in figure 5.

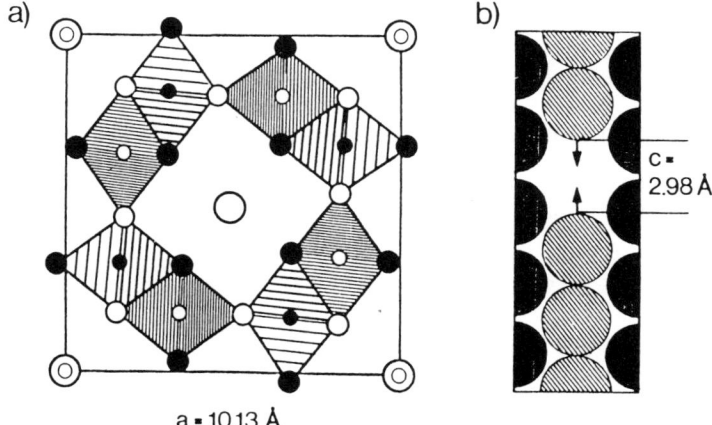

Figure 4: Schematic representation of hollandite–structure
(a) projection of the tetragonal structure parallel to the c–axis: K$^+$–ion (large open circle at the center) O–atoms (medium full and open circles), octahedra–cations (small full and open circles)
(b) distribution of K$^+$–ions (hatched) along the channel. Oxygen atoms are shown in black. The arrows denote the direction of displacements.
(from Rosshirt (1988))

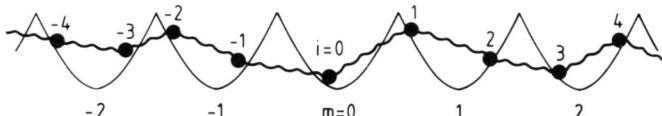

Figure 5: Schematic representation of the model interactions. A harmonic chain of particles interacts with the periodic, piecewise harmonic skeleton potential

Hence the potential energy is given by:

$$V(\{u_i\})=\sum_i [\tfrac{1}{2}(u_{i+1} - u_i)^2+\tfrac{k}{2}(u_i - 2\pi m_i)^2 - 2\pi\mu(m_{i+1}-m_i)] \quad (4a)$$

where

$$m_i = \text{int}\,[(u_i + \pi)/2\pi] \quad (4b)$$

describes the *anharmonic effects* and it also numbers that well of the skeleton potential which contains the i—th particle (see figure 5). For convenience the position u_i of the i—th K—ion is measured in units of $\frac{c}{2\pi} = (q_1)^{-1}$ and the energy is given in units of $C(\frac{c}{2\pi})^2$, with C the elastic constant. The coupling constant k of the skeleton potential is measured in units of C and μ is a tensile force (Lagrange multiplier) to be determined such that:

$$\ell = \frac{c}{2\pi} \lim_{N'-N''\to\infty} [(u_{N'} - u_{N''})/(N' - N'')]$$
$$= c \lim_{N'-N''\to\infty} [(m_{N'} - m_{N''})/(N' - N'')] \quad (5)$$

and therefore

$$\rho^{-1} = \zeta = \ell/c \equiv q_1/q_0 \quad (6)$$

is fixed. Here it is $q_0 = \frac{2\pi}{\ell}$ and the function int(x) denotes the largest integer smaller than or equal to x.

Model (4) is a modified *Frenkel–Kontorova* (FK) *model*, where the cosine–potential (of the skeleton) is expanded around its minima (Vallet et al. (1988)). If the skeleton potential is much stronger than the ion–ion interaction, i.e. k>>1, we expect (4) to be a reasonable approximation of the original FK–model with the cosine–potential. Replacing the ion–ion interaction by a harmonic nearest neighbor coupling may seem to be too crude. Beyeler et al. (1980) have numerically calculated the diffuse part of the static structure factor for a harmonic and a realistic ion–ion coupling, as well as a cosine – and a piecewise parabolic potential. They found that the result does not sensitively depend on the special type of interaction. Therefore we will consider the model (4) as a reasonable description for hollandite. But let us stress an important difference between the smooth cosine–potential and the nonanalytic piecewise parabolic potential. For the former one, Aubry (1978) has proved that a ζ–dependent critical value $k_c(\zeta)$ exists, where the ground state undergoes a transition from an "*analytical*" to a "*nonanalytical*" phase when increasing k. The analytical and nonanalytical regime is characterised by a hull function f which is analytical and nonanalytical, respectively. The physical meaning is the following: in the analytical phase the chain of particles can make a translation without expense of energy in contrast to the nonanalytical phase, where the chain is *pinned*. It is obvious that both regimes will yield different low–temperature properties, since in the analytical regime a zero–frequency mode, a phason, exists, which is absent in the nonanalytic phase. On the other hand the nonanalytic phase is defectable, i.e. defects can be introduced to generate low–energy excitations. This is not possible for the analytical phase. As model (4), a priori, is nonanalytical for *all* k>0, it only can describe the nonanalytical phase. Hence we tacitly assume that K–hollandite belongs to the nonanalytical regime.

3. Ground state and excitations

3.1 METASTABLE CONFIGURATIONS

In this section we present some of the most important properties of the model (4). Due to the piecewise harmonicity of (4), the *metastable configurations* can be exaxtly determied. They are solutions of the nonlinear difference equation:

$$\frac{\partial V}{\partial u_i} \equiv (2+k) u_i - u_{i-1} - u_{i+1} - \phi(u_i) = 0 \tag{7a}$$

where the *nonlinearity* $\phi(u_i)$ is given by

$$\phi(u_i) = 2\pi k\, m_i(u_i) \tag{7b}$$

with $m_i(u_i)$ from (4b).

If the matrix $(\frac{\partial^2 V}{\partial u_i \partial u_j})$ is positive definite, these solutions are metastable. Because V (eq.(4)) is convex for almost all $\{u_i\}$, the metastability is generically guaranteed. With use of the Green function, (7) can be solved analytically (see Aubry (1989) and Reichert et al. (1985) for a similar model). One obtains in the thermodynamic limit:

$$u_i(\{m_j\}) = A \sum_{n=-\infty}^{\infty} \eta^{|i-n|} m_n \tag{8a}$$

with

$$A = 2\pi \frac{1-\eta}{1+\eta} \tag{8b}$$

$$\eta = 1 + \frac{k}{2} - \sqrt{k + (\frac{k}{2})^2} < 1 \tag{8c}$$

which is a solution of (7a), only if the *self-consistency condition:*

$$m_i = \text{int}[(u_i(\{m_j\}) + \pi)/2\pi] \tag{9}$$

is fulfilled. This result has the following physical meaning. For $k \to \infty$, corresponding to $\eta \to 0$, it follows from (8) that:

$$u_i(\{m_j\}) \to 2\pi m_i \quad ,$$

i.e. the i-th particle is in the minimum of the m_i-th well of the skeleton potential (figure 6a). Therefore any *arbitrary occupation configuration* $\{m_i\}$ with $m_{i+1} > m_i$ determines a metastable configuration for $k = \infty$. When decreasing k, the particles will depart from their position at $2\pi m_i$ (figure 6b) and will remain in the m_i-th well, if (9) holds. One can show (see Uhler et al. (1988), appendix B) in case that

the "fluctuations"

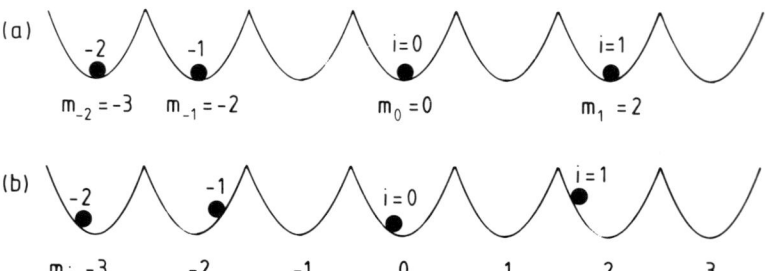

Figure 6. Schematic representation of a metastable configuration for (a) $k = \infty$; (b) $k < \infty$

$$m_{i+1} - m_i \leq r \quad , \quad r \in \mathbb{N} \quad , \quad m_{i+1} \geq m_i , \tag{10}$$

are bounded, that (9) is fulfilled for *all* $\{m_i\}$ consistent with (10), provided $k > k_0(r) \sim r$. This criterion is obvious, since an increase of the fluctuations $(m_{i+1} - m_i)$ increases the elastic forces between particle "i" and "i+1" and in order to keep them in the m_i-th and m_{i+1}-th well, respectively, the coupling k of the skeleton potential has to be large enough.

Substituting (8a) into (4a) and introducing the *pseudo spins*

$$\sigma_i = m_{i+1} - m_i \quad , \quad \sigma_i \geq 0 \tag{11}$$

one obtains (Pietronero et al. (1979), Aubry (1983 a))

$$\mathcal{H}(\{\sigma_i\}) \equiv V(\{u_i(\{m_j\})\})$$

$$= \frac{1}{2} \sum_{i \neq j} J(i-j) \sigma_i \sigma_j - h \sum_i \sigma_i \tag{12a}$$

with

$$J(i-j) = 2\pi A \, \eta^{|i-j|} \tag{12b}$$

and the "magnetic field"

$$h = 2\pi \mu - \frac{1}{2} J(0) \tag{12c}$$

These results are illustrated in figure 7.

The configuration space can be decomposed into cells C_σ uniquely characterized by

$\{m_i\}$, or equivalently by $\{\sigma_i\}$, provided (9) holds. Each cell contains exactly *one* metastable configuration $\{u_i(\{m_j\})\}$ and its energy follows from the Ising Hamiltonian (12a). It is interesting to note that $u_i(\{m_j\})$ cannot take any value, but is restricted to a Cantor set (see e.g. Aubry (1981)).

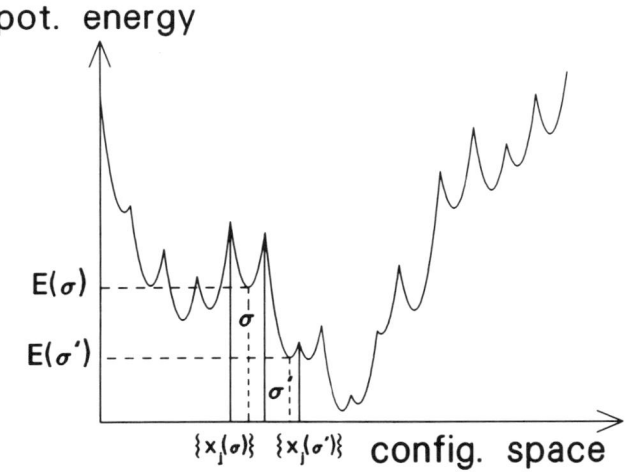

Figure 7: Schematic representation of the potential energy landscape of model (4) in the configuration space. x_j corresponds to u_j and $E(\sigma)$ to $\mathcal{U}(\{\sigma_i\})$.

The consequence of this is illustrated in figure 8. Particles can only sit within the hierarchy of allowed bands.

Figure 8: Hierarchy of forbidden regions (black) which are separated by the allowed bonds (white)

Each allowed band in figure 8, again contains an infinite number of forbidden regions. This fractal–like structure will turn out to be crucial for the hierarchical low–temperature behavior of our model.

3.2 GROUND STATE

It has been proved by several authors (Hubbard (1978), Pokrovsky et al.

(1978), Aubry et al. (1983)) that for the ground state it is

$$\sigma_i^{(0)}(\zeta,\varphi) = \text{int}\,[(i+1)\zeta + \varphi] - \text{int}\,[i\zeta + \varphi] \tag{13}$$

for given incommensurability ratio ζ. It is easy to show that $\sigma_i^{(0)}$ only takes the values n and n+1, if $n < \zeta < n+1$ ($n \in \mathbb{N}$). Without restricting generality, we asssume n = 0 from now on. Then the incommensurate ground state configuration is determined by a quasi–periodic sequence $\{\sigma_i^{(0)}\}$ with $\sigma_i^{(0)} \in \{0,1\}$. In section 4 we will discuss an important property of this sequence, which is its *renormalizability*. For $\zeta = r/s$ with r, s two irreducible positive integers, the corresponding sequence $\sigma_i^{(0)}$ is periodic with period s:

$$\sigma_{i+s}^{(0)} = \sigma_i^{(0)} \quad . \tag{14a}$$

This implies that:

$$u_{i+s}^{(0)} = 2\pi r + u_i^{(0)} \tag{14b}$$

i.e. the ground state configuration is periodic. Its *unit cell* has lengths $2\pi r$ and it contains s particles.

3.3 EXCITATIONS

In order to calculate the low–temperature properties of (4), we have to determine the low–energy excitations. Of course, there are the phonons. But due to the skeleton potential, a phonon gap $E_g = \hbar\,\omega_g = \hbar\,(kC/M)^{1/2}$ exists. M is the mass of the K–ions. Using k=1.28, M=39 amu and C = 0.176 eVÅ^{-2} (see Beyeler et al. (1980)) it follows that E_g is about 60K. Hence for T<<60K the phonons will not much influence the thermal properties.

Besides the phonons there exist two types of configurational excitations which are easily characterized by using the pseudo–spins:

(i) local rearrangements of e.g. the n–th K–ion:

$$\sigma_{n-1} \rightarrow \sigma'_{n-1} = \sigma_{n-1} \pm 1\,;\sigma_n \rightarrow \sigma'_n = \sigma_n \mp 1;\sigma_i \rightarrow \sigma'_i = \sigma_i,\quad i \neq n-1, n \tag{15a}$$

where it must be $\sigma'_j \geq 0$ for all J.

(ii) elementary discommensurations (phase defects):

$$\sigma_i' = \begin{cases} \sigma_i(\zeta,\varphi), & i \leq n \\ \sigma_i(\zeta,\varphi+\Delta\varphi), & i > n \end{cases} \tag{15b}$$

The local rearrangements, which were proposed by v.Löhneysen et al. (1981) and Pietronero et al. (1981), are bound pairs of a discommensuration ($\Delta\varphi > 0$) and an anti–discommensuration ($\Delta\varphi < 0$) (see Aubry (1983 b) and Vallet et al. (1988)). The excitation energy $\Delta E_n = \mathcal{H}(\{\sigma_i'\}) - \mathcal{H}(\{\sigma_i\})$ which follows from (12), is n–dependent and can become arbitrary small for both types of configurational excitations, provided ζ is irrational. For a commensurate ground state (ζ is rational) both configurational excitations exhibit an energy gap. The discommensurations, in contrast to the local rearragements, destroy the long range order. Since for our one–dimensional model no longer range order exists at finite temperatures, one has to consider the discommensurations as elementary excitations. This is similar to a one–dimensional Ising model with ferromagnetic nearest neighbor interactions. It is well known that localized single spin flips in the ferromagnetic ground state configuration are *not* the elementary excitations, but a domain wall between two domains of up – and down–spins.

An empirical approach based on the elementary discommensurations has been used by Vallet et al. (1988) to determine the specific heat. We will not discuss this in more detail, since we have developed a more sophisticated renormalization method which yields more accurate results.

4. Renormalization — Methods

In this section we present a new type of renormalization method which is designed to treat one–dimensional, incommensurate structures at finite temperatures. A vague idea of part of this method for a special value of ζ, can already be found in the paper by Pietronero et al. (1981). The starting point of our renormalization scheme is an important property of the sequence $\{\sigma_i^{(0)}(\zeta,\varphi)\}$, which is its *renormalizability*.

4.1. RENORMALIZABILITY OF THE GROUND STATE

In order to discuss the renormalizability of the sequence $\{\sigma_i^{(0)}(\zeta,\varphi)\}$ we have to introduce some well–known features of irrational (or rational) numbers. Any irrational (or rational) number ζ can be expanded into *a continued fraction*:

$$\zeta = a_0 + 1/[a_1 + 1/[a_2 + 1/[...1/(a_n + \zeta_n)]]] \quad , \tag{16a}$$

where the *positive* integers $a_i (i \geq 1)$ and the remainders ζ_i are recursively defined by

$$a_i = \text{int}(1/\zeta_{i-1}) \tag{16b}$$

$$\zeta_i = 1/\zeta_{i-1} - a_i \quad , \quad i \geq 1 \tag{16c}$$

with the initial values

$$a_0 = \text{int}(\zeta) \quad , \quad \zeta_0 = \zeta - a_0 \quad . \tag{16d}$$

due to our choice $0<\zeta<1$, it is $a_0 = 0$. In case of a rational number ζ, the continued fraction is finite. Truncating (16a) at order n and replacing $(a_n + \zeta_n)$ by p where $1 \leq p \leq a_n$, yields a *rational approximate* $r_{n,p}/s_{n,p}$ of ζ, where $r_{n,p}$ and $s_{n,p}$ follow from the recursion relations:

$$r_{n,p} = r_{n,p-1} + r_{n-1} \quad ; \quad n \geq 1 \quad , \quad 1 < p \leq a_n$$

$$r_{n,1} = r_{n-1} + r_{n-2} \quad ; \quad n > 1 \quad , \quad p=1 \quad \quad (17a)$$

$$r_{n,a_n} \equiv r_n$$

(same for $s_{n,p}$ and s_n) with the initial values

$$r_0 = a_0 = 0, \quad r_{1,1} = a_0 + 1 = 1 \quad ; \quad s_0 = 1 \quad , \quad s_{1,1} = 1 \quad . \quad (17b)$$

$r_{n,p}/s_{n,p}$ for $1 \leq p < a_n$ are called *intermediate convergents* and $r_{n,a_n}/s_{n,a_n} \equiv r_n/s_n$ for $p = a_n$ *principal convergents*. As we will see below, this distinction will get a physical meaning.

Now, let us return to the renormalizability of $\{\sigma_i^{(0)}\}$. Only, the most important steps are given here; for more details the reader may consult Vallet et al. (1986), (1988). One can show that the infinite sequence $\{\sigma_i^{(0)}(\zeta,\varphi)\}$ can be generated by an infinite sequence of two "*words*" (or blocks) $W_{n,p}$ and W_{n-1}. The formation of these "*words*" is determined by the continued fraction expansion, only. $W_{n,p}$ and W_{n-1} follow from the recursion relations:

$$W_{n,p} = \{W_{n,p-1}, W_{n-1}\} \quad ; \quad n \geq 1 \quad , \quad 1 < p \leq a$$

$$W_{n,1} = \{W_{n-1}, W_{n-2}\} \quad ; \quad n > 1 \quad , \quad p=1 \quad \quad (18a)$$

$$W_{n,a_n} \equiv W_n$$

with the initial "*words*"

$$W_0 = \{0\} \quad , \quad W_{1,1} = \{1\} \quad . \quad (18b)$$

$\{W,W'\}$ denotes the composition of both "*words*" W and W'. Note the similarity between (17) and (18). The initial "*words*" W_0 and $W_{1,1}$ consist of one "*letter*" only, which is 0 and 1, respectively. The sequence $\{\sigma_i^{(0)}\}$ can just as good be presented by the sequence of the *two "words"* W_{n-1} and $W_{n,p}$ ($n \geq 1$ and $1 \leq p \leq a_n$) as follows. Let us represent the smaller "*word*" W_{n-1} by a *renormalized* pseudo spin $\sigma^{(n,p)} = 0$ and the larger "*word*" $W_{n,p}$ by $\sigma^{(n,p)} = 1$. Then the sequence $\{\sigma_j^{(n,p)}\}$, containing the same information as $\{\sigma_i^{(0)}(\zeta,\varphi)\}$, follows from the r.h.s. of (13), i.e.

$$\sigma_j^{(n,p)} = \text{int}[(j+1)\zeta_{n,p} + \varphi_{n,p}] - \text{int}(j\zeta_{n,p} + \varphi_{n,p}) \qquad (19a)$$

but with a *renormalized incommensurability ratio*

$$\zeta_{n,p} = \frac{\zeta\, s_{n-1} - r_{n-1}}{(r_{n,p} - r_{n-1}) - \zeta(s_{n,p} - s_{n-1})} \qquad (19b)$$

and a *renormalized phase* $\varphi_{n,p}$ (see Vallet et al. (1988), appendix 2) which will not be explicitly given, since $\varphi_{n,p}$ as well as φ does not play a crucial role. $\zeta_{n,p}$ only involves the original incommensurability ratio ζ, the numerators r_{n-1}, $r_{n,p}$ and denominators s_{n-1}, $s_{n,p}$ of the rational approximates of ζ. Hence, all quantities needed for the *"word"* renormalization of the ground state is determined from a *single* number, which is ζ. Let us illustrate this renormalization method by means of an example. We choose, e.g. $\zeta = \frac{1}{\rho} - 1$ with $\rho = 0.77$ and $\varphi = 0$. From (16b)–(16c) we immediately get:

$$a_0 = 0\,,\ a_1 = 3\,,\ a_2 = 2\,,\ a_3 = 1\,,\ a_4 = 6\,,\ldots$$

and from (13):

$$\sigma_i^{(0)}: \quad \ldots 1000100100100010010001001\ldots \qquad (20)$$

where the cross denotes the position of $i = 0$. Remember, this sequence determines the ground state configuration $\{u_i^{(0)}\}$ uniquely, except for an additive constant, since $\{m_i^{(0)}\}$ can only be determined from (11) up to a constant, e.g. $m_{i=0}^{(0)}$. Figure 9 illustrates the equivalence of (20) and the corresponding ground state configuration. Since $\sigma_i = m_{i+1} - m_i$, σ_i denotes the number of maxima of the

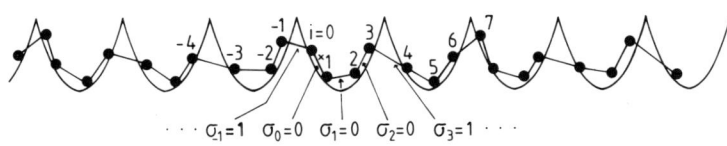

Figure 9: Schematic representation of the ground state configuration $\{u_i^{(0)}\}$ related to the sequence $\{\sigma_i^{(0)}\}$ given by (20). The cross denotes the nearest neighbor bond corresponding to $\sigma_{i=0}^{(0)}$.

skeleton potential between particles "i" and "i+1". Because $\sigma_{i=0}^{(0)} \in \{0,1\}$, there is at most only one maximum between adjacent particles (cf. figure 9).

The sequence (20) already suggests that $\{\sigma_i^{(0)}\}$ can be represented as a sequence of the two "*words*" $\{100\}$ and $\{1000\}$. Using (18) we now can construct the hierarchy of "*words*" W_{n-1}, $W_{n,p}$ and the corresponding renormalized sequence $\{\sigma_j^{(n,p)}\}$:

n = 1:

p = 2 $\qquad W_{1,2} = \{10\}, W_0 = \{0\}$

$\qquad\qquad\quad \sigma_j^{(1,2)}$: .... 100101010010100101...

p = a_1 = 3 $\qquad W_{1,3} \equiv W_1 = \{100\}, W_0 = \{0\}$

$\qquad\qquad\quad \sigma_j^{(1,3)}$: ... 10111011011...

n = 2:
p = 1 $\qquad W_{2,1} = \{1000\}, W_1 = \{100\}$

$\qquad\qquad\quad \sigma_j^{(2,1)}$: ...1001010...

p = a_2 = 2 $\qquad W_{2,2} \equiv W_2 = \{1000100\}, W_1 = \{100\}$

$\qquad\qquad\quad \sigma_j^{(2,2)}$: ...1011...

n = 3:

p = a_3 = 1 $\qquad W_{3,1} \equiv W_3 = \{1000100100\}, W_2 = \{1000100\}$

$\qquad\qquad\quad \sigma_j^{(3,1)}$: ...10...

etc.

Note that the sequences $\{\sigma_j^{(n,p)}\}$ were obtained by the identification W_{n-1} with 0 and $W_{n,p}$ with 1. Of course, $\{\sigma_j^{(n,p)}\}$ can directly by generated from (19) without use of the "*word*" construction.

This explicit example elucidates the mechanism of the renormaliztion method even more:
In the sequence (20) the digit 1 is always isolated, which implies that each 1 is *followed* by 0. Therefore one can use $\{10\}$ (which is just $W_{1,2}$) as a new "*word*" represented by $\sigma_j^{(1,2)} = 1$ and the old "word" $\{0\}$ which is represented by $\sigma_j^{(1,2)} = 0$. This prescription applied to (20) yields the sequence $\{\sigma_j^{(1,2)}\}$ shown above. In $\{\sigma_j^{(1,2)}\}$ still 1 is *followed* by 0. Again $\{1,0\}$ (now 0 and 1 denote $\sigma_j^{(1,2)}$!) can be taken as a new "*word*" $W_{1,3} \equiv W_1$ which equals $\{100\}$ for the original spins $\sigma_i^{(0)}$.

Representing $\{100\}$ in $\{\sigma_i^{(0)}\}$ or $\{10\}$ in $\{\sigma_j^{(1,2)}\}$ by $\sigma_j^{(1,3)} = 1$ and the old "*word*" $\{0\}$ still by $\sigma_j^{(1,3)} = 0$, one obtains the sequence $\{\sigma_j^{(1,3)}\}$ given above. Now, 0 is isolated, i.e. 0 is *preceded* by 1. This situation just occurs for $p = a_1$, $p = a_2$, etc., i.e. for the principal convergents for which $p = a_n$. Taking in $\{\sigma_j^{(1,3)}\}$ again $\{10\}$ as a new "*word*" $W_{2,1}$ (represented by $\sigma_j^{(2,1)} = 1$) which corresponds to $\{1000\}$ for the original spins and the "*word*" $\{1\} = W_1$ corresponding to $\{100\}$ for the original spins (now represented by $\sigma_j^{(2,1)} = 0$), we obtain the sequence $\{\sigma_j^{(2,1)}\}$ shown above. Thus at the stage $n = 2$, $p = 1$, we just obtain the two words $W_1 = \{100\}$ and $W_{2,1} = \{1000\}$ as suggested above. Since in the sequence $\{\sigma_j^{(2,1)}\}$ again 1 is isolated, we can proceed analogous as for $\{\sigma_i^{(1,1)} \equiv \sigma_i^{(0)}\}$. If ζ is an irrational number, this procedure never stops, i.e. the ground state can be renormalized arbitrary often.

It is also instructive to use for ζ the golden ratio $\tau = (\sqrt{5} - 1)/2$ which yields:

$$a_0 = 0, \quad a_i = 1 \quad \text{for all } i \geq 1.$$

Because $a_i = 1$ for all $i \geq 1$, no intermediate convergents exist. The principal convergents are given by the ratio

$$\frac{F_n}{F_{n+1}}$$

of subsequent Fibonacci–numbers F_n where F_n obeys the recursion relations (cf.(17)):

$$F_{n+1} = F_n + F_{n-1}$$

with the initial values $F_0 = 0$, $F_1 = 1$. One can prove that (see e.g. Hardy and Wright (1960), Schroeder (1990))

$$\zeta F_{n+1} - F_n = (-1)^n \tau^{n+1} \tag{21}$$

Using (21) and the identification $r_{n-1} \cong F_{n-1}$, $s_{n-1} \cong F_n$, $r_{n,p} \cong F_n$ and $s_{n,p} \cong F_{n+1}$ we obtain from (19b)

$$\zeta_{n,p} = \zeta = \tau$$

for *all* n and $p = 1$ (since $a_i = 1$ for all $i \geq 1$). This means that the golden number is a *fixed point* of the renormalization equation (19b) and that the sequences $\{\sigma_j^{(n,p)}\}$ are identical up to a phase shift given by $\varphi_{n,p}$. This property has been intensively used to describe physical properties of one–dimensional quasicrystals generated by Fibonacci–sequence $\{\sigma_i^{(0)}(\zeta)\}$ (see e.g. Kohmoto et al. (1983), Ostlund et al. (1983), Schilling (1984).

Finally we mention another interpretation of the "*words*" W_{n-1} and $W_{n,p}$. If we use $\frac{r_{n-1}}{s_{n-1}}$ and $\frac{r_{n,p}}{s_{n,p}}$ for ζ we obtain a *commensurate* ground state with a *periodic* sequence $\{\sigma_i^{(0)}\}$. Due to (14a) it is $\sigma_{i+s}^{(0)} \equiv \sigma_i^{(0)}$ with $s = s_{n-1}$ and $s = s_{n,p}$, respectively. Therefore $\{\sigma_i^{(0)}\}$ is a periodic arrangement of "*unit cells*" $\{\sigma_{i_0+1}, \sigma_{i_0+2}, ..., \sigma_{i_0+s}\}$, $i_0 = 0,...,s-1$. For an appropriate choice of i_0 the unit cell with $s = s_{n-1}$ is identical to the "*word*" W_{n-1} and choosing $s = s_{n,p}$ one obtains $W_{n,p}$.

4.2 RENORMALIZATION OF THE SPIN–HAMILTONIAN

In this section we discuss the renormalizability of the Ising Hamiltonian (12a). In section 4.1 we have shown that the ground state of (4a) and therefore also of (12a) can be represented by the two "*words*" W_{n-1} and $W_{n,p}$. Based on this knowledge, we make the following important *assumption*: there exists a hierarchy of temperatures $T^{(n,p)}$, $n \geq 1$ and $1 \leq p \leq a_n$ such that for $T = O(T^{(n,p)})$ the thermodynamic accessible spin configurations $\{\sigma_i\}$ are given by all arbitrary sequences of both "*words*" W_{n-1} and $W_{n,p}$. Note that the ground state of (12a) is contained within this set of sequences. Representing W_{n-1} and $W_{n,p}$ by the renormalized pseudo spin $\sigma^{(n,p)} = 0$ and $\sigma^{(n,p)} = 1$, respectively, we have to determine the thermodynamical weight of each sequence of words. This can be done as follows (see Vallet et al. (1988), chapter 5). The weight of each $\{\sigma_j^{(n,p)}\}$ can be represented by a *renormalized* spin Hamiltonian $\mathcal{H}^{(n,p)}$. $\mathcal{H}^{(n,p)}$ can always be written as

$$\mathcal{H}^{(n,p)} = \sum_i [J_0^{(n,p)} + J_1^{(n,p)} \sigma_i^{(n,p)} + J_2^{(n,p)} \sigma_i^{(n,p)} \sigma_{i+1}^{(n,p)} + ...] \quad (22)$$

where the dotes describe r–nearest neighbor two–spin interactions with $r > 1$ as well as m–spin interactions between m spins with $m \geq 3$. We anticipate, that for n large enough the higher order interactions can be neglected. The coupling constants $J_\nu^{(n,p)}$ for $\nu = 0,1,2$ are determined by comparison of the *exact* ground state energy of the original Ising–Hamiltonian (12a) with the corresponding approximate energy of the renormalized Hamiltonian (22) for the three *commensurate* spin configurations:

$$...W_{n-1}\ W_{n-1}\ W_{n-1}\ W_{n-1}...$$

$$\ldots W_{n,p} \; W_{n-1} \; W_{n,p} \; W_{n-1} \; W_{n,p} \ldots \qquad (23)$$

$$\ldots W_{n,p} \; W_{n,p} \; W_{n,p} \; W_{n,p} \; W_{n,p} \ldots \quad .$$

Representing W_{n-1} and $W_{n,p}$, respectively, by $\sigma_j^{(n,p)} = 0$ and $\sigma_j^{(n,p)} = 1$, the energies $\mathcal{H}^{(n,p)}$ are easily calculated for the three configurations (23). On the other hand W_{n-1} and $W_{n,p}$ represent the "unit cells" in terms of the original spins. The energy of a commensurate ground state with $\zeta = r/s$ can exactly be calculated (Aubry 1983a). Comparing both results yields for the renormalized nearest neighbor coupling constant:

$$J^{(n,p)} \cong 2\pi k A \, \eta^{s_{n,p+1}} \; ; \quad n \geq 1, \; 1 \leq p \leq a_n \qquad (24)$$

where for $p = a_n$, $s_{n,a_n+1} \equiv s_{n+1,1}$ must be used. We have skipped the subcript 2 for $J^{(n,p)}$. $J_0^{(n,p)}$ and $J_1^{(n,p)}$ are not given here, since they are not important. $J_0^{(n,p)}$ is just an additive constant and $J_1^{(n,p)}$ can be included into the "magnetic field" (Lagrange multiplier) which must be included in order to fix ζ. $s_{n,p+1}$ increases linearly with p (cf. (17a)) and in the average exponentially with n. Therefore the coupling constants $J^{(n,p)}$ decreases very fast with n which yields the hierarchy of well-separated temperatures $T^{(n,p)} = J^{(n,p)}/k_B$. The thermodynamical weight for $T = O(T^{(n,p)})$ is given by:

$$\rho^{(n,p)} = \frac{1}{Z^{(n,p)}} \exp[-\beta(\mathcal{H}^{(n,p)} - h \sum_j \sigma_j^{(n,p)})] \qquad (25a)$$

with the partition function

$$Z^{(n,p)} = \operatorname{tr}\{\exp[-\beta(\mathcal{H}^{(n,p)} - h \sum_j \sigma_j^{(n,p)})]\} \qquad (25b)$$

and the renormalized, nearest neighbor Hamiltonian

$$\mathcal{H}^{(n,p)} = J^{(n,p)} \sum_j \sigma_j^{(n,p)} \sigma_{j+1}^{(n,p)}, \quad \sigma_j^{(n,p)} = 0,1 \; . \qquad (25c)$$

From (5), (6) and (11) it follows that

$$\zeta = \frac{1}{N} \sum_{i=1}^{N} \sigma_i \quad , \qquad (26)$$

where N is the total number of K–ions. For given n and p, fixed ζ implies a fixed renormalized incommensurability ratio $\zeta_{n,p}$ (19b). Therefore for $T = O(T^{(n,p)})$ the "magnet field" h in (25a) has to be determined such that

$$\zeta_{n,p} = \frac{1}{N_0} \sum_{j=1}^{N_0} \langle \sigma_j^{(n,p)} \rangle \quad , \tag{27}$$

where N_0 is the number of "*words*".

5. Results

In this section we will apply the renormalization method in order to calculate the specific heat c(T) and the static structure factor S(q) for *fixed* ζ.

5.1 SPECIFIC HEAT

The specific heat per K–ion for fixed ζ is given by:

$$c(T) = -T \frac{\partial^2 f(T,\zeta)}{\partial T^2} \tag{28}$$

where the free energy per K–ion follows from

$$f(T,\zeta) = -\lim_{N \to \infty} \frac{1}{N} k_B T \ln Z(T,\zeta,N) \tag{29}$$

and

$$Z(T,\zeta,N) = \int d^N p \int d^N u \, \exp\{-\beta [\sum_i \frac{p_i^2}{2M} + V(\{u_i\})]\}. \tag{30}$$

The u–integration in (30) is restricted such that ζ is fixed. Since the configuration space can be decomposed into cells C_σ (see figure 7) it is

$$\int d^N u \, \exp\{-\beta V(\{u_i\})\} = \sum_{\{\sigma_i\}}{}' \int_{C_\sigma} d^N u \, \exp\{-\beta V(\{u_i\})\} \tag{31}$$

where the sum is restricted such that (26) holds. Let

$$u_i = u_i(\{\sigma_j\}) + \epsilon_i \quad , \tag{32}$$

then $V(\{u_i\})$ can be expanded:

$$V(\{u_i\}) = V(\{u_i(\{\sigma_j\})\}) + \frac{1}{2}\sum_{i,j} \frac{\partial^2 V}{\partial u_i \partial u_j}(\{u_i(\{\sigma_j\})\})\, \epsilon_i \epsilon_j. \tag{33}$$

With use of (12a), (31) and (33) we get from (30):

$$Z(T,\zeta,N) = \sum_{\{\sigma_i\}}{}' \exp\{-\beta \mathcal{H}(\{\sigma_i\})\} \int d^N p \int_{C'_\sigma} d^N \epsilon \exp\left\{-\beta \left[\sum_i \frac{p_i^2}{2m} + \frac{1}{2}\sum_{i,j} M_{ij}\epsilon_i \epsilon_j\right]\right\}$$

with:

$$M_{ij} = \frac{\partial^2 V}{\partial u_i \partial u_j}(\{u_i(\{\sigma_j\})\})$$

and C'_σ is the transformed configuration cell C_σ. For temperatures T such that $k_B T \ll E_g$ (the phonon gap), it is

$$\int_{C'_\sigma} d^N\epsilon \exp\{\ldots\} \cong \int_{\mathbb{R}_N} d^N\epsilon \exp\{\ldots\}$$

such that we obtain

$$Z(T,\zeta,N) \cong Z_{\text{spin}}(T,\zeta,N)\, Z_{\text{phonon}}(T,N) \tag{34a}$$

with

$$Z_{\text{spin}}(T,\zeta,N) = \sum_{\{\sigma_i\}}{}' \exp\{-\beta \mathcal{H}(\{\sigma_i\})\}$$

$$Z_{\text{phonon}}(T,N) = \int_{\mathbb{R}_N} d^N p \int_{\mathbb{R}_N} d^N\epsilon \exp\{-\beta[\sum_i \frac{p_i^2}{2m} + \frac{1}{2}\sum_{i,j} M_{ij}\,\epsilon_i\,\epsilon_j]\} \tag{34b}$$

and Z_{phonon} is the contribution of the phonons, which of course can exactly be calculated. Since for $k_B T \ll E_g$ the phonons can be disregarded, this has not to be done. Substituting (34) into (29) we finally get:

$$f(T,\zeta) \cong -\lim_{N\to\infty} \frac{1}{N} k_B T \ln Z_{\text{spin}}(T,\zeta,N) \quad, \tag{35}$$

i.e. the free energy of model (4) is approximated by the free energy of the Ising Hamiltonian (12a).

Now we are in a position where we can take advantage of the renormalization method, since this method provides an approximate scheme for the calculation of $Z_{\text{spin}}(T,\zeta,N)$. In section 4.2 we have found that for $T = O(T^{(n,p)})$ the thermodynamics is given by (25) and (27). Thus the problem is reduced to the calculation of the specific heat for a nearest-neighbor Ising model with fixed magnetization $<\sigma_j^{(n,p)}> = \zeta_{n,p}$. This is an easy exercise, now. The result for $c^{(n,p)}(T,\zeta_{n,p})$ exhibits a maximum (Schottky anomaly) of height $c_0^{(n,p)}$ at a temperature $T_0^{(n,p)}$. Note that $c^{(n,p)}$ is the specific heat per renormalized spin. $c_0^{(n,p)}$ and $T_0^{(n,p)}$ are presented in figure 10. As expected, $c_0^{(n,p)}$ converges to zero when $<\sigma_j^{(n,p)}> = \zeta_{n,p}$ approaches 0 or 1, because in that case the number of available states converges to zero, thereby reducing the entropy. For n large, numerator and denominator in (19b) are small quantities, since ζ is already well approximated by the rational approximate r_{n-1}/s_{n-1} and $r_{n,p}/s_{n,p}$. Then a small change of ζ may change $\zeta_{n,p}$ from about 0.5 to about 0 or 1 which implies a huge change in $c_0^{(n,p)}$. Thus, our renormalization method based on the continued fraction expansion predicts a *high sensitivity* of the specific heat on a small change of the incommensurability ratio $\zeta = \rho^{-1}$ and therefore on a small change of the K-ion concentration ρ. We will come back to this point below.

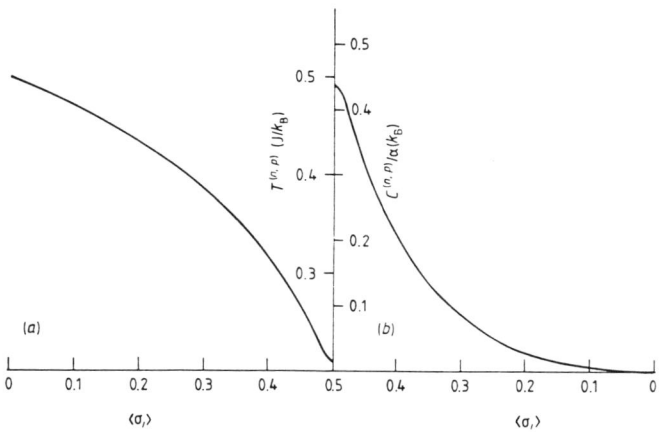

Figure 10 $T_0^{(n,p)}$ in units of $J^{(n,p)}/k_B$ and $c_0^{(n,p)}$ in units of k_B as function of $<\sigma_i^{(n,p)}>$ denoted just by $<\sigma_i>$. Both functions are symmetric with respect to $<\sigma_i^{(n,p)}> = \frac{1}{2}$.

The specific heat c(T) can approximately be presented as a superposition of the Schottky anomalies described by $c^{(n,p)}(T)$:

$$c(T) \cong \sum_{n=1}^{\infty} \sum_{p=1}^{a_n} \alpha_{n,p} \, c^{(n,p)}(T) \qquad (36a)$$

where

$$\alpha_{n,p} = [s_{n-1} + (s_{n,p} - s_{n-1}) \, \zeta_{n,p}]^{-1} \qquad (36b)$$

accounts for the correct weight, since c(T) is given per original spin and $c^{(n,p)}(T, \zeta_{n,p})$ per renormalized spin.

To check the quality of this result we have compared it with an exact numerical calculation of c(T) for $N = 22$ and $\zeta = 3/11 = 1/[3+1/[1+1/2]]$. Both results are shown in figure 11.

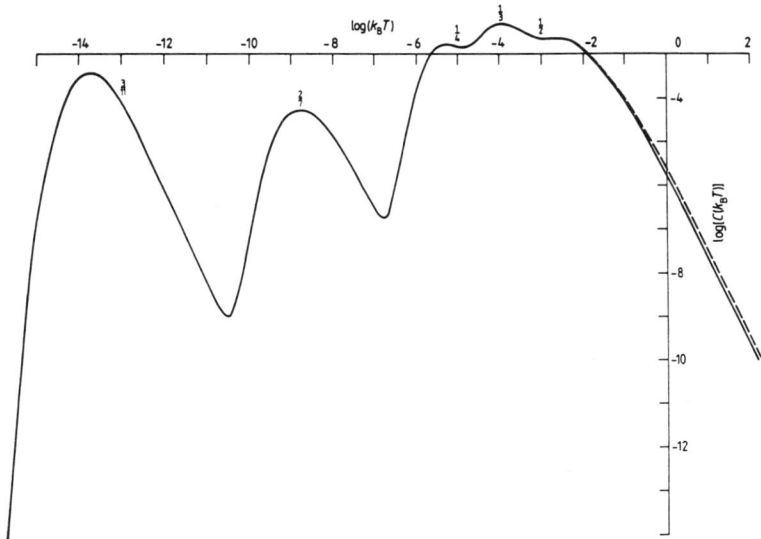

Figure 11: c(T) obtained from (36) (dashed line) and from exact numerical calculation for the Ising hamiltonian (12a) with N=22 and periodic boundary conditions for $\zeta = 3/11$ (solid line). The fractions on top of the maxima give the subsequent rational approximates of ζ.

For low temperatures one cannot distinguish between both results. Note, there are *no* fit parameters. The result in connection with the continued fraction for ζ reveals the following interesting property: the specific heat exhibits a series of *principal* Schottky anomalies which are labelled by n. The principal Schottky anomalies split into *intermediate* Schottky anomalies labelled by $1 \leq p < a_n$ for fixed n. If the continued fraction is finite, i.e. the remainder $\zeta_i = 0$ for $i = n$, then the n-the principal anomaly, only splits into $a_n - 1$ intermediate ones (see Vallet et al. (1988)). For the example in figure 11 it is $a_1 = 3$, $a_2 = 1$, $a_3 = 2$ and $\zeta_3 = 0$. Therefore there are 3 principal anomalies. The first one splits into $a_1 = 3$, the second one into $a_2 = 1$ and the last one into $a_1 - 1 = 1$ intermediate anomalies (cf. figure 11).

Since figure 11 demonstrates that our renormalization method yields good results at least at lower temperatures, we compare our theoretical result with the experimental one by v. Löhneysen et al. (1981), for K–hollandite (figure 12)

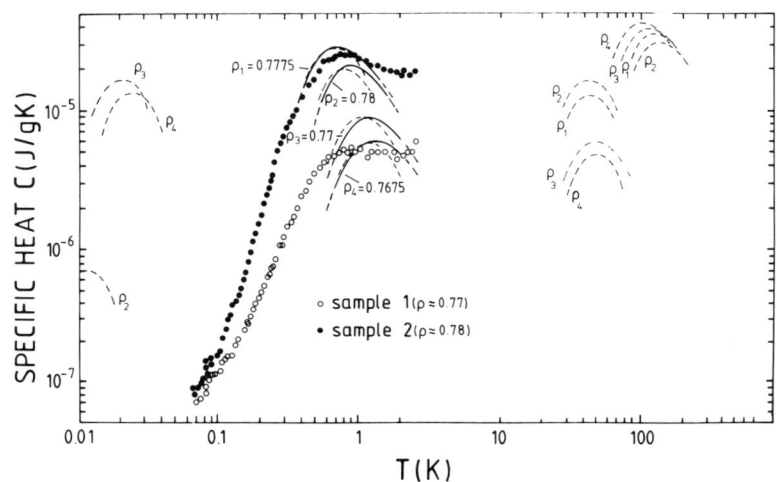

Figure 12. Experimental results for c(T) for K–hollandite with $\rho \approx \rho_3 = 0.7700$ and $\rho \approx \rho_2 = 0.7800$. Theoretical result (dashed lines) from (36) for $\rho_1 = 0.7775$, $\rho_2 = 0.7800$, $\rho_3 = 0.7700$ and $\rho_4 = 0.7675$. Solid lines denote a theoretical result obtained by Pietronero et al. (1981) for the same ρ_i–values. The values for c and k were taken from Pietronero et al. (1981). The exact values for $T_0^{(n,p)}$ and $c_0^{(n,p)}$ were taken from Vallet et al. (1986), Table I. Note the existence of further anomalies at lower and higher temperatures.

First of all two observations can be made from the experimental data:
(i) The results for $\rho \approx 0.77$ and $\rho \approx 0.78$ exhibit a more or less pronounced Schottky anomaly at about 1K.
(ii) the maximum specific heat differs by about a factor of four, though ρ varies about 1.5 % only.

For our theoretical result and also for that of Pietronero et al. (1981), one fit parameter has been used, which is a *common* prefactor to arrange the absolute value for c(T). Thus the relative values of c(T) for the different ρ-values are *not* fitted. The theoretical result by Pietronero et al. (1981) was obtained by a method which has a vague relationship to part of our method. But their approach was restricted to ρ-values with $\rho \approx 3/4$ and the importance of the number–theoretical properties and the renormalizability was not noted. Both theoretical results obtained for the four values ρ_1, ρ_2, ρ_3 and ρ_4 describe the experimental data rather well. But we also mention that both theoretical results yields a specific heat which is about fifty times larger than the experimental result. This has been attributed to the existence of blocking impurities (see v. Löhneysen et al. (1981). Our prediction of a high sensitivity of c(T) on a small change of ρ is consistent with these experimental data. In particular our theory yields about a factor of four for the difference between the maximum specific heat for $\rho = 0.77$ and $\rho = 0.78$, which agrees with the experimental finding. But, since blocking impurities may influence the specific heat, it might be that the observed difference in c(T) for both samples is related to different impurity concentrations. Anyway, our renormalization method for incommensurate structures (d=1) predicts such a high sensitivity and it might be a challenge to search for it by an appropriate experiment.

Besides the Schottky anomalies for $\rho \approx 0.76 - 0.78$ at about 1 K, our method yields anomalies at about 0.035–0.12 K, 55–67 K, 137–145 K and 650–670 K. The anomalies above 50 K may be not observable due to the phonon contribution and the anomalies below 1 K might be modified by quantum effects. Therefore a small temperature window exists only, where experiments are reasonable to test our predictions.

5.2 STATIC STRUCTURE FACTOR

The static structure factor defined by:

$$S(q) = \frac{1}{nN} \sum_{j,k} \langle e^{iq(u_j - u_k)} \rangle \quad (37)$$

is a physical quantity characterizing a structure more directly. Here N is the number of particles and n its density. As before $\langle \ \rangle$ denotes the canonical average. For details of the following calculations the reader is refered to Schilling et al. (1987). With (32) and (33) we obtain from (37):

$$S(q) \cong \frac{1}{nN} \sum_{j,k} \exp[-\frac{1}{2} q^2 \langle (\epsilon_j - \epsilon_k)^2 \rangle_{ph}] \times$$

$$\langle \exp[iq(u_j(\{\sigma_i\}) - u_k(\{\sigma_i\}))] \rangle_{conf.} \quad (38)$$

where $\langle \ \rangle_{ph}$ denotes the canonical averaging over the *phonon degrees of freedom* described by the phonon Hamiltonian

$$\mathcal{H}_{\text{phonon}} = \sum_{j=1}^{N} \frac{p_j^2}{2M} + \frac{1}{2} \sum_{j,k} M_{jk}\, \epsilon_j\, \epsilon_k \tag{39a}$$

and $< \ >_{\text{conf}}$ the *configurational average* defined by:

$$<f(\{\sigma_i\})>_{\text{conf}} = \frac{1}{Z_{\text{spin}}} \sum_{\{\sigma_i\}}{}' f(\{\sigma_i\}) \exp[-\beta \mathcal{H}(\{\sigma_i\})] \tag{39b}$$

with Z_{spin} from (34). Approximating the phonon part by a Debye–Waller factor:

$$\exp[-\tfrac{1}{2} q^2 <(\epsilon_j - \epsilon_k)^2>_{\text{ph}}] \simeq \exp(-2\,W(q)) \tag{40a}$$

where a quantum–mechanical calculation yields

$$2W(q) = -\frac{q^2}{2} \frac{k_B T}{C} \sum_{n=-\infty}^{\infty} [\lambda_n + (\tfrac{1}{2}\lambda_n)^2]^{-1/2} \tag{40b}$$

with:

$$\lambda_n = \frac{(1-\eta)^2}{\eta} + \frac{M}{C}\left(\frac{2\pi}{\hbar\beta} n\right)^2 . \tag{40c}$$

Using the approximation (40a) it follows

$$S(q) \simeq \exp(-2W(q)) \frac{1}{nN} \sum_{j,k} <\exp[iq(u_j(\{\sigma_i\})-u_k(\{\sigma_i\}))]>_{\text{conf}} . \tag{41}$$

Now, the configurational average in (41) can be calculated by means of our renormalization technique. For temperatures $T = O(T^{(n,p)})$ the relevant configurations are sequences of both *"words"* W_{n-1} and $W_{n,p}$. Since both words correspond to unit cells of commensurate structures with $\zeta = \frac{r_{n-1}}{s_{n-1}}$ and $\zeta = \frac{r_{n,p}}{s_{n,p}}$ the position $u_j(\{\sigma_i\})$ can be obtained in the following way (see figure 13 for an illustration). Since for $\zeta = \frac{r}{s}$ the unit cell has length $c\,r$ (in the original length unit) and contains s atoms, the position u_j of the j–th K–ion can be represented as:

$$u_j(\{\sigma_i\}) = u^{(0)} + L_m(\{\sigma_i^{(n,p)}\}) + u^{(\nu)} \tag{42a}$$

where $u^{(0)}$ is the position of the left end of the first unit cell, $u^{(\nu)}$ the position of the j–th atom within the (m+1)–th unit cell and

$$L_m(\{\sigma_i\}) = c \left[r_{n-1} m + (r_{n,p} - r_{n-1}) \sum_{i=1}^{m} \sigma_i^{(n,p)} \right] \quad (42b)$$

is the total length of the m unit cells between the zeroth and the (m+1)–th unit cell. The position $u^{(\nu)}$ within a unit cell of a commensurate structure with $\zeta = r/s$ depend on r/s and on the phase φ. From Aubry (1983a) one obtains

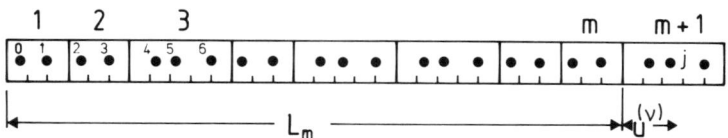

Figure 13: Part of an arbitrary sequence of the two unit cells corresponding to the "*words*" W_{n-1} and $W_{n,p}$. The small vertical lines in each unit cell indicate the period of the skeleton potential. Note that $r_{n-1} = 3$, $s_{n-1} = 2$, $r_{n,p} = 5$ and $s_{n,p} = 3$.

$$u^{(\nu)}(\zeta=r/s,\varphi) = \frac{r}{s}\nu + \varphi$$

$$- \frac{1-\eta}{(1+\eta)(1-\eta^s)} \sum_{\mu=-s+1}^{s} \eta^{|\mu|} \{\frac{r}{s}\nu + \varphi - \text{int}[r(\nu+\mu)/s + \varphi]\}, \quad (43)$$

where $\nu = 1,2,..,s$. In case of $\sigma_i^{(n,p)} = 0$ it is $r = r_{n-1}$, $s = s_{n-1}$, and $r = r_{n,p}$, $s = s_{n,p}$ if $\sigma_i^{(n,p)} = 1$. It can be shown that a reasonable choice for the phase is $\varphi=0$. Here we also note, that the use of (43) (which are the positions in a *commensurate* structure) in (42a) is an approximation already, because generally both unit cells are not arranged periodically. However, the larger the unit cell, the better is this approximation, since the influence of the adjacent cells becomes less important due to the exponentially decaying coupling constant in the Ising Hamiltonian (12a). Substituting (42) into (41) yields finally:

$$S(q) = \exp(-2W(q)) \frac{1}{nN} \sum_{m,m'=1}^{N_0} [|F_{n-1}|^2 < Q_{m,m'} >_{\text{conf}}$$
$$+ |F_{n,p} - F_{n-1}|^2 < \sigma_m^{(n,p)} \sigma_{m'}^{(n,p)} >_{\text{conf}} \quad (44)$$
$$+ F_{n-1}^* (F_{n,p} - F_{n-1}) < \sigma_m^{(n,p)} Q_{mm'} >_{\text{conf}} + \text{c.c}] \quad .$$

$Q_{mm'}$ for $m > m'$ is defined by

$$Q_{mm'} = \exp\{iqc[r_{n-1}(m-m') + (r_{n,p} - r_{n-1})\sum_{i=m'+1}^{m} \sigma_i^{(n,p)}]\} \tag{45a}$$

and for $m < m'$ by $Q_{mm'} = Q^*_{m'm}$. For $m = m'$ it is $Q_{mm} = 1$. The exponent in (45a) is just iq times the total length of the unit cells between the m'-th and (m+1)-th unit cell. The complex factors

$$F_{n-1}(q) = \sum_{\nu=1}^{s_{n-1}} \exp\left[iq\, u^{(\nu)}(r_{n-1}/s_{n-1}, \varphi=0)\right] \tag{45b}$$

$$F_{n,p}(q) = \sum_{\nu=1}^{s_{n-1}} \exp\left[iq\, u^{(\nu)}(r_{n,p}/s_{n,p}, \varphi=0)\right]$$

are the scattering amplitudes (form factors) of both unit cells. The configuration averages in (44) are easily calculated for $T = O(T^{(n,p)})$ by means of the renormalized Ising Hamiltonian $\mathcal{H}^{(n,p)}$. Hence all the quantities entering $S(q)$, (44), are known analytically and even the sum over m and m' can be performed.

Nevertheless the obtained analytical result for $S(q)$ is rather involved and demands for a numerical evaluation. Figure 14 shows the result for K–hollandite with $\rho = 0.77$ and for different temperatures. $S(q)$, or more precise, its diffuse part was measured at room temperature by Beyeler (1976) and is depicted in figure (14a) by the solid line. For an incommensurate structure the reflection peaks (at T = 0K) occur at

$$q_{m,n} = nq_0 + mq_1 \quad ; n,m \in \mathbb{Z} \tag{46a}$$

with

$$q_0 = 2\pi/\ell \, , \quad q_1 = 2\pi/c \tag{46b}$$

and can be labelled by (n,m) (cf. Fig. 14). The experimental result exhibits two main reflections (1,0) and (2,0) and two satellites (−1,2) and (1,1). Beyeler (1976) has suggested a description of his results by a random arrangement of *four* cells containing three, four, five and six atoms. The probability p_i of cell i=1,2,3 and 4 provide three fit parameters. Due to some symmetry, the number of the independent atomic positions within the cells can be reduced to four, which makes in total seven fit parameters. The best fit result is shown in figure 14a as dotted line. There is a reasonable agreement between both results. Our theoretical result involves only η (or k) as fit parameter. Since for given T our renormalization method determines the size and the number of atoms of both unit cells no further fit parameter occurs. Using for C again the value from Pietronero et al. (1981)

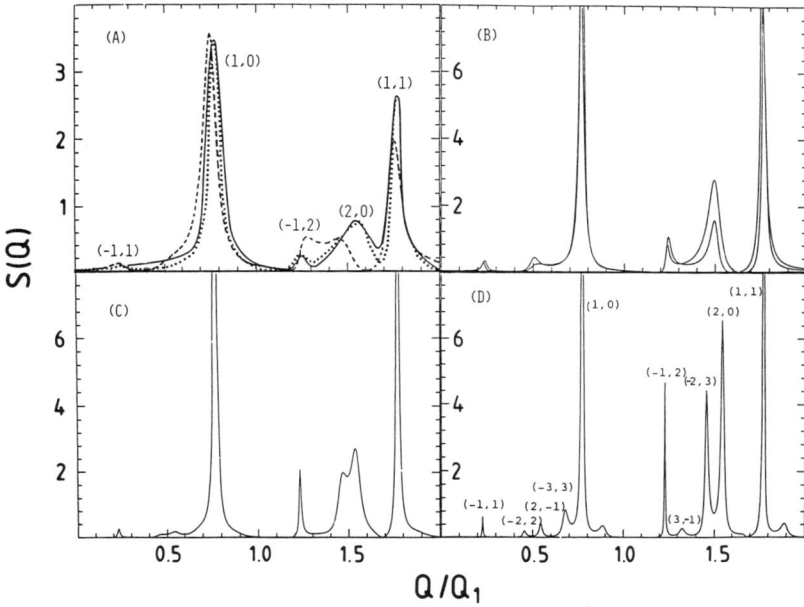

Figure 14: The temperature dependence of S(q) (in arbitrary units) for $\rho = 0.77$. (a) T = 300K: experimental (solid line) and numerical result of Beyeler (1976) (dotted line); our result from the renormalization method is presented by a dashed line. (b) T = 100K: for the upper and lower curve the words $W_0, W_{1,2}$ and $W_0, W_{1,3}$ were used, respectively. (c) T = 10K. (d) = 1k.

and $\eta = 0.34$ (which yields the best fit at T = 300K), we find that both words of our renormalization technique are $W_{1,1}$ and $W_{1,2}$. From the continued fraction expansion of $\zeta = \rho^{-1} \cong 1.2987$ ($\rho = 0.77$) one easily obtains that $r_{1,1} = 2$, $s_{1,1} = 1$ and $r_{1,2} = 3$, $s_{1,2} = 2$. Therefore at room temperature and for $\rho = 0.77$ both unit cells have length 2c and 3c with one and two K–ions, respectively. Our theoretical result is shown as dashed curve in figure 14a. Although our renormalization method reproduces the correct number of reflection peaks which are experimentally observed as well as the additional satellite $(-1,1)$ also existing in the numerical result by Beyeler (1976), the agreement of the intensities and the position of the peaks is not satisfactory. This is not surprising since we expect our method to be good at low temperatures only. A transfer–matrix method in combination with a high–temperature expansion by Geisel (1979) yields in leading order a result which is even worse.

Decreasing the temperature, more and more satellites occur as demonstrated in figure 14a–d. T = 100K is just between both temperatures $T^{(1,2)}$ and $T^{(1,3)}$. Therefore the "*word*" W_0 was used in combination with $W_{1,1}$ and $W_{1,3}$. Both results mainly differ in the intensity (figure 14b).

In figure 15 we present S(q) at T = 1.2K for four different ρ–values. First of all, S(q) for the four ρ–values mainly differ in their intensities. Figure 15a and 15b resemble each other and the same is true for figures 15c and 15d. But both pairs of figures can easily be distinguished from each other due to rather different intensities. The reason for this observation is that ρ_1 and ρ_2 are closer to $r_2/s_2 = 7/9 \cong 0.7777$ and ρ_3 and ρ_4 closer to $r_{3,1}/s_{3,1} \equiv r_3/s_3 = 10/13 \cong 0.7692$. This result again demonstrates the high sensitivity of the measurable quantity S(q) on a small change of ρ, as already found for the specific heat.

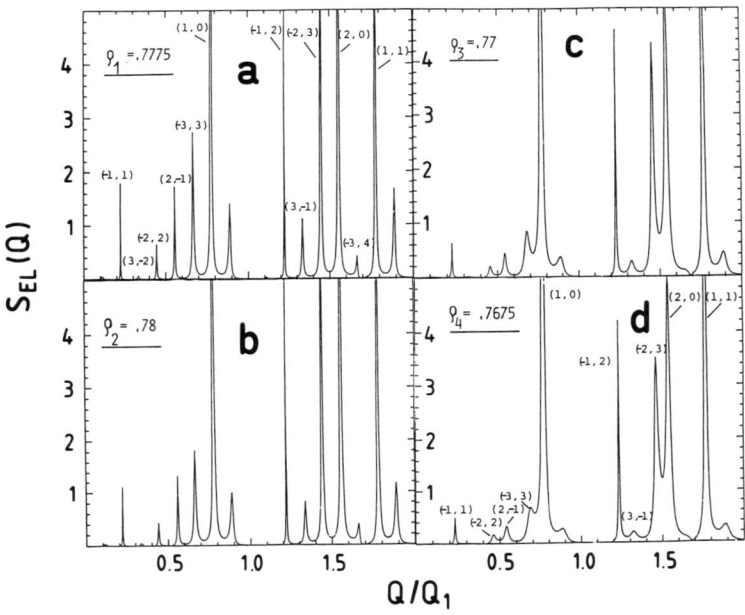

Figure 15: S(q) at T= 1.2K for $\eta = 0.34$ and (a) $\rho_1 = 0.7775$; (b) $\rho_2 = 0.78$; (c) $\rho_3 = 0.77$; (d) $\rho_4 = 0.7675$

6. Summary and Conclusions

In this contribution we have studied the thermal behavior of quasi one–dimensional incommensurate structures. The main emphasis was on the low–temperature specific heat c(T) and the static structure factor S(q).

The theoretical investigations were performed for a Frenkel–Kontorova model with piecewise parabolic (external) potential and for fixed incommensurability ratio ζ. This model describes the nonanalytical regime where the ground state can carry a finite concentration of defects. The metastable defect–configurations are (under some conditions) in a one–to–one correspondence with a sequence $\{\sigma_i\}$ of pseudo spins $\sigma_i = 0,1,2,...,r$ and their energies follow from an Ising–spin Hamiltonian.

Based on the renormalizability of the ground state spin configuration we have found a hierarchy of renormalized nearest neighbor Ising Hamiltonians $\mathcal{H}^{(n,p)}$ which is uniquely determined by the continued fraction expansion of ζ. For $T = O(T^{(n,p)})$, where $k_B T^{(n,p)} = J^{(n,p)}$ is the coupling constant of $\mathcal{H}^{(n,p)}$, the thermodynamical properties of the Frenkel–Kontorova model are well approximated by those of $\mathcal{H}^{(n,p)}$, but with a fixed magnetization which is equal to the renormalized incommensurability ratio ζ.

This method has been applied to calculate $c(T)$ and $S(q)$. $c(T)$ exhibits a hierarchy of Schottky anomalies related to the hierarchy of renormalized coupling constants $J^{(n,p)}$. The comparison with an exact numerical calculation of $c(T)$ for 22 particles and $\zeta = 3/11$ exhibits that the renormalization method yields very good results at low temperatures. Our theoretical results, similarly to those by Pietronero et al. (1981), explain the experimental data for K–hollandite by v.Löhneysen et al (1981) rather well, with exception of the absolute value. The position of the Schottky anomaly at about 1K also favours the discommensurations as elementary excitations and not the local rearrangements of the K–ions as suggested by v.Löhneysen et al. (1981) and Pietronero et al. (1981), since the latter one yield a Schottky anomaly at about 2K using the coupling constants from Pietronero et al. (1981).

The hierarchical behavior of $c(T)$ manifests itself for $S(q)$ in the appearance of more and more satellites with decreasing temperature.

Probably the most interesting result of our renormalization method is the prediction of a high sensitivity of physical quantities on a small change of ζ which corresponds to a small change of ρ, the concentration of the K–ions. Depending on ρ and on T a change of ρ by e.g. one percent may cause a change in $c(T)$ or $S(q)$ by several orders of magnitude. This prediction is in agreement with the experimental results by v.Löhneysen et al. (1981). But due to the existence of blocking impurities this agreement is not conclusive. It would be important to perform new experiments which may clarify this point.

Let us conclude with a physical interpetration of our results. This interpretation is illustrated in figure 16. Let us take $T = O(T^{(n,p)})$. Then the thermodynamically accessible configurations are sequences of W_{n-1} and $W_{n,p}$ (figure 16a). If we *increase* the temperature towards $T^{(n,p-1)}$ a new generation of excitations occur which lead to a dissociation of the larger "*word*" or unit cell $W_{n,p}$ into W_{n-1} and a rest $W_{n,p-1}$ (figure 16b).

Note, this is just the opposite direction of the "*word*"–formation described by (18). $W_{n,p-1}$ is still the larger cell. Increasing the temperature even more, the next generation of higher energy excitations becomes active, which leads to a breaking of $W_{n,p-1}$ into $W_{n-1} \equiv W_{n-1,a_{n-1}}$ and a rest which is W_{n-2} (figure 16c). Note that W_{n-1} is the larger word now. This stage, where the role of the words changes, is just at the principal convergents. Increasing temperature more, the larger "*word*" W_{n-1} will break into W_{n-2} and a rest which is $W_{n-1,a_{n-1}-1}$ and so on. The breaking of the words which occurs in a hierarchical manner as demonstrated in figure 16, can be interpreted as hierarchical "melting". Since we consider a one–dimensional model, there are no singularities in the thermodynamical quantities, like $c(T)$. But the hierarchy of Schottky anomalies are an indication of the "quasi–melting" of the one–dimensional incommensurate structure. The hierarchical "melting" manifests itself in S(q) as disappearance of more and more satellites due to the loss of the long range coherence, because the "rigid" unit cells become smaller and smaller and form a "liquid" phase of cells which loose more and more structure.

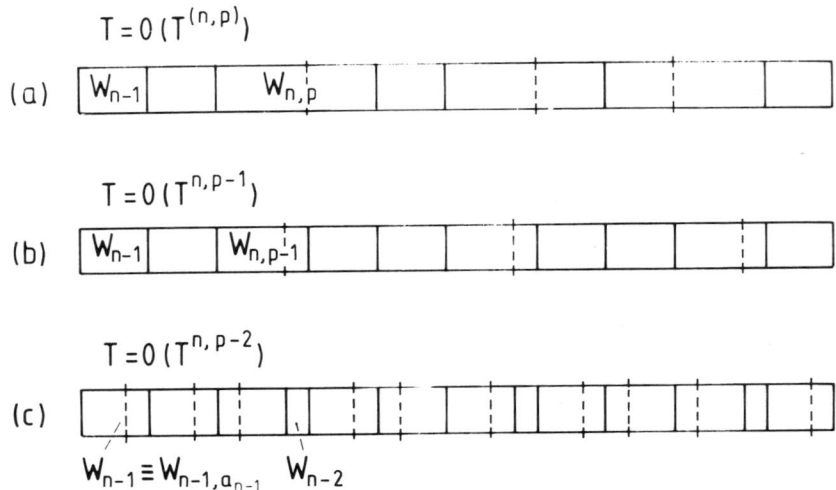

Figure 16: Physical interpretation of the hierarchical thermal behavior as a hierarchical melting. The dashed lines represent the positions where the "*words*" break.

References

Aubry, S. (1978) "The new concept of transition by breaking of analyticity in a crystallographic model" in A.R.Bishop and T.Schneider (eds.), Springer Series in Solid Sciences, Vol.8, Springer, pp. 264–268.

Aubry, S. (1981) "Many defect structures, stochasticity and incommensurability" in R. Balian (ed.), Physics of Defects, North–Holland Publishing pp. 431–451.

Aubry, S. (1983a) "Exact models with a complete devil's staircase", J.Phys.C16, pp. 2497–2508.

Aubry, S. (1983b) "Devil's staircase and order without periodicity in classical condensed matter", J.de Phys.(Paris) 44, pp. 147–161.

Aubry S. and Le Daeron P.Y. (1983) "The discrete Frenkel–Kontonova model and its extensions", Physica 8D, pp. 381–422

Beyeler, H.U. (1976) "Cationic short–range order in the hollandite $K_{1.54}Mg_{0.77}T_{7.23}O_{1.6}$: "Evidence for the importance of ion–ion interactions in superionic conductors", Phys.Rev.Lett.37, pp. 1557–1560.

Beyeler, H.U., Pietronero L. and Strässler S., (1980) "Configurational model for a one–dimensional ionic conductor",Phys.Rev.B22, pp. 2988–3000.

Geisel, T. (1979) "Modulation and incommensurability in a superionic conductor", Solid State Comm.32, pp. 739–743.

Hardy, G.H. and Wright, E.M. (1960) "An introduction to the theory of numbers" Clarendon, Oxford.

Hubbard, J. (1978) "Generalized Wigner lattices in one dimension and some applications to tetracyanoquinodimethane (TCNQ) salts", Phys.Rev.B17, pp. 494–505.

Janssen, T. and Janner, A. (1987) "Incommensurability in crystals", Adv.Phys.36, pp. 519–624.

Kohmoto, M., Kadanoff, L.P. and Tang, C. (1983) "Localization problem in one dimension: Mapping and escape", Phys.Rev.Lett.50, pp. 1870–1873.

v.Löhneysen, H., Schink, H.J., Arnold, W., Beyeler, H.U., Pietronaro, L. and Strässler, S. (1981) "Low–temperature specific–heat anomaly of a one–dimensional ionic conductor", Phys.Rev.Lett.46, pp. 1213–1216.

Ostlund, S., Pandit, R., Rand, D., Schellnhuber, H.J. and Siggia, E.D. (1983) "The 1–D Schrödinger equation with an almost periodic potential", Phys.Rev.Lett.50, pp. 1873–1876.

Pietronero, L. and Strässler, S. (1979) "Anomalous specific heat of a one—dimensional disordered solid", Phys.Rev.Lett.42, pp. 188–191.

Pietronero, L., Schneider, W.R. and Strässler, S. (1981) "Configurational excitations and low—temperature specific heat of the Frenkel—Kontorova model", Phys.Rev.B24, pp. 2187–2195.

Pokrovsky, V.L. and Uimin, G.V. (1978) "On the properties of monolayers of adsorbed atoms", J.Phys.C11, pp. 3535–3549.

Reichert, P. and Schilling, R. (1985) "Glasslike properties of a chain of particles with anharmonic and competing interactions", Phys.Rev.B32, pp. 5731–5745.

Rosshirt, E. (1988) "Röntgen — und Neutronenstreuuntersuchungen von Ordnungsvor — gängen in K—Hollandit im Temperaturbereich von 35K bis 900K", PhD—thesis, Ludwig— Maximilian—Universität, München.

Schilling, R. (1984) "One—dimensional Ising model in an incommensurate field", Phys.Rev.B30, pp. 5190–5194.

Schilling, R. and Aubry, S. (1987) "Static structure factor of one—dimensional non—analytic incommensurate structures", J.Phys.C20, pp. 4881–4889.

Schroeder, M.R. (1990) "Number theory in science and communication", Springer Verlag

Uhler, W. and Schilling, R. (1988) "Model for a glassy adsorbate: Two—level systems and specific heat", Phys.Rev.B37, pp. 5787–5805.

Vallet, F., Schilling, R. and Aubry, S. (1986) "Hierarchical low—temperature behavior of one—dimensional incommensurate structures", Europhys.Letters 2, pp. 815–822.

Vallet, F., Schilling, R. and Aubry, S. (1988) "Low—temperature excitations, specific heat and hierarchical melting of a one—dimensional incommensurate structure", J.Phys.C21, pp.67–105.

CHARGE DENSITY WAVES, PHASE SLIPS, AND INSTABILITIES

S.N. COPPERSMITH
AT&T Bell Laboratories
Murray Hill, NJ 07974
USA

ABSTRACT. Understanding the behavior of sliding charge density waves (CDWs) in the presence of an electric field is a challenging problem because one must understand how the competition between randomness and interactions affects the properties of a nonlinear dynamical system with many degrees of freedom. These lectures describe two aspects of this problem. The first concerns the dynamical generation of defects in CDWs when they are subjected to a uniform electric field. It is shown that amplitude defects, or phase slips, must always occur in the presence of a uniform nonzero electric field for a sample of infinite size. The defect density for real CDWs is estimated and it is shown that phase slips could be present in substantial numbers in even high quality samples. The experimental situation is then addressed, and it is seen that phase slips contribute in an important way to the dynamical response of the CDW in almost all samples. The applicability of this work to other systems such as Wigner crystals and flux lattices in type-II superconductors is discussed. The second concerns the dynamical selection of atypical metastable states when the CDW is subjected to repeated identical voltage pulses. The experimental manifestation of this selection is a synchronization of the CDW current response with the end of the driving pulse.

1. Introduction

Because systems with competing energy scales can display extremely slow relaxations, often experiments probe metastable states that are unrelated to the state that minimizes the free energy. The slow relaxations by which the system goes from these metastable states to the thermodynamic ground state can take years, and thus it makes sense to understand the nonequilibrium state the system reaches when it is subjected to strong driving. In these lectures I will describe differences between equilibrium properties and dynamical properties of a specific experimental system, the sliding charge density wave (CDW).

Some materials undergo a phase transition into a state with a CDW, in which the electron charge density is modulated with a periodicity which need not be simply related to the lattice constant of the material. Several CDW materials have nonlinear current-voltage characteristics, which are ohmic at very small drives, but which display a lower differential resistance for voltages above a reasonably well-defined (and small) threshold voltage. The nonlinear conduction is attributed to motion of the CDW, which contributes substantially to the conductivity at fields above the threshold voltage. Thus, the electrical response can be used to probe the CDW dynamics. A significant advantage to CDWs is the wealth of experiments which have contributed to our knowledge of these systems.[1]

The CDW order parameter is described by an amplitude and a phase; the charge density $\rho(r,t)$ is $\rho(r,t) = \bar{\rho} + \delta\rho(r,t)$, where the modulation of the charge density $\delta\rho(r,t)$ can be written

$$\delta\rho(r,t) = \rho_0(r,t)\cos(Q \cdot r + \phi(r,t)) . \qquad (1)$$

The period of the CDW modulation is $2\pi/Q$; The direction and amplitude of Q is determined by microscopic properties of the material. The amplitude of the order parameter is described by ρ_0 and the phase by $\phi(r,t)$.

In the absence of impurities, the CDW would slide whenever an electric field is applied. However, this behavior is not observed because all experimental samples have impurities. In the presence of impurities, it is energetically favorable for the CDW to distort, and both the amplitude $\rho_0(r)$ and phase $\phi(r)$ of the CDW depend on the position r. When an electric field F is applied to a CDW, the field must overcome this pinning energy before the CDW can move. Thus, these deformations caused by the impurities lead to a nonlinear current-voltage characteristic, where CDW motion appears to set in at a nonzero threshold field F_T.

This paper focusses on the question of how to describe the CDW dynamics in the presence of an applied electric field. The focus is on driven dynamics at zero temperature; thermally-induced relaxations in CDW's are slow enough that this is a very good approximation for the situations addressed. The first lecture discusses the behavior of the CDW in the behavior of a dc voltage. Because the threshold voltage for CDW motion is so low, indicating that the impurity potential is weak compared to the (appropriately normalized) CDW stiffness, one might expect that it is only necessary to consider low-energy long wavelength deformations in order to describe the behavior. However, here it is shown that this expectation is incorrect and one must explicitly include both the phase and amplitude degrees of freedom in order to describe real CDWs. This result shows that driven dynamics are fundamentally different from thermal equilibrium because the application of an electric field causes an instability of a defect-free state that does not arise in the absence of driving. The second lecture describes the response of a CDW to voltage pulses. Experimentally it is observed that the CDW velocity synchronizes with the *end* of the voltage pulse. It is shown that this synchronization arises because the CDW selects out a small subset of all possible configurations, and that this subset consists of states which are in some sense the least stable. Thus, these lectures provide two illustrations that the states that result from deterministic driving differ fundamentally from those that arise in thermal equilibrium.

2. Instability of the Phase Deformation Model

2.1. DEMONSTRATION THAT PHASE SLIPS MUST OCCUR

The CDW has both phase and amplitude degrees of freedom. Long wavelength phase deformations can cost arbitrarily low energy whereas amplitude variations must cost an energy of order the CDW gap energy (of order 1000 K). Since CDW threshold fields are usually less than a volt per centimeter, it is natural to assume that the amplitude degrees of freedom are unimportant and consider a model with phase degrees of freedom only.[2]

This section will focus on a discretized version of the equations of motion for a model including only the phase degrees of freedom of the CDW. These equations can be interpreted in terms of a simple mechanical analog consisting of a system of overdamped particles connected by springs that obey Hooke's law, each in a sinusoidal potential. In addition, a spatially uniform external force (which describes the effects of an applied electric field on the CDW) can be applied. The equations of motion are:

$$\dot{x}_j = k \sum_{\delta \text{ (nn)}} (x_{j+\delta} - x_j) - V_j \sin(x_j - \beta_j) + F \ . \tag{2}$$

The variables x_j describe the positions of the particles, the V_j and β_j describe the pinning potential, k is the spring constant, and F is an external force, which can be viewed as arising from tilting the corrugated surface on which the particles lie. The sum is over nearest neighbors. In one dimension this model is obtained from the continuum equations of motion for the CDW[3] by considering the case of a pinning potential consisting of δ-function impurities, and interpreting the x_j as the phase at the jth impurity.[4] The derivation of the model makes explicit use of the assumption that the phase at the j'th impurity is well defined, which follows if the amplitude degrees of freedom can be neglected.

Now consider the dynamical behavior of the model described by Eqs. (2) for a time-independent F. When the force F is zero, the velocity of each particle for the model described by Eqs. (2) is zero. The system has a number of metastable states that grows exponentially with its size. This fact is most easily seen in the limit $V \gg k$, where it is obvious that moving any given particle to the right or to the left by about 2π yields another metastable state. On the other hand, when $F \gg V$, by summing Eq. (2) over all the particles, one finds that the spatially averaged velocity v is nonzero.

As F is increased, the number of metastable states of Eq. (2) decreases. At a well-defined threshold field F_T there is a transition to a unique moving state, and the behavior near threshold is a dynamic critical phenomenon.[5] The steady state time-averaged velocity, which is the same for each particle, is strictly zero for $F < F_T$ and then rises as $(F - F_T)^\zeta$ for $F > F_T$, where ζ is a dimensionality-dependent critical exponent. A diverging correlation length can be defined. Other quantities with singular behavior at threshold can be found, and the exponents describing these singularities appear to obey scaling relations.

Here it is demonstrated that this picture of the nonlinear conduction, although elegant, has a fundamental flaw.[6] [7] It is shown that for an infinite system, Eq. (2) breaks down when $F > 0$ is constant in time. The argument that shows this is simple and may apply to other systems. In addition to demonstrating that the critical behavior predicted by Eqs. (2) cannot be observed *in principle* in a physical system, the number density of the defects that are generated in practice is estimated[8] and it is shown that it is quite likely that experimental samples are in a regime where the dynamics of the amplitude of the order parameter play an important role. The dynamics are significantly more complicated than the ones described by equations describing phase degrees of freedom only.

Eqs. (2) contain the implicit assumption that the separation between every nearest neighbor pair of particles is bounded, so that

$$|x_{j+\delta} - x_j| \leq S_{max} \qquad (3)$$

for some fixed S_{max} and every j and δ. If this condition is violated, the springs no longer obey Hooke's law and Eqs. (2) is no longer adequate to describe the system's dynamics. (If eq. (3) is not satisfied, the model described by eqs. (2) has unbounded energy density.) When one traces through the derivation of Eqs. (2) from the original CDW equations of motion, it turns out that the condition of bounded particle separations in Eq. (3) corresponds to the condition that the CDW amplitude $\rho_0(\mathbf{r})$ is nonzero everywhere. Here it is shown that eq. (3) is violated using a scaling argument as well as by examining a simpler model where exact calculations are straightforward.

To show $|x_{j+\delta} - x_j|$ is unbounded for the model described by Eqs. (2), examine the threshold state of an infinite system (the state when $F = F_T(\infty)$). Although there are regions that have different local pinning strengths, the configuration adjusts so that the spring forces hold back the regions with weaker pinning. Thus, just at the threshold no particles are moving.

Now examine the spring forces in the threshold state. Since for Eqs. (2) to apply the force exerted by every single spring must satisfy Eq. (3), one is free to choose a region R with less than the typical density of impurities, so that the threshold field of the region $F_T(R)$ is less than $F_T(\infty)$. Choose the region R to have size L, so that its volume $\sim L^d$ and its surface $\sim L^{d-1}$. Define $F_T(R)$ as the threshold field of region R if the springs connecting R to the rest of the system are removed (and perhaps replaced with periodic boundary conditions imposed by rigid bars). Since $F_T(R) < F_T(\infty)$, in the absence of boundary springs R would move with nonzero velocity. Since in the actual system R is not moving, the springs along the boundary of R must exert a force on the region. The magnitude of this spring force scales as $(F_T(\infty) - F_T(R))L^d$. This force is exerted by the L^{d-1} springs on the boundary, so for at least one spring along the boundary one must have

$$\frac{\text{force}}{\text{spring}} \geq (F_T(\infty) - F_T(R))L \quad . \qquad (4)$$

This scaling is easily seen for the special case of an impurity-free region by summing Eqs. (2) over region R. It does not rely on the assumption of linear spring forces; Newton's third law ensures that the spring forces in the interior of R cancel in pairs for any short-ranged interactions between particles.

Using Eq. (4), already one can see that a sufficiently large region with no impurities will have springs stretched more than is admissible ($|x_{j+\delta} - x_j| > S_{max}$ for some j, δ along the boundary). These very rare regions are discussed below. However, first consider typical fluctuations in the threshold field of regions of size L. One expects the fluctuations in $F_T(L)$ to scale as the fluctuations in the number density of impurities in the region. One expects the total number of impurities in the region, $N_I(R)$, to fluctuate by an amount of order $\sqrt{N_I(R)}$, which implies that the number density fluctuations scale as $L^{-d/2}$. This scaling implies that $F_T(\infty) - F_T(L) \sim L^{-d/2}$, so (using Eq. (4)) the force per spring scales as $L^{1-d/2}$.

This argument implies that in one dimension the typical strains scale as $L^{1/2}$, and in two dimensions it is reasonable to expect them to scale as $\log(L)$.

The scaling argument can be shown to apply rigorously for a simpler model first introduced by Mihaly et al.[9] The main simplification of their model is that the nonlinear pinning force in Eq. (2) is replaced by a random coefficient of friction. The equations of motion are:

$$f_j = k \sum_{\delta(nn)} (x_{j+\delta} - x_j) + F - d_j \quad (5)$$

$$\dot{x}_j = f_j \quad \text{if } f_j > 0,$$

$$\dot{x}_j = f_j + 2d_j \quad \text{if } f_j < -2d_j,$$

$$\dot{x}_j = 0 \quad \text{otherwise.}$$

The x_j are the positions of the particles, F is the uniform force, and the d_j describe the friction coefficients. The sum is over nearest neighbors, and k is the spring constant.

For this model, just at threshold $f_j = 0$ for every j, and the threshold force obeys $F_T = <d_j>$, where the brackets denote a spatial average.[7] Since at threshold equations (5) are linear in the x's, it is straightforward to fourier transform them and evaluate $<(x_{j+\delta} - x_j)^2>$ for a system of size L. One indeed finds for this model $<(x_{j+\delta} - x_j)^2>^{1/2} \sim L^{1/2}$ in one dimension and $<(x_{j+\delta} - x_j)^2>^{1/2} \sim \log(L)$ in two dimensions.[10]

In three dimensions, the strain at the boundaries of large regions does not typically diverge as $L \to \infty$, but there will always be a small but nonzero chance that there is a region of size L with no impurities at all.[11] The threshold force $F_T(R)$ for these regions is zero, so that Eq. (4) implies that if $L > kS_{max}/F_T(\infty)$, the bound Eq. (3) is violated.

2.2. ESTIMATING THE PHASE SLIP DENSITY

So far we have only shown that in principle defects must be present in systems of infinite size. However, experiments can be done only on finite-size samples, and one must address the issue of whether the phase slip density is large enough to be experimentally relevant. This question is particularly important to address in the CDW case because CDWs are 3-dimensional systems with threshold voltages for conduction that are very small on the scale of the gap energy, which is the energy required to drive the CDW amplitude to zero and hence describes the energy to create a phase slip. Therefore, if the density of defects were determined by the relative sizes of the threshold voltage and the defect energy, then the defect density would be unobservably small. However, this naive expectation is not correct because CDWs are described by the limit where the pinning involves the collective action of many impurities. It turns out that the dominant defects are stationary regions present in the sliding state of the CDW, and that the density of defects in this regime is of order unity for experimentally relevant parameters.

The estimation of the density of phase slips must explicitly account for the collective nature of CDW pinning.[12] One must calculate the probability of finding a region of a given size with a given impurity concentration, and comparing the impurity pinning forces with the spring forces both in the interior and boundary of the region. Since there is a greater probability of observing a given variation in impurity concentration in small regions, and because as a rule strongly pinned regions are smaller than weakly pinned regions, most phase slips arise at the boundaries of regions that are anomalously strongly pinned (as opposed to more weakly pinned than average).

It is shown here that most of the defects arise at the boundaries of regions that remain stationary while the rest of the CDW is moving. The dominant contribution arises from regions that are basically undistorted, so the density of phase slips is determined by the relative sizes of the bare impurity potential V_{bare} and the phase slip energy E_{ps}. The threshold field itself plays no role in determining the phase slip density.

For these estimates it is adequate to consider the model which includes only phase degrees of freedom. The energy E of the continuum version of the system is given by

$$E = k\int d^d x\, (\nabla \phi(\mathbf{r}))^2 - \int d^d x\, V(\mathbf{r}) \cos(\mathbf{Q}\cdot\mathbf{r} + \phi(\mathbf{r})) \,. \tag{6}$$

These equations describe a d-dimensional system where ϕ is the CDW phase, $V(\mathbf{r})$ describes the impurity potential, and k describes the CDW stiffness. The first term describes the elastic cost of deformations, and the second term describes the effects of $V(\mathbf{r})$, which couples to the CDW charge density.

First recall how one estimates the threshold field F_T of a CDW. One assumes that F_T is proportional to the pinning energy per unit volume that the CDW gains by deforming in the presence of the impurity potential.[2] This pinning energy density is determined by the competition between the CDW elasticity and the impurity potential. In an infinite system an undistorted CDW would have zero energy density because $V(\mathbf{r})$ is random. However, the CDW can adjust to the impurity configuration by distorting, which increases the elastic energy but can lower the impurity contribution. One can estimate the relative sizes of the elastic energy cost and impurity energy gain by considering a region of size L.[13] Because of statistical fluctuations in the impurity potential, an appropriate deformation typically yields an impurity energy gain proportional to the square root of the number of impurities in the region, so that the energy density gain is $V(n_i/L^d)^{1/2}$, where n_i is the impurity density, d is the number of dimensions, and V is the impurity strength. Because the typical distortion will be an amount $1/Q$ accumulating over a distance L, the elastic energy density cost of the distortion scales as $1/L^2$. (Lengths are measured in units of $1/Q$.)

Minimization of the energy as a function of L leads to a pinning energy density of k/ξ_{LR}^2, where the Lee-Rice length $\xi_{LR} \sim (k/V\sqrt{n_i})^{2/(4-d)}$. The energy gain from the impurity potential and the elastic energy cost are the same order of magnitude on the length scale ξ_{LR}.

The preceding argument gives the pinning energy density of a typical region, but one would like to know the distribution of pinning energies. Since the strongly pinned regions are smaller than the weakly pinned ones, the probability of finding a fluctuation in the impurity potential leading to a strongly pinned region is much greater than one would estimate by looking at a region of size the Lee-Rice length.

It is reasonable that on a length scale ξ, the distribution of impurity potential energies is described by a Gaussian with width proportional to the square root of the number of impurities in the region $(n_i \xi^d)^{1/2}$, so that the distribution of impurity energies per unit volume is described by a Gaussian of width $V(n_i/\xi^d)^{1/2}$.[14] The elastic energy cost of the distortion once again scales as Ck/ξ^2, where C is a constant of order unity.

The pinning energy per unit volume of a region ε_{pin} is the sum of the elastic and impurity terms. The probability of observing a value of ε_{pin} in a region of size ξ, $p(\varepsilon_{pin}, \xi)$ is

$$p(\varepsilon_{pin}, \xi) \sim \exp-\left\{(\varepsilon_{pin} + Ck/\xi^2)^2 \xi^d/(n_i V^2)\right\} . \quad (7)$$

The total number of regions with pinning energy density ε_{pin}, $P(\varepsilon_{pin})$, is $\int d\xi\, p(\varepsilon_{pin}, \xi)$.[15] The exponential dependence of the integrand enables one to evaluate the integral using steepest descents, leading to the emergence of a dominant length scale $\tilde{\xi}$ which satisfies

$$\varepsilon_{pin} = C\left(\frac{4}{d} - 1\right)\frac{k}{\tilde{\xi}^2} . \quad (8)$$

Evaluating Eq. (7) at $\tilde{\xi}$ yields

$$-\log(p(\varepsilon_{pin})) \sim \left[\frac{\varepsilon_{pin}}{E_{LR}}\right]^{(2-\frac{d}{2})} , \quad (9)$$

where E_{LR} is the typical (Lee-Rice) pinning energy $E_{LR} \sim \left[\frac{V^4 n_i^2}{k^d}\right]^{\frac{1}{4-d}}$. In three dimensions $-\log(p(\varepsilon_{pin})) \sim \varepsilon_{pin}^{1/2}$; this result means that strongly pinned regions are vastly more probable than in a Gaussian distribution.

The calculation of the phase slip density can be done via a simple generalization of the arguments we just used to obtain the distribution of pinning energies. The new wrinkle on the situation is that phase slips will arise only when the pinning potential is strong enough to overcome not only the elastic forces in the region but also the springs on the boundary.

The calculation proceeds by defining a quantity $\tilde{G}_{pin}$, which is related to the impurity fluctuation energy density ΔV by $\tilde{G}_{pin} = |\Delta V| - k/\xi^2 - kS_{max}/\xi$. Since the fluctuations in ΔV are described by a gaussian of width proportional to $\xi^{-d/2}$, the distribution of $\tilde{G}_{pin}$ for regions of size ξ can be written:

$$p(\tilde{G}_{pin}, \xi) \sim \exp-\left\{(\tilde{G}_{pin} + k/\xi^2 + kS_{max}/\xi)^2 \xi^d/(n_i V^2)\right\} . \quad (10)$$

In three dimensions the quantity in brackets in Eq. (10) is a monotonically increasing function of ξ, so that the dominant contribution comes from the smallest possible values of ξ.

The regions with phase slips are those where the impurities happen to be arranged so that there is substantial pinning energy even when the regions are undistorted. The number of regions with phase slips is determined by the ratio of the bare impurity potential to the phase slip energy. One needs to have a statistical fluctuation leading to a correlation of L^d impurities, where L is determined by the condition $L \geq kS_{max}/V$. Thus one expects the density of phase slips n_{ps} to obey $n_{ps} \sim \exp[-(kS_{max}/V)^d]$, where d is the number of dimensions.

The estimates presented in this section show that the phase slip density is determined by the ratio of the bare pinning potential to the phase slip energy. Theoretical estimates of both these energies are on the order of the CDW gap energy.[1] In addition, it is possible to obtain information about the size of the bare pinning potential experimentally. One can determine the bare pinning potential if one knows both the threshold field and elastic constant. The threshold field is easily measured, and recently the elastic constant in $NbSe_3$ has been measured by X-ray scattering determination of the position-dependent change in CDW wavevector caused by the effects of the strong pinning centers at the contacts of a sample.[16] This experiment yields a value of the elastic constant (and hence bare impurity potential strength) which is consistent with the theoretical estimates. Since the ratio of the bare impurity potential to the phase slip energy is not small, the density of phase slips in experimental samples could easily be nonnegligible. However, one cannot make a definite theoretical statement because the density of phase slips depends exponentially on this ratio.

2.4. EFFECTS OF PHASE SLIPS ON CDW DYNAMICS

The arguments presented above show that a sufficiently large sample of CDW must exhibit phase slips when it is driven by a uniform, time-independent electric field. However, they do not address at all the issue of what happens when phase slips are present. To make progress on this question, one must understand the details of the dynamics at the defects that are generated when the strains get so large that the phase deformation model breaks down. This problem is complex and depends on features of the physical system which did not need to be considered in the phase deformation model.

In this section we discuss dynamics in the presence of defects for CDWs. CDWs differ from other systems such as flux lattices in type-II superconductors and Wigner crystals because CDW wavelengths are not conserved. One can have a moving region completely surrounded by stationary CDW; one creates and removes CDW wavelengths at the edges of the region. This is possible because it is possible to interconvert between normal carriers and electrons in the condensate. (A mechanical analogy to this situation is riding an escalator: both the lower and upper stories are stationary, and the escalator creates stairs at the lower story and destroys them at the upper story.) This feature of the CDW system makes it significantly simpler to analyze than other systems, because one expects the phase-deformation model to apply so long as one restricts consideration to a defect-free region.

It is natural to generalize the equations of motion (2) by replacing the Hooke's law springs by springs described by a force law where the maximum force is bounded:

$$\dot{x}_i = k \sum_{\delta, \, nn} f_{spring}(x_{i+\delta} - x_i) - V_i \sin(x_i - \beta_i) + F \quad , \tag{11}$$

where the spring forces f_{spring} are now bounded, $|f_{spring}(x_{i+\delta} - x_i)| < k \, S_{max}$ for all i. Arguments similar to those given in section 2 can be used to prove that the velocity of any model of this type must be inhomogeneous for any nonzero force F. One finds two regions with substantially different pinning strengths and averages the analog of Eqs. (2) over the interiors of the two regions. Since the interaction term is smaller by a factor of $1/L$ than the other terms in the equation, the difference in the pinning strength leads directly to a difference in the velocities of the two regions.

We now address the question of how the presence of defects affects the behavior of the model near the threshold. In one and two dimensions, one expects the CDW to break up into disconnected pieces moving at different velocities, implying that the threshold behavior present in Eqs. (2) is completely destroyed. For $d > 2$ a typical region does not have diverging strain at its boundary, so one expects a connected region to start moving all together at a well-defined value of the force. However, when this region starts to move, a nonzero density of regions with stronger than typical pinning will remain stationary; the arguments showing this are exactly analogous to those presented above for the anomalously weakly pinned regions. The implications of these defects for the dynamical behavior depends on their microscopic dynamics.

The simplest case to analyze is a "rubber band" model where the spring force is linear in the particle separations up to a critical separation S_{max}, after which the spring force is zero. For this case it is clear that the spring force exerted at the boundary of a moving region and a stationary region is zero, since the separation between neighboring particles is unbounded at long times. Therefore, when a region breaks free, the spring force changes discontinuously as a function of driving force, leading to a jump in the velocity of the region. In three dimensions, since at the threshold a nonzero fraction of springs break at the same value of the force, the velocity must jump. However, a different model where the spring force is linear in the particle separations up to a critical separation S_{max}, after which the spring force remains at the value kS_{max} most likely leads to a continuous velocity characteristic with a nonanalyticity at a well-defined threshold. Thus, there is no symmetry argument requiring the transition to a moving state to be continuous; one must consider the dynamics of the depinning process.

In real CDWs the dynamics at the phase slips are complicated. One must understand the interconversion between normal carriers and CDW as well as the interactions between the slow phase deformation modes and the fast amplitude modes. Understanding these dynamics remains a challenging problem.

2.5. ROLE OF PHASE SLIPS IN EXPERIMENTS

Most of our knowledge of dynamics in the presence of defects is indirect. Nonetheless, there is good evidence that phase slips play an important role in the dynamics of almost all samples. Although at this stage it is unclear whether the phase slips arise from macroscopic sample inhomogeneities or from the mechanism discussed in this paper, there is evidence that no samples can be described by the Fukuyama-Lee-Rice model in the regime where the phase correlation (or Lee-Rice) length is much smaller than the sample size.

One clearcut feature is that a sample which has two or more regions with different time-averaged velocity must have accompanying phase slips. Therefore, experimental demonstration of inhomogeneous velocity implies the presence of phase slips. Experimentally, inhomogeneous current (which implies the presence of phase slips) has been demonstrated unambiguously in $K_{0.3}MoO_3$.[17] The only material where the current flow is homogeneous enough that the possibility of transport without phase slips can be contemplated (ignoring phase slips that must occur at the electrical contacts) is $NbSe_3$. Although almost all samples exhibit substantial rounding of the current-voltage characteristic,[18] there are a small fraction of samples (about 1 in 10^4) where the threshold is very sharp and where the low-frequency temporal fluctuations in the spatially-averaged CDW velocity are small.[19] The presence of a sharp threshold field is correlated with uniform cross-sectional area, which indicates that there are significant finite size effects which are not included in the model. Even ignoring this complication, when combined ac and dc fields are applied to these samples, the response is significantly more complex than that predicted using the phase deformation model.[20]

A second, more indirect, experimental probe into the the relevance of phase slips is broad band noise, which is observed above threshold in almost all samples.[21] [22] The usefulness of broad band noise as a probe arises because it is known that it does not occur in the phase-deformation model of CDWs[23] [24] In samples with a thickness step that leads to two regions with different velocities, the broadband noise has been shown to be correlated with the presence of the step.[25] However, the question remains whether there are sources of broad band noise other than macroscopic velocity inhomogeneity. It has been shown that in many $NbSe_3$ samples the instantaneous CDW velocity appears substantially more homogeneous than the time-averaged value,[26] indicating that noise sources other than spatial velocity inhomogeneity may be present. Once again, these noise sources are not included in the phase deformation model, so it is reasonable to speculate that they may be identified with phase slips.

In real CDW materials phase slips can arise not only from the mechanism described here but also from sample inhomogeneities (e.g., clumping of impurities) and from thickness variations, as well as because of nonuniform electric fields. However, this issue is semantic, because the arguments here show that an amplification of the variations of the pinning potential occurs (based on surface to volume ratios), so that one is merely asking whether the observed strong-pinning behavior is "intrinsic" or an amplification of weak pinning behavior.

2.6. DEFECTS IN OTHER SYSTEMS

Two other systems for which the considerations in this paper are relevant are magnetic flux lattices in type-II superconductors and Wigner crystals. If one considers only long-wavelength acoustic modes, both these systems are described by equations of motion that are very similar to the CDW phase-deformation model described in section 2 of this paper.[27] Therefore, one must again ask whether including only long wavelength deformations is justified, and if it is not, how the dynamics are affected by the presence of short-wavelength defects (grain boundaries).

The arguments of section II which show that a model including only long wavelength deformations is unphysical apply to any system where the interactions are short ranged (or screened).[28] However, there is an essential difference between the flux lattice and Wigner crystal on one hand and the CDW on the other hand because CDW wavelengths are not conserved whereas flux lines and electrons are. This difference means that in the latter two systems it is not possible to have an isolated moving region--particles must flow into and out of the region. Therefore, even though a CDW of infinite size has nonzero velocity for any $F > 0$, a flux lattice or a Wigner can have strictly zero velocity at a nonzero value of the force. However, this does not mean that the phase deformation model can be used to describe the behavior near threshold. In two dimensions one expects the system to break up into disconnected pieces,[29] and even in three dimensions, since the dominant defects are stationary regions inside a moving medium, and these are allowed in all the systems, one must account explicitly for the plastic flow around the stationary regions if one hopes to understand the response near the onset of motion.

The nonlinear response of flux lines in type-II superconductors is a particularly interesting system to investigate the importance of defects, because the energy cost of creating a defect can be varied by varying the magnetic field. At fields very close to the upper critical field H_{c2}, the nonlinear current-voltage characteristic of the flux lattice in $NbSe_2$ undergoes a systematic evolution that might be indicative of an increasing role of defects as their energy cost is reduced.[30]

In these proceedings Bob Willett describes experiments that probe an insulating state of a very low-density two-dimensional electron system, which is most commonly interpreted as a pinned Wigner crystal. The Wigner crystal is composed of electrons which couple to an applied electric field, and therefore one can probe its nonlinear dynamics in ways analogous to those used in the CDW system. The two-dimensionality of the system makes it likely that grain boundaries should be more important than in a three-dimensional system. Estimates of the Lee-Rice length in this system[31] are of order a lattice constant, which means the energy cost of creating a grain boundary is of order the impurity pinning potential. Therefore it is reasonable to expect substantial creation of defects (grain boundaries) and significant velocity inhomogeneity and rounding of the threshold in this system. Experimentally measured nonlinear current-voltage characteristics in these systems appear quite rounded, and the onset of nonlinear conduction appears correlated with a large increase in broad band noise.[32]

2.7. SUMMARY

To summarize, in this lecture it was shown that the dynamics of a CDW in the infinite volume limit cannot be described using a model that includes only long wavelength deformations. Including the effects of the short-wavelength amplitude degrees of freedom in the model fundamentally changes the dynamical behavior of the model. There is strong evidence that phase slips play an important role in many experiments.

3. The Pulse Duration Memory Effect

This lecture describes some joint work with P.B. Littlewood,[33] which concerns a peculiar memory effect that was first seen experimentally.[34] [35] Figure 1(a) shows experimentally measured current traces that are induced by repeated voltage pulses. For each pulse length, the velocity is rising just as the pulse ends. When the length of the pulse is changed, the velocity characteristic undergoes substantial rearrangement, but after a few pulses of the new duration, the velocity rise at the end of the pulse is again observed.

First, note that this experiment does not violate causality. The CDW "knows" when the end of a pulse is coming because it assumes it has the same duration as the preceding pulse. Therefore, we are left only with the question of why the CDW "cares" about the length of the pulse--why the current oscillations are synchronized with it.

Now one expects the velocity of the chain to exhibit a transient oscillation at the start of a pulse. When one particle moves in a periodic potential, its velocity oscillates as it goes down the washboard. When many particles are coupled together, each one has an oscillating velocity. In steady state the oscillations are out of phase and the spatially-averaged velocity of the infinite system is independent of time. However, at the beginning of a pulse each particle starts off at the bottom of a potential well, and hence initially the oscillations are in phase.

However, at the end of the pulse this type of argument does not apply. In fact, any deterministic equation describing a single degree of freedom in a periodic potential cannot exhibit this effect. The velocity characteristic of the particle is entirely determined by the equation of motion and the initial conditions. Since the initial conditions are known ($x = \dot{x} = \ddot{x} = \ldots = 0$), there is no freedom left to adjust the velocity at the end of the pulse.

Peter Littlewood and I demonstrated that Eqs. (2) can be used to understand this effect.[36] Although Eqs. (2) are deterministic, there is freedom to adjust the velocity characteristic because

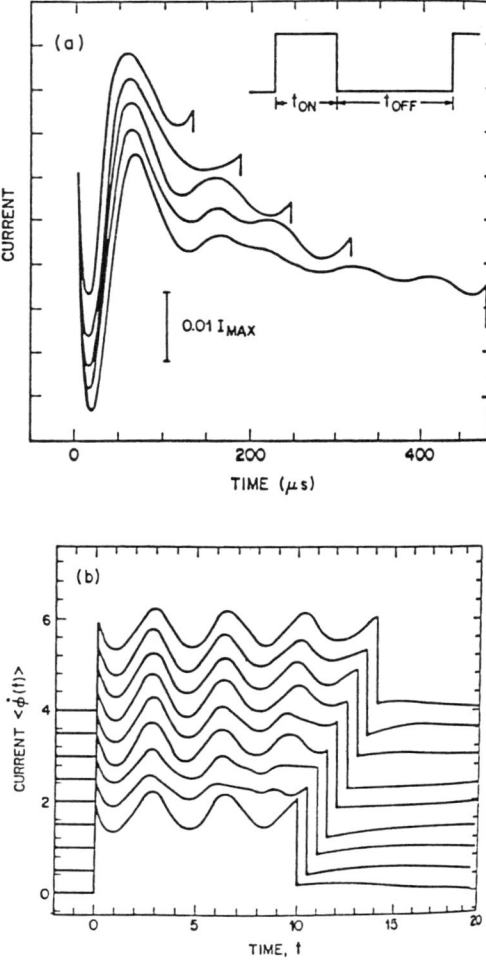

Figure 1. Current oscillations in response to a square-wave driving field of about 10 times the threshold (see inset) in $K_{0.3}MoO_3$ at 45 K (from Fleming and Schneemeyer[34]).
(b) Current oscillations observed in numerical simulations of a modification of eqs. (2) (described in ref. 35) for a one-dimensional system with 50 particles. In both experiment and simulation the pulse length is varied for fixed pulse height; different curves have their vertical axes offset for clarity. Both experiment and simulation show a tendency to have velocity increasing at the end of a pulse.

there are many metastable states available to provide different initial conditions. Just at the beginning of a pulse, the CDW is in a metastable state, so the pulse application can be viewed as a mapping from one metastable state to another. Thus, the velocity synchronization arises from the CDW choosing some subset of possible initial conditions.

Our first step was to do a simulation of a modification of Eqs. (2) (the modification is unimportant, but I want to be historically accurate), the results of which are shown in Fig. 1(b). One can see that although the synchronization is not perfect, there is a tendency for the velocity to be rising at the end of a pulse. Thus it seems that Eqs. (2) have a feature that leads to the pulse duration memory effect. So the problem was now to understand what features in Eqs. (2) lead to the effect.

The understanding we reached involves the following logic. By taking a spatial average of the time derivative of Eqs. (2), one sees that increasing velocity implies that $<V_j \cos(x_j - \beta_j)> < 0$ at the end of a pulse. Thus, the velocity increase implies that a preponderance of particles are at local maxima of their potentials. We claim that the potential maxima are selected because they separate the basins of attraction for different metastable states.

One then needs to understand why the system ends up at the boundary of a basin of attraction. It turns out that the mechanism is closely related to why one always finds a lost item at the last place one looks--once it is found the search ends. Similarly, the CDW is searching for a fixed point--a configuration that is invariant when a pulse is applied. (This evolution occurs because faster-moving regions move ahead and then feel spring forces that make them move more slowly.) As soon as a fixed point is reached, the system stops evolving. Hence the time evolution results in the observation of fixed points that are the least stable in the sense that they are the closest to configurations that are not invariant under application of a pulse.

Thus, the essential features leading to the organization are:

1. Many inequivalent metastable states, which leads to many fixed points of the mapping between the metastable configurations. This fixed point must lie in a fairly compact region in the configuration space.

2. A source of feedback that causes the system to evolve slowly towards its set of fixed points as the map is iterated.

When the system starts out, generically it is far from a fixed point, but the iteration of the mapping causes the system to evolve towards its region of fixed points. The system eventually reaches a configuration that is on the boundary of the region of fixed points. This state need not be a typical fixed point, but by definition, no further evolution occurs. Thus, the dynamics cause the system to select out a particular subset of the permissible configurations.

The simplest model that exhibits this effect is one particle in a sinusoidal potential connected by a spring to the point $x = 0$.[37] The equation of motion is:

$$\dot{x} = -kx - V \sin x + F(t) \ . \qquad (12)$$

The particle position is x, the potential strength is V, k is the spring constant (which is assumed to be much less than V), and $F(t)$ once again describes repeated square wave force pulses of magnitude F and duration T. The spring breaks the symmetry so that the system has many inequivalent metastable states.

The effect of applying repeated pulses to this one particle system is pictured in figure 2, where the evolution is shown for parameter values $F = 20$, $V = 5$, $k = 0.05$, and $T = 0.25$. The first pulse causes the particle to jump over into the next well. The spring force retarding the motion then increases by $\sim 2\pi k$. After many pulses the spring force increases so much that the particle

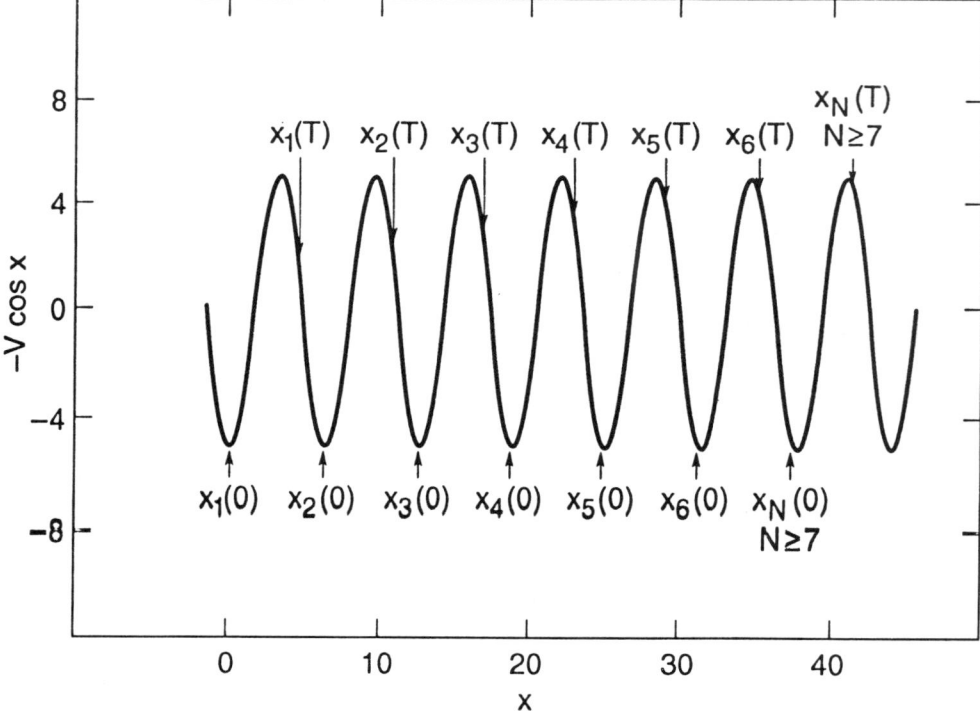

Figure 2. Dynamical evolution for the one particle model described by the equation of motion (12) with $F=20$, $V=5$, $k=0.05$ and $T=0.25$. When the particle starts out at $x=0$, a pulse causes it to jump over into the next well. The spring force retarding the motion then increases by $\sim 2\pi k$. After many pulses the spring force increases enough that the particle just barely fails to jump over a well top. This configuration yields an increasing velocity at $t=T$. It is invariant under application of a pulse and hence repeats indefinitely.

just barely fails to jump over a well top. This configuration yields an increasing velocity at $t=T$. It is invariant under application of a pulse and hence repeats indefinitely.

We still needed to show that this mechanism still applies when many particles are coupled together. Heuristically, one can view the one-particle equation as representing the motion of each particle relative to the mean field motion of the system as a whole. However, we were not able to construct an analytically solvable model with many degrees of freedom which displays the effect. Instead, we constructed a system of coupled maps which has the ingredients described above and studied its evolution numerically.

The mapping we used can be visualized as consisting of two steps, one analogous to the change in configuration caused by application of a field, and the second analogous to the relaxation to a metastable state that occurs when the field is off. The defining equations are

$$y_j^n = t(k(x_{j+1}^n - 2x_j^n + x_{j-1}^n) + F - d_j), \quad (13a)$$

$$x_j^{n+1} = \text{int}(y_j^n + \frac{1}{2}), \quad (13b)$$

with periodic boundary conditions $x_1 = x_{N+1}$. The variable x_j^n describes the position of the jth particle at the beginning of the nth pulse, whereas y_j^n is the position of the jth particle just as the nth pulse is turned off. The variables d_j are reminiscent of the pinning potential in Eqs. (2), and they are taken to random numbers uniformly distributed between 0 and 1. The mapping thus depends on the three parameters t, k, and F, with the notation chosen to suggest analogy with the coupled differential equations (2).

This system of maps is similar to those that can be derived directly from the differential equations (2) in the limit $F \gg V \gg k$.[38] The first step of the map mimics the behavior of the system while the pulse is on, and the second step approximates the effects of turning the pulse off. However, the arguments leading to organization do not depend on the details of the equations of motion (2), but rather on the presence of many metastable states and weak feedback. Therefore, it is perhaps more useful to view the mapping as an example of a dynamical system with these features rather than as a model of charge density waves.

Since the second step of the mapping is discontinuous when $\text{frac}(y_j)$, the fractional part of $y_j = \frac{1}{2}$, this point marks the boundary of a basin of attraction of a fixed point. Thus, the organization manifests itself for the maps when a preponderance of y_j's have fractional part close to ½.

Figure 3 shows the the distribution of $\text{frac}(y_j)$ for a system of 1000 particles in bins of size 0.001 at the end of different numbers of iterations. The map defined by Eqs. (13) was iterated with parameters $t = 1$, $F = 2$, and $k = 0.001$. Initially the $\text{frac}(y_j)$'s are uniformly distributed in the unit interval, but as the map is iterated a pronounced peak in the distribution grows at $\text{frac}(y) = 1/2$.

The peak shows no signs of disappearing as the number of particles is made larger. Thus, increasing the number of degrees of freedom does not destroy the tendency to organization. The peak emerges well before the system reaches a fixed point, for even after 5000 iterations the system is still evolving for the conditions shown. One can envision a process where longer and longer length scales have homogenized their velocities, with piling up at boundaries occurring at each stage. One might hope to describe this process with a mode-coupling theory.

Thus, a simple yet puzzling experimental result can be explained as a self-organization process in a dynamical system with many degrees of freedom. We suspect that the mechanism applies to a broad class of multivariate dynamical systems, but investigating the generality of the phenomenon remains a challenging problem for the future.

To summarize, a model of sliding CDW's provides a simple yet interesting example of a dynamical system with many degrees of freedom. The model has strains that diverge in the limit of infinite size in the presence of a time-independent force (which implies the model discussed in this paper is unphysical and must be modified). When large force pulses are applied to a CDW, it displays organized behavior which can be understood in terms of a selection mechanism for fixed points of a multivariate dynamical system. In the future we hope to understand whether the ideas used to understand CDW's can be applied to other nonlinear dynamical systems with many degrees of freedom.

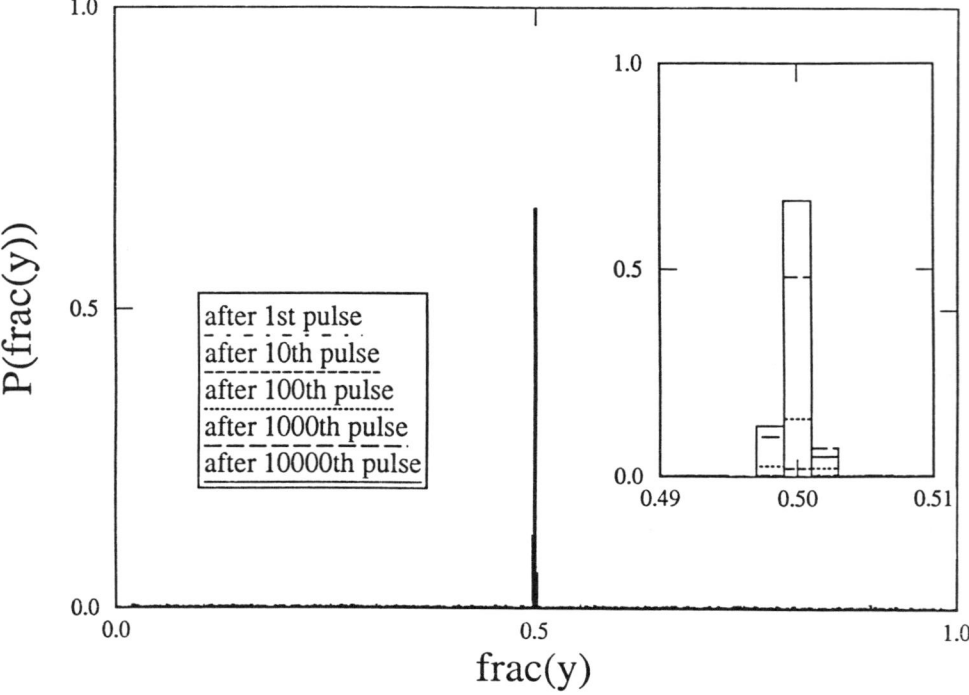

Figure 3. Histogram of the fractional part of y_j after 1 (dash-dotted line), 10 (dashed line), 100 (dotted line), 1000 (long-dashed line), and 10000 (solid line) iterations of the map [Eqs. (13)] with 1000 degrees of freedom with parameter values t=1, F=2, and k=0.001. The quantity P(frac(y)) is the fraction of the y's with fractional parts whose values are within a bin of size 0.001. The boundary of every basin of attraction is at $y = 1/2$. The accumulation of y's at half-integer values is clearly visible.

References

1. For a review of CDWs, see, e.g., G. Grüner, Rev. Mod. Phys. **60**, 1129 (1988).

2. H. Fukuyama and P.A. Lee, Phys. Rev. B**17**, 535 (1978); P.A. Lee and T.M. Rice, Phys. Rev. B**19**, 3970 (1979).

3. L. Sneddon, M.C. Cross, and D.S. Fisher, Phys. Rev. Lett. **49**, 292 (1982).

4. The relation between the equations of motion used here and the original CDW equations of motion is discussed in L. Pietronero and S. Strassler, Phys. Rev. B **28**, 5683 (1983); P.B. Littlewood in *Charge-Density Waves in Solids,* ed. L.P. Gor'kov and G. Gruner (Elsevier, Amsterdam, 1989); D.S. Fisher, Phys. Rev. B. **31**, 1396 (1985).

5. See, e.g., D.S. Fisher, Phys. Rev. B **31**, 1396 (1985); S.N. Coppersmith and D.S. Fisher, Phys. Rev. A **38,** 6338 (1988); P. Sibani and P.B. Littlewood, Phys. Rev. Lett. **64**, 1305 (1990); P.B. Littlewood and C.M. Varma, Phys. Rev. B **36**, 480 (1987); J.B. Sokoloff, Phys. Rev. B **31,** 2270 (1985); A. Middleton and D.S. Fisher, Phys. Rev. Lett. **66**, 92 (1991); C.R.

Myers and J.P. Sethna, unpublished; O. Narayan and D.S. Fisher, Phys. Rev. B**46**, 11520 (1992).

6. S.N. Coppersmith, Phys. Rev. Lett. **65**, 1044 (1990).

7. S.N. Coppersmith and A.J. Millis, Phys. Rev. B**44**, 7799 (1991).

8. S.N. Coppersmith, Phys. Rev. B**44**, 2887 (1991).

9. L. Mihaly, M. Crommie, and G. Gruner, Europhys. Lett. **4**, 103 (1987).

10. Similar calculations have been done in the context of oscillator entrainment; see, H. Sakaguchi et al., Prog. Theor. Phys. **77**, 1005 (1987); S.H. Strogatz and R.E. Mirollo, J. Phys. A **21**, L699 (1988) and Physica D **31**, 143 (1988).

11. The argument involving very rare fluctuations is related to arguments used for random magnets; R.B. Griffiths, Phys. Rev. Lett. **23**, 17 (1969); M. Randeria, J. Sethna, and R. Palmer, Phys. Rev. Lett. **54**, 1321 (1985).

12. The methods used are similar to those used to calculate the density of states of impurity states in the band tails of disordered semiconductors; see B.I. Halperin and M. Lax, Phys. Rev. **148**, 722 (1966); J. Zittartz and J.S. Langer, Phys. Rev. **148**, 741 (1966).

13. Y. Imry and S.-k. Ma, Phys. Rev. Lett. **35**, 1399 (1975).

14. This result can be demonstrated simply for a simplified model.[8]

15. M. Randeria, J. Sethna, and R. Palmer, Phys. Rev. Lett. **54**, 1321 (1985).

16. D. DiCarlo, E. Sweetland, M. Sutton, J.D. Brock, and R.E. Thorne, Phys. Rev. Lett. **70**, 845 (1993).

17. See, e.g., P. Segransan et al., Phys. Rev. Lett. **56**, 1854 (1986).

18. See, e.g., S. Bhattacharya, M.J. Higgins, and J. P. Stokes, Phys. Rev. Lett. **63**, 1508 (1989).

19. J. McCarten, D.A. DiCarlo, M.P. Maher, T.L. Adelman, and R.E. Thorne, Phys. Rev. B**46**, 4456 (1992).

20. M.J. Higgins, A.A. Middleton and S. Bhattacharya, Bull. A.P.S. **38**, 383 (1993), and preprint.

21. S. Bhattacharya, J.P. Stokes, M.O. Robbins, and R. A. Klemm, Phys. Rev. Lett. **54**, 2453 (1985); M.O. Robbins, J.P. Stokes, and S. Bhattacharya, Phys. Rev. Lett. **55**, 2822 (1985).

22. M.S. Sherwin and A. Zettl, Phys. Rev. B**32**, 5536 (1985).

23. P.B. Littlewood, Phys. Rev. B**33**, 6694 (1986).

24. A.A. Middleton, Ph. D. Thesis, Princeton University (1990); A.A. Middleton, Phys. Rev. Lett. **68**, 670 (1992).

25. M.P. Maher, T.L. Adelman, J. McCarten, D.A. DiCarlo, and R.E. Thorne, Phys. Rev. B**43**, 9968 (1991).

26. S. Bhattacharya et al., Phys. Rev. Lett. **59**, 1849 (1987); G.L. Link and G. Mozurkewich, Solid State Commun. **65**, 15 (1988).

27. See, e.g., A. Schmid and W. Hauger, J. Low Temp. Phys. **11**, 667 (1973).

28. The arguments in this paper do not imply an instability of a model including long-wavelength deformations only in systems with unscreened long-ranged interactions if the impurity potential is weak enough. However, this situation is significantly more complicated than that described by the phase deformation model because in such a system shear and rotation cost much less energy than compression, and one must account for this when considering the dynamics of the long-wavelength modes.

29. In this volume Henrik Jensen describes numerical simulations demonstrating that plastic flow is important in two-dimensional systems. O. Pla and F. Nori (private communication) have also obtained good numerical evidence that inhomogeneous conduction occurs in two dimensions for a model of flux lines with short-range interactions.

30. S. Bhattacharya and M. Higgins, preprint.

31. See, e.g., B.G.A. Norman, P.B. Littlewood, and A.J. Millis, Phys. Rev. B**46**, 3920 (1992).

32. Y.P. Li, T. Sajoto, L.W. Engel, D.C. Tsui, M. Shayegan, Phys. Rev. Lett. **67**, 1630 (1991).

33. S.N. Coppersmith and P.B. Littlewood, Phys. Rev. B **36**, 311 (1987). See also C. Tang et al., Phys. Rev. Lett. **58**, 1161 (1987).

34. R.M. Fleming and L.F. Schneemeyer, Phys. Rev. B **33**, 2930 (1986).

35. S.E. Brown, G. Gruner, and L. Mihaly, Solid State Commun. **57**, 165 (1986).

36. It can be shown that Eqs. (2) need not have unbounded strains when $F(t)$ takes the form of repeated pulses. In addition the synchronization occurs even if the CDW has broken up into several pieces. Thus, studying Eqs. (2) is adequate to understand the effect.

37. S.N. Coppersmith, Physics Letters A **125**, 473 (1987).

38. S.N. Coppersmith, Phys. Rev. A **36**, 3375 (1987).

UNIVERSALITY IN COMMENSURATE-INCOMMENSURATE PHASE TRANSITIONS

B. HU and J. SHI
Department of Physics
University of Houston
Houston, TX 77204-5506

B. LIN
Department of Physics
Case Western Reserve University
Cleveland, OH 44106-7079

ABSTRACT. To investigate the problem of universality in commensurate-incommensurate phase transitions, we have studied certain generalized Frenkel-Kontorova models with different interparticle and external potentials. Many new and interesting features have been found.

1. Introduction

One of the most commonly used models in the study of commensurate-incommensurate phase transitions is the Frenkel-Kontorova (FK) model. The standard FK model consists of a chain of particles connected by harmonic springs subjected to an external sinusoidal potential:

$$H = \sum_i \left[\frac{1}{2}(x_{i+1} - x_i - \mu)^2 + \frac{k}{(2\pi)^2}(1 - \cos 2\pi x_i) \right]. \tag{1}$$

Here x_i is the position of the ith atom, μ the natural length of the spring, and k the ratio of the strengths of the external and interparticle potentials. In most of the earlier studies, the FK model was treated in the continuum approximation. An entirely new approach was pioneered by Aubry,[1] who studied the discrete model directly via the powerful Kolmogorov-Arnold-Moser (KAM) theorem. In this approach, a crucial connection is made between the FK model and the standard map. To see this connection, let us look at the equilibrium condition of Eq. (1):

$$\frac{\partial H}{\partial x_i} = 0, \tag{2}$$

which gives

$$x_{i+1} - 2x_i + x_{i-1} - \frac{k}{2\pi}\sin x_i = 0. \tag{3}$$

Defining $y_i = x_i - x_{i-1}$, one can rewrite this second-order difference equation into a set of first-order equations:

$$y_{i+1} = y_i + \frac{k}{2\pi} \sin x_i,$$
$$x_{i+1} = x_i + y_{i+1}, \qquad (4)$$

which is seen to be just the standard map. This connection enables one to make use of the KAM theorem to gain new insights into the FK model.

The mean particle separation (or "winding number") is defined as

$$\omega = \lim_{(n-n') \to \infty} \frac{x_n - x_{n'}}{n - n'}. \qquad (5)$$

If ω is a rational number, the particles are always pinned (corresponding to a periodic orbit). However, if ω is an irrational number, then there exists a critical k_c such that when $k < k_c$, the system is unpinned; yet when $k > k_c$, the system is pinned.[2] In the KAM language, the invariant torus has broken up into a cantorus. This transition, called by Aubry[1] a "transition by breaking of analyticity," is very similar to a phase transition and exhibits very interesting critical behaviors.

One of the most important problems in the study of critical behavior is that of universality. To investigate this problem, we have studied two generalized FK models: in one of them, the interparticle potential is changed from harmonic to Toda; in the other, the external potential is changed from sinusoidal to one with an arbitrary inflection point. Their critical behaviors are found to be markedly different. Due to limitation of space, we can only very briefly summarize our findings and refer the interested reader to more detailed accounts published elsewhere.[3,4]

2. Toda Model

The interparticle Toda potential[5] is given by

$$W(x_{i+1} - x_i) = \frac{1}{\beta^2}\left[e^{-\beta(x_{i+1}-x_i-\mu)} - 1\right] + \frac{1}{\beta}(x_{i+1} - x_i - \mu). \qquad (6)$$

β gives a measure of anharmonicity: as $\beta \longrightarrow 0$, the Toda potential becomes harmonic; as $\beta \longrightarrow \infty$, the potential becomes that of a hard-rod. We have studied[4] both the local and global scaling behaviors of the Toda model.

2.1. LOCAL SCALING BEHAVIOR

Near the critical point k_c the critical behavior of various physical quantities can be characterized by a set of critical exponents.

2.1.1. *Phonon Gap*. The phonon spectrum gives the eigenmodes of small vibrations of the particles around their equilibrium positions. The phonon gap Ω is the lowest frequency in the spectrum. It can easily be shown[2] that a state is pinned if and only if $\Omega \neq 0$. For $k < k_c$, the state is unpinned and $\Omega = 0$; as $k > k_c$, a gap appears

$$\Omega(k) \sim (k - k_c)^\chi, \qquad (7)$$

which defines the phonon gap exponent χ.

2.1.2. *Coherence Length.* The coherence length ξ measures the distance over which a perturbation propagates when one of the particles is moved from its equilibrium position. It can be shown that the coherence length is simply the reciprocal of the Lyapunov exponent γ, $\chi = \gamma^{-1}$. Below k_c, $\gamma = 0$; above k_c,

$$\gamma(k) \sim (k - k_c)^\nu, \tag{8}$$

which defines the coherence length exponent ν.

2.1.3. *Peierls-Nabarro Barrier.* The Peierls-Nabarro (PN) barrier is the minimal energy needed to translate continuously the chain on the external potential. Below k_c, $E_{PN} = 0$; above k_c,

$$E_{PN}(k) \sim (k - k_c)^\psi, \tag{9}$$

which defines the Peierls-Nabarro barrier exponent ψ.

We have computed[3] these critical exponents for $\omega = (\sqrt{5} - 1)/2$. In the Toda model, they are, within numerical accuracy, all equal to those of the standard model: $\chi = 1.02$, $\nu = 0.99$, $\psi = 3.00$. Therefore, the Toda model should belong to the same universality class as the standard model.

2.2. GLOBAL SCALING BEHAVIOR

We found[3] that the reflection symmetry present in the phase diagram of the standard model is lost in the Toda model due to its exponential form of interaction. The phase diagram also depends on the anharmonicity parameter β: as β increases, the Arnold tongues swing more and more to the left. We have also studied[3] the Devil's staircase and its multifractal properties. Both its singularity spectrum $f(\alpha)$ and generalized dimension D_q are found to depend on β. Therefore, the global scaling behavior of the Toda model is different from that of the standard model.

3. Inflection Model[4]

In this model, we change the external sinusoidal potential to one with an arbitrary inflection point:

$$V(\theta_i) = \frac{1}{2}\theta_i^2 \left(1 - \frac{2}{z+1}|2\theta_i|^{z-1}\right), \quad \theta_i = x_i \pmod{1}, \tag{10}$$

where $\theta_i \in [-\frac{1}{2}, \frac{1}{2})$ and $V(\theta)$ is periodically extended, $V(\theta_i) = V(\theta_i + 1)$. This potential is so designed that it has an adjustable degree of inflection z at $\theta_i = 0$. The motivation for studying such a potential came from our previous studies of the circle map[8] and the nonanalytic twist map.[9] In the circle map, we found that the scaling exponents α and δ depend on z. In other words, z serves as a universality criterion. In the nonanalytic twist map, we found that the scaling exponents depend on z for $2 \leq z < 3$. Moreover, for $z > 3$ the KAM torus can reappear after it has disappeared, and this process can recur many, even infinitely many, times. When the study was extended to the FK model, it was not surprising that similar phenomena have also been observed. Our major findings are the following:

1. The critical exponents χ, ν, and ψ all depend on the degree of inflection z for $2 \leq z < 3$, but remain the same as those of the standard model for $z > 3$. The degree of inflection thus plays a role quite similar to that of dimensionality in second-order phase transitions, with $z = 2$ and 3 corresponding respectively to the lower and upper critical dimensions.

2. For $z > 3$, a sequence of pinning and depinning transitions, reflecting the disappear and reappear of the KAM torus, occurs. This phenomenon may be of relevance to physical systems exhibiting a sequence of metal-insulator transitions.

Despite its deceptively simple structure, the FK model exhibits an extremely rich and interesting critical behavior. The study reported here represents but a first step towards understanding universality in the FK model, a more extensive and in-depth study is being undertaken.

Acknowledgments

This work was supported in part by the University of Houston President's Research and Scholarship Fund and the ROC National Research Council. One of us (BH) would like to thank Drs. C. K. Hu, H. L. Yu, and Felix Lee for inviting him to visit the Institute of Physics of Academia Sinica and the Department of Physics of Tsinghua University. Another one of us (JS) would like to acknowledge the award of a SSC National Fellowship.

References

1. S. Aubry, Phys. Rep. **103**, 127 (1984), and references therein.
2. N. Coppersmith and D. S. Fisher, Phys. Rev. B **28**, 2566 (1983).
3. B. Lin and B. Hu, J. Stat. Phys. **69**, 1047 (1992).
4. J. Shi and B. Hu, Phys. Rev. A **45**, 5455(1992).
5. A. Milchev, Phys. Rev. B **33**, 2062 (1986); **38**, 2808 (1988).
6. M. Peyrard and S. Aubry, J. Phys. C **16**, 1593 (1983); L. de Seze and S. Aubry, *ibid* **17**, 389 (1984).
7. O. Biham and D. Mukamel, Phys. Rev. A **39**, 5326 (1989).
8. B. Hu, A. Valinia and O. Piro, Phys. Lett. A **144**, 7 (1990).
9. B. Hu, J. Shi and S. J. Kim, J. Stat. Phys. **62**, 631 (1991); J. Shi and B. Hu, Phys. Lett. A **156**, 267 (1991).

GLASSY BEHAVIOR OF THE CHARGE/SPIN DENSITY WAVE GROUND STATE

K. BILJAKOVIĆ
Institute of Physics of the University of Zagreb
P.O. Box 304
41001 Zagreb
Croatia

ABSTRACT. The density wave systems are very good candidates for glassy-forming systems as they show many metastable states and high cooperativity. Charge density wave compounds are shown to exhibit a glassy behaviour over very wide temperature range and in very different energy-time windows. Special attention is given to the evidence for the glassy state in the charge density wave ground state at very low temperatures, and to almost similar effects recently found in spin density wave compounds. These results place density wave systems into a special group of glasses with some very unique features among the found universalities.

1. Introduction

Almost 40 years ago Fröhlich (1954) and Peierls (1955) independently introduced the concept of charge density wave. A one-dimensional metal with strong electron-phonon coupling has a pronounced inclination to lower its energy by a distortion of the lattice and the opening of a gap at the Fermi surface, consequently inducing a transition into an insulating state (called a Peierls transition). The term "charge density wave" (CDW) reflects the main characteristic of the new collective mode -the lattice and the electron charge density form a new periodic structure, with a wavelength λ that is longer than the original lattice period a. If λ/a is not a simple rational fraction, then the CDW is incommensurate with the underlying structure, otherwise it is commensurate.

In the 70 s a CDW ground state was shown to exist first in two-dimensional, then in quasi-one-dimensional compounds. But only in 1976 the idea of a sliding conductivity (when the entire CDW, if incommensurate, can slide through the lattice, hereby contributing to the electric conductivity) envisioned by Fröhlich (1954), was demonstrated for the first time in measurements of dc transport properties in $NbSe_3$ [Monceau et al 1976] and later in the ac conductivity [Ong and Monceau (1977) and Grüner et al (1980)]. Two mechanisms have been involved for explaining the existence of a force sufficiently large to dislodge the CDW and to enable it to slide. The first one considers the CDW commensurate with the lattice and therefore pinned by the lattice potential [Lee et al 1974]. The second one is the phase locking of the incommensurate CDW to impurities [Fukuyama and Lee 1978, Lee and Rice (1979)], a phenomenon very analogous to static friction. Depending on the individual pinning strength of each impurity the CDW can be pinned either strongly or weakly.

Very often the CDW wavelength exhibits only a small deviation from low-order commensurability. This quasi-commensurate state can be described as commensurate domains separated by topological defects : phase slips in 1D, domain walls or discommensurations (DCs) in 2D and dislocation loops (DLs) in 3D [McMillan (1976), Lee and Rice (1979)]. The CDW superlattice clamped between defects suffers either a compression or a dilatation which implies that DCs bear an electric charge. It involves the importance of underlying screening mechanism.

The properties of the collective conductivity state induced by a sliding CDW are now relatively well established and described in many review articles, proceedings of conferences and books [separate list is given after the references]. In this contribution I am dealing with the glassy behaviour of CDW systems.

Because of the strong interaction with impurities, the CDW cannot be described by a unique ground state. The static properties of the CDW as assumed in the Fukuyama-Lee-Rice (FLR) model [Lee and Rice (1979)] show strong similarities with those of an XY magnet in the random field. The random nature of the CDW ground state is reflected by the occurrence of many metastable states. There is a large number of experiments which have demonstrated the existence of these metastable states.

I do not intend to give a detailed review of the glassy behaviour of CDW systems, however in chapter 2 some illustrative and representative experimental results and the main ideas of some theoretical models (inserted into interpretation) will be presented which might be related to the specific field of our investigation. I will try to specify the different energy-time windows which characterize the existing experiments. In chapter 3 experiments of energy relaxation will be described in the range of smallest energy (few K) and longest time (seconds to few days) window which will give some new, additional properties of the CDW ground state. Finally, in chapter 4, I will point out the similarities in low temperature relaxational processes in spin-density wave (SDW) and CDW systems. As the dynamics in both types of compounds shows an exceptionally rich spectrum of exotic structural, transport and mechanical properties, density wave (DWs) systems must take a special place among the other, very diverse systems exhibiting a glassy behaviour.

2. Metastability

Already the popular FLR theory for sliding CDW predicts the possibility of many metastable states. Similar to the random-field XY model, the processes involving thermal hopping over the free energy barriers separating such states might be the dominant relaxational mechanism at long times (or/and at low temperatures). The relaxational behaviour depends on the distribution of pinning strengths or, in other words, on the broad distribution of localized low-energy (phase excitations) modes [Littlewood and Rammal (1988)]. The model of an elastic string pinned by impurity potential with quenched random phases, analyzed from a microscopic point of view, gives also the relaxation dynamics coupled in a hierarchical way [Pietronero 1989]. The existence of low-energy excitations (LEE) in CDWs and their high cooperativity together with the sudden change in the Hamiltonian or sudden application of a potential can give rise to the infrared divergence which has been proposed as the unique property at the origin of the anomalously slow relaxational processes in glasses [Ngai and Lin (1981)]. A more detailed and complete theoretical overview is given in this issue by S. Coopersmith.

Now, I am going to give some of very many different experimental evidences of the existence of metastable states but limited to the linear conducting regime, with an applied electric field whenever possible lower than the threshold field and to look at the states closer to the CDW ground state.

2.1. MEMORY EFFECTS, HYSTERESIS AND LONG TIME RESISTIVITY RELAXATION

Gill (1981) first pointed out that the CDW has a transient response to current pulses which depends on the sample history, i.e. on the polarity of the proceeding pulse. This effect now called "pulse-sign memory" is accepted as the ample evidence that CDWs are not rigid.

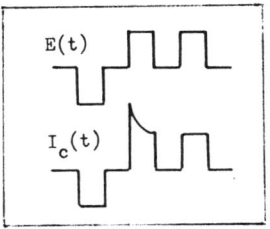

Figure 1. The overshoot phenomenon. The response $I_c(t)$ to a pulsed field $E(t)$ [from Gill (1984)].

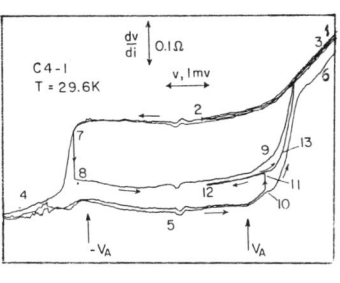

Figure 2

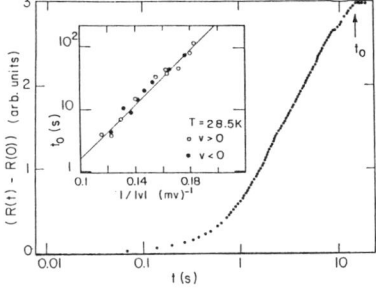

Figure 3

Figure 2. Electrical hysteresis in NbSe$_3$ shown as differential resistance R vs. voltage V [from Ong et al (1984)].

Figure 3. Variation of R vs. logt for a field V in NbSe$_3$ corresponding to the vertical line 7-8 in Fig. 2. Inset shows the variation of t_o -the time when the relaxation is interrupted at the value $R(t_o)$.

Metastable states are believed to be also the natural origin of low-field or/and temperature resistivity hysteresis. As shown first by Gill (1983) in TaS$_3$ and later confirmed in some other CDW systems, the differential resistance has an unusual dependence on the electrical history. Similar hysteretic feature of R-V curves (as on Fig. 2) have been also found in R-T curves. In a more sophisticated experiment by Duggan et al (1985), it has been shown that the relaxation of the resistance on the hysteretic loop (Fig. 2) is via a unique interrupted relaxational process. In contrast to complicated structure of the R-V hysteresis the dynamics shown on Fig. 3 is more simple and it has been shown that the observed relaxational process R vs. t can be fitted (for $t < t_o$) to the stretched-exponential form with the stretching parameter $\beta = 0.8$ [see after Eq. (2)]

and relaxation time of few seconds. Moreover the time t_0 at which the relaxation is interrupted obeys $t_0(T,E) \sim \tau_0 \exp(A/TE)$ law, i.e. in the underlying Arrhenius law the apparent activation energy scales with the applied field. It gives very interesting identical temperature and field dependence which put closer the mechanism of depinning and the thermally activated relaxational processes (author note). Ong et al (1985) have made conclusion that hidden in the CDW problem should be a single process (conversion) with time scales spanning 10^6 s to 10^{-6} seconds and that similar effects observed in very different time windows are glimpses of the same phenomenon.

The dc conductivity corresponding to a metastable state created by a quick heat treatment was found to be time dependent in the time domain of $10-10^5$ sec and a simple model, based on the analogy with spin glasses, has been proposed by G. Mihaly and L. Mihaly (1984).

2.2. DIELECTRIC RELAXATION

The ac response of the CDWs has been intensively studied. In general, the frequency range has been divided into three regimes [Fleming and Cava (1989)] which can be also identified on Fig. 4.

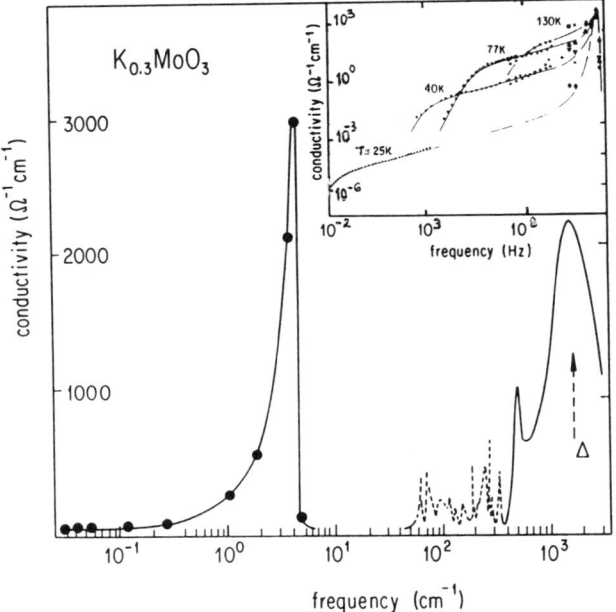

Figure 4. Frequency-dependent conductivity $\sigma(\omega)$ in the CDW state of $K_{0.3}MoO_3$. The inset shows the low-frequency response [from Gorkov and Grüner (1989) and Degiorgi and Grüner (1991)].

I region (far-infrared) between the pinned mode and the Peierls gap.
II region (10-100 GHz) dominated by the pinned phase mode of the CDW predicted by Lee, Rice and Anderson.
III region (dc-10 MHz) termed the dielectric relaxation regime where the dynamics is described as the decay of an induced polarization due to elastic deformation of the CDW [explained by Littlewood (1987)].

In this low frequency range (the inset of Fig. 4) the dielectric response can be phenomenologically described by Havriliak and Negami (1966) formula [Fleming and Cava (1979)]:

$$\varepsilon(\omega) = \varepsilon_\infty + \frac{\varepsilon_0 \varepsilon_\infty}{[1+(i\omega\tau)^{1-\alpha}]^\beta} \quad (1)$$

with ε_0 static and ε_∞ high-frequency value of dielectric constant. Eq. 1 represents the distribution of modes around τ, and α and β characterizing respectively the width and the asymmetry of this distribution. The temperature variation (60 K < T < 100 K) of the mean relaxation time τ of blue bronze [Cava et al (1984)] indicates that the CDW relaxation process displays Arrhenius behaviour with an activation energy of 829 K, very close to Peierls gap. The dielectric constant ε_0 increases with decreasing temperature. By 60 K Re $\varepsilon(\omega)$ develops a "cusp" characteristic of glassy systems.

The low-frequency loss peak has been explained as the consequence of the screening of localized pinned collective oscillations with a wide distribution of corresponding modes [Fig. 5, Littlewood (1987)]. The dynamical screening is proportional to the single particle resistivity and consequently the main relaxation time of this distribution has been expected to increase in thermally activated manner, with an activation energy in the range of the Peierls gap, as really it has been found experimentally. The theory assumed that the CDW is in equilibrium, a state which can be reached only on time scales much longer than the damping time which strongly increases at very low temperatures. Even if the same model [Littlewood and Rammal (1988)] gives very detailed analysis of glassy relaxation, distinguishing four different physical regimes explored with increasing time, it does not suppose any singular temperature.

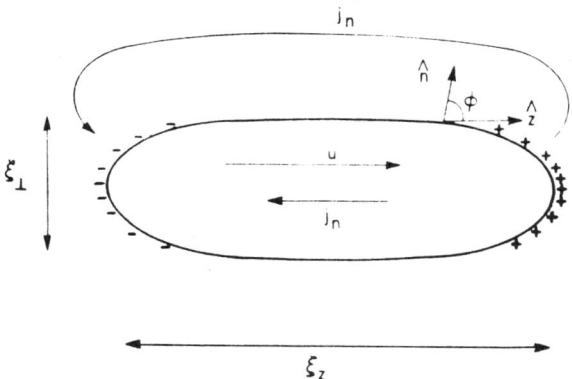

Figure 5. Schematic picture of a localized mode as a uniform displacement u of the CDW over a region $\xi_z \sim \mu m$ showing the induced polarization charges and the normal current "backflow" [from Littlewood (1987)].

However the properties of the CDW ground state are drastically changed at low temperatures. The behaviour of the CDW at very low frequencies may be better studied by applying step-like perturbations and recording the time-dependent relaxation processes in the similar manner to that used for spin glasses (SG). It has been shown that at lower temperatures, the polarization at zero external field exhibits a very weak relaxation. In addition some very recent measurements of the dielectric response at lower frequencies and at lower temperatures give indication of strong slowing down. Below 40 K and for frequencies ω < 100 kHz the real part of dielectric response of TaS3 exhibits a large peak which increases and moves to lower temperatures with decreasing frequency. It is a reminiscence of the cusp like dielectric or magnetic susceptibility of different glasses and spin glasses. So far the observed divergence of ε'(T, ω) has been explained as a transition of the CDW into a new glassy-like state at T ~ 13 K [Nad' and Monceau (1993)]. Similar features have been also found in blue bronze [Fig. 6, Yang and Ong (1991)].

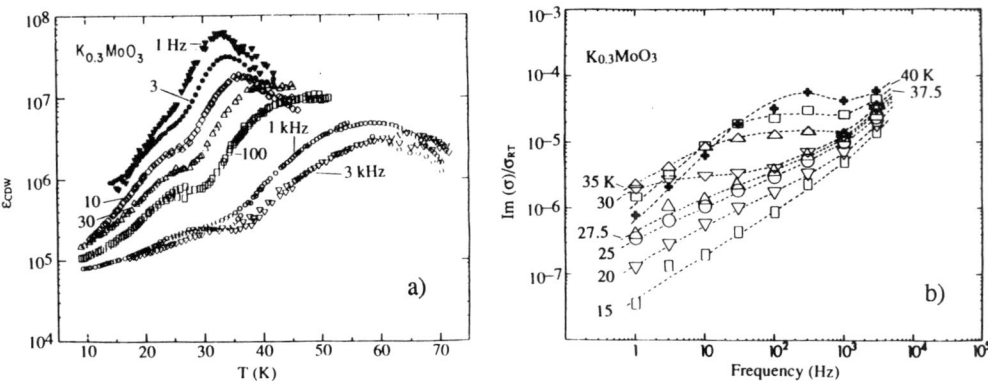

Figure 6. a) Temperature dependence of the real part of the dielectric response $\varepsilon'(\omega) = \varepsilon_{CDW}$ of the pinned CDW in $K_{0.3}MoO_3$. b) The frequency dependence of $Im\sigma \sim \omega\varepsilon_1$ at temperatures 15-40 K for the same sample [from Yang and Ong (1991)]

The analysis of the frequency dependence of the maximum (Fig. 6) gives again activated CDW dynamics but now with lower activation energy (700 K < Peierls gap value) at lower temperatures (30 < T < 70). However it is possible from the same data to get that ε_{CDW} diverges at temperatures close to 28 K. Moreover it can be noticed that the discontinuity in imaginary part of ac conductivity (Fig. 6b) occurs at approximately the same temperature.

As shown above, the temperature region in CDWs where glassy behaviour has been demonstrated was a very broad one and the single particle (Peierls) gap has been accepted as the highest barrier which is directing the activated dynamics of the complex glassy relaxation although some ambiguities remained evident regarding the activated dynamics at low temperatures. But, of course, a nice theoretical argument concerning the dimensionality exists [Littlewood and Rammal (1988)]. The scale of characteristic height of barriers between metastable states is set by the pinning energy. For a large correlation length - Lee-Rice length this characteristic barrier height will be much larger than k_BT in dimensions d > 2, but much smaller for d < 2. Consequently, the thermal effects should be much stronger in one and two dimensions than in three dimensions.

Although it has been pointed out that the freezing develops gradually, it seems to be very difficult to connect the "high" temperature behaviour with the low temperature (T < 1 K) relaxational behaviour with activation energies of 1-2 K. These energies have been estimated from very slow energy relaxation in specific heat measurements where ergodicity breaking with evolving "aging" effects have been recently established [Biljakovi ć et al (1989, 1991)]. In fact the very recent dielectric measurements of Nad' and Monceau (1993) at low frequencies and at low temperatures are the missing link between these different temperature-frequency windows discussed above.

3. Low Temperature Thermodynamical Properties

As in other disordered material (glasses and spin glasses or polymers) the CDW metastable states are also expected to contribute to the thermodynamical properties at very low temperature. Indeed very rich spectra of the effects revealing the glassy nature of the CDW ground state has been found in low temperature (T < 10 K) specific heat experiments, where additional excitations (LEE) to "regular" phonons contribute to the specific heat C_p, which are somewhat reminiscent of two-level systems (TLS) in structural glasses. On the other hand, and in a different way as for the case of glasses [Loponen et al (1982)], C_p becomes strongly time dependent below T ~ 0.5 K. This was firstly demonstrated by a non-exponential decay of the energy relaxation [Biljaković et al (1989)]. Figure 7a shows two features of the excess heat capacity, which we ascribed to the CDW phase excitations, compared with the specific heat of the structurally akin system with no CDW showing regular Debye behaviour (Fig. 7b). Fig. 7c is the example for the comparison with a typical glassy system such as SiO_2.

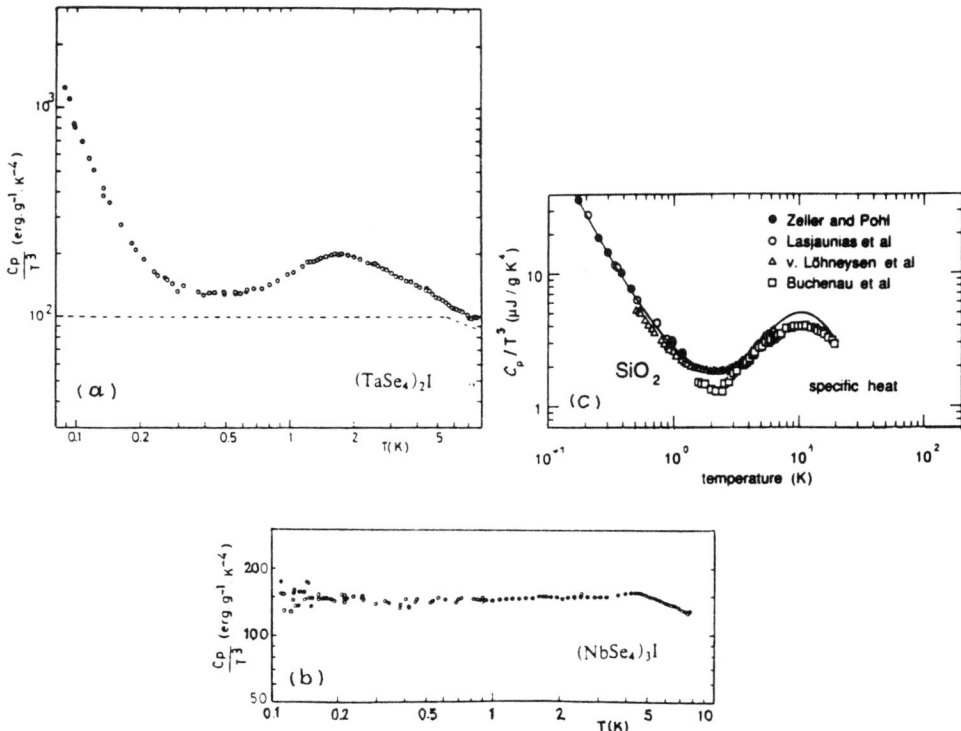

Figure 7. a) Specific heat of $(TaSe_4)_2I$ divided by T^3 shows two characteristic features. b) Specific heat of $(NbSe_4)_3I$ divided by T^3 has a "regular" Debye behaviour [from Biljaković et al (1986)]. c) Specific heat of SiO_2 : C_p/T^3 vs. T shows the comparison to the model calculation [from Gil et al (1993)].

3.1. SPECIFIC HEAT BELOW 1 K

Although the characteristic peak in C_p/T^3 is one of mostly intriguing and attractive features (for testing different models) of C_p at low temperatures, I will essentially consider the temperature range below 1 K, mainly because it is also the temperature region in which the slow relaxational processes take part.

The technique we are using in our investigation of low-temperature thermal properties of CDW systems is a transient heat pulse technique in which the specific heat is calculated from the decay of the temperature increment after a heat pulse, as $\Delta T(t) = \Delta T_0 \exp(-t/\tau)$ (of 10 % or less) (Fig. 9) with $\tau = CR$ as described in Fig. 8. In the case when the system does not have any slow relaxing modes, the relaxation is mainly defined by the thermal link (what is also the experimental limit of our technique). In the glassy state, when an internal relaxation due to LEE is present, this simple electrical equivalent circuit should be used.

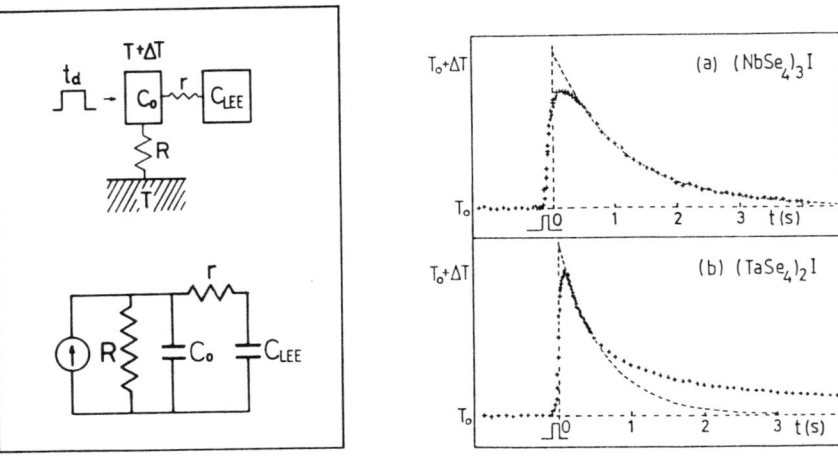

Figure 8   Figure 9

Figure 8. Schematic representation of the method with the equivalent electrical circuit where R represents the thermal link, C_o is the regular part of specific heat (addenda and phonons), and the part which depends on time C_{LEE} coupled to the phonons through the thermal resistance r giving rise to the internal time constant $\tau_{int} = rC_{LEE}$.

Figure 9. Thermal transients after a short heat pulse at T = 0.11 K in structurally similar compounds : (NbSe$_4$)$_3$I without a CDW ground state and (TaSe$_4$)$_2$I exhibiting a Peierls transition at T_p = 265 K. $\Delta T/T_o$ is about 10 %. The dashed line is the exponential decay.

In addition, the kinetics and consequently the value of the heat capacity which can be calculated from the total heat release, is dependent upon the duration of the thermal perturbation t_d. In fig. 10 this time dependence of C_p is shown in the case of o-TaS$_3$ below T = 0.5 K where there appears the large deviation from the phonon contribution, characteristic of the presence of LEE. The maximum value (which is about 10 times larger than in the case of the short-pulse regime at the lowest temperature) is obtained after a duration of perturbation of about $t_d = 10$ hours, when the system has reached the thermodynamical equilibrium (see 3.2.2.). The progressive increase of $C_p(t)$ indicates how heat is poured out into more and more low-energy modes. A more precise analysis of the time-dependence of the different excitations (e.g. the coupling between Ta nuclei and LEE) is under the way, but in the condition of short pulses this extra-contribution obeys a smooth temperature power-law dependence ($\Delta C_p \sim T^\nu$ with $\nu < 1$), as it has been measured in numerous CDW compounds (fig. 11), which reflects a slow energy dependence in the corresponding energy range for the density of states n(E) [Lasjaunias et al (1990)].

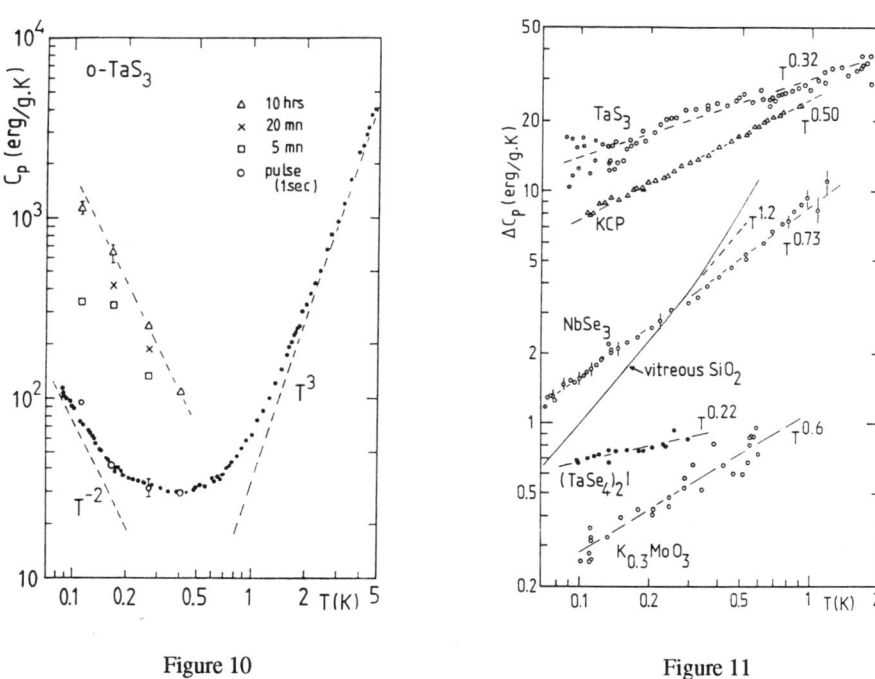

Figure 10   Figure 11

Figure 10. Total specific heat of o-TaS$_3$, measured on different time spans. (•) usual transient technique, i.e. response to a heat pulse of duration of 1 sec or less. Other symbols (Δ, x, □, ○) : C_p is calculated by the integration of total energy release through the heat link, following a thermal perturbation of different durations, from 1 sec to several hours. Maximum values correspond to the thermodynamical equilibrium state reached by the LEE [Lasjaunias et al, to be published].

Figure 11. Residual specific heat (measured by the transient heat pulse technique on a time span of 1-10 sec) after subtraction of the contribution of phonons, of hyperfine nuclear term (in case of TaS$_3$), of residual electronic term (in case of NbSe$_3$). ΔC_p obeys a power law T^ν, with exponent $\nu \lesssim 1$, and the amplitude remains in the range of glassy materials (see the contribution of vitreous silica, varying as $T^{1.2}$, drawn for comparison) [from Lasjaunias et al (1990)].

3.2. LONG TIME RELAXATION OF ENERGY

In the temperature region where LEE give a dominant contribution to the specific heat it is observed in almost all cases that the relaxation of the system, after it has been excited by the small heat perturbation ($\Delta T_0/T \lesssim 10$ %), has a long thermal drift. This deviation from the exponential decay is the manifestation of a new relaxational processes which are slowing down when temperature has been decreased or the perturbation prolonged. It is a typical feature of the glassy state.

Detailed investigation of the long time relaxation of the energy shows four distinct features:
(i) a non-exponential decay,
(ii) a dependence on the time during which the thermal perturbation has been applied,
(iii) the evolution of the relaxation indicating a cross-over between a non-equilibrium state and the thermodynamical equilibrium state when the system has been allowed to "age",
(iv) the relaxation is thermally activated with the activation energy of ~ 1 K (depending on the duration of the perturbation).

3.2.1. *Non-Exponential Decay*. The response $\Delta T(t)$ to two different types of heat perturbations have been studied : (i) to a heat pulse no longer than 1 s and (ii) to a heat flow of different durations t_d (playing the role of a waiting time t_w by analogy to spin glasses) up to longer than a day (~ 10^5 sec) [Biljaković et al (1989)].

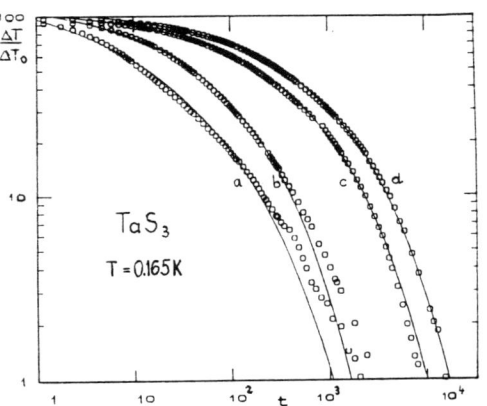

	t_d(s)	α	β	τ(s)
a	0.01	0.064	0.41	30
b	10	0.033	0.50	80
c	4×10^4	0.014	0.48	410
d	1.3×10^5	0.011	0.49	800

Figure 12. Variation of $\Delta T/\Delta T_0$ as function of time in a log-log plot for TaS$_3$ at T = 0.165 K and for different waiting times. Fitting parameters (for Eq. 2) are given in the table.

The decay can be accounted for by a stretched exponential function with the pre-exponential factor which mainly corresponds to an initial decay and obeys a power law regime (Fig. 12).

$$\frac{\Delta T}{\Delta T_0} = A \left(\frac{t}{\tau_p}\right)^{-\alpha} \exp\left[-\left(\frac{t}{\tau_p}\right)^{\beta}\right] \qquad (2)$$

Both, α and β parameters are temperature dependent and while α depends drastically on time, β does only slightly. For TaS$_3$ β drops continuously from a value very close to 1 at 0.6 K down to 0.25 at 0.1 K as shown in Fig. 13. Together with the initial β(T) dependence, this hints that thermally activated processes, in contrast to tunneling processes, are the origin of the time dependent specific heat in glasses.

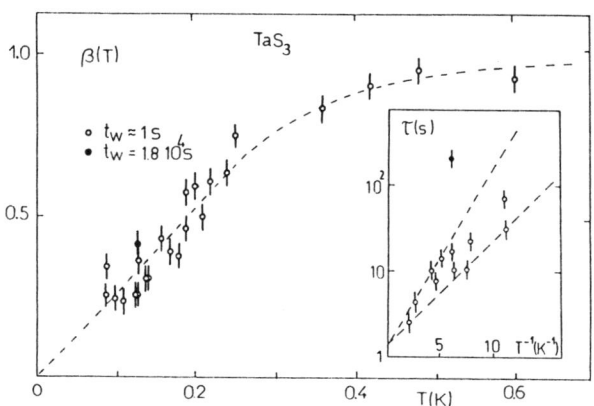

Figure 13. Temperature dependence of parameters β and τ_p defined in Eq. 1 mainly for short pulses. Only τ_p changes with the waiting time, β does not.

3.2.2. *Aging Effect and Cross-Over to Equilibrium*. Together with polymers the SGs show very pronounced aging properties. The response of the system after a perturbation has been switched of depends on the time this perturbation has been applied [Lungren L. *et al* (1983) and Alba M. *et al* (1986)]. This effect obviously exists in CDWs as already shown on fig. 12. We assume that ΔT(t)/ΔT(0) may be expressed in terms of a distribution of relaxation times g(lnτ) as :

$$\frac{\Delta T(t)}{\Delta T(0)} = \int_{t_{co}}^{\infty} g(\ln\tau) \exp(-t/\tau) d\ln\tau \qquad (2)$$

where t_{co} is a short-time cut-off ; g(lnτ) can be obtained from the partial differentiation of the relaxation function with respect to lnt [Lungren *et al* (1983)] :

$$g(\ln\tau) \simeq \frac{d[\Delta T(t)/\Delta T(0)]}{d\ln t} \qquad (3)$$

The logarithmic derivative of ΔT(t) drawn in fig. 14a and b shows a peak in the relaxation rate at τ_p. The pronounced increase of the aging effect when T is decreased is also clearly presented, as well as the change in the shape of the relaxation rate. τ_p increases with t_d indicating that the system has reached its thermodynamical equilibrium when τ_p saturates at long t_d as shown on fig. 15 for two CDW systems.

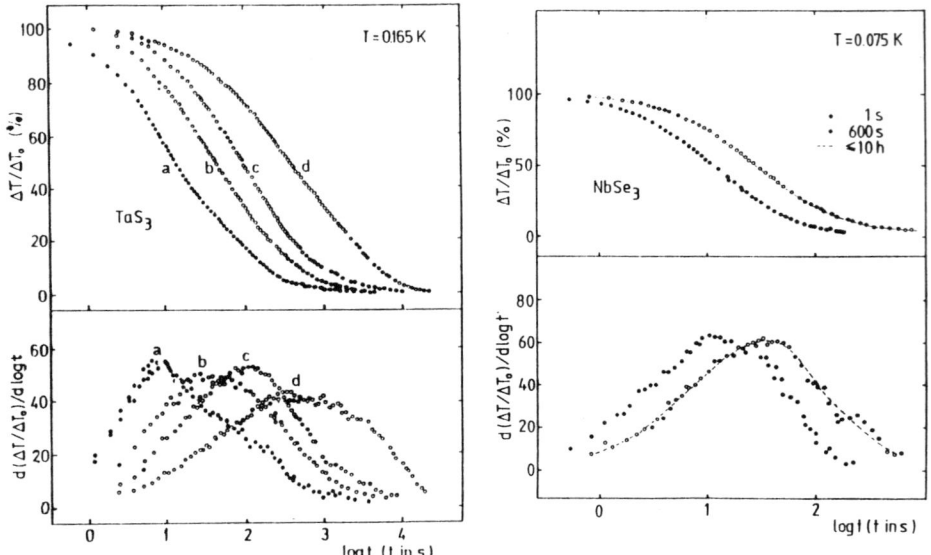

Figure 14. Aging effects for a) TaS$_3$ at T = 0.165 K and b) NbSe$_3$ at T = 0.075 K. In NbSe$_3$ the saturation occurs at much shorter times than in TaS$_3$.

The non-equilibrium properties reflected in the aging effect are much more pronounced in TaS$_3$ (fig. 14a) and (TaSe$_4$)$_2$I than in NbSe$_3$ (fig. 14b) where they appear only below 150 mK. Thus for TaS$_3$ at T = 0.1 K (fig. 15) the behaviour is similar to that of spin glasses with the maximum t_w (10 hours) which remain smaller or equal to the longest relaxation time. To our knowledge, it is the first evidence reported in glassy materials for such a *cross-over between non-equilibrium and equilibrium states*.

Different predictions for the divergence of the relaxation time at the phase transition temperature T_c have been given by different models for disordered systems. For a finite T_c, in the hypothesis of critical slowing down, τ diverges as

$$\tau/\tau_0 \sim (1-T_c/T)^{-z\nu} = \xi^z , \qquad (5)$$

where ν is the critical exponent for the correlation length ξ and z the critical dynamical exponent. Acceptable fits for TaS$_3$ would give very small T_c ($T_c \lesssim 0.03$ K) and huge $z\nu$ ($z\nu > 30$) and respectively $T_c \lesssim 0.01$ K and $z\nu \gtrsim 25$ for NbSe$_3$. Alternatively an Arrhenius law $\tau = \tau_0 \exp(W/kT)$ gives comparable fits. This is not surprising when viewed in the spirit of Souletie's argument (1988) that expression (5) remains analytic (it can be expanded in terms of T_c/T at all temperatures larger than T_c). In the limit of $T > T_c$, $\tau/\tau_0 \sim 1+z\nu T_c/T+ ...$, where $z\nu T_c$ exists and remains finite. Therefore the Arrhenius law is simply the limit of zero critical temperature of Eq. (5) with the activation energy given by $W = z\nu T_c$. Figure 16 where we have drawn the variation of τ_{max} as a function of 1/T, shows that an activated behavior is followed for TaS$_3$ with $W \sim 0.8$-1 K (similar value is obtained for (TaSe$_4$)$_2$I) and $W \sim 0.25$ K for NbSe$_3$. When the system is not allowed to reach equilibrium, for example, for short pulses of 1 s in TaS$_3$, the activation energy is almost 3 times lower (dashed line in Fig. 16). Thus the effect of aging yields a deepening of the local minima in the energy configuration.

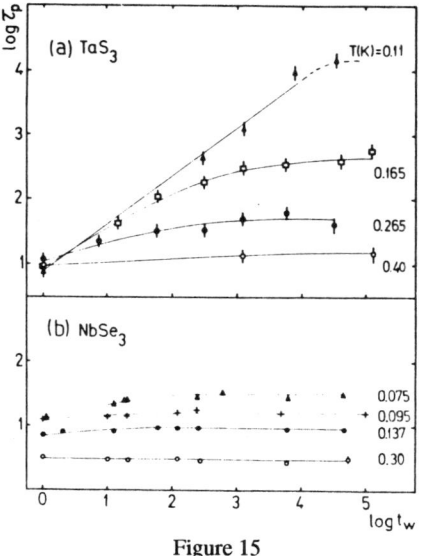
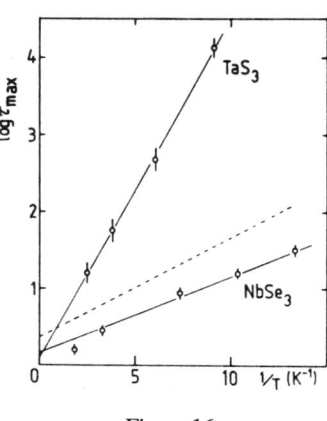

Figure 15   Figure 16

Figure 15. Variation of the peak in the relaxation rate (fig. 14) as a function of the waiting time $t_w = t_d$ at several temperatures. The cross over between non-equilibrium and equilibrium state occurs when the variation of τ_p flattens off with t_w.

Figure 16. Variation of τ_{max} : the saturated value of τ_p in fig. 2, as a function of 1/T, shows activated behaviour.

These results show that, below 0.5 K, CDW internal degrees of freedom are still active with activation energies of ~ 1 K. Defects in the CDW superstructure, phase slips in 1D, dislocation loops in 3D, might be at the origin of the long time, low T, energy relaxation. One can thus think about a structure of a network of dislocations of non-uniform size similar to the ramified fractal lattice on percolation clusters with similar dynamics. The time constant of these large objects is expected to be high, which may account for the τ_0 value of a few seconds in the Arrhenius behaviour (fig. 16).

4. Long Time Relaxation in SDW Systems

Without entering deeper into the problem, I will only quote Overhauser (1960) who said that the spin-density wave can be thought of most easily by considering it to be the sum of two CDWs, with antiparallel spins and opposite phases, therefore keeping the total charge density of the electron gas uniform in space.

Recent measurements in the spin density wave (SDW) state of $(TMTSF)_2PF_6$ (TMTSF is tetramethyltetraselenofulvalene) showed results very similar to those in CDWs. A set of nontrivial properties could not be explained in the context of a simple model of SDW state : thus, the electrical conductivity carried by the sliding SDW has been shown to suddenly drop below 2.6 K [Tomić et al (1991) and Kriza et al (1991)], the low frequency relaxation dynamics reflecting the internal SDW deformations has been reported to freeze out below 2 K [Mihaly G. et al(1991)] ;

anomalies in the spin lattice relaxation ($1/T_1$) have been reported at ~ 3 K and 1.9 K with a non-exponential relaxation below 3 K [Takahashi *et al* (1991) and Wang *et al* (1993)]. Finally the specific heat shows a jump at approximately the same temperature with out of equilibrium features at the lower temperatures [Lasjaunias *et al* (1992)].

Specific heat measurements integrate all possible degrees of freedom (excitations) if the system is allowed to reach equilibrium. If this condition cannot be fulfilled because of diverging relaxation times, the underlying thermodynamics are masked by kinetic effects. As in the case of our previous studies on CDW materials, we have observed the progressive deviation to the exponential decay below 0.6 K announcing new, very slow relaxational times originating from the LEE typical for glassy states as represented in Fig. 17.

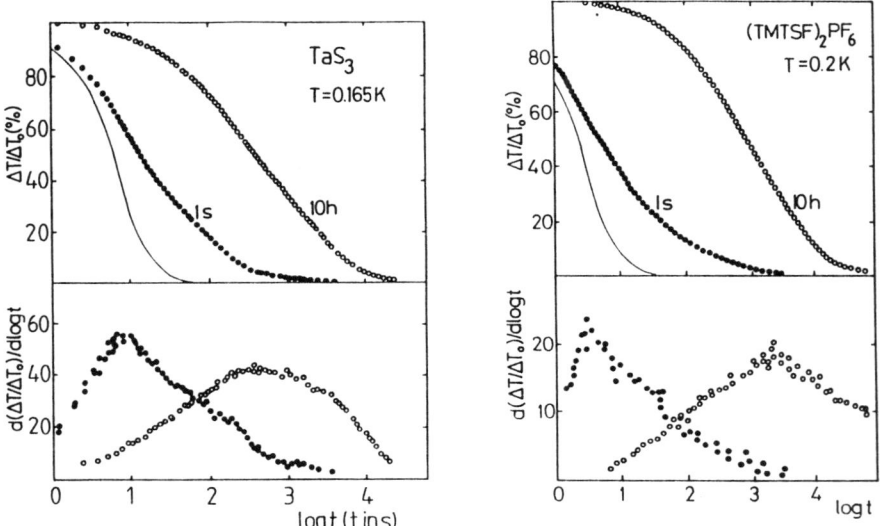

Figure 17. The relaxation of the total specimen (Fig. 8) after the small heat perturbation (upper curves) of TaS_3 (CDW) and PF_6 (SDW) and its derivative (lower curves) for two extreme conditions ; (•) short pulses (~ 1 s) and (o) long perturbation (10 h). The lines correspond to exponential decay with the time constant defined by the maximum in the relaxation rate. The shape of relaxation rate gives information on the distribution of relaxation time. It broadens with the duration of the perturbation and translates to longer times (aging effect). The effects are more pronounced in PF_6.

We obtain even more pronounced "aging" effects in $(TMTSF)_2PF_6$ than in CDW systems, especially TaS_3, for which these effects were the strongest. The position of the peak at τ_{max} in the energy relaxation rate is shifted towards larger times when the duration of perturbation (or waiting time t_w) is increased. However these aging effects saturate beyond an ergodic time t_e [τ_{max} saturates at t_e] revealing a cross-over between a non-equilibrium ground state and a thermodynamical equilibrium state as in CDWs.

Similarly, the variation of t_e with temperature obeys an Arrhenius law (upper part of Fig. 18), that shows that the system rather follows activated dynamics of the remaining very low LEE modes with an activation energy $W \simeq 1.8$ K (to be compared to $W \sim 1$ K for TaS_3).

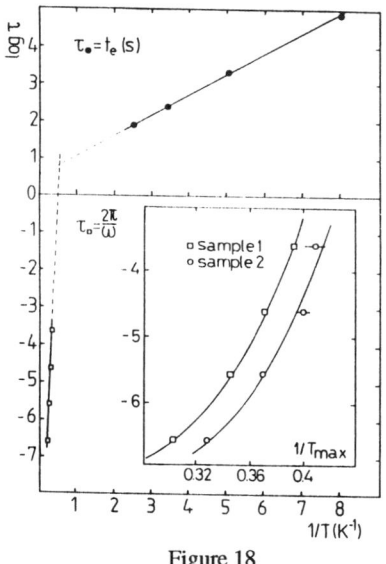

Figure 18

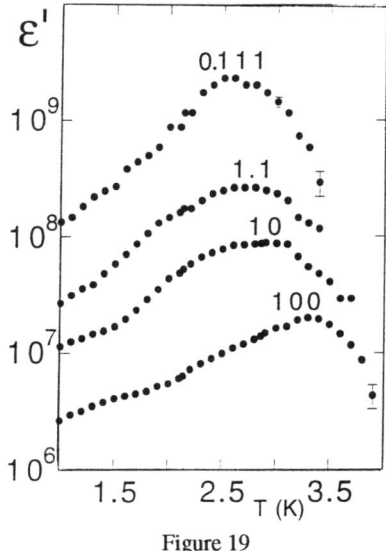

Figure 19

Figure 18. Variation of relaxation time τ in dielectric (□) and energy relaxation (•) measurements as a function of T^{-1} in $(TMTSF)_2PF_6$. For dielectric relaxation the data $\tau = \nu^{-1}$ correspond to maxima in the real part of the dielectric constant as shown in Fig. 1 for $\nu = \omega/2\pi$. Relaxation time for energy relaxation $\tau = t_e$ is this one at the thermodynamical equilibrium when aging effects are saturated. When extrapolated, both relaxation processes merge at $T_g \sim 2$ K. The inset shows on an extended scale the dielectric relaxation time for sample 1 and 2. The full curve is the Volger-Fulcher or critical slowing down law (with parameters given in the text).

Figure 19. Temperature variation of the real part of the dielectric constant ε' of $(TMTSF)_2PF_6$ at frequencies given in kHz.

As for CDWs, the critical slowing down analysis can be done and the estimated CSD exponent is $z\nu > 100$ (!!) with $T_c \lesssim 0.01$ K.

In CDWs, the temperature range of observed glassy dynamics was very broad ($T < T_p$; with $T_p \gtrsim 100$ K). Regarding the relatively squeezed temperature region (in PF_6 $T_p \simeq 12$ K), a possible glassy transition should appear very close, or inside our temperature range. However one of the features of C_p of PF_6 measured below the SDW transition was the small step (~ 15 %) at $T \sim 3$ K, temperature below which C_p becomes time dependent [Lasjaunias et al (1992)]. Is it possible to interpret this "anomaly" as the calorimetric glassy transition in the SDW phase ? We found also the decoupling of fast and slow C_p modes ; the fast ones associated with the underlying crystal lattice and the slow ones originating from the LEE typical for glassy states [Biljaković et al, to be published]. The answer to the question which arises naturally -"Is there a real glass transition above this temperature range where the very slow modes dominate ?" we tried to find it in the dielectric relaxation.

The peaks we observe in dielectric constant, shown in Fig. 19, and their dependence with frequency and temperature are reminiscent of very similar dependences measured in various glassy-like materials as spin-glasses [the review ; Binder and Young (1986)], orientational glasses [the review ; Höchli et al (1990)] and also CDW glasses [Nad' and Monceau (1993)]. In all these materials, the divergence of ϵ' occurs when some characteristic temperature T_g is approached. The "cusps" in ϵ' in Fig. 19 are the onset of "blocking" of the cluster or domain size which occurs for $1/\tau \simeq \omega$ when the system is not able to follow the perturbation any more. The dependence with temperature of the mean relaxation time τ is drawn in the lower part of Fig. 18, and with an extended scale in the inset of Fig. 18. The frequency sensitivity of the temperature T_m of the maxima has been widely used as the criterion to classify the different glasses : while an Arrhenius law is predicted for non-interacting clusters, the slight curvature in the log $\tau = f(1/T_m)$ plot drawn in the inset of Fig. 18 indicates that either the phenomenological Vogel-Fulcher (VF) law :

$$\tau/\tau_0 \simeq \exp[(E_a/k(T_m-T_0)] \tag{6}$$

or the critical slowing down (CSD), Eq. 5, should be valid. VF law implies a characteristic local ordering temperature T_0 arising from interactions between the clusters with the largest relaxation time. The relaxation is controlled by energy barriers of order E_a. CSD behaviour is associated with a real transition at T_c where τ diverges. From VF one gets $10^{-9} < \tau_0 \leq 10^{-8}$ s, $8 < \epsilon_a < 10$ K, $1.8 < T_0 < 2$ K and from CSD : $10^{-10} < \tau_0 \leq 10^{-9}$ s, $6 < z\nu < 8$ and $2.2 < T_c < 2.3$ K. Experiments at lower frequency are underway to better differentiate between both behaviours.

The separation or bifurcation [Rössler (1991)] between two different relaxation processes at low temperatures in (TMTSF)$_2$PF$_6$ is clearly visible in Fig. 18. If the measured dependence of relaxation time with T are both extrapolated, the cross-over between both processes would occur around 2 K, temperature we can tentatively define as the glassy transition temperature T_g.

5. Glass Transition ?

The low temperature experimental data of dielectric constant and specific heat of one CDW prototype system, TaS$_3$ and one SDW prototype system, PF$_6$, substantiate the separation into two distinct dynamical regions. The dipolar relaxation dynamics in these DWs shows strong slowing down and points towards a "static" glass transition temperature ($T_g \sim 2$ K in PF$_6$ and $T_g \sim 13$ K in TaS$_3$). The pronounced time dependent specific heat shows enormously increased time needed for achieving the thermodynamic equilibrium. This very slow relaxational process we attribute to the β process, i.e. the secondary process in the glassy state (there exists certain confusion in using α and β in the terminology of field of glasses).

We are now going to briefly discuss these results in the general frame of the glassy transition. It is widely accepted that slow, secondary, β relaxation with activation energy following an Arrhenius law is a basic feature of the glass transition [Goldstein (1960)]. In contrast to the fast, α, process, the β process also persists below T_g. At higher temperatures the β process merges into the α one when its mean correlation time reaches that of the α process. In this context we can propose the same scenario for SDW's as for CDW's [Biljaković et al (1989,1991)] as there are no essential differences between DW's except that there is no lattice distortion in SDW.

The picture of viscous flow was the source of the prediction of secondary relaxations as a universal property of liquids [Goldstein (1969)]. It seems that β relaxations are extrinsic to the glassy state, arising from degrees of freedom that fortuitously remain active below T_g, but are not relevant to the degrees of freedom (types of motion) that give rise to the viscous elastic response of a "liquid", and whose freezing out gives rise to the experimental glass transition.

Although long time existing model of CDW motion relative to the electron fluid (incompressible) suggested by McMillan (1975) and further elaborated (for collective transport of the charge with the moving CDW) by Littlewood (1987) and Littlewood and Rammal (1988) gave the evidence of low-frequency response, it was never brought to the point when some other, low temperature glassy transition has to be discussed except the Peierls transition by itself. The electrical resistance of the motion of the CDW relative to the electron fluid has been accepted as the dominant mechanism of energy dissipation so that the damping parameter, or the "viscosity" scales with the normal resistivity, $\gamma \sim \rho_n$. The freezing out of screening currents has as the consequence the increased rigidity of the condensate. Considering only the elastic deformation of the CDW this kind of model can give the explanation for the gradual freezing of CDW below Peierls transition but it cannot account for the very slow relaxational dynamics with very low activation energies in the temperature range where ergodicity breaking with evolving "aging" effects have been established. Variation of the CDW amplitude should actually be strong in the vicinity of DLs which delimit the domains but the amplitude excitations have been totally ignored in the elastic model. As we proposed [Biljaković et al (1991)], the network of DLs, coarsening in the freezing process, might give to the system another possibility to relax.

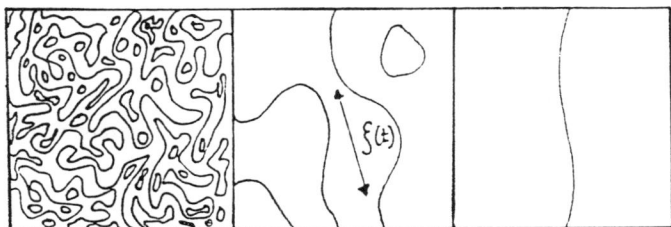

Figure 20. Coarsening of a quench-cooled Ising model. The growth in the time of the typical domain size follows a power-law : $\xi(t) \sim t^p$ [from Bruinsma (1990)].

Following the description in the phase space, it may be said that the landscape of the restricted phase space into which the system sinks during the glass transition can be represented as a deep well with a rough and corrugated bottom. These minima correspond to degrees of freedom not affected by the glass transition which are associated to topological defects in the DW superstructure such as dislocation loops and/or disclinations. Certainly a dislocation loop cannot jump as a whole rigid entity ; there must be a sequential or cooperative character to the process involving only localized portions of any loop.

Glass-forming systems have been ranged in 2 or 3 different classes of universality characterized by $zv \sim 6\pm1$, $zv \sim 20\pm5$ and $zv \rightarrow \infty$ [Souletie (1990)]. Above ~ 2 K the dielectric relaxation processes in PF_6 can be described with $zv \sim 7$. We showed that at lower temperatures energy relaxation implies $zv \gtrsim 100$, thus revealing a cross-over from "fragile" ($zv \sim 7$) to "strong" behaviour. Similar, but not so drastic cross-over has been found in TaS_3, as the dielectric relaxation was characterized by $zv \sim 15$ and the value obtained from equilibrium relaxation time of specific heat was $zv \gtrsim 30$. Very likely is the change of dimensionality of the system on approaching the critical temperature. It will be of great interest to classify DWs regarding their dynamical behaviour and to find out the relation between the "fragility" of the system, the intensity of underlying electron-phonon interactions and screening phenomena.

Besides the glass transition temperature T_g, there exists a temperature T_A which marks a transition between two regimes of different relaxation mechanism [Fischer *et al* (1992)] and especially the appearance of cooperatively rearranging domains (CRD) in which the elementary units are trapped after loosing their degrees of freedom. The apparent change of dynamics in DW systems happens below MHz range which might imply in analogy with Fischer *et al* (1992) the characteristic length (CRD) of few hundred nanometers. This allows us to characterize the DW's as mesoscopic glasses putting them together with colloidal systems and polymers which are also showing very long living metastable states.

The apparent similarity of the very slow relaxation in DW systems with that one in spin glasses deserves some consideration too because on the one hand it is striking to find the characteristics of SG in a system with long range SDW order and on the other hand to know that an attempt has been made to treat the metallic SG as a short-range "frustrated" SDW [Werner (1990)]. A recent model of aging in disordered systems allows us to quantify this effect [Bouchaud (1992)]. Our measurements in SDWs as those previously in CDWs are the first to clearly show a cross-over from a non-equilibrium to an equilibrium regime in the aging phenomenon which has been designated by Bouchaud (1992) as "interrupted aging". Our results indicate that the ergodic time scale in DWs is much shorter than in SG. The ergodic time corresponds to the time for which all accessible metastable states in the corrugated bottom have been visited and so it is related to the number of available states and it is a measure of the "complexity" of the system. In SG and DWs the number of metastable states increases when the temperature is lowered. In PF_6 the ergodic time corresponds to the maximum of the equilibrium relaxation rate t_e (Fig. 18). This time reaches a value of 10^5 s at 0.120 K while in SG compounds as AgMn or CrIn this ergodic time is ~ 10^6 s near T_g [Bouchaud *et al* (1993)]. The "complexity" of DWs will only reach that of SG at very low temperature i.e. below 0.1 K.

The very unique features of DW systems overlap with similar properties of very diverse set of glassy materials ranging from spin glasses to polymers but they also show some particularities which might help to better understanding of glassy behaviour. DW system present a very good model system for reduced dimensionality, competing time and energy scales.

Acknowledgements. - This work has been growing up in very stimulating collaboration with my colleagues P. Monceau and J.C. Lasjaunias in CRTBT (CNRS) in Grenoble, where also all experiments have been performed. I thank J. Souletie for introducing me the universalities, B. Castaing, R. Currat, E. Lorenzo-Diaz, F. Nad', J. Odin, M. Saint-Paul and other colleagues from the same laboratory for long and illuminating discussions and their hospitality during my frequent visits. I am grateful to E. Vincent, J. Hamman and J.P. Bouchaud for helping me to better understand the spin glasses and for bringing our fields closer. Some of results would not be so nicely fitted without the help of the students D. Starešinić and S. Slijepčević. I also appreciate the moral help and support of all my colleagues at Institute of Physics of the University of Zagreb, my home institution. Without the great help and self-denying of Danièle Devillers this paper would not be finished on time. I also thank A. Briggs for making the final corrections.

I remember so often Dubravka Janda my professor of philosophy in XIV Gymnasium "25. Maj" in Zagreb who inspired me with her presentation of the problem of Subject and Object to study Physics. She passed away last year.

REFERENCES

Alba, M., Ocio, M. and Hamman, J. (1986), Europhys. Lett., 2, 45.

Biljaković, K., Lasjaunias, J.C., Zougmoré, F., Monceau, P., Levy, F., Bernard, L. and Currat, R. (1986), Phys. Rev. Lett., 57, 1907.

Biljaković, K., Lasjaunias, J.L., Monceau, P. and Levy, F. (1989), Europhys. Lett. 8, 771.

Biljaković, K., Lasjaunias, J.L., Monceau, P. (1989), Phys. Rev. Lett., 62, 1512.

Biljaković, K., Lasjaunias, J.L., Monceau, P. (1991), Phys. Rev. B, 43, 3117.

Biljaković, K., Lasjaunias, J.L., Monceau, P. (1991), Phys. Rev. Lett., 67, 1902.

Binder, K. and Young, A.P. (1986), Review of Modern Physics, 58, 801.

Bouchaud, J.P. (1992), J. Phys. I, 2, 1705.

Bouchaud, J.P., Vincent, E. and Hammann, J., to be published.

Bruinsma, R. (1990), in the book 6 (eds. Baeriswyl *et al*), 209.

Cava, R.J., Fleming, R.M., Littlewood, P., Rietman, E.A., Schneemeyer, L.F. and Dunn, R.G. (1984), Phys. Rev. B, 30, 3228.

Duggan, D.M., Jing, T.W., Ong, N.P. and Lee, P.A. (1985), Phys. Rev. B, 32, 1397.

Fischer, E.W., Donth, E., Steffen, W. (1992), Phys. Rev. Lett., 68, 2344, and references therein.

Fleming, R.M. and Grimes, C.C. (1979), Phys. Rev. Lett., 42, 1423.

Fleming, R.M. and Cava, R.J. (1989), in the book 1 (ed. C. Schlenker), 259.

Fröhlich, H. (1954), Proc. R. Soc. A, 223, 296.

Fukuyama, A. and Lee, P.A. (1978), Phys. Rev. B, 17, 535.

Gil, L., Ramos, A., Bringer, A. and Buchenau, U. (1993), Phys. Rev. Lett., 70, 182.

Gill, J.C. (1981), Solid State Commun., 39, 1203.

Gill, J.C. (1983) in Proceedings of the International Symposium of Non-Linear Transport and Related Phenomena in Inorganic Quasi One-Dimensional Conductors, Hokkaïdo University, 139.

Gill, J.C. (1984), in book 1 (eds. Hutiray *et al*), 377.

Goldstein, M. (1969), J. Chem. Phys., 51, 3728.

Gor'kov, L.P. and Grüner, G. (1989), in book 4 (ed. L.P. Gor'kov and G. Grüner), 1.

Grüner, G., Tippie, L.C., Sanny, J., Clark, W.G. and Ong, N.P. (1980), Phys. Rev. Lett., 45, 935

Havriliak, S. and Negami, S. (1966), J. Polym. Sci. C, 14, 99.

Höchli, H.T., Knorr, K., Loidl, A. (1990), Adv. Phys., 39, 405.

Kriza, G., Quirion, G., Traetteberg, O. and Jérôme, D. (1991), Europhys. Lett., 16, 585.

Lasjaunias, J.C., Biljaković, K. and Monceau, P. (1990), Physica B, 165&166, 893.

Lasjaunias, J.C., Biljaković, K., Monceau, P. and Bechgaard, K. (1992), Solid State Commun., 84, 297.

Lee, P.A., Rice, T.M. and Anderson, P.W., Solid State Commun., 14, 703.

Lee, P.A. and Rice, T.M. (1979), Phys. Rev. B, 19, 3970.

Littlewood, P.B. (1987), Phys. Rev. B, 36, 3108.

Littlewood, P.B. and Rammal, R. (1988), Phys. Rev. B, 38, 1675.

Loponen, M.T., Dynes, R.C., Narayanamurti, V. and Garno, J.P. (1982), Phys. Rev. B, 25, 1161.

Lungren, L., Svedlindh, P., Mordblad, P. and Beckman, O. (1983), Phys. Rev. Lett., 51, 911.

Mc Millan, W.L. (1976), Phys. Rev. B, 14, 1496.

Mc Millan, W.L. (1975), Phys. Rev. B, 12, 1197.

Mihaly, G., Kim, Y. and Grüner, G. (1991), Phys. Rev. Lett., 67, 2713.

Monceau, P., Ong, N.P., Portis, A., Meerschaut, A. and Rouxel, J. (1976), Phys. Rev. Lett., 37, 602.

Monceau, P., Richard, J. and Renard, M. (1980), Phys. Rev. Lett., 45, 43.

Nad', F. and Monceau, P. (1993), to appear in Solid State Commun.
Ngai, K.L. and Lin Fu-Sin (1981), Phys. Rev. B, 24, 1049, references therein.
Ngai, K.L., private communication.
Ong, N.P. and Monceau, P. (1977), Phys. Rev. B, 16, 3443.
Ong, N.P., Duggan, D.D., Kalem, C.B. and Jing, T.W. (1985), in book 1 (eds. Hutiray et al), 387.
Overhauser, A.W. (1960), J. Phys. Chem. Solids, 13, 71.
Peierls, R.E. (1955), Quantum Theory of Solids, Oxford University Press, London.
Pietronero, L. (1989), in Progress on Electron Properties of Solids (eds. R. Girlando et al), Kluwer Academic Publishers, p. 239.
Rössler, E. (1991), J. Non-Cryst. Solids, 131-133, 242.
Souletie, J. (1990), J. Phys. France, 51, 883.
Souletie J. (1988), J. Phys. (Paris), 49, 1211.
Takahashi, T., Harada, T., Kobayashi, Y., Kanoda, K., Suzuki, K., Murata, K. and Saito, G. (1991), Synth. Metals, 41, 3985.
Tomić, S., Cooper, J.R., Kang, W., Jérôme, D. and Maki, K. (1991), J. Phys. I (France), 1, 1603.
Werner, S.A. (1990), Comments Cond. Mat. Phys., 15, 55.
Wong, W.H., Hanson, M.E., Alavi, B., Clark, W.G. and Hines, W.A. (1993), Phys. Rev. Lett., 70, 1882.
Yang, Jie and Ong, N.P. (1991), Phys. Rev. B, 44, 7912.

BOOKS

1. "Charge-density waves in solids" (1984), Proceedings of the International Conference, Budapest (1984), Lecture Notes in Physics, eds. G. Hutiray and J. Solyom, Springer-Verlag, Berlin.
2. "Electronic Properties of inorganic quasi-one-dimensional compounds" (1985), Part I - Theoretical ; Part II - Experimental, Physics and Chemistry of Materials with Low-Dimensional Structure, ed. P. Monceau, D. Reidel Publishing Company, Dordrecht.
3. "Low-dimensional conductors and superconductors" (1987), NATO ASI Series B : Physics, Vol. 155, eds. D. Jérôme and L.G. Caron, Plenum Press, New York.
4. "Low-dimensional electronic properties of molybdenum bronzes and oxides" (1989), Physics and Chemistry of Materials with Low-Dimensional Structures, Vol. 11, ed. C. Schlenker, Kluwer Academic Publishers, Dordrecht.
5. "Charge-density waves in solids" (1989), Modern Problems in Condensed Matter Sciences, Vol. 25, eds L.P. Gor'kov and G. Grüner, North-Holland, Amsterdam.
6. "Applications of statistical and field theory methods to condensed matter" (1990), NATO ASI Series B : Physics, Vol. 218, eds. Baeriswyl et al, Plenum Press, New York.

The Critical Behavior of 1-d Charge Density Waves

ZEEV OLAMI
Department of Chemical Physics
Weizmann Institute of Science
Rehovot Israel

ABSTRACT A theory for the statics and domain structure of a coarse grained model for Charge Density Wave is presented. Arguments for exponents and exponent relations are given.It is shown that in the case of elastic instabilities the correlation exponents are smaller then the naive exponents 2/d. The relaxation is connected with the statics of this model in and below the critical point.It is shown that below and in the critical point the system does not develop any strain singularities. There is a diverging discontinuity between the width of the surface in the moving state and in the critical point. Phase slips are expected to have an effect only above the critical point in the moving state.

1 Introduction

The dynamics of charge density waves under an applied field has received a lot of attention [1]. Those systems have a set of interesting properties. The depinning transition is a dynamical phase transition where the polarization and correlation length diverge as the system approaches the depinning transition. Below the depinning transition the system displays many properties which are reminiscent of glassy systems. Many metastable states with a large distance between them in phase space exist in this system. In the reaction to stretched field slow relaxations are observed and hysteretic reaction of the polarization and the current are observed experimentally [2],[3].

The usual models for this transition are derived from the Fukuyama Lee Rice hamiltonian [4]. The critical behaviour below the transition is poorly understood though it was studied numerically by several authors [5],[6] . It is evidently useful to study a coarse grained and Mihally et. al. and studied later also by Parisi and Pietronero[9] [10]. The model is defined by the following equations:

$$\frac{d\phi_i}{dt} = \theta(F_i)F_i F_i = -2\phi_i + \phi_{i-1} + \phi_{i+1} + F - \mu_i \tag{1}$$

The phases ϕ_i can be considered as coarse grained versions of the original phases in the usual models. h_i is a local random variable with a constant probability distribution between o and 1. F is the applied field and θ is the a Heavyside function ($\theta(x) = 1$ for a positive x and 0 for negative x.

Argument for the relevance of this model can be found in both references. It is clearly the limit of small elastic constants in the FLR model. However, this models miss the essential nonstabilities and jumping above the transition. However below the transition when the individual phases move a large distance this model is a good approximation. Therefore we mainly focus on those properties in this paper.

We analyze the reaction of an initially pinned system to fields smaller then the critical field. We estimate the different critical exponents connect them to the polarization exponents and present numerical results to complete the set of exponents and to check the validity of the exponent relations. We then discuss the temporal relaxation and relate it to the static characterization of the model.

2 The statics of the model

We first briefly discuss the dynamics above the depinning transition. It can be easily shown that $F_c = \langle \mu_i \rangle$ and that the current scales linearly with the field above the transition. The phase differences in this steady state can be derived using the fact that the force has to be the same all over the lattice. Using this the phase difference in the point i is

$$\delta_i = \phi_{i+1} - \phi_i = \sum_{i=1}^{i} \mu_j - \langle \mu \rangle \qquad (2)$$

Notice that those phase differences are not dependent on the driving force in the depinned steady state running state. The relative positions are always the same in relation to each other above the transition. A much weaker result was proved by Middleton lately for the usual dynamical rules [13]. The introduction of a field reduces the phase space volume of the metastable states. A randomly chosen state will be far from this reduced state so we expect to see similar results near the transition for all randomly chosen states. The selected state will not be related to any minimal energy consideration or to any ground state energy. It will usually be the first stable state the system reaches under this force.

It was noted by Coppersmith that eq. (2) implies a singularity of the local strain with system size [11]. It can be easily shown that the local strain scales as $L^{1/2}$. This indicates that there is a critical system size beyond which corrections because of phase slips will have to be introduced. Similar effects are seen above the transition because of the boundary conditions. We show in this paper that for the one dimensional model and probably in higher dimensions those singularities will not exist below and in the transition point. The physical scenario for this kind of models is that phase slips will be part of the transport process but will appear only above the threshold point in the moving state. So the physical picture below the critical point is valid for any system size.

The main physical process we want to understand is the reaction of an initially pinned system to a field below the depinning transition. When a field is introduced on a pinned system a certain part of the system will be depinned. Initially the pinned points will be determined by the condition $F - \mu_i < 0$. Since the elastic interaction can depin those points the clusters can grow until their growth will finally be frozen by their pinned boundary points. As the field approaches the critical value the final sizes of the depinned domains will diverge. However below F_c the domain growth is always stopped by the boundary points.

The natural way to describe the system below F_c, after it relaxed to the final frozen state, is by the domain number distribution function $N_f(r, L)$, where r is the domain size, $f = F_c - F$ and L is the system size. We assume that this function have the following scaling form near the transition:

$$N_f(r, L) = L f^\lambda r^{-\tau} f(r/\xi(f)) \qquad (3)$$

where the correlation length diverges near the transition as $\xi(f) \sim f^{-\nu}$. Since throughout the process the total length of the depinned domains is proportional to L the following exponent relation is found for $1 < \tau < 2$. The average cluster size will scale in this case as: $\langle r \rangle \sim f^{-\lambda}$.

We first estimate the critical exponents for this configuration. To do this we first prove that the average sum of the local field fluctuation $h_i - \rangle h \langle$ in a single unpinned domain is not dependent on it's average size. We define this quantity as

$$\mu_{sc} = 1/N_d \sum_{unpinned-points} \mu_i - \langle \mu \rangle$$

We first note that the total sum of this fluctuation over pinned and unpinned domains is zero. Therefore

$$\sum_{unpinned-domains} (\mu_i - F_c) + \sum_{pinned-domains} (\mu_i - F_c) = 0$$

. Since the pinned domains do not change dramatically in size during the growth process we can estimate the second term to be $N_d R_p W_f$, where R_p is the scale of the pinned clusters, N_d is their number (equal to the number of unpinned domains) and W_f is the scale of the distribution width. This implies

$$\mu_{sc} = R_p W_f \qquad (4)$$

This enable us to calculate the average size of a cluster below F_c. The average force on cluster of size $\langle r \rangle$ is $RW_f - f\langle r \rangle$. Since this average should be larger then zero this implies that

$$\langle r \rangle \sim f^{-1} \qquad (5)$$
$$\lambda = 1.0 \qquad (6)$$

We now discuss the constraints for the stability of a single depinned domain. A domain is a free string held in it's two sides by a pinned point where the force cannot be larger then the pinning forces. This elastic problem can be easily solved. We get two constraints on the random variables μ_i in the domain.

$$\sum_0^N \mu_i < c \qquad (7)$$

$$0 > \sum_0^N \mu_i/N > -c \qquad (8)$$

where c has the scale of the width of the distribution function, and the sum is done inside the domain, between the two boundary points. Since this is the first metastable state the system reaches the cluster size is determined by the first time those two conditions are fulfilled. Therefor the exponent is defined by the first return probability of those conditions.

This first return probability scales with the exponent 1.333 ± 0.005 which is consistent with the value $4/3$ (this was found in computer simulations of this first return problem). Therefore we get $\tau = 4/3$. Using this we can calculate the scaling of the correlation length. We get $\nu = 1.5$. Notice that the naive argument would indicate that $\nu = 2$ because of the following argument given in [10]. The fluctuations in the pinning force scale as $r^{1/2}$ and should be smaller then the pinning force. However it is obvious from the stability requirements that the characteristic cluster will be larger then this estimate. Therefor the estimates given before for limits on the correlation exponent are wrong [10]SCA

Another exponent that characterizes this transition is the polarization exponent $P = \sum /phi_i \sim f^{-\delta}$ To derive the polarization scaling we first have to estimate the scaling of the domains total movement $\sum_{domain} \phi_i$ versus it's size. To do this we write a mean field continuous equation for the phase in the depinned cluster.

$$\frac{d\phi}{dt} = \Delta \phi + <\mu> \qquad (9)$$

with the fixed boundary conditions $\phi(-l) = \phi(l) = 0$. The obvious final state solution is $\phi(x) = \frac{<\mu>}{2}(x^2 - L^2)$. So the total polarization scales as $P(l) \sim <\mu> l^3$. However since we showed before that $\mu_s c$ is size independent we can estimate $<\mu> \sim 1/l$ and therefor:

$$P(r) \sim r^D, D = 2.0 \qquad (10)$$

The same scaling can be derived analytically using eq. ( ). It can be easily shown that the sum of the phases in a depinned domain is proportional to $\sum i^2 \mu_i$ inside the domain. Using the double condition on this sum it can be shown that the sum is proportional the the square of the domain size consistent with the previous result and simulations. The random sums inside the clusters are no longer dominated by the gaussian fluctuations.

Using this scaling result we can relate the polarization to the other scaling exponents:

$$P = \int_0^{\xi(f)} N_f(r, L) dr \sim f^{\nu(D-1)} \qquad (11)$$

Using the previous results we can estimate the exponents in the critical point where the correlation length is in a scale L. It can be easily shown that the number of free domains scale as $L^{1/3}$ and that the total polarization scale as L^2. This is in contrast to the scaling above the critical point which is $L^{2.5}$. This obviously indicates that there are no singularities in the phase differences below and in the critical voltage.

We made a numerical check of those predictions. we find that the exponents are consistent with our theoretical predictions.

3 The temporal behavior

We now discuss the temporal behavior of this model we first give some numerical results and then derive the different scaling exponents. The relaxation after the introduction of a field was claimed to be a stretched exponential form [9][10]. Careful numerical simulations prove this function to be :

$$J(t) = \sum_{i=1}^{L} \frac{d\phi_i}{dt} = t^{-\beta(f)} exp(\frac{-t^{1/2}}{\xi'(f)}) \qquad (12)$$

The long range relaxation is exponential

$$J(t) \sim exp(\frac{-t}{\xi'^2}) \qquad (13)$$

In the critical point $J(t) = t^{-\beta_0}$, $\beta_0 = 1/4$,

$$\langle r(t) \rangle \sim t^{\beta'} \qquad (14)$$

, and $\beta_0 = \beta'$.

We first explain the dynamics at the critical point. The behaviour in the critical point is related to the domain growth. As stated before , in the beginning of the growth process about half of the systems phases are free to move ,forming small domains of a characteristic size of unity. During the growth domains merge by depinning of boundary points reducing the total number of domains in the system. The current is given by

$$J(t) = \sum_{depinned domains} (\mu_i - F_c) - \sum_{boundary} phi_i$$

As we have already shown in eq.(4) the first term is proportional to the number of depinned domains. The second term is also proportional to the domain number. Therefore the current is proportional to the domain number and obviously the numerical estimate $\beta_0 = \beta'$ is exact. To derive those exponents we have to understand the domain growth. As the growth proceeds most of the domains are frozen. Only the big domains at the edge of the distribution continue to grow. We assume that the upper edge of the distribution scales as t^a. Since the frozen part of the distribution will still be defined by the static exponent we get:

$$-\beta - a(\tau - 2) = 0.$$

The polarization is dominated by the largest clusters so $P = t^{2a}$. Since also $P = t^{(1-beta)}$ we get an additional exponent relation, $1 - \beta - 2a = 0$. Solving both exponents we get: $\beta = \frac{1}{4}$ and $a = \frac{3}{8}$. As a check of those arguments we measured the scaling of the largest clusters getting $r_{max} = t^{0.4 \pm 0.03}$, which is consistent with our estimates.

The stretched exponential dependence of the relaxation was claimed to be related to a static distribution of clusters in [10]. This assumption is unfounded. The stretched exponential relaxation is related to the relaxation of the individual largest domains. To understand it we again use equation (7) to analyze the relaxation of a single domain. It can easily be shown that :

$$J(t) = \int_{-l}^{l} exp(\frac{-t}{s^2}) ds$$

To estimate the behaviour of this integral we note that after a time t we can estimate that all modes up to $s = t^{1/2}$ have relaxed. Using this we can estimate :

$$J(t) \sim (l - t^{1/2})$$

and therefor for short times we get:

$$J(t) \sim exp(-t^{1/2}/l) \tag{15}$$

Since the largest clusters will be the only active ones in the end of the growth process we expect the behaviour then to be $l = /xi$. When the time scales will be larger $t > \xi^2$ a simple exponential relaxation will be observed with a time constant proportional to ξ^2. This indicates that $\xi = \xi'$.

Therefor three time domains will be observed in the temporal relaxation. The first region is related to the domain growth and is characterized by a power law relaxation. We don't have a detailed understanding why the effective exponents change. The next stage is a stretched exponential relaxation of the largest frozen clusters. The final state after times of the scale of ξ^2 is a simple exponential relaxation with a time constant proportional to ξ^2. We checked that indeed the scaling of those scales with f is the same and that the stretched exponential time constant is proportional to the correlation length.

4 Conclusion

In this paper we explored theoretically and numerically various features of the spatio-temporal behaviour of CDW using this simplified model. We showed how one can derive a detailed understanding of the correlations and the temporal relaxation using a more detailed description the domain structure and growth.

This model is, as we stated simpler then the usual FLR models used to describe CDW. s enables us to get a detailed understanding of it's properties. However there are strong indications that this model and the usual CDW model belong to the same universality class. The argument that we gave for cluster stability can be used to the FLR models. The same dynamics s observed near the threshold[10]. Initial simulation show that the same domain scaling and disribuions are observed. It is not clear wether the phase disordered dominated FLR models simulated by Another interesting feature in this model is that the .A Middleton [7] will display the same universality class. The results seem to indicate a crossover in the scaling. correlation length here is very far from the naive expectation of 2.0.

I thank P.Bak S. Maslov and M. Paszcuzki for critical reading of this manuscript. I appreciate the support and hospitality of Brookhaven National Laboratory. This work was supported by the Division of Basic Energy Sciences, USDOE under contract no. DE-AC02-76CH00016.

References

[1] For a summary see G. Grunner Rev. Mod. Phys.**60** ,1129 (1988)

[2] Z. Wang and N.P. Ong Phys. Rev. Lett. **58**, 2375 (1987)

[3] G. Kriza and L. Mihally Phys. Rev. Lett. **56**, 2529 (1986)

[4] Fukuyama and P.A. Lee Phys. Rev. B **17**, 535 (1978) P.A. Lee and T.M. Rice Phys. Rev. B **19**, 3970 (1979)

[5] S. Sibani and P.B. Littlewood Phys. Rev. Lett. **64**, 1305 (1990)

[6] A.A. Middleton and D.S. Fisher Phys. Rev. Lett. **66**. 92 (1991)

[7] L. Pietronero and S.Strassler Phys. Rev. B **28**, 5863 (1984)

[8] L. Mihally ,M. Crommie and G. Grunner Eur. Phys. Lett. **4**, 103 (1987)

[9] I. Webman Phil. Mag. 56,743 (1987) L. Parisi and L. Pietronero Eur. Phys. Lett. **16**, 321 (1991) Physica **A 179** 1(1991) Almost all the exponents relevant to this paper which are given in those references are wrong, However I. Webman gives the proper phenomenology for this model and states the problem correctly.

[10] S. Coppersmith Phys. Rev. Lett. **65**, 1044 (1990)

[11] A.A. Middleton Phys. Rev. Lett. **68**. 670 (1992)

[12] J.T. Chayes,L. Chayes, D.S. Fisher and T. Spenser Phys. Rev. Lett. **57**, 2999 (1986) A.A. Middleton and D.S. Fisher 1992

ELECTRON SOLIDIFICATION IN TWO DIMENSIONS

R. L. WILLETT
AT&T Bell Laboratories
600 Mountain Avenue
Murray Hill, NJ 07974

ABSTRACT. Several years of intense experimental investigations have left the question of quantum electron solidification in two dimensions open. We re-examine the problem by first reviewing the fundamental concepts of 2D solid formation, then look carefully at a full range of experiments performed on GaAs/AlGaAs heterostructures in the extreme quantum limit. Finally we focus on recent experiments using surface acoustic waves that determine the dynamical conductivity of the 2D electron system, and discuss these findings at small filling factors.

The topic of electron solidification in two dimensions will be discussed in this report. First a brief background review will be given, introducing both classical and quantum electron solids and the experimental systems in which they are explored. Emphasis is placed on the proposed quantum solid in the experimental system of the GaAs/AlGaAs heterostructure. The second section covers in some detail past experiments and results examining the question of electron solidification in heterostructures. These methods range from standard d.c. transport to optical probe to radio-frequency experiments. The final section will cover our recent studies of the extreme quantum limit in 2D using surface acoustic waves. The mechanics of this technique will be reviewed and the results discussed in light of the controversy of whether electrons do indeed order into a lattice at large B fields.

1. Background

The conditions for electron solid formation in the 2D electron system (2DES) are dictated by the competing energy scales of the system; potential energy, kinetic energy, thermal energy, and last but not least the disorder potential. These scales are defined by the system in which the 2DES resides; there are several types: a) electrons on helium, b) GaAs/AlGaAs heterostructures, and c) silicon MOSFETs. In each case confinement of the electrons perpendicular to the interface is achieved electrostatically with only the lowest electronic subband populated at low temperatures in order to study a "pure" 2D system. As seen in Fig. 1, for example, some extension in the z direction will occur.

The relative scale of potential to kinetic energy is important in determining the transition to electron solid. For sufficiently large Γ electron solidification will occur where

$$\Gamma = <V>/<K.E.>; \quad <V> = (\pi n)^{1/2} e^2/\varepsilon$$

$$<K.E.> = k_B T \text{ for } T \to \infty$$

$$<K.E.> = \pi \hbar^2 n/m_c \equiv E_F \text{ for } T \to 0$$

and where n is the electron areal density. Physical parameters for the 2D electron systems mentioned above are displayed in Table 1 and allow evaluation of the Γ parameter. The thermal energy limits of $T \to \infty$ or $T \to 0$ determine whether the system is in the classical or quantum region, and this will be discussed below.

The classical 2D electron solid region is defined where

$$E_F \ll k_B T \text{ and} <V> > k_B T$$

so that

$$\Gamma = [(\pi n)^{1/2} e^2/\varepsilon]/k_B T .$$

Such a system is electrons on He (Fig. 1). Classical freezing was first observed by Grimes and Adams [1] using an RF excitation of the electrons to detect ordering (Fig. 2). They observed the melting transition at $\Gamma = 131 \pm 7$, with the experimentally extracted phase diagram shown in Fig. 3. This classical melting transition is believed to be mediated by the unbinding of pairs of electron lattice defects, as suggested by Kosterlitz and Thouless (KT) [2] and elaborated by Halperin and Nelson [3]. The potential to kinetic energy ratio predicted for melting in this theory is $\Gamma_{KT} = 125$.

For sufficiently high electron density, the 2DES will no longer behave classically. This fundamental idea led to theoretical extension [4] of the proposed phase diagram to that qualitatively shown in Fig. 4. We can now consider the quantum 2D electron solid. In this case the temperature is low so that $E_F \gg k_B T$, in which case the kinetic energy is determined by the density so that

$$\Gamma = (me^2)(\pi n)^{-1/2}/\hbar^2 \varepsilon \equiv r_s .$$

The Wigner transition [5] is then realized as melting occurs when one raises the density (as opposed to the classical case where increasing the density induced freezing). Numerical calculations suggest that in order to observe melting, the mean inter-particle separation in effective Bohr radii, r_s, must be approximately 37 for zero temperature [6]. These theoretical results support a phase diagram which must look at least qualitatively like that displayed in Fig. 4, with the Wigner transition occurring as one increases density at $T = 0$.

As can be seen by examining Table 1, the quantum phase transition ($T = 0$) is difficult to experimentally access. On bulk He $r_s \approx 3 \times 10^3$ for $n = 10^7 \text{cm}^{-2}$. In semiconductor structures the r_s is about 3 with density $n \approx 10^{11} \text{cm}^{-2}$. Constraints in the available density range in each system prohibit reaching $r_s \sim 50$.

Table I. Physical parameters for three different 2DES.

	electrons on He	Si-MOS	GaAs/Al$_x$Ga$_{1-x}$As
E_B (meV)	0.7	5-50	20-40
$\langle z \rangle$ Å	110	30	50-100
n_c (cm^{-2})	$10^5 - 10^9$	$10^{11} - 10^{13}$	$10^{10} - 10^{12}$
E_F (meV)	$10^{-7} - 10^{-3}$	1-50	0.5-40
m^*/m_m	1.0	0.19	0.066
τ (sec)	10^{-7}	10^{-12}	$10^{-12} - 10^{-11}$
μ (cm^2/v$_s$)	10^7	10^3	$10^5 - 10^7$
ε	1.0	11	14

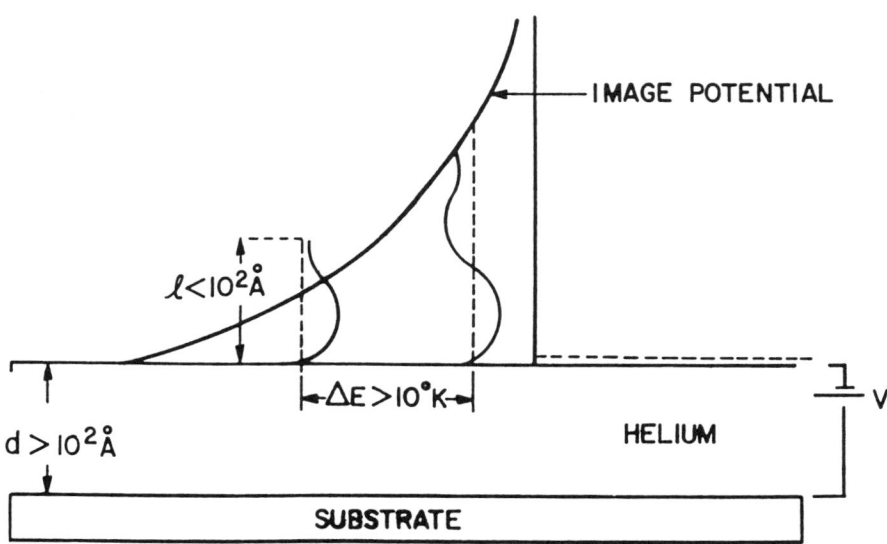

Fig. 1. Schematic of configuration for electrons trapped in a metal-helium-vacuum system. The wave functions perpendicular to the interface are drawn in. The static voltages V fixes the average charge density. Ref. 15.

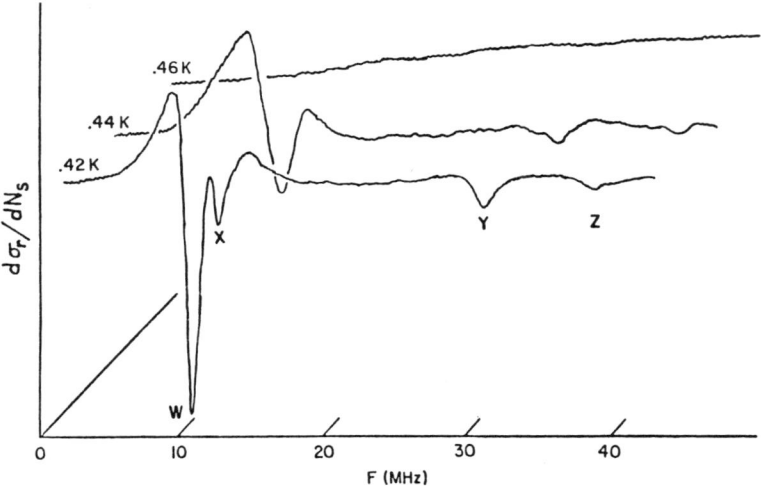

Fig. 2. Experimental traces showing coupled plasmon-ripplon resonances at an areal density of $\approx 4.4 \times 10^8$ electrons/cm^{-2}. These resonances disappear when the crystal melts at T = 0.457K. Ref. 1.

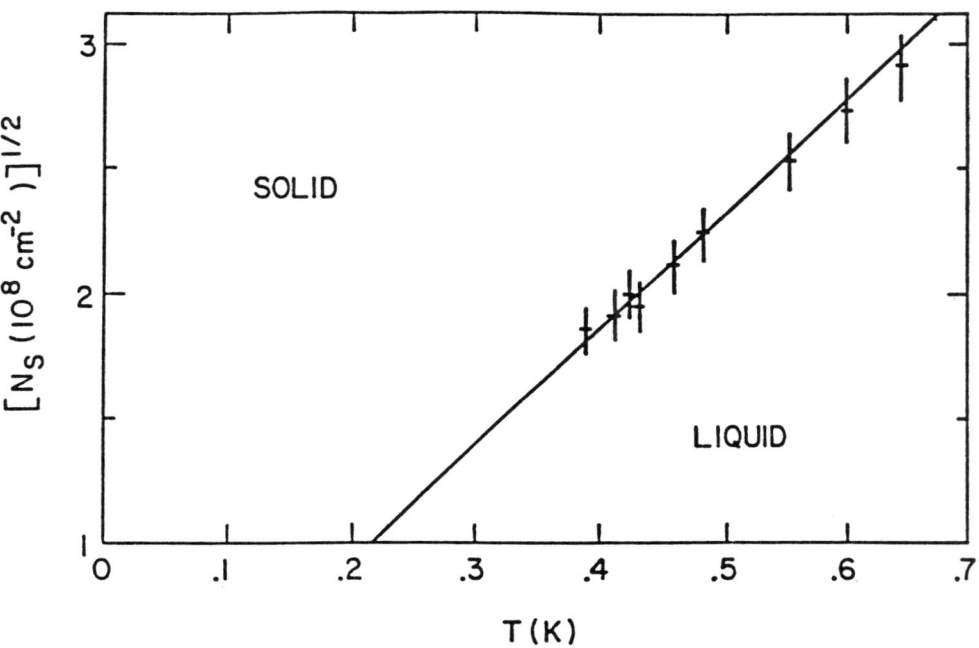

Fig. 3. Data points denote the melting temperatures measured at various values of the electron areal density n_s. Along the line the quantity Γ has a value 131. Ref. 1.

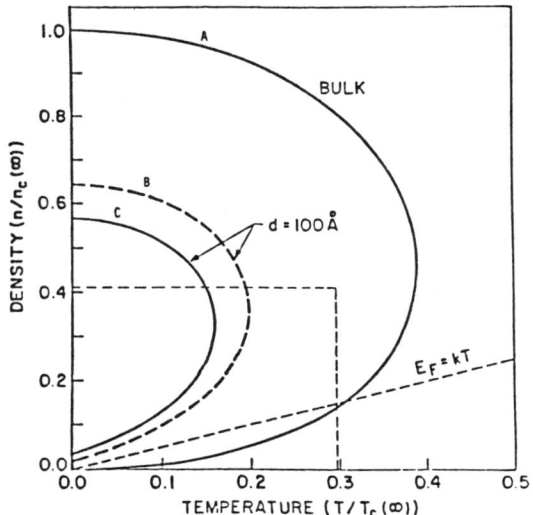

Fig. 4. Qualitative phase diagram for 2D electron system. Curve A is for bulk He, B and C for 100Å films on dielectric substrates with $\varepsilon = 10$ and $-\infty$ (metal) respectively. Ref. 15.

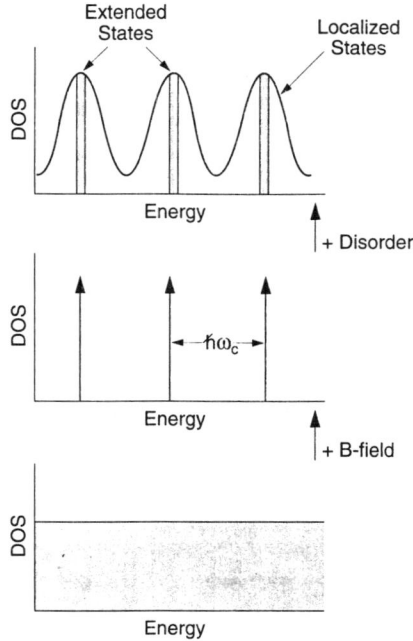

Fig. 5. Density of states for a non-interacting 2D electron system without and with a magnetic field, and the effect of disorder.

Since the appropriate r_s is difficult to reach in the known systems, another means must be explored to reduce the kinetic energy, leading us to consider the 2DES in a large magnetic field. In 2D without application of a magnetic field the density of states is constant in energy with $dn/dE = m^*/2\pi\hbar^2$ up to the Fermi level (Fig. 5) for a noninteracting system. Application of a magnetic field quenches the kinetic energy of the system, resulting in a set of Landau levels separated in energy by $\hbar\omega_c$ and being described as delta functions for $\omega_c\tau = \infty$. The degeneracy of each LL is eB/h. These Landau levels are of course broadened in the real world by disorder. In this picture extended states at the center of the Landau level are separated by localized states between Landau levels. The position of the Fermi level is described in this scheme by the filling factor $\nu \equiv E_F/\hbar\omega_c = ne/hB$. From this one can see that at constant density sweeping B field will sweep through the Landau levels.

Interactions in the $B \neq 0$ 2DES are critically dependent upon disorder. The pertinent energy scales in the system are those of the thermal energy $k_B T$, the disorder potential $<V_d>$, and the interaction energy $<V_i>$. In order to examine the interaction energy we must satisfy the condition that $k_B T$ and $<V_d> \ll <V_i>$ presuming we have quenched the kinetic energy ($\hbar\omega_c \gg k_B T$).

We will now closely examine the experimental 2DES used in the remainder of this review, the GaAs/AlGaAs heterostructure, and will specifically address how these energetic constraints are met. The constraint of minimal disorder is severe and has been achieved in GaAs/AlGaAs heterostructures using the scheme shown in Fig. 6. These structures are grown using molecular beam epitaxy. In this structure, the electrons reside at the GaAs/AlGaAs interface. These electrons are transferred to the interface from Si donors located a distance from the interface in sheet dopant layers. This modulation doping [7] provides a spacer between the 2DES and the ionized donors, thus reducing scattering from these charges. These donor sheets can still scatter the 2D electrons, but their remoteness and the sheet nature of the doping cause only long wavelength potential fluctuations [8]. Residual disorder exists at the 2D electron layer in the form of interface roughness and intrinsic impurities in the GaAs. Only by improved vacuum during the growth process can the latter be improved. The relative mobilities are defined by $\mu = e\tau/m$; this is the parameter of significance in assessing heterostructure quality. Structures with $\mu \geq 10^7 \text{cm}^2/\text{Vs}$ have been achieved using the above schemes.

The effects of disorder in heterostructures are readily observable, and in fact dictate the results of the most common probe, simple d.c. transport. If the disorder is sufficiently strong to overwhelm electron-electron interactions, the single particle interplay with disorder is still observable: this is the integral quantum Hall effect [9], (Fig. 7). This effect is characterized by zeros or minima in longitudinal resistance ρ_{xx} and concomitant plateaus in the transverse resistance ρ_{xy} at integral filling factors. A single particle picture easily describes this effect. As the magnetic field is swept, the Fermi level traverses through the delocalized and localized states of the Landau levels. When the delocalized states are traversed, the Hall resistance changes linearly with field as in any metal. In the localized states however, the Hall resistance is pinned at quantized values $(h/e^2)(1/m)$, m an integer, while the longitudinal resistance (conductance) drops to zero as no conduction is possible at the Fermi level. As disorder in the system is increased,

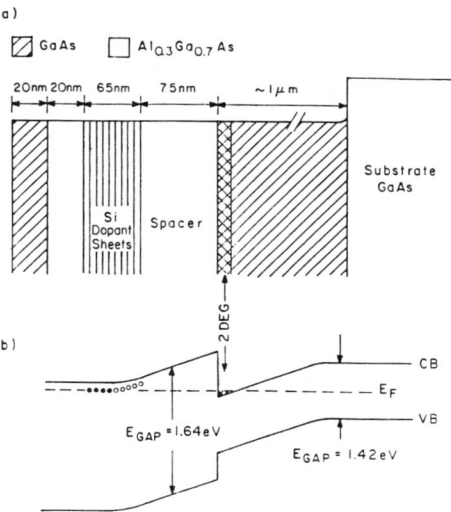

Fig. 6. High mobility GaAs/AlGaAs heterostructure; a) cross-section and b) energy diagram.

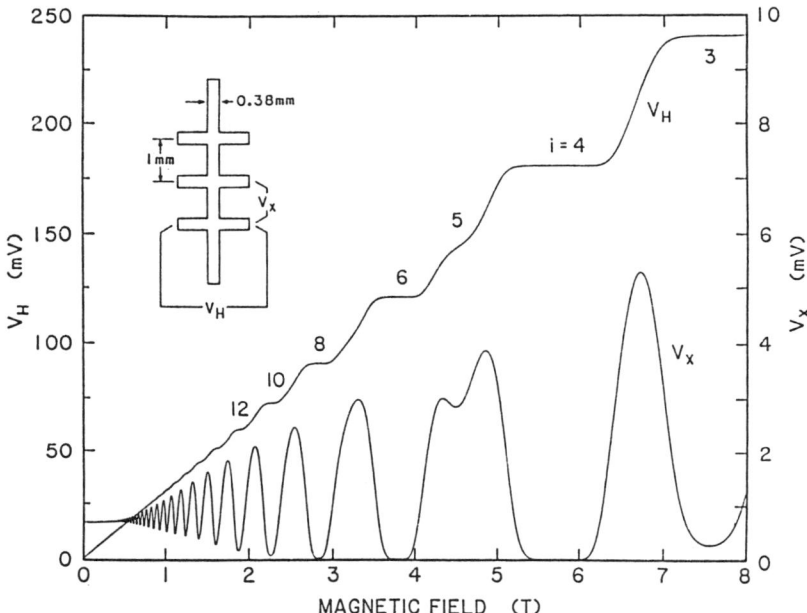

Fig. 7. Chart recordings of V_H and V_x vs. B for a GaAs/AlGaAs heterostructure cooled to 1.2K. The source-drain current is 25.5 μA and $n = 5.6 \times 10^{11}/cm^2$. Cage et al. IEEE Trans. Instrum. Meas. IM-34, 301 (1985).

the Hall plateaus widen as more of the total states are within the localized bands. Transport in the integral states is activated with respect to the temperature, reflecting the energy gap defining that filling factor ($\Delta = \hbar\omega_c$ for ν = even, $\Delta = g\mu_0 B$ for ν = odd at the spin gaps).

If disorder is decreased and magnetic field is increased such that $\nu \leq 2$, the vestiges of electron-electron interactions can now be observed. An effect wholly similar to the IQHE in simple transport appears, the fractional quantum Hall effect [10] (FQHE). Rather than Hall resistance plateaus at $(e^2/h)(1/m)$, these plateaus appear at $(e^2/h)(1/\nu)$ with ν an odd denominator fraction (Fig. 8). Belying the trivial similarity of the effects is the underlying mechanism of the FQHE. No simple single particle picture would predict an energy gap at $\nu = 1/3$! The effect can be traced to electron-electron interactions, with the system apparently resistant to minor density (filling factor) changes, manifesting as energy gaps at odd denominator filling factors. The principle FQHE states at $\nu = 1/m$, m an odd integer are described as liquid like, incompressible collective electron states using Laughlin's [11] variational wave function:

$$\psi_m(z_1 \cdots z_N) = \prod_{i<j}(z_i - z_j)^m e^{(-1/4)\Sigma_k |z_k|^2}$$

where $z_i = x_i + iy_i$ is the complex position of the electron. This wave function represents an incompressible, constant density state at $\nu = 1/m$, with charge carriers $e^* = e/m$. A finite energy is required to add magnetic flux/electrons to the system. This picture was expanded to describe the full series of odd denominator fractional states $\nu = p/(2p+1)$ by considering how the elementary excitations of the FQHE interact [12]. Just as the electrons are believed to condense into principle FQHE states, at sufficiently high densities of the quasiparticle excitations, these too may condense to form new liquid-like states. Thus the 1/3 state may, upon decrease of B, enter into a new state at $\nu = 2/5$ and so on.

This picture of electron correlations may, in fact, be substantially modified by a new description of the extreme quantum limit. In this model by Halperin, Reed, and Lee [13] a Chern-Simons gauge field Fermion may occur at $\nu = 1/\text{even}$. These particles may then dictate interactions in the extreme quantum limit; the principle series of fractional Hall states at $\nu = p/(2p+1)$ will merely represent the IQHE for the particles with B referenced to $\nu = 1/2$.

Given these apparent manifestations of electron-electron interactions, the true observation of electron solidification has been and still is eagerly anticipated. This liquid to solid transition has received little theoretical concentration, although predictions of the filling factor at which the transition occurs have been made. To properly parametrize the transition, one can construct a potential to kinetic energy ratio considering the 2D system in an applied B field;

$$(e^2/\varepsilon_0)\pi n^{1/2}/\hbar\omega_c \equiv \nu r_s < 1$$

Using this criteria, the larger r_s for fixed ν, the more the solid is favored. At sufficiently small ν for any r_s the solid will be favored; Lam and Girvin [14] proposed that the solid is lower in energy than the liquid for $\nu < 1/7$ even for $r_s = 0$.

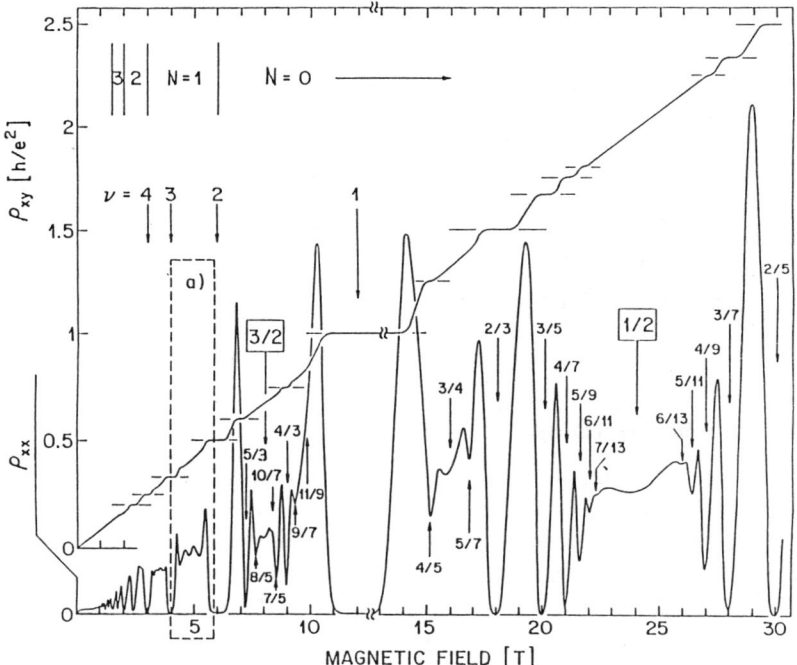

Fig. 8. Overview of diagonal resistivity ρ_{xx} and Hall resistivity ρ_{xy} of a high mobility heterostructure. Filling factors ν and Landau levels N are indicated. R. L. Willett et al., Phys. Rev. Lett. 59, 1776 (1987).

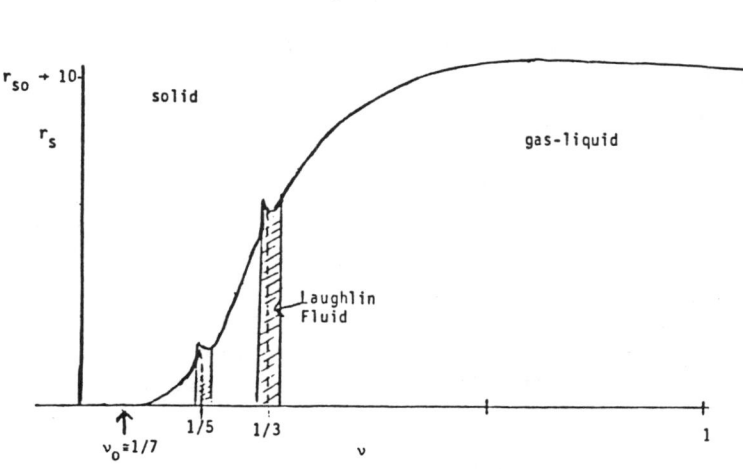

Fig. 9. Qualitative phase diagram at T = 0 for a 2-D electron gas in the lowest Landau level ($\nu \equiv E_F/\hbar\omega_c$). Ref. 15.

These painfully general criteria for solid formation were at least put into the written word by Platzman [15]. He generated phase diagrams of the magnetic field induced Wigner solid considering first the FQHE to solid transition (Fig. 9). For T = 0 he guessed that the ν = 1/3 Laughlin state would solidify at $r_s \simeq 10$. The remainder of the curve reflects the notion that as ν is decreased, the solid is favored for increased r_s. The finite temperature phase diagram was likewise generated in an ad-hoc fashion resulting in the qualitative phase diagram of Fig. 10. The limits demonstrated here represent $r_s = \infty$, where magnetic field has no effect on the classical Kosterlitz-Thouless melting curve at $\Gamma = 131$. *Trends* are shown for decreasing r_s: finite r_s decreases the solid regime while large fields enhance the solid regime.

2. Past Electron Solid Experiments

In the previous section, a simple rational was presented for the phase diagram of a 2DES in a magnetic field, but the effect of disorder as related to solid formation was not discussed. The role of disorder is actually the central issue in experimental tests of presumed 2D solids. This is due to the fact that it is difficult to experimentally differentiate between magnetic localization and electron solidification. As such, theoretical efforts to incorporate the disorder into any model are invaluable but simultaneously complex. One does not know a-priori whether the disorder is a perturbation on the solidification or that, in the other extreme, the ordering is a perturbation on magnetic localization. Since the systems that concern us here (GaAs/AlGaAs heterostructures) are known to support electron correlation in the limit of high B field and low temperature (the FQHE), it is then natural to conclude that the disorder *may not* dominate the correlations. Given this framework of guarded optimism, experimental studies were pursued with the understanding that solidification must be demonstrated conclusively before assigning any properties to the solid. The debate between features of magnetic localization and solidification will be addressed throughout presentation of the data.

Very early studies of the FQHE encountered apparent magnetic freeze-out at small filling factors in samples where indications of the principle FQHE states (1/3, 1/5) were present [16]. No systematic study of this phenomenon was undertaken. Sample quality improved dramatically before this issue was revisited, which allowed a direct comparison between low disorder (high mobility) and high disorder (low mobility) systems [17]. Such a comparison is displayed in Figures 11 and 12. In Fig. 11 small filling factor longitudinal transport is shown for three samples of different mobilities. All three demonstrate insulating behavior at their highest respective B fields, but also demonstrate different strengths of the FQHE. The lowest mobility sample supports the FQHE only down to ν = 1/3, yet the highest mobility sample can support the 1/5 FQHE.

The nature of the insulating phase in each sample is characterized by the temperature dependence of the transport coefficients. Figure 12 shows $\ln \rho_{xx}$ versus 1/T for each in the insulating range, with the highest mobility showing distinctly different transport from the lower μ systems. The lower μ samples do not demonstrate an activated transport, but

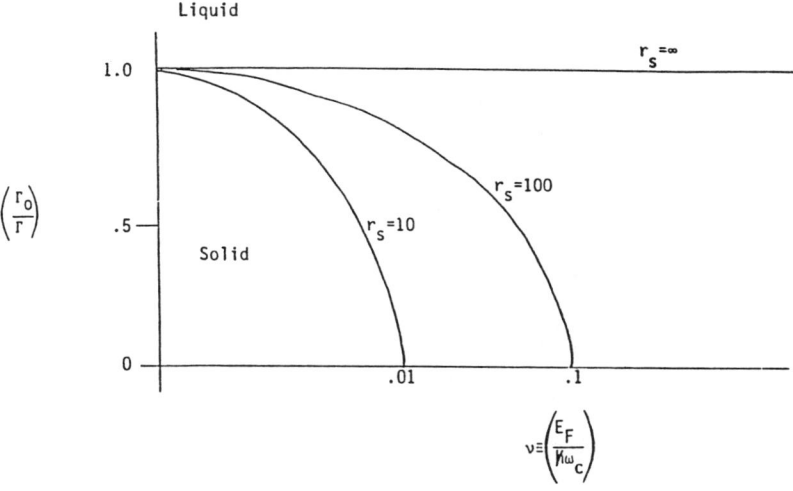

Fig. 10. Qualitative effects of r_s and magnetic field on the solidification boundary of a 2D electron gas.

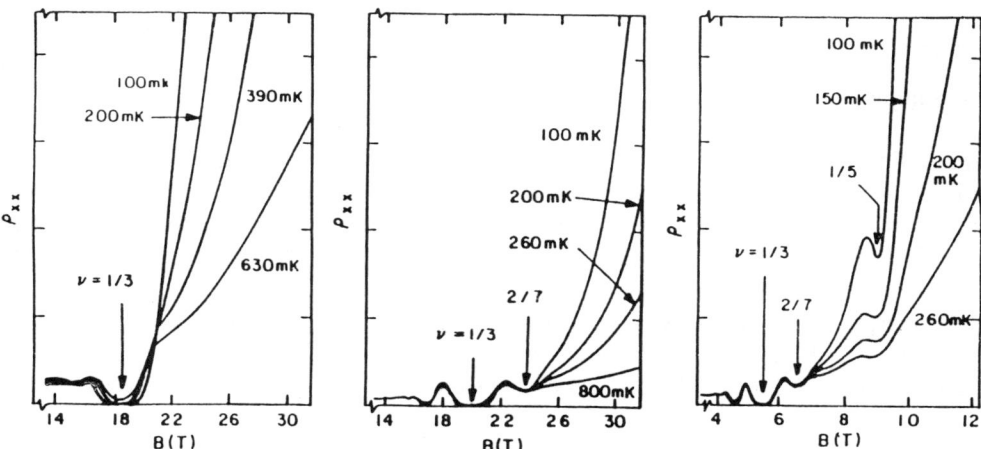

Fig. 11. Longitudinal resistivities of three samples with mobilities from left to right of 5×10^5 cm^2/V-sec, 1×10^6 cm^2/V-sec, and 2×10^6 cm^2/V-sec $\equiv$ sample A. Ref. 18.

rather a temperature dependence more characteristic of a hopping conduction; $\rho_{xx} \sim \exp(\alpha/T^{1/2})$. The highest mobility sample demonstrates an activated transport throughout the insulating range; $\rho_{xx} = \exp(E_A/T)$, with E_A specific to the magnetic field value. The most interesting aspect of this result is that as mobility has since been increased by an order of magnitude, this general finding of activated transport at small ν has been consistently observed. Perhaps this indicates a new regime of disorder with a single quantitative response in the transport.

Further analysis of the highest mobility system transport data presented above reveals an interesting picture suggestive of electron solidification. Figure 13 shows the filling factor dependence of the activation energies. The results can be generalized as $E_A = E_o - \alpha\nu$ with E_o and α dependent upon the sample density. Higher densities show larger E_A in the large B limit - the classical limit. In the lower B limit the higher density has onset at smaller ν. Comparison should be made between these results and the heuristic picture of electron solidification. In this picture, as seen in these results, for the large B limit higher density should increase the Coulomb interaction and thus increase the energy scale (E_A) of the system excitations. The onset of the activated behavior should reflect the liquid to solid transition in the *quantum* limit (Fig. 9). As described previously, decreasing r_s in this range should promote solidification at a smaller filling factor. Both of these principles are observed experimentally in Fig. 13.

These results demonstrate a qualitative agreement with what is expected in electron solidification phase diagrams, yet these findings are rather non-specific. Following these results, Chui and Esfarjani [19] addressed finite temperature transport of a pinned 2D electron lattice. They argued that for a periodic 2D electron lattice in a B field, thermally activated bound dislocation pairs contribute to the conductivity from which an activation energy E_A can be deduced. Near the melting point, this activation energy E_A approaches zero as $E_A = E_0(1-\nu/\nu_0)$. They calculated E_0 from the change in the zero-point energy of the phonons as a dislocation pair is created and find argreement to within 30% of the experimental results in Fig. 13.

As sample quality has improved the standard transport measurements in the small filling factor range have generally continued to display activated behavior, but with an interesting process [20] occurring in the vicinity of ν = 1/5. Unlike the ρ_{xx} results of Fig. 11 where the ν = 1/5 state does not develop completely, in systems with lower disorder the ground state at ν = 1/5 is determined to be an incompressible quantum liquid (Fig. 14). This is demonstrated by vanishing resistivity ρ_{xx} as the temperature T → 0. At filling factors below ν = 1/5 as well as in a narrow region above ν = 1/5, ρ_{xx} diverges exponentially as T → 0. This contrasts with the T dependence at any larger ν. These results suggest that the quantum liquid at ν = 1/5 is surrounded by a insulating phase, perhaps, but not definitely that of an electron solid. This reentrant phase picture is supported by the calculations showing that the energetic cusp at ν = 1/5 may be lower in energy at the liquid state than the surrounding solid energy (Fig. 14 inset). It can be argued that as magnetic field is increased, if magnetic localization or freeze-out is responsible for the insulating phase then it should be unfavorable for a correlated liquid state to be energetically preferred at even higher B field: once "frozen-out", the carriers won't be able to recondense into a correlated state. However, one can imagine that

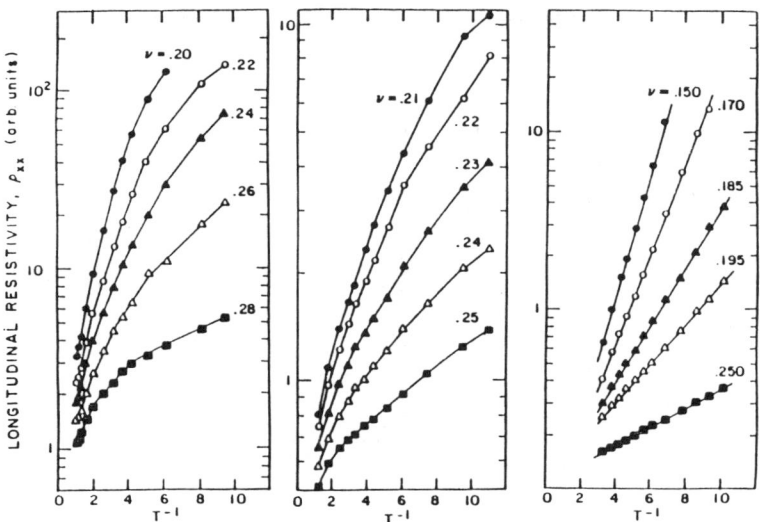

Fig. 12.   Temperature dependence of longitudinal resistivity ρ_{xx} at five different filling factors in the insulating range for the samples of Fig. 11. Ref. 18.

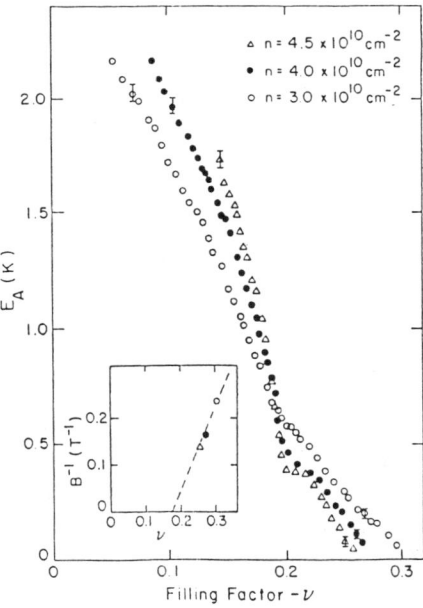

Fig. 13.   Filling factor dependence of energy E_A at three different carrier densities for sample A of Figs. 11 and 12. Note the crossover in the data at $\nu \approx 0.19$. The inset shows extrapolation of onset ($E_A = 0$) to infinite B using $\nu = nh/eB$. Ref. 17.

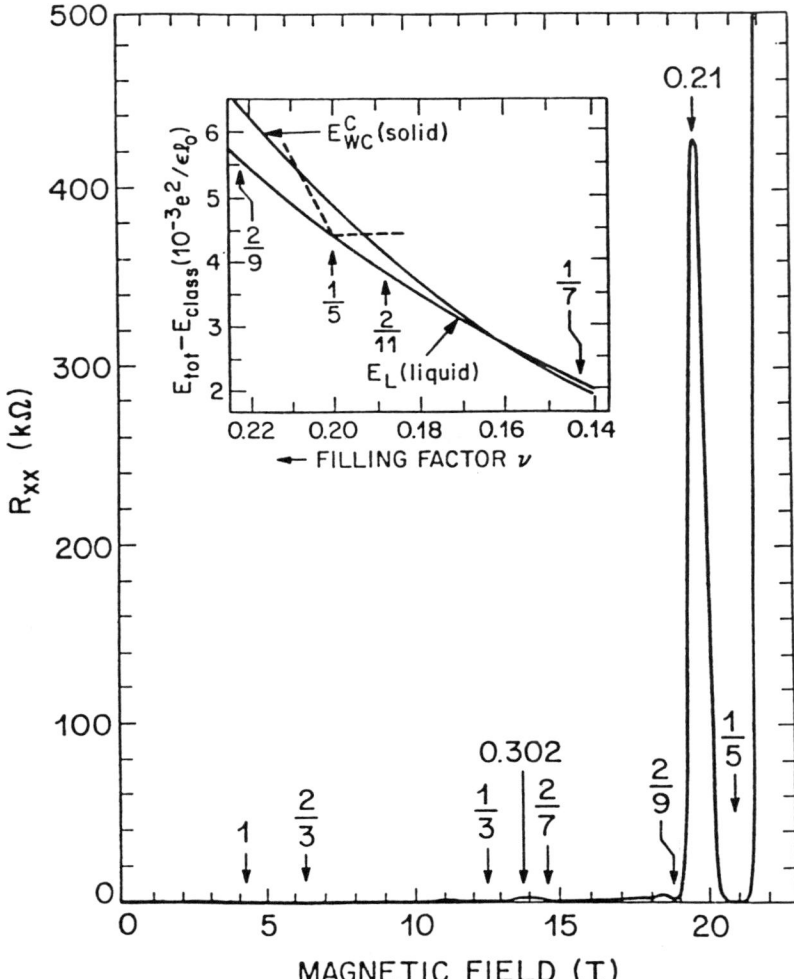

Fig. 14. Diagonal resistance R_{xx} vs. magnetic field at $T \approx 90$ mK. Data are taken on a square sample so that $\rho_{xx} = \alpha R_{xx}$ with $\alpha \sim 1$. At $\nu = 1/5$, $\rho_{xx} \to 0$ indicating that the $\nu = 1/5$ quantum liquid forms the ground state. The resistivity ρ_{xx} in the sharp-spike at $\nu \sim 0.21$ and for all $\nu \leq 1/5$ is rising exponentially on lowering the temperature. All FQHE features at lower magnetic field are well developed but practically invisible on this scale. Inset: Result of a calculation for the total energy per flux quantum of the solid (E^c_{wc}) and interpolated $1/m$ quantum liquids (E_L) as a function of filling factor (Ref. 14). A classical energy ($E_{class} = 0.78213 \, \nu^{-1/2}$) is subtracted for clarity. The dashed lines represent the cusp in the total energy of the liquid at $\nu = 1/5$. Its extrapolation intersects the solid at $\nu \sim 0.21$ and 0.19 suggesting *two* transitions from quantum liquid to quantum solid around $\nu = 1/5$. Ref. 20.

locally a deformable liquid may actually be able to accommodate intrinsic disorder and therefore still allow condensation. As such this reentrant phase picture does not define the presence of a solid.

Further transport experiments, I-V studies, have attempted to specifically address electron solid properties in the 2DES. In known charge density wave systems, discontinuities [21] in I-V measurements have been an indicator of a pinned, ordered solid. Upon application of a sufficiently large electric field the solid will be depinned and slide over the impurities. The results of the first such studies [22] are shown in Fig. 15 and Fig. 16. Indeed as one moves into the insulating phase discontinuities in the differential resistance are observed, and in fact multiple discontinuities can be seen. The values of the depinning voltages can be related to a correlation length via simple models [23, 24]: these results would suggest a correlation length $L_0 \approx (0.02\ e/(\varepsilon E_T))^{1/2}$ of the order of 100 electron lattice spacings. While it is inviting to assume that this type of analysis and these results are definitive descriptions of a solid, in truth other mechanisms must be considered. Filamentary current paths in the system could well result in I-V discontinuties since the Hall voltage across these filaments could cause breakdown across the path, appearing as discontinuities in the differential resistance. With this caveat in place, further I-V studies must be examined for specific findings that allow determination of properties of an electron solid only.

Further I-V studies of the 2DES have revealed provocative findings which are again consistent with the electron solid picture but not necessarily specific to it [25]. In higher mobility samples, strongly nonlinear I-V were again observed and in this case low frequency noise was noted around the $\nu = 1/5$ state for depinning fields (Fig. 17). The noise is characteristic of a weakly pinned e-solid around 1/5, with the noise peaking and then vanishing in the insulating phase for $\nu < 1/5$. This regime was consequently described as a "Wigner Glass" since the disorder apparently overwhelms the ordering, quenching the freedom to slide and produce noise.

In other experiments [26], two-level random switching was observed at a step on the nonlinear I-V curve above the conduction threshold (Fig. 18). These results were interpreted as an insulating phase that can support two conduction configurations, corresponding to two sliding states in the electron solid picture, which switch between one another through tunneling.

While these findings in sum may suggest an electron solid none defines the system as supporting ordering. In all these results there are no findings, such a narrow-band noise, which could be used to determine solidification by measuring the only internal parameter that is known a-priori — the electron separation $a_0 \sim n^{-1/2}$.

Beyond I-V measurements, other more recent transport studies have discussed possible electron solidification. Using a 2D system comprised of holes [27] rather than electrons insulation in the small filling factor limit was explored using a different energy scale since the hole effective mass is substantially larger than that of the electron. Whereas the electron system showed reentrant behavior around $\nu = 1/5$, the hole system displayed reentrance around $\nu = 1/3$ (Fig. 19). This finding is consistent with increased Landau-level mixing which would promote e-solid formation at larger filling factors. A

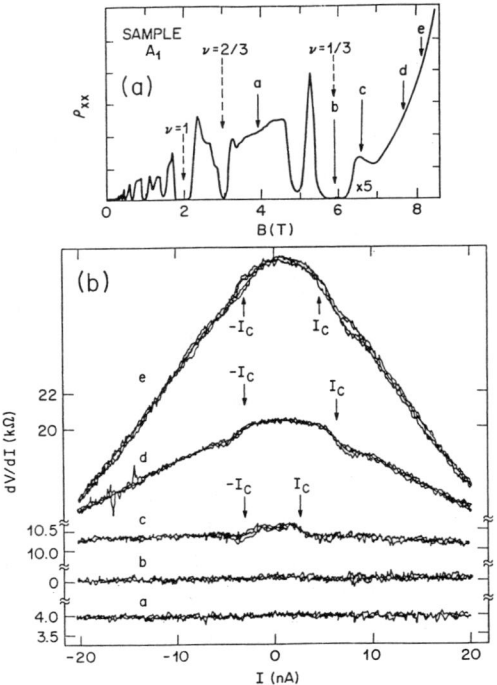

Fig. 15. Longitudinal resistivity vs. magnetic field (a) and differential resistance vs. excitation current (b) for a high mobility heterostructure B. Field values at which differential resistance is measured in (b) are marked a through e (a). The temperature is 80 mK. Ref. 22.

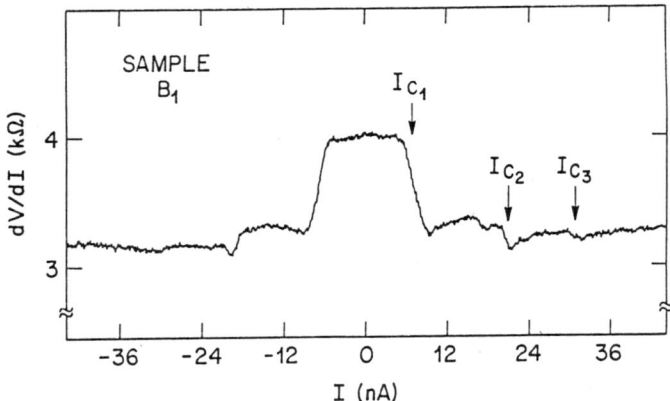

Fig. 16. Differential resistance vs. excitation current in sample C of density 6×10^{10} and $\mu_s = 1.2 \times 10^6$ cm^2/V-sec at B = 8 Tesla (ν=0.31) and T = 80K. Ref. 22.

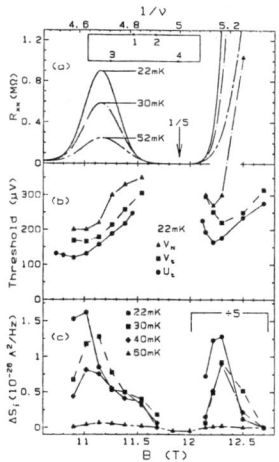

Fig. 17. B-field dependences of (a) diagonal resistivity R_{xx}, (b) threshold voltages U_t from four terminal, V_t, two-terminal, and V_N noise measurements. (c) Noise increase above the threshold defined by $\Delta S_i \equiv S_i(V=1\text{ mV}) - S_i(V=0)$, averaged between 500 and 775 Hz. Data points are connected to guide the eye. The sample geometry is shown in top of (a). Ref. 25.

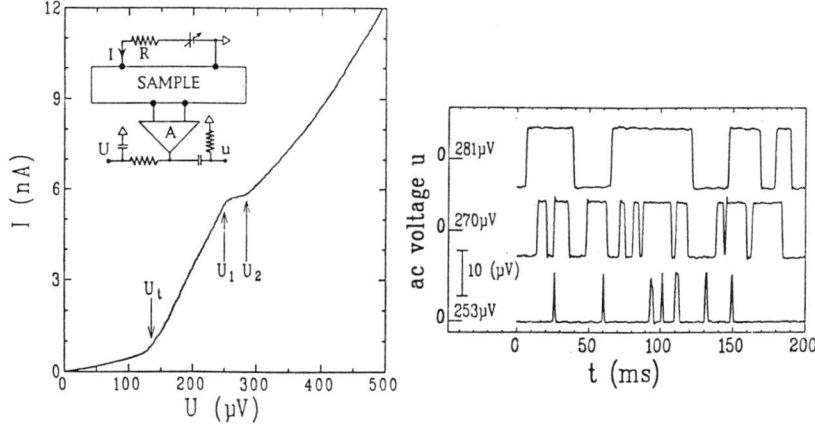

Fig. 18. Four terminal d.c. I-V measured at B = 11.2T ($\nu \approx 0.21$) and T = 19 mK. $U_t \approx 140$ μV is the threshold voltage. $U_1 \approx 250$ μV and $U_2 \approx 280$ μV correspond to the beginning and end of the step on the I-V. Two-level random switching is observed on the step for $U_1 < U < U_2$. The sample geometry and circuit are shown in the inset. In lower panel are typical traces of the ac component of voltage as a function of time, measured at B = 11.2T ($\nu \approx 0.21$) and T = 19 mK for three biases: U = 253 μV (close to U_1), U = 281 μV (close to U_2) and U = 270 μV (between U_1 and U_2). Ref. 26.

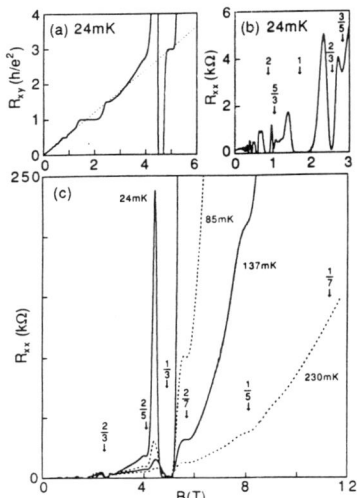

Fig. 19. Details of magnetotransport coefficients in a 2D-hole system: a) the Hall resistance R_{xy}, b) the low field R_{xx}, and c) the temperature dependence of R_{xx}. Ref. 27.

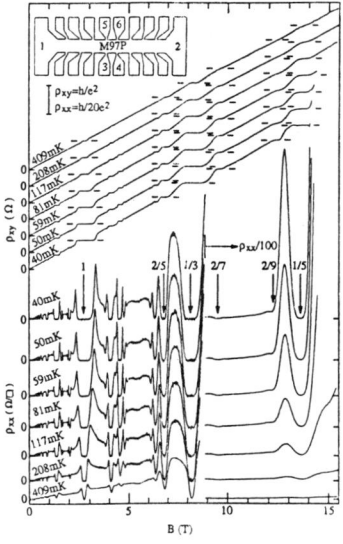

Fig. 20. Temperature dependence of ρ_{xx} 1 → 2,5–6 and ρ_{xy} 1 → 2,4–6 after averaging the +B and −B traces. The data show $\rho_{xy} = 5h/e^2$ at $\nu = 1/5$, $\rho_{xy} = 9h/2e^2$ at $\nu = 2/9$, and $\rho_{xy} \approx B/ne$ at the $\nu \sim 0.21$ insulating phase peak and for ν slightly smaller than 1/5. The inset shows the geometry of the Hall bar sample M97P and the contacts used in the measurements. Ref. 28.

localization model to combat this interpretation is not immediately obvious.

A new result which may well be the most damning evidence against magnetic freeze out as the cause of the insulating phase is a careful measurement of the Hall resistance [28] near $\nu = 1/5$. In the insulating phase for $\nu < 1/5$ and $\nu > 1/5$, ρ_{xy} is found to be *normal* (= B/ne) and independent of T (Fig. 20). This directly rules out magnetic freeze-out since the carrier density should drop in this model.

A technique which recently has been employed by several investigators is photoluminescence detection [29, 30] of the 2D electron gas in the extreme quantum limit. These experiments are difficult due to the low luminescence yield of the 2D gas with respect to the background substrate response. As such the results in this area are very empirical, with features in luminescence generally associated with presumed properties of the electron solid. Shown in Fig. 21 is a result [29] reported in an interesting structure using a δ-doped monolayer of acceptors (Be atoms) located 250 Å away from the GaAs/AlGaAs interface. This structure promotes greater overlap of the electron and hole wavefunctions resulting in larger luminescence yield. A new line appears in the radiative-recombination spectrum of the 2D electrons as filling factor is reduced and at low temperatures. This finding is attributed to a "crystallization effect", with the new line occurring at lower energy than the principle line since the electron solid has a lower ground state energy. Other studies have likewise reported additional lines in luminescence [30] at low ν and T, but certainly further work is needed to clarify precisely what is probed in these experiments.

In overviewing experimental studies of the electron solid, an important concern are studies that probe low energy excitations of the solid. It is best to first review in detail the theoretical expectations. The problem of pinning and conductivity of 2D charge-density waves in magnetic fields was examined by Fukuyama and Lee [31]. In the limit of no disorder and at small wave vector q the excitations of the system are split into two magnetophonon branches, $\omega_\pm$

$$\omega_+ \approx \omega_c + \omega_p^2/2\omega_c, \quad \omega_- \simeq \omega_t \omega_p/\omega_c$$

with

$$\hbar\omega_c = \hbar eB/m, \quad \omega_p^2 = 2\pi e^2 n_s q/\varepsilon m$$

and $\omega_t = (\mu/mn_s)^{1/2} q$, the sheer modulus = μ. The lower frequency mode describes propagation of a shear wave in the solid, with precise delineation of this mode a definitive indication of solid formation. This is the mode of experimental interest. In order to realistically examine this low energy excitation the effects of disorder must be considered. The first consequence of disorder is pinning of the solid with consequent ω gap formation at q = 0. Beyond this, the mode is expected to be broadened (in ω) because of the disorder. Figure 22 schematically shows a qualitative estimate of the mode structure (without showing the mode broadening), displaying a q = 0 gap and eventual $\omega \sim q^{3/2}/B$ at large q. The exact q dependence of the mode below the pure shear propagation ($q^{3/2}$) is in question. Within this picture one could expect that for an ω sweep at fixed q through the mode a broad resonance will be observed with the mode width on the order of the peak frequency at small q. This resonance will manifest itself

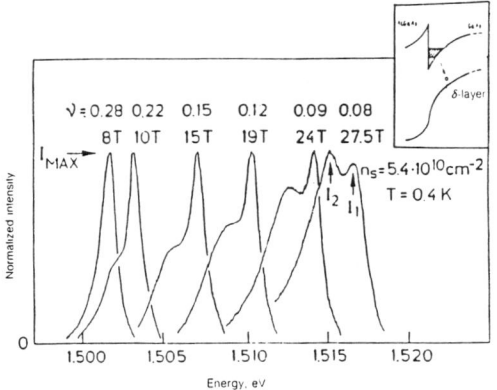

Fig. 21. Spectra of the radiative recombination of 2D electrons with photoexcited holes in a monolayer of acceptors. The spectra were recorded in various magnetic fields for a sample with a 2D-electron density $n_s = 5.4 \times 10^{10}$ cm^{-2} at T = 400 mK (the spectra have been normalized to the same peak intensity). The inset shows an energy diagram of the heterostructure and the optical transition of interest. Ref. 29.

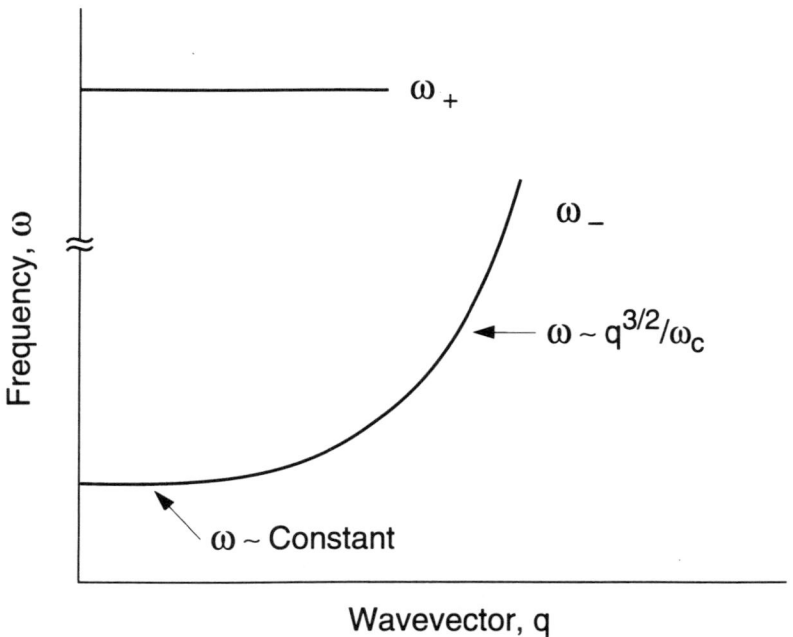

Fig. 22. Schematic dispersion relations for excitations in a pinned 2D solid. Note that broadening of the modes in ω is not represented.

as a broad conductivity peak centered at the pinning frequency. These expectations will be reviewed again when examining the surface acoustic wave experiments.

A set of experiments that are important to examine are the RF meander line studies [32-34] on the presumed electron solid. The experiment involves an RF transmission meander line placed on top of the heterostructure. The RF signal transmission through the line is effected by the interaction of the line's E-field and the 2D electron gas as dictated by the 2DES conductivity. In Fig. 23, the absorption spectrum for a set of ν and T conditions are shown with the principle finding being a broad absorption complex. Because the broad absorption disappears with decreased B field and increased temperature the feature is stated to reflect electron solidification. This assertion may or may not be true, but one can certainly assign the response to a B and T dependent change in the system conductivity as known from the d.c. transport measurements. Within the large absorption complex, smaller peaks spaced ~0.1 GHz can be seen. These features were claimed to be representative of a $q^{3/2}$ or a $q^{1/2}$ mode dependence. Unfortunately, the spacing of these peaks must change with B field in any dispersive mode representing a solid, which is clearly not supported by the raw data shown in the figure. It would be inviting to assign the overall complex or broad absorption peak to an excitation of the electron solid, but the data show that as B is decreased, the mode frequency decreases; again this is contrary to the expected dispersion f(q)/B. In all likelihood the broad resonance represents only an overall resistance change in the system and therefore an impedance change in the line. Reflections in the lines could occur with the frequency spacing of the small peaks shown due to line impedance mismatches. As such these RF results provide questionable information concerning the 2D electron system at small filling factors.

Given this review of d.c. transport, optics, and meander line RF experiments, we will now turn to an experimental system which has proven itself to accurately measure the RF conductivity in the 2DES.

3. Surface Acoustic Wave Experiments

As outlined in the previous sections, a large number of experimental efforts have been undertaken to search for proof of existence and properties of the quantum electron solid in 2D, with the serious limitation that little is known theoretically about this phase. While the general nature of the liquid to solid transition may be understood to the level that a phase diagram can be contrived, the only other experimentally accessible characteristic should be the systems' ability to support a low energy shear mode. The effect of disorder on this magnetophonon propagation leaves the specific dispersion relation for this excitation somewhat open to debate (see Fig. 22). Moreover, since this excitation promises to be the most specific indicator of electron solid properties, or of properties of the 2DES at small ν in general, it is worth examining experimentally. For this reason we have used surface acoustic waves to probe the 2DES. The following section outlines our studies, with first an introduction to the technique.

The surface acoustic wave (SAW) technique [35] allows determination of the dynamical conductivity $\sigma_{xx}(q,\omega)$ [36] of the 2DES whose q and ω are the wavevector and frequency of sound, and the sound velocity $v_s = \omega/q$. The technique works as follows. Both ends of a typically 3 × 5 mm heterostructure are etched, leaving a mesa containing the 2DES about 3 mm square. Interdigital SAW transducers are patterned lithographically and evaporated directly onto the ends of the samples (Fig. 24). The finger spacing of the transducer determines the SAW frequency/wavelength, with each transducer able to produce multiple harmonics — typically the fundamental, third and fifth harmonics. This corresponds to each transducer being able to examine several points on the sound dispersion curve (Fig. 23). The full dispersion line can be tested by making multiple transducers producing different fundamental and higher harmonic waves.

The SAW are produced in one transducer using a pulse of <1 μsec duration. A voltage across the transducer finger produces a lattice distortion in the GaAs since this material is piezoelectric. For a voltage at frequency $\omega = 2\pi v_s / \lambda_{fingers}$ a traveling wave is produced. This wave traverses the surface of the sample, travels through the mesa where it interacts with the 2DES, and is detected at the other transducer. Amplitude and frequency change of the transmitted wave are measured using standard boxcar integration and homodyne detection. The traveling time of the SAW pulse from one transducer to the other allows clean detection of the transmitted pulse since transients in the system due to the initial pulse have died out by the time the pulse reaches the receiving transducer.

The conductivity of the 2DES effects the SAW amplitude and velocity and this measured response in the sound properties allows determination of the sheet conductivity $\sigma_{xx}(q,\omega)$. In this simple model, a relaxation time of the 2DES is contained within the sheet conductivity such that

$$\Gamma = (\alpha^2/2) q (\sigma_{xx}/\sigma_m)/(1+(\sigma_{xx}/\sigma_m)^2)$$

and

$$\frac{\Delta v}{v} = (\alpha^2/2)/(1+(\sigma_{xx}/\sigma_m)^2)$$

where amplitude $\sim \exp(-\Gamma x)$, x is the path length of the sound, α is the effective piezoelectric coupling coefficient $\alpha^2/2 = 3.2 \times 10^{-4}$, $\sigma_m = v_s(\varepsilon_0 + \varepsilon_s) = 3.5 \times 10^{-7}$ ($\Omega^{-1}/\square$) and ε_0, ε_s are the dielectric constants of the vacuum and semiconductor. From this one can see that the amplitude is minimized for $\sigma_{xx}(q,\omega) = \sigma_m$ and that the technique is sensitive to conductivities centered around σ_m (Fig. 25).

The relaxation model outlined above holds for a single relaxation time parametrized by a real conductivity $\sigma_{xx}(q,\omega)$. However, since both amplitude and velocity are measured it is possible to deduce a *complex* conductivity $\sigma_{xx}(q,\omega) = \sigma' + i\sigma''$ if indeed a complex conductivity is present. Then the expressions above (1 and 2) take the form

$$\Gamma = (\alpha^2/2) q \, \sigma'/[(1+\sigma'')^2 + (\sigma')^2]$$

and

$$\frac{\Delta v}{v} = (\alpha^2/2)[1+\sigma'']/[(1+\sigma'')^2 + (\sigma')^2]$$

where $\sigma' = \text{Re } \sigma_{xx}(q,\omega)/\sigma_m$ and $\sigma'' = \text{Im } \sigma_{xx}(q,\omega)/\sigma_m$. One can trivially determine that an imaginary component is present by mapping the measured Γ versus $\Delta v/v$, which

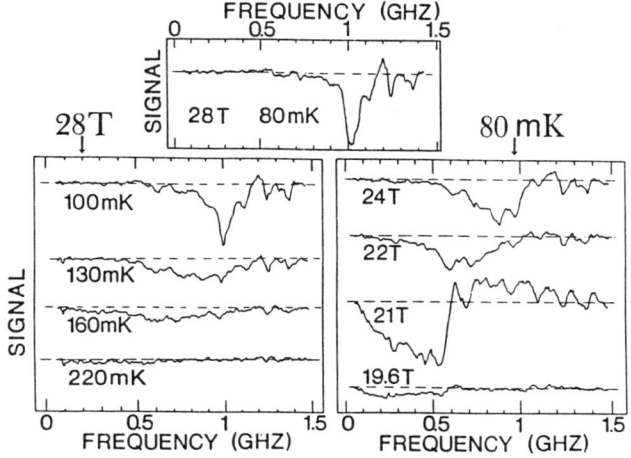

Fig. 23. Evolution of the meander line radio-frequency absorption spectrum with temperature at constant field and with field at constant temperature: The density $n = 9.4 \times 10^{11}$ cm^{-2}, filling factor $\nu = (3.92\text{T})/H$, reduced temperature $t = T/(560 \text{ mK})$. Ref. 32.

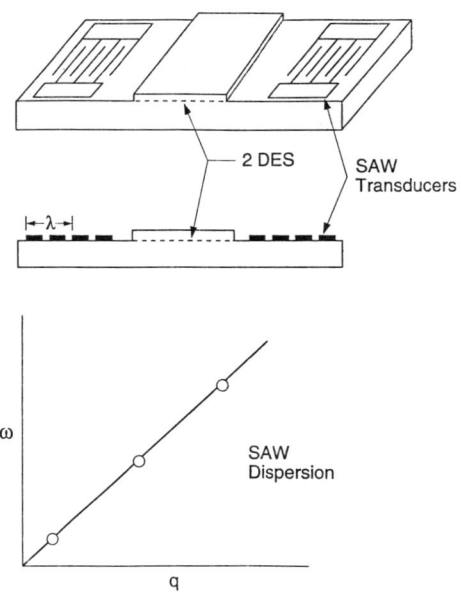

Fig. 24. Schematic representation of surface acoustic wave (SAW) setup. The typical dimension between transducers is ~1 mm. The SAW dispersion is shown below, marking the fundamental frequency (corresponding to λ) and two higher harmonics.

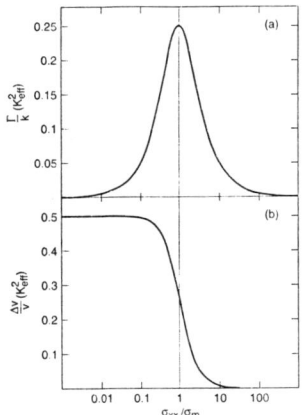

Fig. 25. Attenuation coefficient Γ per unit wavevector k (a) and change in sound velocity $\Delta v/v_0$ (b) in units of K_{eff}^2 as a function of sheet conductivity σ_{xx} of the 2DES which is assumed to be located on top of the crystal surface. Ref. 35.

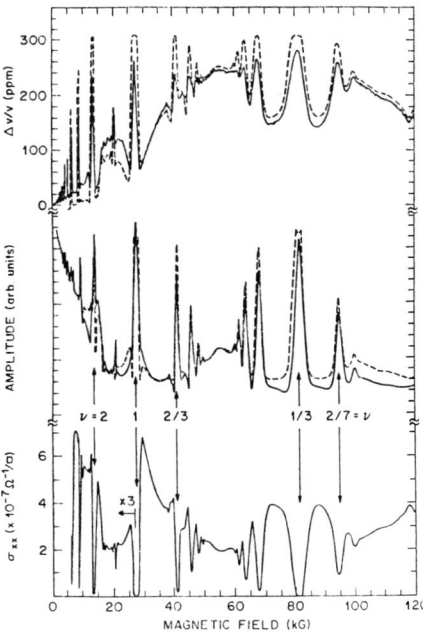

Fig. 26. Conductivity, SAW amplitude, and SAW velocity shift vs. magnetic field at 160 mK and 235 MHz. Solid lines are measured values, and dashed lines are results of the measured d.c. conductivities used in expressions (1) and (2) with $\sigma_m = 4 \times 10^{-7}$ Ω^{-1}. Ref. 37.

should reveal a semi-ellipse if the conductivity is purely real and not complex. As such, the measured SAW $\Delta v/v$ and Γ provide a direct measurement of the 2DES conductivity $\sigma_{xx}(q,\omega)$ at the frequency and wavevector of the SAW. Experimentally the d.c. conductivity can be simultaneously measured, allowing direct comparison between the d.c. and a.c. σ_{xx} of the system.

Measurements [37] of the SAW transmitted amplitude and velocity and simultaneous d.c. conductivity are shown in Fig. 26 for a high mobility heterostructure at dilution refrigerator temperatures. The SAW amplitude and velocity show increases when the d.c. conductivity drops: for example at $v = 1/3$. This represents piezoelectric stiffening, since at low conductivity the 2DES cannot relax to short-out the piezoelectric field of the SAW and so the sound velocity remains high. The major feature to note from this result is that the conductivity features seen in the d.c. are also seen in the SAW amplitude and frequency. The d.c. conductivity is converted to amplitude and $\Delta v/v$ using the relaxation model (relations 1 and 2) with a good agreement between the measured SAW results and the d.c. transport derived $\Delta v/v$ and SAW amplitude. This means that at this SAW frequency and wavevector, the $\sigma_{xx}(q,\omega)$ is roughly equivalent to the d.c. σ_{xx} over the magnetic field range displayed here. Clearly the SAW technique is a sensitive measure of 2DES conductivity, and we are now poised to examine the small filling factor range.

At small v, in the vicinity of $v = 1/5$ where the insulating phase can be seen in d.c. transport and has been examined by other techniques (section II), the SAW amplitude and velocity demonstrate a marked discrepancy [38] between σ_{xx} (d.c.) and $\sigma_{xx}(q,\omega)$. Figure 27 shows that on the high B field side of $v = 1/5$ the longitudinal resistivity grows extremely large. At these magnetic field values the SAW measurements give much lower sound velocity shift values than expected from the d.c. transport. This means that over this filling factor range the high frequency conductivity is much larger than the d.c. conductivity as determined using the relaxation model. The raw SAW attenuation results are consistent with this picture; immediately beyond 1/5 the attenuation is small and nearly saturated, as expected for large, high frequency conductivity.

In order to understand the SAW results at small v, we take advantage of the measurements of both amplitude and velocity to derive both real and imaginary high frequency conductivity. If an excitation mode is indeed present in the 2DES and the SAW frequency and wavevector range cross the mode peak then it is expected that the real part of the conductivity will be peaked at the mode center and the imaginary part of the conductivity will go through zero (Fig. 28).

The real and imaginary contributions to the conductivity can be seen immediately in a phase plot of the measured SAW attenuation versus velocity. As noted above, the phase plot will give a semicircle for only real conductivity and deviations from the semicircle represent imaginary conductivity. Figure 29 shows phase plots for a single magnetic field point in the presumed electron solid range: in these plots the temperature is swept from the highest to lowest values. The highest frequency plot (2.67 GHz) shows data lying on the semicircle, and as T is lowered (moving to smaller $\Delta v/v$) a large deviation from purely real conductivity is apparent. Over the entire temperature range of the 0.93 GHz measurement the purely real conductivity curve is traced. At the lowest

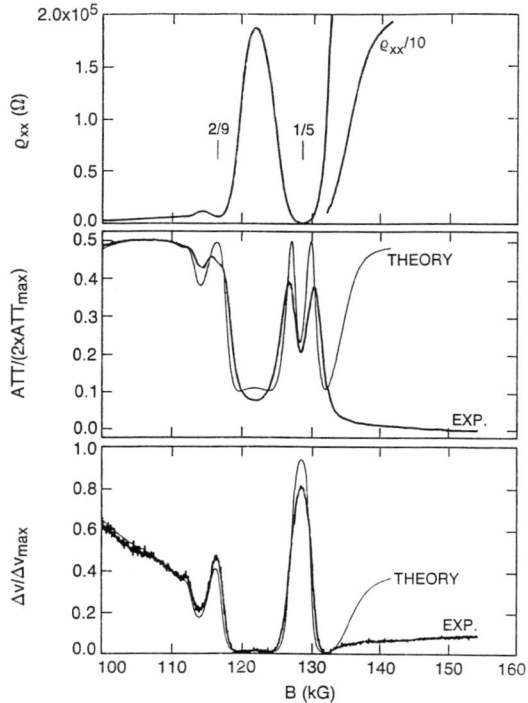

Fig. 27. Magnetic field dependence of d.c. resistance ρ_{xx}, normalized SAW attenuation, and velocity shift in the small filling factor range at 80 mK and 235 MHz. The theory lines are calculated from expressions 1 and 2 using ρ_{xx} (d.c.). Ref. 38.

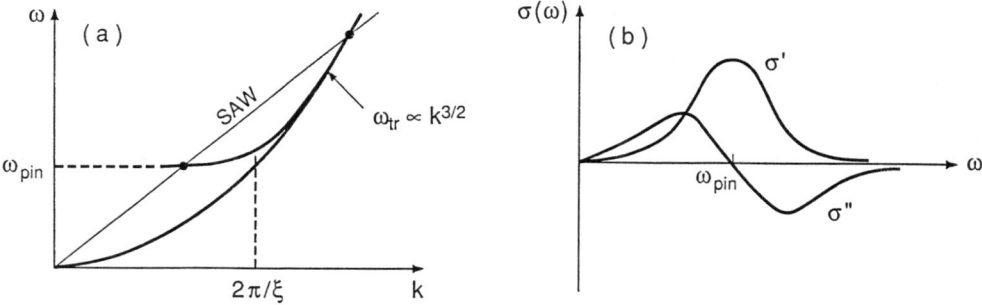

Fig. 28. a) Dispersion relation of surface acoustic waves (SAW) and collective modes of a disordered Wigner crystal. b) Real and imaginary parts of conductivity near the crossing of the pinning mode and SAW dispersion line.

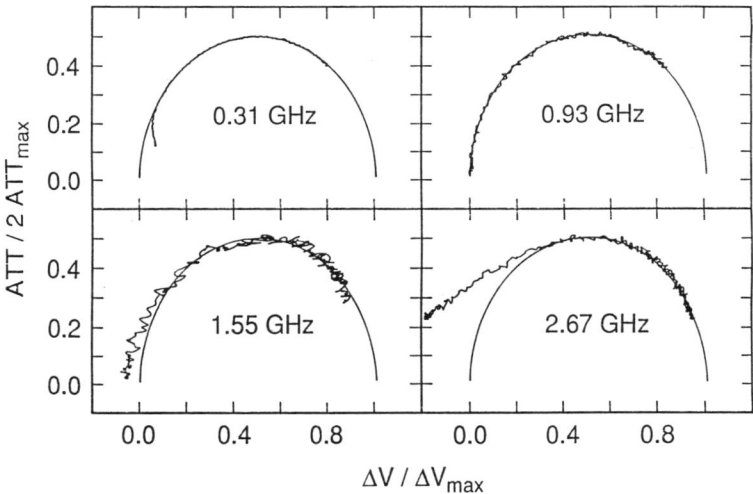

Fig. 29. Phase plots of sound attenuation vs. velocity shift at four different frequencies for the same sample at a constant B($\nu=0.176$). The theoretical line based on Eqs. (1) and (2) with $\sigma'' = 0$ forms a semicircle. Ref. 38.

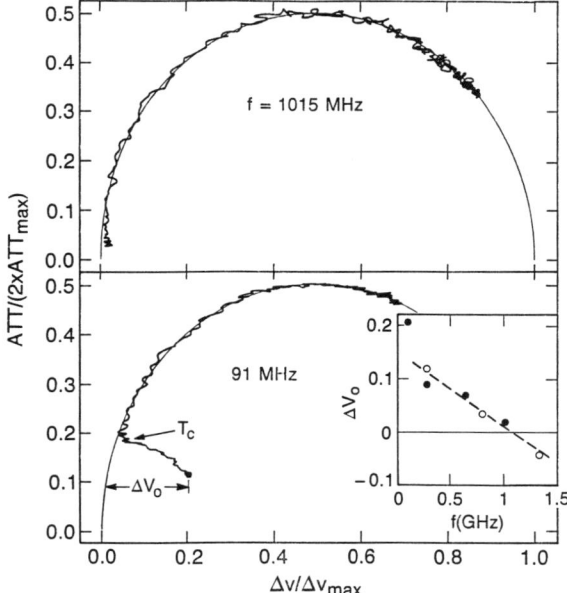

Fig. 30. A phase plot of SAW attenuation vs. velocity shift at two different frequencies for sample D at a constant B($\nu=0.167$). The theoretical line based on Eqs. (1) and (2) forms a semicircle. In the inset, Δv_0 is plotted as a function of frequency for sample D(○) and sample E (●, $\nu=0.140$). Ref. 38.

frequency a deviation from purely real σ_{xx} is again seen at low T, but with a sign opposite to that of the high frequency SAW. Another set of phase plots is shown in Fig. 30 for the same sample. In the low freqency (91MHz) plot at the low T end a substantial inward deviation occurs, signalling a sizable positive imaginary contribution. The deviation initiates at a distinct temperature T_c. As in Fig. 29 (but not shown here), intermediate frequencies demonstrated intermediate values of Δv_0, which we define as the lowest temperature deviation of the SAW velocity from the real conductivity semicircle. These data suggest that as the SAW frequency is increased, at lowest temperature a large imaginary conductivity component exists and decreases in magnitude, crosses though zero, then increases with opposite sign as the frequency increases further. Values of Δv_0 are plotted versus frequency in the inset to Fig. 30, showing that at a frequency of about 1 GHz the imaginary conductivity crosses through zero, which is characteristic of a mode crossing.

Given that the imaginary contribution to the conductivity has now been established, the real part of the conductivity can be extracted by again studying temperature sweeps at constant magnetic field values in the insulating regime. Figure 31, lower panel, shows the temperature sweeps for four SAW frequencies and d.c. transport converted to real conductivity. At high temperatures, no frequency dependence is seen in σ_{xx}. At T_c, the frequency dependence becomes apparent. In the low temperature limit, the conductivity increases from the lowest frequency, is a maximum near 1 GHz, then decreases at higher frequencies. This pattern is repeated for all transducers sets and samples examined. This behavior indicates a peak in the real conductivity at ~1 GHz, which further corroborates the model of a conductivity mode crossing.

The properties of the mode crossing are consistent in general with the presumed properties of a broad pinning mode in a disordered electron solid, as outlined in section II. To further appreciate the features of this mode and relate it to the electron solid, we can study it's form as a function of filling factor and temperature. Fig. 32 shows the real mode peak for a set of temperatures at one filling factor value. Clearly no mode is present at high temperatures, ≥ 250 mK. As T is lowered the conductivity at $\omega = 0$ decreases while at 1GHz the peak increases. Note that the mode center does not appear to shift substantially as T is lowered. The mode itself at all temperatures is somewhat assymetric, with a substantial high frequency tail. Recall that ω increases linearly with q for SAW measurements so that in the electron solid picture of the mode crossing the SAW dispersion crosses the mode at a not-constant q (see Fig. 28). The highest frequency/wavevector tail of the mode may therefore be a prelude to a *re-crossing* of the shear mode, beyond the low q, pinned range, in the range of wavevector dependence when the true shear properties can be examined.

The filling factor dependence of the real conductivity mode is shown in Fig. 33. Note as B is increased the magnitude of the mode increases. It is significant that in this data for the different ν the mode peak does not shift appreciably — this shows that the principle mode crossing is q-independent since any q-dependent mode necessarily will have a cyclotron frequency dependence (see Section II). At the high frequency mode tail interesting behavior is noted for the filling factors beyond $\nu = 1/5$. For higher B the conductivity apparently *drops*, as expected for a q dependent mode since $\omega \sim f(q)/\omega_c$.

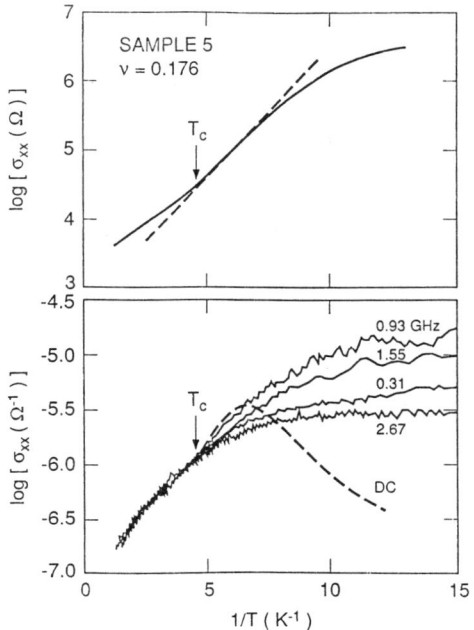

Fig. 31. In the upper panel is a semilog plot of d.c. resistance vs. inverse temperature. In the lower panel is a similar plot of rf conductivity including d.c. conductivity as a broken line.

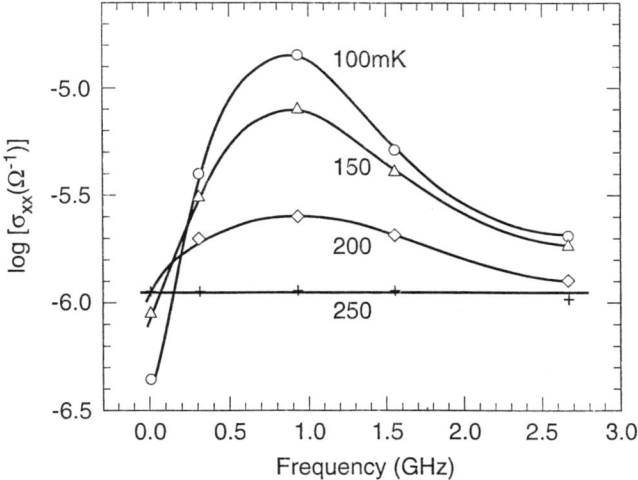

Fig. 32. Finite frequency/wavevector real conductivity mode crossing at filling factor $\nu = 0.160$ for a set of five different temperatures as measured with SAW. Note formation and peaking of the mode at ~0.9 GHz for T < 250 mK.

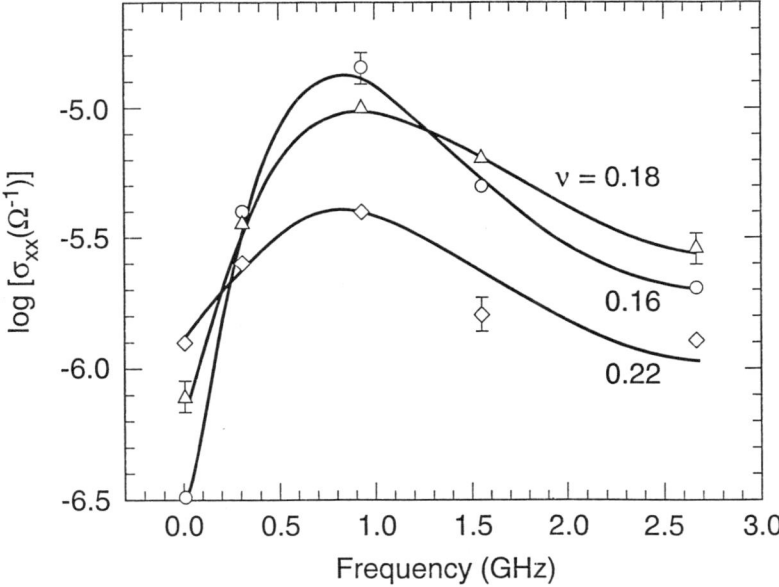

Fig. 33. Finite frequency/wavevector real conductivity mode crossing at T = 100 mK for three different magnetic field values. Note that the mode peak does not move appreciably for the range of filling factor displayed.

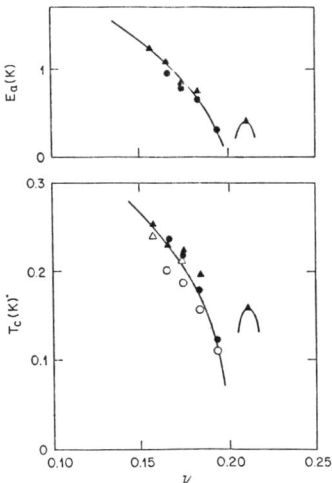

Fig. 34. Activation energy and transition temperature of electron solid vs. filling factor measured in 2 different samples (circles and triangles). The open and closed symbols are determined from rf- and dc- measurements, respectively. T_c is defined for both in Fig. 31.

As in the case of the temperature dependence of the mode crossing shown above, this high frequency feature may indicate penetration of the q-dependent range of the dispersion within the model of the electron solid. Clearly the data displayed here offer only intimations to this behavior, since the tendencies outlined are within the experimental errors shown in the figures. The data do indicate that further study is necessary in the high frequency SAW range beyond that already explored.

If the model of electron solid formation is used to explain the SAW results then one can estimate the correlation length ξ of the solid from the pinning frequency. In the long wavelength $k < 2\pi/\xi$ regime, the pinning frequency is constant and scales like $\omega_p \omega_t/\omega_c$ where $\omega_p^2 = 2\pi ne^2 k/\epsilon_p m^*$ and the shear mode frequency $\omega_t^2 = \eta k^2/nm^*$ are determined at the wavevector $k = 2\pi/\xi$ and the cyclotron frequency $\omega_c = eB/m^*$. By estimating the shear modulus $\eta = 4k_B T_c n$ for the transition temperature $T_c = 200$ mK we obtain $\xi = 1$ μm from our our measured pinning frequency. This value of ξ is about 20 lattice spacings of the electron solid but smaller than the wavelength $\lambda_{min} \approx 2.5$ μm used at 1.0 GHz. This provides self-consistency of our analysis by giving a constant pinning frequency over the range of wavelengths where the mode peak appears to be B field independent. Our estimate of ξ is shorter than the coherence length of 10 μm calculated within the weak pinning model from the depinning threshold electric field [39]. Because the samples used in both sets of experiments seem to be comparable in quality, the difference presumably comes from different models used in estimating the correlation length. In our case, the shortest wavelength data at $\lambda \sim 1.0$ μm clearly rule out a 10 μm correlation length.

Finally, one may use the SAW data, particularly the stark frequency dependent conductivity, to generate a phase diagram [40] of a presumed electron solid. Temperature sweeps at many filling factors for different SAW frequencies (see Fig. 31) allowed determination of T_c as a function of ν. These data are compiled in Fig. 34, lower panel, and demonstrate the mode onset observed at $\nu < 1/5$ and at $\nu \geq 1/5$. This is consistent with the re-entrant behavior noted in d.c. transport in previous experiments. For these same samples, activation energies of the ρ_{xx} were measured and plotted as E_a in Fig. 34 upper panel. Note that the same general behavior or qualitative properties can be derived from the simple transport results — both are consistent with the presumed electron solid phase diagram.

In closing this section on SAW studies and the experimental review of all techniques, a cautionary note is warranted. No experimental evidence exists to date, including that of the SAW experiments, that definitively indicates electron solid formation. While the data in total may be viewed from the perspective of this picture and be found to be consistent with the electron solid model, as is almost always the case no experiment absolutely proves a theory to be correct. The SAW data show promise for further testing of the insulating phase and probably provide the most specific picture of the phase to date. However, a new class of experiments which can directly measure the electron ordering, such as scattering experiments or direct electron imaging, are probably needed to sort out this interesting but as yet unsolved question.

It is a pleasure to acknowledge my collaborators in this work: H.W. Jiang at M.I.T., Dan Tsui at Princeton University, and Horst Stormer, Mikko Paalanen, Rene Ruel, Loren Pfeiffer, Ken West, and Kirk Baldwin, all at A.T.&T. Bell Laboratories.

References

[1] C.C.Grimes and G.Adams, PRL **42**, 795 (1979).

[2] M.Kosterlitz and D.Thouless, J. Phys. C **6**, 1181 (1973).

[3] B.Halperin and D.Nelson, Phys. Rev. B **19**, 2457 (1979).

[4] P.M.Platzman and H.Fukuyama, Phys. Rev. B **10**, 3150 (1974).

[5] E.P.Wigner, Phys. Rev. **45**, 1002 (1934).

[6] D.Ceperly, Phys. Rev. B **18**, 3126 (1979).

[7] H.L.Stormer, Surf. Sci. **132**, p. 529 (1980).

[8] R.Lassnig, Solid State Commun., **65**, 765 (1988).

[9] K.vonKlitzing, G.Dorda, and M.Pepper, Phys. Rev. Lett. **46**, 494 (1980).

[10] D.C.Tsui, H.L.Stormer, and A.C.Gossard, Phys. Rev. Lett. **48**, 1559 (1982).

[11] R.B.Laughlin, Phys. Rev. Lett. **50**, 1395 (1983).

[12] F.D.M.Haldane, Phys. Rev. Lett. **51**, 605 (1983).

[13] B.I.Halperin, P.A.Lee, N.Read, Phys. Rev. B **47**, 7312 (1993).

[14] P.K.Lam, S.M. Girvin, Phys. Rev. B **30**, 473 (1984).

[15] P.M.Platzman, "The Physics of the 2 Dimensional Electron Gas", NATO ASI series, Plenum 1987.

[16] E.E.Mendez, L.L.Chang, M.Heiblum, L.Esaki, M.Naughton, U.Martin, and J.Banks, Phys. Rev. B **30**, 7310 (1984).

[17] R.L.Willett, H.L.Stormer, D.C.Tsui, L.N.Pfeiffer, R.W.West, and K.W.Baldwin, Phys. Rev. B **38**, 7881 (1988).

[18] R.L.Willett, H.L.Stormer, D.C.Tsui, L.N.Pfeiffer, and K.W.West, "High Magnetic Fields in Semiconductor Physics", Proceedings of the Wurzburg High Magnetic Field Conference, Springer-Verlag, 1988.

[19] S.T.Chui and K.Esfarjani, Phys. Rev. Lett. **66**, 652 (1991).

[20] H.W.Jiang, R.L.Willett, H.L.Stormer, D.C.Tsui, L.N.Pfeiffer, and K.W.West, Phys. Rev. Lett. **65**, 633 (1990).

[21] J.P.Stokes, A.N.Bloch, A.Jamossy, and G.Garner, Phys. Rev. Lett. **52**, 372 (1984).

[22] R.L.Willett, H.L.Stormer, D.C.Tsui, L.N.Pfeiffer, K.W.West, M.Shayegan, M.J.Santos, and T.Sajoto, Phys. Rev. Lett. B **40**, 6432 (1989).

[23] H.Fukuyama and P.A.Lee, Phys. Rev. B **17**, 535 (1978).

[24] L.Bonsall and A.A.Maradudin, Phys. Rev. B **15**, 1959 (1977).

[25] Y.P.Li, J.Sajoto, L.W.Engel, D.C.Tsui, and M.Shayegan, Phys. Rev. Lett. **67**, 1630 (1991).

[26] Y.P.Li, T.Sajoto, L.W.Engel, D.C.Tsui, and M.Shayegan, Phys. Rev. B **47**, 9933 (1993).

[27] M.Santos, Y.W.Suen, M.Shayegan, Y.P.Li, L.W.Engel, and D.C.Tsui, Phys. Rev. Lett. **68**, 1188 (1992).

[28] T.Sajoto, Y.P.Li, L.W.Engel, D.C.Tsui, and M.Shayegan, Phys. Rev. Lett. **70**, 2321 (1993).

[29] H.Buhmann, W.Joss, K.von Klitzing, I.V.Kukushkin, G.Martinez, K.Ploog, and V.B.Timofeev, JETP Lett. **52**, 306 (1990).

[30] R.G.Clark, Phys. Soc. T **39**, 45 (1991); E. M. Goldys et al., Phys. Rev. B **46**, 7957 (1992).

[31] H.Fukuyama and P.A.Lee, Phys. Rev. B **18**, 6245 (1978).

[32] E.Y.Andrei, G.Deville, D.C.Glattli, and F.I.B.Williams, E.Paris, and B.Etienne, Phys. Rev. Lett. **60**, 2765 (1988).

[33] H.L.Stormer and R.L.Willett, Phys. Rev. Lett. **62**, 972 (1989).

[34] H.L.Stormer and R.L.Willett, Phys. Rev. Lett. **68**, 2104 (1992).

[35] A.Wixworth, J.Scriba, M.Wassermeier, J.P.Kotthaus, G.Weimann, W.Schlapp, Phys. Rev. B **40**, 7874 (1989).

[36] A.Wixworth, J.P.Kotthaus, and G.Weimann, Phys. Rev. Lett. **56**, 2104 (1986).

[37] R.L.Willett, M.A.Paalanen, R.R.Ruel, K.W.West, L.N.Pfeiffer, and D.J.Bishop, Phys. Rev. Lett. **65**, 112 (1990).

[38] M.A.Paalenan, R.L.Willett, P.B.Littlewood, R.R.Ruel, K.W.West, L.N.Pfeiffer, and D.J.Bishop, Phys. Rev. B **45**, 17342 (1992).

[39] F.I.B.Williams, et al. Phys. Rev. Lett. **66**, 3285 (1991).

[40] M.A.Paalanen, R.L.Willett, R.R.Ruel, P.B.Littlewood, K.W.West, and L. N. Pfeiffer, Phys. Rev. B **45**, 13784 (1992).

TWO–DIMENSIONAL WIGNER SOLID VERSUS INCOMPRESSIBLE QUANTUM LIQUID

GUENTHER MEISSNER
Theoretische Physik,
Universitaet des Saarlandes
66041 Saarbruecken,
Germany

ABSTRACT. A unified many–body approach is shown to allow for studying the nature of two condensed phases in high magnetic fields: a 2D Wigner solid (*WS*) of the guiding centers of Coulomb interacting electrons and an incompressible quantum liquid (*IQL*) of Bose condensed charge–vortex composites. A competition between the two phases is found to give rise to a re–entrance behavior, if the mean number of magnetic flux quanta per electron exceeds a certain critical value.

1. Introduction

The main purpose of this contribution is to discuss some predictions of a unified many–body approach for a liquid and a solid phase, competing in a plane, where strong correlations of the considered many–body system of Coulomb–interacting electrons (charge: $Q = -e$) are of significance in the limit of high magnetic fields being applied perpendicular to that plane. Due to the well–known massive degeneracy associated with 2D free–electron motion of cyclotron frequency ω_c, the electron–electron interaction within the lowest Landau level plays an important role. In actual systems disorder potentials, as e.g. at hetero–junction interfaces, may also be essential [1]. The highly correlated motion of these electrons at sufficiently small random disorder, however, should favor the formation of the two novel phases of condensed matter at temperatures (T) low enough compared to the cyclotron energy, i.e., $k_B T \ll \hbar \omega_c$, and at magnetic fields (B) sufficiently high such that the mean number $\nu^{-1} = B/(\phi^0 n_e)$ of magnetic flux quanta $\phi^0 = ch/e$ per electron exceeds one (electron density: n_e).

The *IQL*–phase of Bose condensed charge–vortex composites $e_p \phi_q^0$ for certain rational filling factors $\nu = p/q < 1$ (with p and q denoting mutual primes and q usually being odd) exhibits the fractional quantum Hall effect (FQHE), i.e.: a quantized Hall conductance $\sigma_{12} = \nu\, e^2/h$ is accompanied by minima in the longitudinal conductance σ_{11}, surviving moderate disorder [2].

The *WS*–phase with a lattice–periodic structure of the guiding centers of the electrons, minimizing their Coulomb repulsion $e^2/\epsilon r_L$ repeatedly at non–rational filling factors below a certain critical value ν_c, exhibits a non–vanishing shear modulus $\mu > 0$ ($r_L = (c\hbar/eB)^{1/2} = (\hbar/m\omega_c)^{1/2}$: Larmor radius; m: effective electron mass; ϵ: background dielectric constant). In the dilute electron ($\nu \to 0$) or high–magnetic field ($r_L \to 0$) limit, the ground–state of that sort of a re–entrant system may finally be identified with the triangular electron lattice formed by a classical 2D Wigner crystal of lattice constant $d = (\sqrt{3}\, n_e/2)^{-1/2}$ [3]. A random potential will induce a finite correlation length ξ in the Wigner solid, whose shear stiffness μ remains meaningfull, however, as long as $\xi/d \gg 1$ [4].

2. Outline of the many–body approach

The fundamental quantity for calculating ground–state energies $E(\nu)$ and collective excitations $\omega(k)$ in our many–body approach [5] is the wave–vector and frequency dependent response function

$$\chi_{\Delta\Delta}(k,\Omega=\omega+i\eta) = \frac{i}{\hbar} \int_0^\infty dt\, e^{i\Omega t} <[\Delta(k,t),\Delta(-k,0)]> \tag{1}$$

for non–commuting density fluctuations $\Delta(k) = \Sigma_\ell \exp(-ik\cdot X(\ell))$ of the guiding centers $X(\ell)$ with non–commuting Cartesian components, being reflected in the commutation relations $[\Delta(k),\Delta(k')] = -2i \sin(k_\alpha \epsilon_{\alpha\beta} k'_\beta r_L^2/2)\, \Delta(k+k')$. Closely related are the spectral function $\chi''_{\Delta\Delta}(k,\omega)$, being given by the discontinuity of the complex response function $\chi_{\Delta\Delta}(k,\Omega)$ at the real frequency axis, and the static structure factor $S_{\Delta\Delta}(k) = <\Delta(k)\Delta(-k)>/N_e$ of N_e electrons.

Using the concept of broken magnetic translation invariance, an implicit equation for the melting temperature $T_M(\nu)$ as a function of the filling factor $\nu = 2\pi r_L^2 n_e$ can finally be derived quite generally from a homogeneous integral equation, holding for the covariant derivative $D <\bar{\rho}(r)>$ of the expectation value of that part of the electron density operator

$$\bar{\rho}(r) = \int d^2k\, \exp(ik\cdot r - k^2 r_L^2/4)\, \Delta(k)/2\pi \tag{2}$$

which no longer contains the fast cyclotron motion. The quantity A in $D = (\partial/\partial r + iQ\hbar A/c)$ denotes the vector potential of the magnetic field $\vec{B} = \nabla \times A$. The competition between Coulomb repulsion and exchange attraction apparently is a sufficient condition for this system to become unstable against broken magnetic translation invariance at non–rational filling factors $\nu < \nu_c$. Due to a specific kind of particle–hole symmetry in the lowest Landau level, the Fourier transform of the effective interaction between two guiding centers

$$v_{eff}(k) = \frac{1}{2}[v(k) - \frac{r_L^2}{2\pi}\int d^2k'\, e^{i k \cdot k' r_L^2} v(k')], \tag{3}$$

with the repulsive part $v(k) = (2\pi e^2/\epsilon k)\, exp(-k^2 r_L^2/2)$, indeed reveals this sort of competition, i.e. $v_{eff}(k) < 0$, if $kr_L > 1$.

3. Summary of results

Based on the outlined many–body approach, predictions concerning the ground–state energy $E(\nu)$, the collective excitations $\omega(k)$, and the phase seperation curve $T_M(\nu)$ of the two competing phases are conceivable.

The critical filling factor ν_c can be estimated from comparing ground–state energies calculated both for the solid and the liquid phase, i.e. $E_S(\nu_c) = E_L(\nu_c)$. Our estimate obtained for $\nu_c \simeq 0.202 \gtrsim 1/5$ together with reentrant aspects around $\nu = 1/5$ seems to be in reasonable agreement with experimental findings [1]. For the WS–phase the one–loop approximation of the high–magnetic field expansion

$$E_S(\nu)/N_e = \{-0.78213\,\nu^{1/2} + 0.24101\,\nu^{3/2} + 0.02992\,\nu^{5/2} + ...\}(e^2/\epsilon r_L)$$

has been used, containing the static lattice energy, the harmonic zero–point energy, and anharmonic corrections [6]. A fit of Monte–Carlo simulations with Laughlin's wave function [8] was used for $E_L(\nu)/N_e$ in the IQL–phase.

Concerning the phase separation curve $T_M(\nu)$, we would just like to mention, that the familiar mean–field result [7], $k_B T_M/(e^2/\epsilon r_L) = 0.557\nu(1-\nu)$, is easily recovered in our approach as the leading approximation, replacing, e.g., a generalized particle–hole interaction of the lowest Landau level by v_{eff} of Eq. (3). Strong correlations could be included by evaluating the polarization operator in single–mode approximation rather than in Hartree–Fock approximation.

Results for collective excitations $\omega(k)$ in the absence of disorder have extensively been reviewed before [5]. The finite gap $\omega_L(k=0) \neq 0$ of the magneto–rotons in the IQL–phase versus gapless behavior $\omega_S(k \to 0) \sim \mu\, k^{3/2}$ of the magneto–phonons (Goldstone mode) in the WS–phase in the long–wavelength limit has been shown to hold rigorously. Thus

$$\omega_S^2(k) = \det|M_{\alpha\beta}(k,0)|\,\omega_c^{-2} = \frac{2\pi e^2}{m^2 \omega_c^2 \epsilon}\,\mu\, k^3 + 0(k^4) \tag{4}$$

where $M_{\alpha\beta}(k,0) = 2\pi e^2 n_e k_\alpha k_\beta/(k\epsilon m) + Z_{\alpha\beta\gamma\delta}k_\gamma k_\delta + 0_{\alpha\beta}(k^4)$ denotes the static self–energy and $\mu = m\, n_e\, Z_{1122}$ is the isothermal shear modulus. In the dilute electron $(\nu \to 0)$ limit, e.g., $\mu = 0.25406\, e^2 n_e^{3/2} \epsilon^{-1}$ [3].

The possibility for the liquid–solid transition to occur via softening of the magneto–roton minimum (at certain finite wave vectors in the excitation spectrum $\omega_L(k)$ becoming a reciprocal lattice vector in the WS–phase) can also be shown to exist rather generally. This could have implications on attempts to interprete optical studies [9] of Wigner cristallization in the presence of defects. Due to random disorder a gap ω_p will open up at $k=0$ in the magneto–phonon dispersion, since the pinning force gives rise to additional restoring forces $m\omega_p^2 \delta_{\alpha\beta}$ which have to be added to the expression of the static self–energy in the absence of disorder. Hence, the correlation length $\xi = (2\pi e^2 \mu/\omega_c^2 m^2 \epsilon)^{1/3}\, 2\pi/\omega_p^{2/3}$ of the Wigner solid is finally determined by its pinning frequency ω_p.

References

[1] Jiang, H.W., R.L. Willet, H.L. Stoermer, D.C. Tsui, L.N. Pfeiffer, and K.W. West (1990), Phys. Rev. Lett. **65**, 633.
[2] Tsui, D.C., H.L. Stoermer, and A.C. Gossard (1982), Phys. Rev. Lett. **48**, 1599; Laughlin, R.B. (1983), Phys. Rev. Lett. **50**, 1395.
[3] Meissner, G., H. Namaizawa, and M. Voss (1976), Phys. Rev. B **13**, 1370; Bonsall, L. and A.A. Maradudin (1977), Phys. Rev. B **15**, 1959.
[4] Fisher, D.S., Lectures on: *Low Temperature Phases, Ordering and Dynamics in Random Media*.
[5] Meissner, G. (1993), Physica B **184**, 66.
[6] Meissner G. and U. Brockstieger (1987), in G. Landwehr Ed. (Springer, Berlin, Heidelberg) *Series in Solid State Sciences* **71**, 85.
[7] Fukuyama, H., P.M. Platzman, and P.W. Anderson (1979) B **19**, 5211.
[8] Levesque, D., J.J. Weis, and A.H. MacDonald (1984), B **30**, 1056.
[9] See, e.g., Buhmann, H., W. Joss, K. von Klitzing, I.V. Kukushkin, A.S. Plaut, G. Martinez, K. Ploog, and V.B. Timofeev (1991), Phys. Rev. Lett. **66**, 926.

DOMAIN GROWTH AND COARSENING

A. J. BRAY
Department of Theoretical Physics
The University of Manchester
Manchester M13 9PL
United Kingdom

ABSTRACT. The theory of phase ordering dynamics – the growth of order through domain coarsening when a system is quenched from the homogeneous phase into a broken-symmetry phase – will be reviewed, and recent developments discussed. Interest will focus on the scaling regime that develops at long times after the quench. How can one determine the growth laws that describe the time-dependence of characteristic length scales, and what can be said about the form of the associated scaling functions? Particular attention will be paid to systems described by more complicated order parameters than the simple scalars usually considered, e.g. vector and tensor fields. The latter are needed to describe phase ordering in nematic liquid crystals, on which there have a number of recent experiments. The study of topological defects (domain walls, vortices, strings, monopoles) provides a unifying framework for discussing coarsening in these different systems.

1. Introduction

Systems quenched from a disordered phase into an ordered phase do not order instantaneously. Instead, the length scale of ordered regions grows with time as the different broken symmetry phases compete to select the equilibrium state.

To fix our ideas, consider the ferromagnetic Ising model. Figure 1 shows the spontaneous magnetization as a function of temperature. The arrow indicates a temperature quench, at time $t = 0$, from an initial temperature T_I above the critical point T_C to a final temperature T_F below T_C. At T_F there are two equilibrium phases, with magnetization $\pm M_0$. Immediately after the quench, however, the system is in an unstable disordered state corresponding to equilibrium at temperature T_I. The theory of phase ordering kinetics is concerned with the dynamical evolution of the system from the initial disordered state to the final equilibrium state.

Part of the fascination of the field, and the reason why it remains a challenge more than 30 years after the first theoretical papers appeared, is that, in the thermodynamic limit, final equilibrium is never achieved! This is because ergodicity is broken in the ordered phase, and the longest relaxation time diverges with the system size. Instead, a network of domains of the equilibrium phases develops, and the typical length scale associated with these domains increases with time t. This situation is illustrated in Figure 2, which shows a Monte Carlo simulation of a two-dimensional Ising model, quenched from $T_I = \infty$ to $T_F = 0$. Inspection of the time sequence will convince the suggestible reader that we have here a *scaling* phenomenon – the domain patterns at later times look statistically similar to those at earlier times, apart from a global change of scale. This 'dynamic scaling hypothesis', which will be formalized later, has played an important role in the development of the theory.

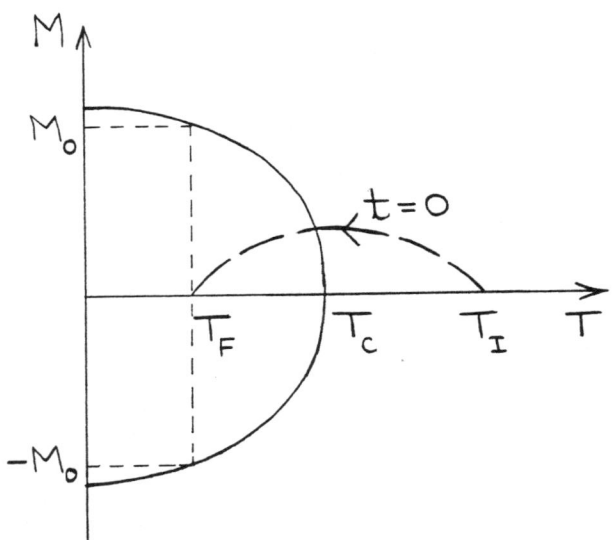

Figure 1. Spontaneous magnetization of the Ising model as a function of temperature. The arrow indicates a quench, at time $t = 0$, between temperatures T_I and T_F.

For pedagogical reasons, we have introduced domain-growth in the context of the Ising model, and will continue to use magnetic language for simplicity. The same phenomenon has been studied for many decades, however, by metallurgists, in the context of binary alloys, where it is known as 'Ostwald ripening'. A similar phenomenon occurs in the phase separation of fluids under cooling, known as 'spinodal decomposition', although in that case the phase separation is accelerated by the gravitational field, which severely limits the temporal duration of the scaling regime. The gravitational effect can be avoided by using density-matched binary liquids. All of the above systems, however, contain an extra complication not present in the Ising ferromagnet. This is most simply seen by mapping an AB alloy onto an Ising model. If we represent an A atom by an up spin, and a B atom by a down spin, then the *equilibrium* properties of the alloy can be modelled very nicely by the Ising model. There is one important feature of the alloy, however, that is not captured by the Ising model with conventional Monte-Carlo dynamics. Flipping a single spin in the Ising model corresponds to converting an A atom to a B atom (or vice versa), which is inadmissible. The dynamics must conserve the number of A and B atoms separately, i.e. the magnetization (or 'order parameter') of the Ising model should be *conserved*. This will influence the form of the coarse-grained equation of motion, as discussed in section 2, and lead to slower growth than for a non-conserved order parameter.

In all the systems mentioned so far, the order parameter (e.g. the magnetization of the Ising model) is a scalar. In the last few years, however, there has been increasing interest in systems with more complex order parameters. Consider, for conceptual simplicity, a planar ferromagnet, in which the order parameter is a vector confined to a plane. After a quench into the ordered phase, the magnetization will point in different directions in different regions of space, and singular lines (vortex lines) will form at which the direction is not well defined. These 'topological defects' are the analog of domain walls for the scalar systems. We shall find that, quite generally, an understanding of the relevant topological defects in the system, combined with the scaling hypothesis, will take us a long way towards understanding the forms of the growth laws and scaling functions for phase ordering in a large variety of systems.

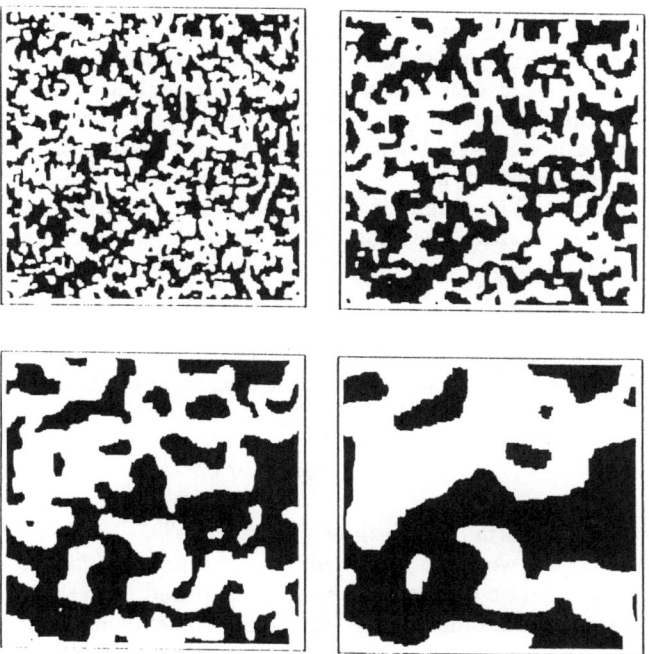

Figure 2. Growth of domains at $T = 0$ for a $2-d$ Ising model after 5, 15, 60 and 200 Monte Carlo steps per spin following a quench from $T = \infty$. The system size is 256×256. (From J. G. Kissner, Ph. D. thesis, University of Manchester, 1992).

2. Dynamical Models

It is convenient to set up a continuum description in terms of a coarse-grained order-parameter field ('magnetization density') $\phi(\mathbf{x}, t)$, which we will initially take to be a scalar field. A suitable Landau free-energy functional to describe the ordered phase is

$$F[\phi] = \int d^d x \, [\frac{1}{2} (\nabla \phi)^2 + V(\phi)] \,, \qquad (1)$$

where the 'potential' $V(\phi)$ has a double-well structure, e.g. $V(\phi) = (1 - \phi^2)^2$. We will take the minima of $V(\phi)$ to occur at $\phi = \pm 1$, and adopt the convention that $V(\pm 1) = 0$.

The two minima of V correspond to the two equilibrium states, while the gradient-squared term in (1) associates an energy cost to an interface between the phases.

In the case where the order parameter is not conserved, an appropriate equation for the time evolution of the field ϕ is

$$\begin{aligned}\partial\phi/\partial t &= -\delta F/\delta\phi \\ &= \nabla^2\phi - V'(\phi)\,,\end{aligned} \qquad (2)$$

where $V'(\phi) \equiv dV/d\phi$. Eq. (2) says that the rate of change of ϕ is proportional to the gradient of the free-energy functional in function space. This equation provides a suitable coarse-grained description of the Ising model, as well as alloys that undergo an order-disorder transition on cooling through T_C, rather than phase separating. Such alloys form a two-sublattice structure, with each sublattice occupied predominantly by atoms of one type. In Ising model language, this corresponds to antiferromagnetic ordering. The magnetization is no longer the order parameter, but a 'fast mode', whose conservation does not significantly impede the dynamics of the important 'slow modes'.

When the order parameter is conserved, as in phase separation, a different dynamics is required. In the alloy system, for example, it is clear physically that A and B atoms can only exchange locally (not over large distances), leading to diffusive transport of the order parameter, and an equation of motion of the form

$$\begin{aligned}\partial\phi/\partial t &= \nabla^2 \delta F/\delta\phi \\ &= -\nabla^2[\nabla^2\phi - V'(\phi)]\,,\end{aligned} \qquad (3)$$

which can be written in the form of a continuity equation, $\partial_t\phi = -\nabla\cdot j$, with current $j = -\lambda\nabla(\delta F/\delta\phi)$. In (3), we have absorbed the transport coefficient λ into the timescale (and a similar kinetic coefficient has been absorbed in (2)).

Eqs. (2) and (3) are sometimes called the Time-Dependent-Ginzburg-Landau (TDGL) equation and the Cahn-Hilliard equation respectively. A more detailed discussion of them in the present context can be found in the review by Langer [1]. The same equations with additional Langevin noise terms on the right-hand sides are familiar from the theory of critical dynamics, where they are 'model A' and 'model B' respectively in the classification of Hohenberg and Halperin [2], a nomenclature we will adopt.

The absence of thermal noise terms in (2) and (3) indicates that we are working at $T = 0$. A schematic Renormalization Group (RG) flow diagram for T is given in Figure 3, showing the three RG fixed points at 0, T_C and ∞, and the RG flows. Under coarse-graining, temperatures above T_C flow to infinity, while those below T_C flow to zero. We therefore expect the final temperature T_F to be an irrelevant variable (in the scaling regime) for quenches into the ordered phase. This can be shown explicitly for systems with a conserved order parameter [3]. The thermal fluctuations at T_F simply renormalize the bulk order parameter and the surface tension of the domain walls. Similarly, any short-range correlations present at T_I should be irrelevant in the scaling regime, i.e. all initial temperatures are equivalent to $T_I = \infty$. Therefore we will take the *initial conditions* to represent a completely disordered state. For example, one could choose the 'white noise' form

$$\langle \phi(\mathbf{x},0)\, \phi(\mathbf{x}',0) \rangle = \Delta\, \delta(\mathbf{x}-\mathbf{x}') , \tag{4}$$

where $\langle \cdots \rangle$ represents an average over an ensemble of initial conditions. The above discussion, however, indicates that the precise form of the initial conditions should not be important, as long as only short-range spatial correlations are present.

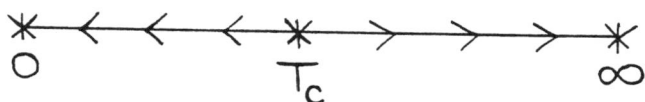

Figure 3. Schematic RG flow diagram, indicating that all $T_I > T_C$ are equivalent to $T = \infty$, and all $T_F < T_C$ to $T = 0$.

The challenge of understanding phase ordering dynamics, therefore, can be posed as finding the nature of the late-time solutions of deterministic differential equations like (2) and (3), subject to random initial conditions. A physical approach to this formal mathematical problem is based on studying the structure and dynamics of the topological defects in the field ϕ. This is approach that we will adopt. For scalar fields, the topological defects are just domain walls.

2.1 DOMAIN WALLS

It is instructive to first look at the properties of a flat equilibrium domain wall. From (2) the wall profile is the solution of the equation

$$d^2\phi/dg^2 = V'(\phi) , \tag{5}$$

with boundary conditions $\phi(\pm\infty) = \pm 1$, where g is a coordinate normal to the wall. We can fix the 'centre' of the wall (defined by $\phi = 0$) to be at $g = 0$ by the extra condition $\phi(0) = 0$. Integrating (5) once, and imposing the boundary conditions, gives $(d\phi/dg)^2 = 2V(\phi)$. This result can be used in (1) to give the energy per unit area of wall, i.e. the surface tension, as

$$\sigma = \int_{-\infty}^{\infty} dg\, (d\phi/dg)^2 = \int_{-1}^{1} d\phi\, [2V(\phi)]^{1/2} . \tag{6}$$

The profile function $\phi(g)$ is sketched in Figure 4. For $g \to \pm\infty$, linearizing (5) around $\phi = \pm 1$ gives

$$1 \mp \phi \sim \exp(-[V''(\pm 1)]^{1/2}|g|) , \qquad |g| \to \infty , \tag{7}$$

i.e. the order parameter saturates exponentially fast away from the walls. It follows that the excess energy is localized in the domain walls, and that the driving force for the domain growth is the wall curvature, since the system energy can only decrease through a reduction in the total wall area.

2.2 THE ALLEN-CAHN EQUATION

The existence of a surface tension implies a force per unit area, proportional to the total curvature, acting at each point on the wall. The calculation is similar to that of the excess pressure inside a bubble. Consider, for example, a spherical domain of radius R, in three dimensions. If the force per unit area is F, the work done by the force in decreasing the radius by dR is $4\pi F R^2 dR$. Equating this to the decrease in surface energy, $8\pi\sigma R dR$, gives $F = 2\sigma/R$. For model A dynamics, this force will cause the walls to move, with a velocity proportional to the local curvature.

This argument was formalized by Allen and Cahn [4], who noted that, close to a domain wall, one can write $\nabla\phi = (\partial\phi/\partial g)_t \, \hat{g}$, where $\hat{g}$ is a unit vector normal to the wall (in the direction of increasing ϕ), and so $\nabla^2\phi = (\partial^2\phi/\partial g^2)_t + (\partial\phi/\partial g)_t \nabla\cdot\hat{g}$. Noting also the relation $(\partial\phi/\partial t)_g = -(\partial\phi/\partial g)_t\,(\partial g/\partial t)_\phi$, (2) can be recast as

$$-(\partial\phi/\partial g)_t\,(\partial g/\partial t)_\phi = (\partial\phi/\partial g)_t \nabla\cdot\hat{g} + (\partial^2\phi/\partial g^2)_t - V'(\phi)\ . \tag{8}$$

Assuming that, for gently curving walls, the wall profile is given by the equilibrium condition (5), the final two terms in (8) cancel. Noting also that $(\partial g/\partial t)_\phi$ is just the wall velocity v (in the direction of increasing ϕ), (8) simplifies to

$$v = -\nabla\cdot\hat{g} = -K\ , \tag{9}$$

where $K \equiv \nabla\cdot\hat{g}$ is the total curvature.

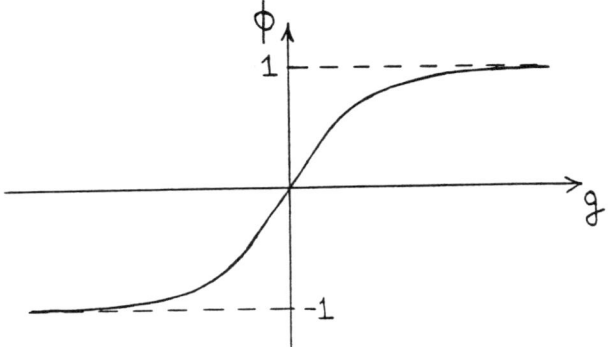

Figure 4. Domain-wall profile function $\phi(g)$ (schematic).

Equation (9) is an important result, because it establishes that the motion of the domain walls is determined (for non-conserved fields) purely by the local curvature. In particular, the detailed shape of the potential is not important: the main role of the double-well potential $V(\phi)$ is to establish (and maintain) well-defined domain walls. (Of course, the well depths must be equal, or there would be a volume driving force). We shall exploit this insensitivity to the potential, by choosing a particularly convenient form for $V(\phi)$, in section 5.

A second, and related, feature of (9) is that the surface tension σ (which does depend on the potential) does not explicitly appear. How can this be, if the driving force on the walls contains a factor σ? The reason is that one also needs the *friction constant* per unit area of wall, η. The equation of motion for the walls in this dissipative system is $\eta v = -\sigma K$. Consistency with (9) requires $\eta = \sigma$. In fact, η can be calculated independently, as follows. Consider a plane wall moving uniformly (under the influence of some external driving force) at speed v. The rate of energy dissipation per unit area is

$$\begin{aligned} dE/dt &= \int_{-\infty}^{\infty} dg \, (\delta F/\delta\phi) \, \partial\phi/\partial t \\ &= -\int_{-\infty}^{\infty} dg \, (\partial\phi/\partial t)^2 \,, \end{aligned} \qquad (10)$$

using (2). The wall profile has the form $\phi(g,t) = f(g-vt)$, where the profile function f will, in general, depend on v. Putting this form into (10) gives

$$dE/dt = -v^2 \int dg \, (\partial\phi/\partial g)^2 = -\sigma v^2 \,, \qquad (11)$$

where (6) was used in the final step, and the profile function $f(x)$ replaced by its $v = 0$ form to lowest order in v. By definition, however, the rate of energy dissipation is the product of the frictional force ηv and the velocity, $dE/dt = -\eta v^2$. Comparison with (11) gives $\eta = \sigma$. We conclude that the Allen-Cahn equation is completely consistent with the idea that domain growth is driven by the surface tension of the walls.

2.3 THE SCALING HYPOTHESIS

Although originally motivated by experimental and simulation results for the structure factor and pair correlation function [5-7], for ease of presentation it is convenient to introduce the scaling hypothesis first, and then discuss its implications for growth laws and scaling functions. Briefly, the scaling hypothesis states that there exists, at late times, a single characteristic length scale $L(t)$ such that the domain structure is (in a statistical sense) independent of time when lengths are scaled by $L(t)$. It should be stressed that scaling has not been proved, except in some simple models [8,9]. However, the evidence in its favour is compelling (see, e.g., Figure 5).

We shall find, in section 4, that the scaling hypothesis, together with a result derived in section 3 for the tail of the structure factor, is sufficient to determine the form of $L(t)$ for most cases of interest.

2.4 GROWTH LAWS

The scaling hypothesis suggests a simple intuitive derivation of the 'growth laws' for $L(t)$. For model A, we can estimate both sides of the Allen-Cahn equation (9). If there is a single characteristic scale L, then the wall velocity $v \sim dL/dt$, and the curvature $K \sim 1/L$. Equating and integrating gives $L(t) \sim t^{1/2}$ for non-conserved scalar fields.

For conserved fields (model B), the argument is more subtle. We shall follow the approach of Huse [10]. The point is that, due to the requirement to conserve the total magnetization, the domain walls cannot move independently. In fact the wall velocity is not proportional to the force (i.e. the local curvature), since a uniform force would have no effect, but to its gradient, $v \sim \nabla K$. Estimating $v \sim dL/dt$, $K \sim 1/L$ and $\nabla \sim 1/L$, gives $L \sim t^{1/3}$ for model B. A much more compelling argument for this result will be given in section 4. We note, however, that the result is in agreement with a large body of computer simulations [10,11], as well as a RG treatment [3]. In the limit that one phase occupies an infinitesimal volume fraction, the original Lifshitz-Slyozov-Wagner theory convincingly demonstrates a $t^{1/3}$ growth. This calculation, whose physical mechanism is the evaporation of material (or magnetization) from small droplets and condensation onto larger droplets, is outside the scope of the present lectures, however, and the reader is referred to the original papers [12], or the reviews by Gunton et al. [13] and Langer [1], for details.

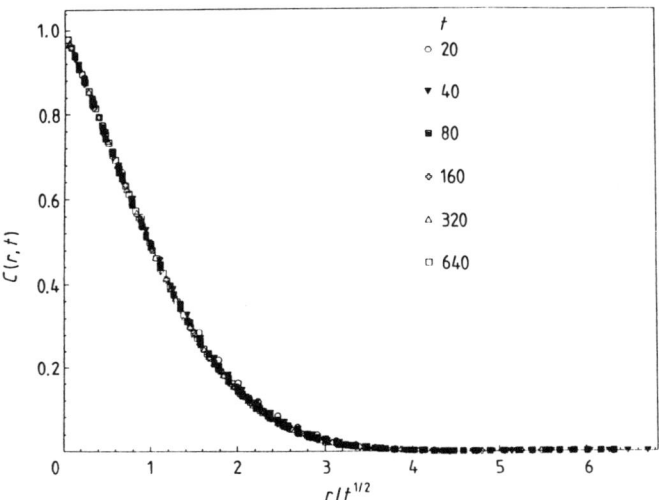

Figure 5. Scaling function $f(x)$ for the pair correlation function of the $2-d$ Ising model with nonconserved order parameter. The time t is the number of Monte Carlo steps per spin. (From K. Humayun and A. J. Bray, J. Phys. A **24**, 1915 (1991)).

2.5 SCALING FUNCTIONS

Two commonly used probes of the domain structure are the equal-time pair correlation function $C(\mathbf{r},t) = \langle \phi(\mathbf{x}+\mathbf{r},t)\,\phi(\mathbf{x},t)\rangle$, and its Fourier transform, the equal-time structure factor, $S(\mathbf{k},t) = \langle \phi_{\mathbf{k}}(t)\,\phi_{-\mathbf{k}}(t)\rangle$. Here angle brackets indicate an average over initial conditions. The structure factor can, of course, be measured in scattering experiments. The existence of a single characteristic length scale, according to the scaling hypothesis, implies that the pair correlation function and the structure factor have the scaling forms

$$\begin{aligned} C(\mathbf{r},t) &= f(r/L)\,, \\ S(\mathbf{k},t) &= L^d\,g(kL)\,, \end{aligned} \qquad (12)$$

where d is the spatial dimensionality, and $g(y)$ is the Fourier transform of $f(x)$. Note that $f(0)=1$, since (at $T=0$) there is perfect order within a domain. The scaling forms (12) are well supported by simulation data and experiment. As an example, Figure 5 shows the scaling plot for $f(x)$ for the 2-D Ising model, with $x = r/t^{1/2}$.

For future reference, we note that the different-time correlation function, $C(\mathbf{r}, t, t') = \langle \phi(\mathbf{x} + \mathbf{r}, t) \phi(\mathbf{x}, t') \rangle$, can also be written in scaling form. A simple generalization of (12) gives

$$C(\mathbf{r}, t, t') = f(r/L, r/L') ,\qquad(13)$$

where L, L' stand for $L(t)$ and $L(t')$. The *autocorrelation* function, $A(t) = C(\mathbf{0}, t, t')$ is therefore a function only of the ratio L'/L. For $L \gg L'$ it has the form $A(t) \sim (L'/L)^\lambda$, where the exponent λ, first introduced by Fisher and Huse in the context of non-equilibrium relaxation in spin glasses [14], is a non-trivial exponent associated with phase ordering kinetics [15]. It has recently been measured in an experiment on twisted nematic liquid crystal films [16].

In the following section, we explore the forms of the scaling functions in more detail. For example, the linear behaviour of $f(x)$, for small scaling variable x in Figure 5, is a generic feature for scalar fields, both conserved and non-conserved. We shall see that it is a simple consequence of the existence of 'sharp' (in a sense to be clarified), well-defined domain walls in the system. A corollary that we shall demonstrate is that the structure factor scaling function $g(y)$ exhibits a power-law tail, $g(y) \sim y^{-(d+1)}$ for $y \gg 1$, a result known as 'Porod's law' [17,18]. In section 4 we shall show that this result, and its generalization to more complex fields, together with the scaling hypothesis, are sufficient to determine the growth law for $L(t)$.

3. Topological Defects

The domain walls discussed in the previous section are the simplest form of 'topological defect', and occur in systems described by scalar fields [19]. They are surfaces, on which the order parameter vanishes, separating domains of the two equilibrium phases. A domain wall is topologically stable: local changes in the order parameter can move the wall, but cannot destroy it. For an isolated flat wall, the wall profile function is given by the solution of (5), with the appropriate boundary conditions, as discussed in section 2.1 (and sketched in Figure 4). For the curved walls present in the phase ordering process, this will still be an approximate solution locally, provided the typical radius of curvature L is large compared to the intrinsic width (or 'core size'), ξ, of the walls. (This could be defined from (7) as $\xi = [V''(1)]^{-1/2}$, say). The same condition, $L \gg \xi$, ensures that typical wall separations are large compared to their width.

Let us now generalize the discussion to vector fields. The '$O(n)$ model' is described by a vector field $\vec{\phi}(\mathbf{x}, t)$, with a free energy functional $F[\vec{\phi}]$ that is invariant under global rotations of $\vec{\phi}$. A suitable generalization of (1) is

$$F[\vec{\phi}] = \int d^d x \, [\frac{1}{2}(\nabla \vec{\phi})^2 + V(\vec{\phi})] ,\qquad(14)$$

where $(\nabla \vec{\phi})^2$ means $\sum_{i=1}^d \sum_{a=1}^n (\partial_i \phi^a)^2$ (i.e. a scalar product over both spatial and 'internal' coordinates), and $V(\vec{\phi})$ is 'mexican hat' (or 'wine bottle') potential, such as

$(1-\vec{\phi}^2)^2$, whose general form is sketched in Figure 6. It is clear that $F[\vec{\phi}]$ is invariant under global rotations of $\vec{\phi}$ (a continuous symmetry), rather than just the inversion symmetry ($\phi \to -\phi$, a discrete symmetry) of the scalar theory. We will adopt the convention that V has its minimum for $\phi^2 = 1$.

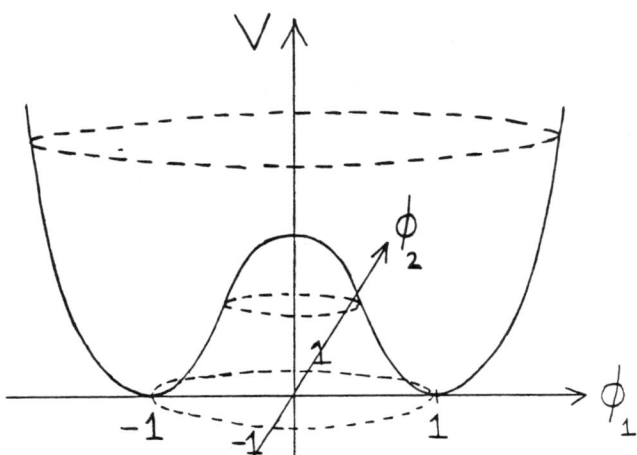

Figure 6. The 'Mexican hat' potential for the $O(n)$ model. The case $n = 2$ is illustrated.

For non-conserved fields, the simplest dynamics (model A) is a straightforward generalization of (2), namely

$$\partial\vec{\phi}/\partial t = \nabla^2\vec{\phi} - dV/d\vec{\phi} . \tag{15}$$

For conserved fields (model B), we simply add another $(-\nabla^2)$ in front of the right-hand side.

Stable topological defects for vector fields can be generated, in analogy to the scalar case, by seeking stationary solutions of (15) with appropriate boundary conditions. For the $O(n)$ theory in d-dimensional space, the requirement that all n components of $\vec{\phi}$ vanish at the defect core defines a surface of codimension $d-n$ (e.g. a domain wall is a surface of codimension $d - 1$: the scalar theory corresponds to $n = 1$). The existence of such defects therefore requires $n \leq d$. For $n = 2$ these defects are points ('vortices') for $d = 2$ or lines ('strings', or 'vortex lines') for $d = 3$. For $n = 3$, $d = 3$ they are points ('hedgehogs', or 'monopoles'). The field configurations for these defects are sketched in Figure 7(a)-(d). Note that the forms shown are radially symmetric with respect to the defect core: any configuration obtained by a global rotation is also acceptable. For $n < d$, the field $\vec{\phi}$ only varies in the n dimensions 'orthogonal' to the defect core, and is uniform in the remaining $d - n$ dimensions 'parallel' to the core.

For $n < d$, the defects are spatially extended. Coarsening occurs by a 'straightening out' (or reduction in typical radius of curvature) as sharp features are removed, and by the shrinking and disappearance of small domain bubbles or vortex loops. These processes reduce the total area of domain walls, or length of vortex line, in the system. For point defects ($n = d$), coarsening occurs by the mutual annihilation of defect-antidefect pairs. The antidefect for a vortex ('antivortex') is sketched in Figure 7(e). Note that the antivortex in *not* obtained by simply reversing the directions of the arrows in 7(b): this would correspond to a global rotation through π. Rather, the vortex and antivortex have different 'topological charges': the fields rotates by 2π or -2π respectively on encircling the defect. By contrast, an antimonopole *is* generated by reversing the arrows in 7(d): the reversed configuration cannot be generated by a simple rotation in this case.

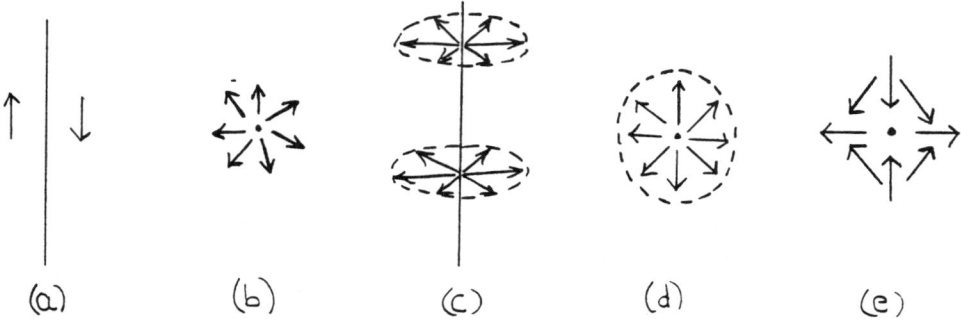

(a)   (b)   (c)   (d)   (e)

Figure 7. Types of topological defect in the $O(n)$ model: (a) domain wall ($n = 1$) (b) vortex ($n = 2 = d$) (c) string ($n = 2, d = 3$) (d) monopole ($n = 3 = d$) (e) antivortex.

For the radially symmetric defects illustrated in 7(b)-(d), the field $\vec{\phi}$ has the form $\vec{\phi}(\mathbf{r}) = \hat{r}\, f(r)$, where $\hat{r}$ is a unit vector in the radial direction, and $f(r)$ is the profile function. Inserting this form into (15), with the time derivative set to zero, gives the equation

$$\frac{d^2 f}{dr^2} + \frac{(n-1)}{r}\frac{df}{dr} - \frac{(n-1)}{r^2} f - V'(f) = 0 , \qquad (16)$$

with boundary conditions $f(0) = 0$, $f(\infty) = 1$. Of special interest is the approach to saturation at large r. Putting $f(r) = 1 - \epsilon(r)$ in (16), and expanding to first order in ϵ, yields

$$\epsilon(r) \simeq \frac{(n-1)}{V''(1)} \frac{1}{r^2} , \qquad r \to \infty . \qquad (17)$$

This should be contrasted with the exponential approach to saturation (7) for scalar fields. Defining a 'core size' ξ through $\epsilon \simeq 1 - \xi^2/r^2$ for large r gives $\xi = [(n-1)/V''(1)]^{1/2}$ for $n > 1$.

The presence of topological defects, seeded by the initial conditions, in the system undergoing phase ordering has an important effect on the 'short-distance' form of the pair correlation function $C(\mathbf{r}, t)$, and therefore on the 'large-momentum' form of the structure factor $S(\mathbf{k}, t)$. To see why this is so, we note that, according to the scaling hypothesis, we would expect a typical field gradient to be of order $|\nabla\vec{\phi}| \sim 1/L$. At a distance r from a defect core, however, with $\xi \ll r \ll L$, the field gradient is much larger, of order $1/r$ (for a vector field), because $\vec{\phi} = \hat{r}$ implies $(\nabla\vec{\phi})^2 = (n-1)/r^2$. Note that we require $r \gg \xi$ for the field to be saturated, and $r \ll L$ for the defect field to be largely unaffected by other defects (which are typically a distance L away). This gives a meaning to 'short' distances ($\xi \ll r \ll L$), and 'large momenta' ($L^{-1} \ll k \ll \xi^{-1}$). The large field gradients near defects leads to a non-analytic behaviour at $x = 0$ of the scaling function $f(x)$ for pair correlations, as we shall now see.

3.1 POROD'S LAW

We start by considering scalar fields. Consider two points $\mathbf{x}$ and $\mathbf{x} + \mathbf{r}$, with $\xi \ll r \ll L$. The product $\phi(\mathbf{x})\phi(\mathbf{x}+\mathbf{r})$ will be -1 if a wall passes between them, and $+1$ if there is no wall. Since $r \ll L$, the probability to find more than one wall can be neglected. The calculation amounts to finding the probability that a randomly placed rod of length r cuts a domain wall. This probability is of order r/L, so we estimate

$$\begin{aligned} C(\mathbf{r},t) &\simeq (-1) \times (r/L) + (+1) \times (1 - r/L) \\ &= 1 - 2r/L, \qquad r \ll L. \end{aligned} \tag{18}$$

The factor 2 in this result should not be taken seriously.

The important result is that (18) is non-analytic in $\mathbf{r}$ at $\mathbf{r} = 0$, since it is linear in $r \equiv |\mathbf{r}|$. Technically, of course, this form breaks down inside the core region, when $r < \xi$. We are interested, however, in the scaling limit defined by $r \gg \xi$, $L \gg \xi$, with $x = r/L$ arbitrary. The nonanalyticity is really in the scaling variable x.

The nonanalytic form (18) implies a power-law tail in the structure factor, which can be obtained from (18) by simple power-counting:

$$S(\mathbf{k},t) \sim \frac{1}{L\,k^{d+1}}, \qquad kL \gg 1, \tag{19}$$

a result known universally as 'Porod's law'. It was first written down in the general context of scattering from two-phase media [17]. Again, one requires $k\xi \ll 1$ for the scaling regime. Although the k-dependence of (19) is what is usually referred to as Porod's law, the L-dependence is equally interesting. The factor $1/L$ is simply (up to constants) the total area of domain wall per unit volume, a fact appreciated by Porod, who proposed structure factor measurements as a technique to determine the area of interface in a two-phase medium [17]. On reflection, the factor $1/L$ is not so surprising. For $kL \gg 1$, the scattering function is probing structure on scales much shorter than than the typical interwall spacing or radius of curvature. In this regime we would expect the structure factor to scale as the total wall area.

This observation provides the clue to how to generalize (19) to vector (and other fields) [20,21]. The idea is that, for $kL \gg 1$, the structure factor should scale as the total volume of defect core. Since the codimension of the defects is $d - n$, the amount of defect per unit volume scales as L^{-n}. Extracting this factor from the general scaling form (12) yields

$$S(\mathbf{k},t) \sim \frac{1}{L^n\,k^{d+n}}, \qquad kL \gg 1, \tag{20}$$

for the $O(n)$ theory, a 'generalized Porod's law' [22-24].

Equation (20) was first derived from approximate treatments of the equation of motion (15) for nonconserved fields [22-24]. These derivations, however, obscured the key role of topological defects. The above heuristic derivation suggests that the result is in fact very general (e.g., it should hold equally well for conserved fields), with extensions beyond simple $O(n)$ models. The appropriate techniques, which also enable the amplitude of the tail to be determined, were developed by Bray and Humayun [21].

3.2 EXACT AMPLITUDES

The qualitative arguments of the previous subsection can be made precise [21], and the *amplitude* of the $k^{-(d+n)}$ Porod tail obtained in terms of the density of defect core, ρ_{def}, which scales as L^{-n}. The basic result is a generalization of (18), in which the leading singular contribution to $C(\mathbf{r},t)$ is a term of order $|\mathbf{r}|^n$ for n odd (or n real, in a continuation of the theory to real n), and $|\mathbf{r}|^n \ln|\mathbf{r}|$ for n even. This in turn implies a power law tail $k^{-(d+n)}$ in $S(\mathbf{k},t)$. We illustrate the method for the case of point defects ($n = d$), and refer the reader to Bray and Humayun [21] for the general case ($n \leq d$).

3.2.1 Point Defects ($n = d$)

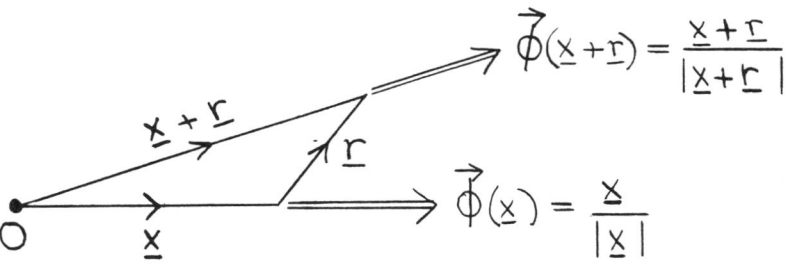

Figure 8. Field configuration near a point defect in the $O(n)$ model.

Consider the field $\vec{\phi}$ at points $\mathbf{x}$ and $\mathbf{x} + \mathbf{r}$ in the presence of a point defect at the origin. We consider the case where $|\mathbf{x}|$, $|\mathbf{x} + \mathbf{r}|$ and $|\mathbf{r}|$ are all small compared to a typical inter-defect distance L, but large compared to the defect core size ξ. Then the field at the points $\mathbf{x}$ and $\mathbf{x} + \mathbf{r}$ is saturated in length (i.e. of unit length) and not significantly distorted by the presence of other defects. Moreover, the field can be taken, up to a global rotation, to be directed radially outward from the origin, as illustrated in Figure 8. Thus

$$\vec{\phi}(\mathbf{x}) \cdot \vec{\phi}(\mathbf{x} + \mathbf{r}) = \frac{\mathbf{x} \cdot (\mathbf{x} + \mathbf{r})}{|\mathbf{x}||\mathbf{x} + \mathbf{r}|} . \tag{21}$$

With $\mathbf{r}$ held fixed we average (21) over all possible relative positions of the point defect, i.e. over all values of $\mathbf{x}$ within a volume of order L^n around the pair of points, with the appropriate probability density ρ_{def}. Focussing on the *singular* part of the correlation function we obtain

$$C_{sing}(\mathbf{r},t) = \rho_{def} \int^L d^n x \left(\frac{\mathbf{x}.(\mathbf{x}+\mathbf{r})}{|\mathbf{x}||\mathbf{x}+\mathbf{r}|} - \text{analytic terms} \right). \quad (22)$$

The 'analytic terms' in (22) serve to converge the x-integral at large $|\mathbf{x}|$, and allow us to extend the integral over all space. We include as many terms in the expansion of (21) in powers of $\mathbf{r}$ as are necessary to ensure the convergence of the integral. When n is even, there is a residual logarithmic singularity. This case can be retrieved from the general n result by taking a suitable limit (see below).

The integral (22) (extended over all space) is evaluated in [21]. The result is

$$C_{sing} = n\,\pi^{n/2-1}\,B\left(\frac{n+1}{2},\frac{n+1}{2}\right) \Gamma\left(-\frac{n}{2}\right) \rho_{def} |\mathbf{r}|^n, \quad (23)$$

where $\Gamma(x)$ is the gamma function, and $B(x,y) = \Gamma(x)\Gamma(y)/\Gamma(x+y)$ is the beta function. Note that the dependence on $|\mathbf{r}|$ can be extracted simply by a change of variable in (22).

The pole in the $\Gamma(-n/2)$ factor for even values of n signals a contribution of the form $|\mathbf{r}|^n \ln(|\mathbf{r}|/L)$ to C_{sing} for those cases. To see how this comes about, we set $n = 2m + \epsilon$ in (23), with m an integer, and take the limit $\epsilon \to 0$. The leading pole contribution, proportional to $\epsilon^{-1} (r^2)^m$, is analytic in $|\mathbf{r}|$ and therefore does not contribute to C_{sing}. The $O(1)$ term (in the expansion in powers of ϵ) generates the logarithmic correction from the expansion of $|\mathbf{r}|^{2m+\epsilon}$. Since, however, the result for the structure factor tail is completely smooth as a function of n, we will not give a separate detailed discussion for even n.

3.2.2 General Case, $n \leq d$

The general case is slightly more complicated than for point defects, and we will simply quote the result, derived in ref. [21]:

$$C_{sing} = \pi^{n/2-1} \frac{\Gamma(-n/2)\Gamma(d/2)\Gamma^2((n+1)/2)}{\Gamma((d+n)/2)\Gamma(n/2)} \rho_{def} |\mathbf{r}|^n, \quad (24)$$

which reduces to (23) for $n = d$.

We remarked in the previous section that for even n the leading singularity is the form of $r^n \ln r$, but the Fourier transform $S(\mathbf{k},t)$ of $C_{sing}(\mathbf{r},t)$ has the same form for even, odd and real n, so we will not consider the even n case explicitly.

3.2.3 The Structure Factor

It remains to Fourier transform (24) to obtain the tail of the structure factor. Simple power counting on (24) gives immediately the power-law tail $S(\mathbf{k},t) \sim k^{-(d+n)}$. To derive the *amplitude* we employ the integral representation

$$\Gamma(-n/2)|\mathbf{r}|^n = \int_0^\infty du\, u^{-n/2-1} \{\exp(-ur^2) - \text{analytic terms}\}, \quad (25)$$

where 'analytic terms' indicates, once more, as many terms in the expansion of $\exp(-ur^2)$ as are necessary to converge the integral. These terms will not contribute to the tail of the Fourier transform, and can be dropped once the transform has been taken. The Fourier transform of (25) is, therefore,

$$\int_0^\infty du\, u^{-n/2-1} \int d^d r \exp(-ur^2 - i\mathbf{k}\cdot\mathbf{r})$$
$$= \pi^{d/2} \int_0^\infty du\, u^{-(d+n)/2-1} \exp(-k^2/4u)$$
$$= \pi^{d/2} \Gamma\left(\frac{d+n}{2}\right) \left(\frac{2}{k}\right)^{d+n}. \tag{26}$$

Inserting the remaining factors from (24) gives the final result

$$S(\mathbf{k},t) = \frac{1}{\pi}(4\pi)^{(d+n)/2} \frac{\Gamma^2((n+1)/2)\Gamma(d/2)}{\Gamma(n/2)} \frac{\rho_{def}}{k^{d+n}}. \tag{27}$$

We note that this expression is smooth as n passes through the even integers. The generality of the result should be noted: it is independent of any details of the dynamics, e.g. whether the order parameter is conserved or non-conserved, and holds independently of whether the scaling hypothesis is valid. We note that, as well as providing an exact result against which to test approximate theories, Eq. (27) can also be used to *determine* the defect density experimentally. Measurements of the tail amplitude in numerical simulations are in agreement with (27) for both scalar [21] and vector [25] fields.

In the following section, the generalized Porod's law (27), combined with the scaling hypothesis, will be used to derive the growth laws for $L(t)$ in a variety of situations [26].

4. Growth Laws

Although the growth laws for both nonconserved and conserved scalar systems, and conserved fields in general, have been derived by a number of methods, there has up until now been no simple, general technique for obtaining $L(t)$. In particular, the growth laws for non-conserved vector fields have, until recently, been somewhat problematical. Here we describe a very general approach, recently developed by Bray and Rutenberg (BR) [26], to obtain $L(t)$ consistently by comparing the global rate of energy change to the energy dissipation from the local evolution of the order parameter. BR derive growth laws for $O(n)$ models, but their results can be applied to other systems with similar defect structures.

The BR approach is based on the energy dissipation that occurs as the system relaxes towards its ground state. They evaluate the energy dissipation by considering the topological defects, which dominate the dynamics when they exist, and compare the global rate of energy change to the energy dissipation from the local evolution of the order parameter. For systems with a single characteristic scale, describing correlations as well as dissipation, they self-consistently determine the growth law, $L(t)$.

The results are summarized in table 1 below. Note that conservation laws (μ) are irrelevant to the growth law below a value determined by the number of components (n) of the order parameter. At the marginal values, logarithmic factors are introduced. The growth laws obtained are independent of the spatial dimension of the system.

We begin by writing down the equation of motion for the Fourier components $\vec{\phi}_{\mathbf{k}}$:

$$\partial_t \vec{\phi}_{\mathbf{k}} = -k^\mu (\partial F/\partial \vec{\phi}_{-\mathbf{k}}), \tag{28}$$

The conventional non-conserved (model A) and conserved (model B) cases are $\mu = 0$ and $\mu = 2$, respectively.

Integrating the rate of energy dissipation from each Fourier mode, and then using the equation of motion (28), we find

$$\begin{aligned} d\epsilon/dt &= \int_{\mathbf{k}} \left\langle (\partial F/\partial \vec{\phi}_{\mathbf{k}}) \cdot \partial_t \vec{\phi}_{\mathbf{k}} \right\rangle \\ &= -\int_{\mathbf{k}} k^{-\mu} \left\langle \partial_t \vec{\phi}_{\mathbf{k}} \cdot \partial_t \vec{\phi}_{-\mathbf{k}} \right\rangle, \end{aligned} \tag{29}$$

where $\epsilon = \langle F \rangle /V$ is the mean energy density, and $\int_{\mathbf{k}}$ is the momentum integral $\int d^d k/(2\pi)^d$. We will relate the scaling behavior of both sides of (29) to that of appropriate integrals over the structure factor, $S(\mathbf{k},t)$. Either the integrals converge and the dependence on the scale $L(t)$ can be extracted using the scaling form (12), or a divergent contribution dominates the integral. Any divergence will occur because of small-scale structure at $kL \gg 1$, since no structure has formed yet for $kL \ll 1$. This small-scale structure leads to the generalized Porod law (20) for the structure factor.

We first calculate the scaling behavior of the energy density, ϵ, which is captured by that of the gradient term in (1):

$$\begin{aligned} \epsilon &\sim \left\langle (\nabla \vec{\phi})^2 \right\rangle \\ &= \int_{\mathbf{k}} k^2 \, L^d \, g(kL), \end{aligned} \tag{30}$$

where we have used the scaling form (12) for the structure factor. If the integral converges, then a simple change of variables yields $\epsilon \sim L^{-2}$. If the integral diverges for $kL \gg 1$, then we use Porod's law (20) and impose a cutoff at $k \sim 1/\xi$, where ξ is the defect core size. This gives [23]

$$\begin{aligned} \epsilon &\sim L^{-n} \xi^{n-2}, & n &< 2, \\ &\sim L^{-2} \ln(L/\xi), & n &= 2, \\ &\sim L^{-2}, & n &> 2. \end{aligned} \tag{31}$$

We see that the energy is dominated by the defect core density, $\rho_{\text{def}} \sim L^{-n}$, for $n < 2$, by the defect field at all length scales between ξ and L for $n = 2$, and by variations of the order parameter at scale $L(t)$ for $n > 2$.

We now evaluate the right side of (29) in a similar way. Using the scaling assumption for the two-time function, $\langle \vec{\phi}_{\mathbf{k}}(t) \cdot \vec{\phi}_{-\mathbf{k}}(t') \rangle = k^{-d} g(kL(t), kL(t'))$, which is the Fourier transform of (13), we find

$$\langle \partial_t \vec{\phi}_{\mathbf{k}} \cdot \partial_t \vec{\phi}_{-\mathbf{k}} \rangle = \left. \frac{\partial^2}{\partial t \partial t'} \right|_{t=t'} \langle \vec{\phi}_{\mathbf{k}}(t) \cdot \vec{\phi}_{-\mathbf{k}}(t') \rangle$$
$$= (dL/dt)^2 L^{d-2} h(kL). \tag{32}$$

When the momentum integral on the right of (29) converges, we obtain, using (32), $d\epsilon/dt \sim -(dL/dt)^2 L^{\mu-2}$. However, if the integral diverges we need to know the behaviour of (32) for $kL \gg 1$.

In this regime, where the important structure is on scales short compared to the typical interdefect spacing, the scaling functions are determined by the field of a single defect [20] and the defect field will be comoving with the defect core,

$$\partial_t \vec{\phi} = \vec{\omega} \times \vec{\phi} - \mathbf{v} \cdot \nabla \vec{\phi}. \tag{33}$$

Here, $\mathbf{v}$ is the velocity of the defect and $\vec{\omega}$ represents rotations for vector fields. For $kL \gg 1$, the velocity term dominates because the rotation scales as $\omega \sim (dL/dt)/L$ while the gradient term involves the short-scale structure, $\mathbf{v} \cdot \nabla \sim (dL/dt)k$. In estimating $v \sim (dL/dt)$, BR are assuming that dissipation occurs primarily by the evolution of defects at scale $L(t)$. Using $\partial_t \vec{\phi}_{\mathbf{k}} \sim (dL/dt) k \, \vec{\phi}_{\mathbf{k}}$ and Porod's law (20), we have

$$\langle \partial_t \vec{\phi}_{\mathbf{k}} \cdot \partial_t \vec{\phi}_{-\mathbf{k}} \rangle \sim L^{-n} k^{-(d+n-2)} (dL/dt)^2, \qquad kL \gg 1. \tag{34}$$

Using (34) in the right of (29) shows that the integral is convergent for $kL \gg 1$ when $n + \mu > 2$. Otherwise the integral is dominated by k near the upper cut-off $1/\xi$. This gives

$$\int_k k^{-\mu} \langle \partial_t \vec{\phi}_{\mathbf{k}} \cdot \partial_t \vec{\phi}_{-\mathbf{k}} \rangle \sim L^{-n} \xi^{n+\mu-2} (dL/dt)^2, \qquad n+\mu < 2,$$
$$\sim L^{-n} \ln(L/\xi)(dL/dt)^2, \qquad n+\mu = 2,$$
$$\sim L^{\mu-2}(dL/dt)^2, \qquad n+\mu > 2. \tag{35}$$

In what cases does dissipation occur primarily at scales $L(t)$, as BR assume? For $n > 2$, the energy density (31), and hence dissipation, is dominated by variations at scale $L(t)$. For $n \leq 2$, with $d > n$, the energy density is proportional to the defect core volume (with an extra factor $\ln(L/\xi)$ for $n = 2$, see (31)), but BR argue that dissipation is still dominated by defect structures with length scales of order $L(t)$. For these cases the rate of energy dissipation at a length scale l is given by the rate at which the core volume disappears at that scale. So the total energy dissipation is given by

$$d\epsilon/dt \sim \int dl \, n(l) \, \dot{l} \, l^{d-n-1}, \tag{36}$$

where $\dot{l}$ is the rate of evolution of defect structures of scale l, $n(l)$ is the number density of features of scale l, and $\dot{l} \, l^{d-n-1}$ is the rate of change of the defect core volume at scale l. (Again, for $n = 2$ the integrand in (36) contains an extra factor $\ln(l/\xi)$). But $j(l) \sim n(l)\dot{l}$ is just the number flux of defect features at scale l, and imposing the continuity equation we see that $\dot{N} \sim j(0)$, where N is the average density of defects. N scales as a volume, $N \sim 1/L^d$, and so $\dot{N}$ is a constant for times of order $L/\dot{L}$. This implies that $j(l)$ is constant for $l \ll L$, in order to provide a constant rate of defect extinction. Hence the contribution to the energy dissipation from structures with scale $l \ll L$ is $d\epsilon/dt \sim \int_0^L dl\, l^{d-n-1}$. The convergence of this integral at small l for $d > n$ implies that structures with scales $l \sim L(t)$ dominate the energy dissipation.

For $n = d$ with $n < 2$, the marginal case in the above argument, the energy density (31) is dominated by the core energy of point defects and dissipation is dominated by defect pairs annihilating. Since the dissipation occurs at separations $l \sim \xi \ll L$ we do not expect the BR approach to cover these cases. In fact, since the bulk energy density does not depend on defect separations $l \gg \xi$, we expect the system to be disordered, with an equilibrium density of defects, at any non-zero temperature. At $T = 0$ we expect slow growth laws that depend on the details of the potential $V(\vec{\phi})$. These cases, $n = d$ with $n < 2$, are at their lower critical dimension.

The $2d$ XY model is outside the scope of this treatment since a more detailed argument [27] shows that dissipation occurs significantly at all length scales between ξ and $L(t)$. Naively, we might expect scaling violations for this case. Yurke et al. [28] suggest a growth law of $L \sim (t/\ln t)^{1/2}$ for non-conserved order-parameters, while simulations by Mondello and Goldenfeld [29] find $L \sim t^{1/4}$ for the conserved case.

In the cases where dissipation occurs at the scale of $L(t)$, comparing the rate of energy dissipation between (35) and the time derivative of (31) gives an equation relating dL/dt and $L(t)$, which can be integrated to give $L(t)$. The results are summarized in table 1.

$L(t)$	$n < 2$	$n = 2$	$n > 2$
$n + \mu < 2$	$t^{1/2}$	–	–
$n + \mu = 2$	$(t/\ln t)^{1/2}$	$t^{1/2}$	–
$n + \mu > 2$	$t^{1/(n+\mu)}$	$(t \ln t)^{1/(2+\mu)}$	$t^{1/(2+\mu)}$

Table 1
Growth law of the length scale $L(t)$ for various number of components, n, and conservation laws, μ. Note that $n = d$ is excluded for $n \leq 2$.

For non-conserved fields ($\mu = 0$), we find $L \sim t^{1/2}$ for all systems (with $d > n$ or $n > 2$). Leading corrections in the $n = 2$ case are interesting: the $\ln L$ factors in (31) and (35) will in general have different effective cutoffs, of order the core size ξ. This leads to a logarithmic correction to scaling, $L \sim t^{1/2}(1 + O(1/\ln t))$, and may account for the smaller exponent (~ 0.45) seen in simulations of $O(2)$ systems [30,31,25].

For conserved fields ($\mu > 0$) our results agree with the the RG analysis [3], with additional logarithmic factors for the marginal cases $n = 2$ and $n + \mu = 2$. Note that the conservation law is only relevant for $n + \mu \geq 2$. Therefore for vector fields ($n \geq 2$), any $\mu > 0$ is sufficient to change the growth law, while for scalar fields ($n = 1$) the conservation law is irrelevant for $\mu < 1$, in agreement with the RG analysis [3]. Simulations by Siegert and Rao [32] for $n = 2$, with $d = 3$ and $\mu = 2$, obtain growth exponents slightly over 1/4, which may be difficult to distinguish from the predicted $L \sim (t \ln t)^{1/4}$ behaviour.

Since systems without topological defects ($n > d + 1$) will have convergent momentum integrals for $kL \gg 1$, we obtain $L \sim t^{1/(2+\mu)}$ for these cases. We can also apply this result to systems with topological textures ($n = d + 1$), even though the appropriate Porod's law is not known. Since defects with $n > d$ must be spatially extended and without a core, they will have a smaller large-k tail to their structure factor $S(\mathbf{k}, t)$ than any defects with cores. So for $n > 2$, when the energy dissipation clearly occurs at length scales of order $L(t)$ (see (31)) and the momentum integrals for defects with cores converge, our results should apply. As a result, table 1 will apply for *any* system, except perhaps those with $d \leq n \leq 2$ [33] Of course, whether the structure factor scales, as in (12), for systems with textures is an open question.

The strength of this approach is that it can be applied to systems with more complicated order parameters than n-component vectors. The details of the energy functional (1) are unimportant [34]. All we need is the existence of some 'elastic energy', an effective conservation law (μ), and the defect structure if any. The derivation is independent of the initial conditions, and so, e.g., applies equally to critical and off-critical quenches *as long as the system scales at late times*. We simply choose a Porod's law (20) to represent the dominant defect type, which is the one represented by the smallest n. For example, in bulk nematic liquid crystals, the existence of string defects leads to (20) with $n = 2$, which with no conservation law implies a $L \sim t^{1/2}$ growth law, consistent with recent experiments [35] and simulations [36].

In summary, we have shown how growth laws for phase ordering can be obtained assuming only the dynamic scaling hypothesis and a generalized Porod's law for the large-momentum tail of the structure factor. By focusing on the total energy dissipation, we do not need to consider the defect dynamics explicitly. This results in a powerful yet simple method that addresses many cases of interest.

5. Scaling Functions for Non-Conserved Fields

The calculation of scaling functions, e.g. the pair correlation scaling function $f(x)$ (see Eq. (12)), has been a long-standing challenge. A number of approximate scaling functions have been proposed for non-conserved fields, but in my view none of them is completely satisfactory. The most physically appealing approach for scalar fields is that of Ohta, Jasnow and Kawasaki (OJK) [37], which starts from the Allen-Cahn equation (9) for the interfaces. Below we will briefly review the OJK method, as well as an earlier approach by Kawasaki, Yalabik and Gunton (KYG) [38], and more recent work by Mazenko [39,40]. Finally we discuss in detail a new approach [41] which has the virtue that it can, in principle, be systematically improved.

5.1 APPROXIMATE THEORIES FOR $f(x)$

5.1.1 The OJK Theory

A common theme in the approximate theories of scaling functions is the replacement of the physical field $\phi(\mathbf{x},t)$, which is ± 1 everywhere except at domain walls, where it varies rapidly, by an auxiliary field $m(\mathbf{x},t)$, which varies smoothly in space, through a non-linear function $\phi(m)$ with a 'sigmoid' shape (such as $\tanh m$). In the OJK theory [37], the dynamics of the domain walls themselves, defined by the zeros of m, are considered. The normal velocity of a point on the interface is given by the Allen-Cahn equation (9), $v = -K = -\nabla\cdot\mathbf{n}$, where K is the total curvature, and $\mathbf{n} = \nabla m/|\nabla m|$ is a unit vector normal to the wall. This gives $v = \{-\nabla^2 m + n_a n_b \nabla_a \nabla_b m\}/|\nabla m|$. In the frame moving with the interface, $dm/dt = 0 = \partial m/\partial t + \mathbf{v}\cdot\nabla m$. But $\mathbf{v}\cdot\nabla m = v|\nabla m|$ so $v = -(\partial m/\partial t)/|\nabla m|$. The two expressions for v can be used to eliminate v and obtain

$$\partial m/\partial t = \nabla^2 m - n_a n_b \nabla_a \nabla_b m , \tag{37}$$

where $\mathbf{n} = \nabla m/|\nabla m|$. OJK simplified this non-linear equation by replacing $n_a n_b$ by its spherical average δ_{ab}/d to obtain the diffusion equation

$$\partial m/\partial t = D \nabla^2 m , \tag{38}$$

with diffusion constant $D = (d-1)/d$. This can be readily solved, for gaussian initial conditions with correlator $\langle m(\mathbf{x},0)m(\mathbf{x}',0)\rangle = \Delta\delta(\mathbf{x}-\mathbf{x}')$, to give $\langle m(1)m(2)\rangle = \{\Delta/(8\pi Dt)^{d/2}\}\exp(-r^2/8Dt)$, where '1' and '2' represent space points. In the scaling regime, one can replace the function $\phi(m)$ by $\text{sgn}(m)$, because the walls occupy a negligible volume fraction. Then

$$C(12) = \langle \text{sgn}\,m(1)\,\text{sgn}\,m(2)\rangle = (2/\pi)\sin^{-1}(\gamma) , \tag{39}$$

$$\gamma = \langle m(1)m(2)\rangle/[\langle m^2(1)\rangle\langle m^2(2)\rangle]^{1/2} = \exp(-r^2/8Dt) . \tag{40}$$

The gaussian average over the field m required in (39) is standard (see, e.g. [42]). Eqs. (39) and (40) define the 'OJK scaling function'. Note that (apart from the trivial dependence through D) it is independent of the spatial dimension d. We will present arguments that it becomes exact in the large-d limit. The OJK function fits experiment and simulation data very well. As an example, we show the function $f(x)$ for the $d = 2$ scalar theory in Figure 9.

5.1.2 The KYG Method

An earlier approach, due to Kawasaki, Yalabik and Gunton (KYG) [38], building on still earlier of Suzuki [43], was based on an approximate resummation of the direct perturbation series in the non-linearity for the quartic potential $V(\phi) = (1/4)(1-\phi^2)^2$. In terms of the mapping $\phi(m) = m/(1+m^2)^{1/2}$, it gives, instead of (38),

$$\partial m/\partial t = \nabla^2 m + m , \tag{41}$$

which gives an exponential growth superimposed on the diffusion. After the replacement $\phi(m) \to \text{sgn}(m)$, however, this drops out and the scaling function (9) is recovered (but with $D = 1$).

The KYG method has the virtue that it can be readily extended to vector fields [22,44]. Eq. (41) is again obtained, with but m replaced by a vector auxiliary field $\vec{m}$, with $\vec{\phi} = \vec{m}/(1 + \vec{m}^2)^{1/2}$. At late times, $\vec{\phi} \to \hat{m}$, a unit vector, and $C(12) = \langle \hat{m}(1) \cdot \hat{m}(2) \rangle$. Taking gaussian initial conditions for $\vec{m}$, the resulting scaling function is given by Eq. (59) below [22], with γ again given by (40). The same scaling function was obtained independently by Toyoki [23]. We will call it the 'BPT scaling function'.

Both (38) and (41) suffer from the weakness that (for scalar fields) the width of the interface changes systematically with time. Since $\phi(m)$ is linear in m for small m, and $|\nabla \phi|$ is fixed (by the interface profile function) in the interface, we expect $\langle (\nabla m)^2 \rangle = \text{const.}$. Eqs. (38) and (41), however, give $\sim t^{-(d+2)/2}$ and $\sim \exp(2t)/t^{(d+2)/2}$ for this quantity, corresponding to increasing and decreasing interface widths respectively. This problem can be solved in an ad-hoc way [42] by introducing an extra term $h(t)m$ in (38), with $h(t)$ chosen by the requirement $\langle (\nabla m)^2 \rangle = \text{const.}$, giving $h(t) \simeq (d+2)/4t$ at late times. The scaling function (39), however, is unaffected by this 'Oono-Puri extension' of OJK.

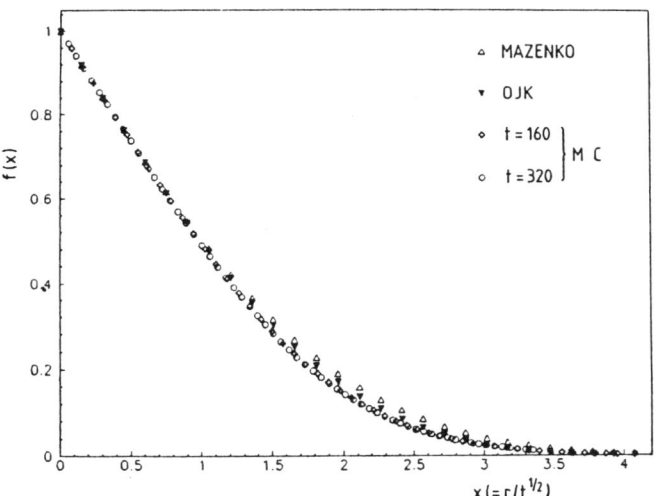

Figure 9. Scaling function $f(x)$ for the $2 - d$ Ising model, showing Monte-Carlo data from Figure 5, and the approximations of OJK [37] and Mazenko [39]. The scale lengths $L(t)$ for the theoretical curves were chosen to give the correct slope in the linear 'Porod' regime at small x. (From K. Humayun and A. J. Bray, Phys. Rev. B **46**, 10594 (1992)).

5.1.3 Mazenko's Method

The recent theory of Mazenko [39,40] handles the interface width in a more natural way. Mazenko's approach combines a clever choice for the function $\phi(m)$ with the minimal assumption that the field m is gaussian. Specifically $\phi(m)$ is chosen to be the equilibrium interface profile function, defined by $\phi''(m) = V'(\phi)$, with boundary conditions $\phi(\pm\infty) = \pm 1$, $\phi(0) = 0$. The field m then has a physical interpretation, near walls, as a coordinate normal to the wall. With this choice, (2) becomes $\partial_t \phi = \nabla^2 \phi - \phi''(m)$. Multiplying by ϕ at a different space point and averaging over initial conditions gives $(1/2)\partial_t C(12) = \nabla^2 C(12) - \langle \phi''(m(1)) \phi(m(2)) \rangle$. If m is assumed to be a gaussian field, the final term can be expressed in terms of $C(12)$ itself, to give [39]

$$(1/2)\,\partial_t C = \nabla^2 C + a(t)\gamma dC/d\gamma \,, \qquad (42)$$

with $C(\gamma)$ given by (39), and γ still given by the left part of (40), but no longer simply $\exp(-r^2/8t)$. In (42), $a(t) = \langle m^2 \rangle^{-1}$. Using (39) for $C(\gamma)$ gives $\gamma dC/d\gamma = (2/\pi)\tan[(\pi/2)C]$. Then (42) becomes a closed non-linear equation for C. For a scaling solution, one requires $a(t) = a_0/t$ for large t in (42). The constant a_0 is fixed by the requirement that the large-distance behaviour of C be physically reasonable [39]. The virtues of the Mazenko approach are (i) only the assumption that the field m is gaussian is required, (ii) the scaling function has a non-trivial dependence on d (whereas, apart from the trivial dependence through D, (39) and (40) are independent of d), and (iii) the non-trivial behaviour of *different-time* correlation functions [15] emerges in a natural way [45]. In practice, however, the shape of the scaling function $f(x)$ differs very little from that of the OJK function given by (39) and (40). Both are in good agreement with experimental data and numerical simulations. The Mazenko function for $d = 2$ is included in Figure 9. The Mazenko approach can be also be used for vector fields [24] and, with some modifications, for conserved scalar [40] and vector [46] fields.

5.2 A SYSTEMATIC APPROACH

All of the above treatments suffer from the disadvantage that they invoke an uncontrolled approximation at some stage. Very recently, however, a new approach has been developed [41] which recovers the OJK and BPT scaling functions in leading order, but has the advantage that it can, in principle, be systematically improved.

5.2.1 Scalar Fields

For simplicity of presentation, we will begin with scalar fields. The TDGL equation for a non-conserved scalar field $\phi(\mathbf{x}, t)$ is given by Eq. (2). We recall that, according to the Allen-Cahn equation (9), the detailed form of $V(\phi)$ should not be important, a fact that we will exploit.

Following Mazenko [39], we define the function $\phi(m)$ by the equation

$$\phi''(m) = V'(\phi) \,, \qquad (43)$$

where primes indicate derivatives, with boundary conditions $\phi(\pm\infty) = \pm 1$. We have noted that $\phi(m)$ is just the equilibrium domain-wall profile function, with m playing the role of the distance from the wall. Therefore, the spatial variation of m near a domain wall is completely smooth (in fact, linear). The additional condition $\phi(0) = 0$ locates the center of the wall at $m = 0$. Figure 10 illustrates the difference between ϕ and m for a cut through the system. Note that, while ϕ saturates in the interior of domains, m is typically of order $L(t)$, the domain scale. Rewriting (2) in terms of m, and using (43) to eliminate V', gives

$$\partial_t m = \nabla^2 m - \frac{\phi''(m)}{\phi'(m)} \left(1 - (\nabla m)^2\right). \tag{44}$$

For general potentials $V(\phi)$, Eq. (44) is a complicated non-linear equation, not obviously simpler than the original TDGL equation (2). For reasons discussed in section 2.2, however, we expect the scaling function $f(x)$ to be *independent* both of the detailed form of the potential and of the particular choice for the distribution of initial conditions. Physically, the motion of the interfaces is determined by their *curvature*. The potential $V(\phi)$ determines the domain wall *profile*, which is irrelevant to the large-scale structure. For example, for a single spherical domain of one phase, in a sea of the other phase, the Allen-Cahn equation (9) gives $\partial_t R = -(d-1)/R$ for the radius R of the domain, independent of the details of the potential. This equation for R can be verified directly from (2) for R large compared to the width ξ of the wall [1].

Similarly, the initial conditions determine the early-time locations of the walls, which should again be irrelevant for late-stage scaling properties. For example, in Mazenko's approximate theory, both the potential and the initial conditions drop out from the equation for $f(x)$.

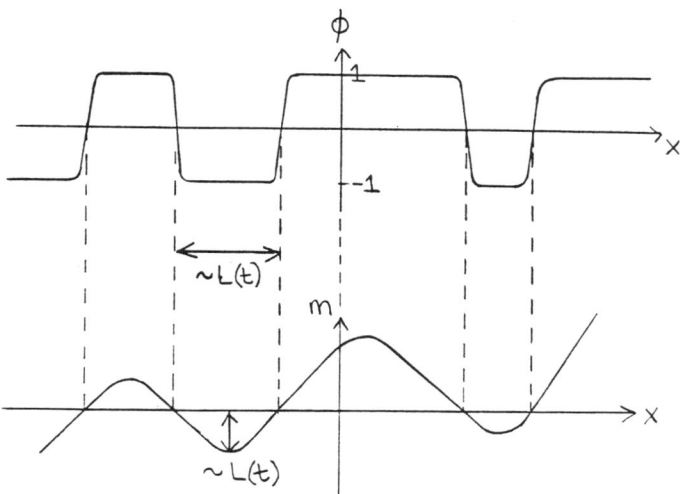

Figure 10. Spatial variation (schematic) of the order parameter field ϕ and the auxiliary field m, defined by Eq. (43).

The key step in the present approach is to exploit the notion that the scaling function should be independent of the potential (or, equivalently, independent of the wall profile) by choosing a particular $V(\phi)$ such that Eq. (44) takes a much simpler form (Eq. (48)). Specifically we choose the domain-wall profile function $\phi(m)$ to satisfy

$$\phi''(m) = -m\,\phi'(m). \tag{45}$$

This is equivalent, via (43), to a particular choice of potential, as discussed below. First we observe that (45) can be integrated, with boundary conditions $\phi(\pm\infty) = \pm 1$ and $\phi(0) = 0$ to give the wall profile function

$$\phi(m) = (2/\pi)^{1/2} \int_0^m dx \, \exp(-x^2/2) = \text{erf}(m/\sqrt{2}) \,, \tag{46}$$

where erf(x) is the error function. Also, (43) can be integrated once, with the zero of potential defined by $V(\pm 1) = 0$, to give

$$V(\phi) = (1/2)(\phi')^2 = (1/\pi)\exp(-m^2) = (1/\pi)\exp(-2[\text{erf}^{-1}(\phi)]^2) \,, \tag{47}$$

where $\text{erf}^{-1}(x)$ is the inverse function of erf(x). In particular, $V(\phi) \simeq 1/\pi - \phi^2/2$ for $\phi^2 \ll 1$, while $V(\phi) \simeq (1/4)(1-\phi^2)^2|\ln(1-\phi^2)|$ for $(1-\phi^2) \ll 1$ [47].

With the choice (45), Eq. (44) reduces to the much simpler equation

$$\partial_t m = \nabla^2 m + (1 - (\nabla m)^2) \, m \,. \tag{48}$$

This equation, though still non-linear, represents a significant simplification of the original TDGL equation. We believe, however, on the basis of the physical arguments discussed above, that it retains all the ingredients necessary to describe the universal scaling properties.

We now proceed to show that the usual OJK result is recovered by simply replacing $(\nabla m)^2$ by its average (over the ensemble of initial conditions) in (48), and choosing a gaussian distribution for the initial conditions. In order to make this replacement in a controlled way, however, and to facilitate the computation of corrections to the leading order results, we systematize the treatment by attaching to the field m an internal 'color' index α which runs from 1 to N, and generalize (48) to

$$\partial_t m_\alpha = \nabla^2 m_\alpha + \left(1 - N^{-1} \sum_{\beta=1}^N (\nabla m_\beta)^2\right) m_\alpha \,.$$

Eq. (48) is the case $N = 1$. The OJK result is obtained, however, by taking the limit $N \to \infty$, when $N^{-1}\sum_{\beta=1}^N (\nabla m_\beta)^2$ may be replaced by its average. In this limit (48) becomes (where m now stands for one of the m_α)

$$\partial_t m = \nabla^2 m + a(t) \, m \tag{49}$$
$$a(t) = 1 - \langle(\nabla m)^2\rangle \,, \tag{50}$$

a self-consistent *linear* equation for $m(\mathbf{x},t)$.

It is interesting that the replacement of $(\nabla m)^2$ by its average in (48) is also justified in the limit $d \to \infty$, where d is the number of spatial dimensions, because $(\nabla m)^2 = \sum_{i=1}^d (\partial m/\partial x_i)^2$. If m is a gaussian random field (and the self-consistency of this assumption follows from (49) – see below) then the different derivatives $\partial m/\partial x_i$ at a given point x are independent random variables, and the central limit theorem gives, for $d \to \infty$, $(\nabla m)^2 \to d\langle(\partial m/\partial x_i)^2\rangle = \langle(\nabla m)^2\rangle$, with fluctuations of relative order $1/\sqrt{d}$. While this approach is not so simple to systematize as that adopted above, it seems very likely that our leading order results become exact for large d.

As discussed above, we will take the initial conditions for m to be gaussian, with mean zero and correlator (in Fourier space)

$$\langle m_{\mathbf{k}}(0)\,m_{-\mathbf{k}'}(0)\rangle = \Delta\,\delta_{\mathbf{k},\mathbf{k}'}\,, \qquad (51)$$

representing short-range spatial correlations at $t=0$. Then m is a gaussian field at all times. The solution of (49) is $m_{\mathbf{k}}(t) = m_{\mathbf{k}}(0)\exp(-k^2 t + b(t))$, where $b(t) = \int_0^t dt'\,a(t')$. Inserting this into (50) yields

$$a(t) \equiv db/dt = 1 - \Delta \sum_{\mathbf{k}} k^2\,\exp(-2k^2 t + 2b)\,.$$

After evaluating the sum one obtains, for large t (where the db/dt term can be neglected), $\exp(2b) \simeq (4t/\Delta d)(8\pi t)^{d/2}$, and hence $a(t) \simeq (d+2)/4t$. This form for $a(t)$ in (49), arising completely naturally in this scheme, reproduces exactly the 'Oono-Puri extension' to the OJK theory, an ad-hoc addition to OJK's diffusion equation for m, designed to keep the wall-width finite as $t \to \infty$ [42].

The explicit result for $m_{\mathbf{k}}(t)$, valid for large t, is

$$m_{\mathbf{k}}(t) = m_{\mathbf{k}}(0)\,(4t/\Delta d)^{1/2}\,(8\pi t)^{d/4}\,\exp(-k^2 t)\,, \qquad (52)$$

from which the equal-time two-point correlation functions in Fourier and real space follow immediately:

$$\langle m_{\mathbf{k}}(t)\,m_{-\mathbf{k}}(t)\rangle = (4t/d)(8\pi t)^{d/2}\exp(-2k^2 t)\,, \qquad (53)$$
$$\langle m(1)\,m(2)\rangle = (4t/d)\exp(-r^2/8t)\,, \qquad (54)$$

where '1', '2', are a shorthand for space-time points $(\mathbf{r}_1,t)$, $(\mathbf{r}_2,t)$, and $r = |\mathbf{r}_1 - \mathbf{r}_2|$.

We turn now to the evaluation of the correlation function of the original fields ϕ. Since, from (54), m is typically of order $\sqrt{t}$ at late times it follows from (46) that the field ϕ is saturated (i.e. $\phi = \pm 1$) almost everywhere at late times. As a consequence, the relation (46) between ϕ and m may be simplified to $\phi = \text{sgn}(m)$ as far as the late-time scaling behavior is concerned. Thus $C(12) = \langle \text{sgn}(m(1))\,\text{sgn}(m(2))\rangle$. The calculation of this average for a gaussian field m proceeds just as in the OJK calculation. The OJK result, (39) and (40), (with $D=1$) is recovered. The present approach, however, makes possible a systematic treatment in powers of $1/N$. The work involved in calculating the next term is comparable to that required to obtain the $O(1/n)$ correction to the $n=\infty$ result for the $O(n)$ model [48].

5.2.2 Vector Fields

For vector fields, the TDGL equation is given by (14), where $V(\vec{\phi})$ is the usual 'mexican hat' potential with ground-state manifold $\vec{\phi}^2 = 1$. This time we introduce a *vector* field $\vec{m}(\mathbf{x},t)$, related to $\vec{\phi}$ by the vector analog of (43), namely [24]

$$\nabla_m^2 \vec{\phi} = \partial V/\partial \vec{\phi}\,, \qquad (55)$$

where ∇_m^2 means $\sum_{a=1}^n \partial^2/\partial m_a^2$ for an n-component field. We look for a radially symmetric solution of (55), $\vec{\phi}(\vec{m}) = \hat{m}\,g(\rho)$, with boundary conditions $g(0) = 0$,

$g(\infty) = 1$, where $\rho = |\vec{m}|$ and $\hat{m} = \vec{m}/\rho$. Then the function $g(\rho)$ is the defect profile function for a topological defect in the n-component field, with ρ representing the distance from the defect core. In terms of $\vec{m}$, the TDGL equation for a vector field reads

$$\sum_b \frac{\partial \phi_a}{\partial m_b} \frac{\partial m_b}{\partial t} = \sum_b \frac{\partial \phi_a}{\partial m_b} \nabla^2 m_b + \sum_{bc} \frac{\partial^2 \phi_a}{\partial m_b \partial m_c} \nabla m_b \cdot \nabla m_c - \nabla_m^2 \phi_a . \tag{56}$$

Just as in the scalar theory, we can attach an additional 'color' index α $(= 1,\ldots,N)$ to the vector field $\vec{m}$, such that the theory in the limit $N \to \infty$ is equivalent to replacing $\nabla m_b \cdot \nabla m_c$ by its mean, $\langle (\nabla m_b)^2 \rangle \delta_{bc}$ in (56). Noting also that $\langle (\nabla m_b)^2 \rangle$ is independent of b from global isotropy, (56) simplifies in this limit to

$$\sum_b \frac{\partial \phi_a}{\partial m_b} \frac{\partial m_b}{\partial t} = \sum_b \frac{\partial \phi_a}{\partial m_b} \nabla^2 m_b - \nabla_m^2 \phi_a \left(1 - \langle (\nabla m_1)^2 \rangle \right) , \tag{57}$$

where m_1 is any component of $\vec{m}$. Finally, this equation can be reduced to the linear form (49), with m replaced by $\vec{m}$, through the choice $\nabla_m^2 \phi_a = -\sum_b (\partial \phi_a / \partial m_b) m_b$ or, more compactly, $\nabla_m^2 \vec{\phi} = -(\vec{m} \cdot \nabla_m) \vec{\phi}$, to determine the function $\vec{\phi}(\vec{m})$. Substituting the radially symmetric form $\vec{\phi} = \hat{m} g(\rho)$ gives the equation

$$g'' + \left(\frac{n-1}{\rho} + \rho\right) g' - \frac{n-1}{\rho^2} g = 0 , \tag{58}$$

a generalization of (45), for the profile function $g(\rho)$, with boundary conditions $g(0) = 0$, $g(\infty) = 1$. The solution is linear in ρ for $\rho \to 0$, while $g(\rho) \simeq 1 - (n-1)/2\rho^2$ for $\rho \to \infty$. The potential $V(\vec{\phi})$ corresponding to this profile function can be deduced from (55), though we have been unable to derive a closed form expression for it. Note that we are making here the natural assumption that scaling functions are independent of the details of the potential for vector fields, as well as for scalar fields.

For the vector theory, Eqs. (49) and (50) hold separately for each component of the field. Taking gaussian initial conditions, with correlator (51), yields $a(t) \simeq (d+2)/4t$ again, giving (54) for each component. The final step is the evaluation of the two-point function $C(12) = \langle \vec{\phi}(1) \cdot \vec{\phi}(2) \rangle$. Since $|\vec{m}|$ scales as $\sqrt{t}$, we can replace the function $\vec{\phi}(\vec{m})$ by $\hat{m}$ at late times. Then $C(12) = \langle \hat{m}(1) \cdot \hat{m}(2) \rangle$ in the scaling regime. The required gaussian average over the fields $\vec{m}(1), \vec{m}(2)$ yields finally [22-24]

$$C(12) = \frac{n\gamma}{2\pi} \left[B\left(\frac{n+1}{2}, \frac{1}{2}\right) \right]^2 F\left(\frac{1}{2}, \frac{1}{2}; \frac{n+2}{2}; \gamma^2\right) , \tag{59}$$

where $B(x,y)$ is the beta function, $F(a,b;c;z)$ the hypergeometric function, and $\gamma \equiv \gamma(12)$ is given once more by (40) (with $D = 1$). Eq. (59) is the BPT scaling function. Again, it can be systematically improved by expanding in $1/N$.

It is easy to show that (59) contains the singular term of order r^n (with an additional logarithm for even n) that generates the Porod tail (20) in the structure factor. This feature was effectively built into the theory through the mapping $\vec{\phi}(\vec{m})$.

In Figure 11, we compare the BPT scaling function with numerical simulation results [25], both for the pair correlation function and the structure factor. In making the fit, γ was replaced by $\exp(-r^2/L(t)^2)$ and $L(t)$ adjusted separately at each time to give the best fit of the data to the theory for $C(\mathbf{r},t)$. The structure factor plots, on a log-log scale, confirm the existence of the Porod tail (20) in the data.

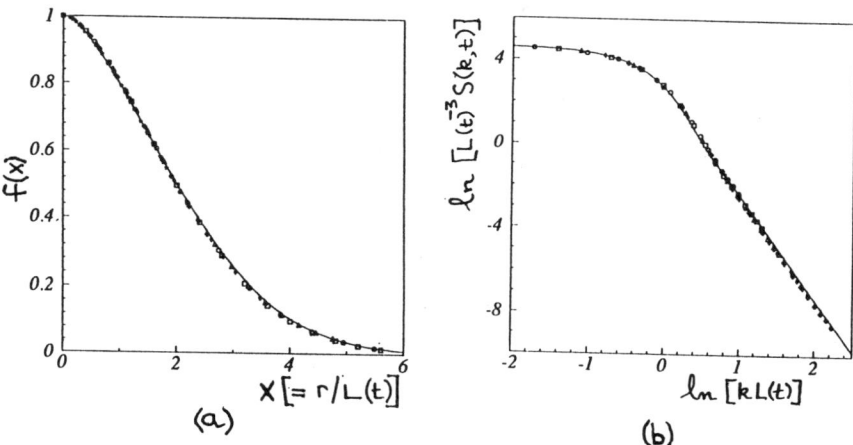

Figure 11. Comparison of the BPT function (59) with numerical simulation data [25] for the case $d = 3$, $n = 2$. (a) the real space scaling function $f(x)$ (b) log-log scaling plot for the structure factor. The slope of the structure-factor tail on the log-log plot is -5, in agreement with (20).

6. Discussion

In these lectures I have reviewed our current understanding of the dynamics of phase ordering, and discussed some recent developments. The concept of topological defects provides a unifying framework for discussing the growth laws for the characteristic scale, and motivates approximate treatments of the pair correlation function. To keep things as simple as possible, I have concentrated on the simplest dynamical models, namely models A and B of the Hohenberg-Halperin classification [2]. These are adequate for the discussion of simple systems such as dynamic Ising models and certain alloy systems. Phase separation in fluids, however, is more complicated since coupling of the order parameter to hydrodynamic modes must be explicitly included [49]. Even in fluid systems, however, the Cahn-Hilliard equation (3) provides a reasonable description for early and intermediate times.

The most important consequence of the presence of topological defects in the system is the 'generalized Porod law', equation (20), for the large kL tail of the structure factor. This power-law tail, whose existence has long been known for scalar systems, has recently been observed in computer simulations of various vector systems [25,31,50]. It should be stressed that the form of the tail depends only on the nature of the dominant topological defects. In nematic liquid crystals, for example, the presence of disclinations [19], or '1/2-strings', implies a structure factor tail described by (20) with $n = 2$ [51], i.e. a k^{-5} tail for bulk systems. This tail has been seen in simulations [36,51], and is not inconsistent [51] with experimental results [35].

The Porod law (20), together with the scaling hypothesis, leads to a powerful and general technique for deriving growth laws [26]. The results are summarized in table 1. Again, the technique is more general than the simple $O(n)$ models to which it has been applied here. Nematic liquid crystals, for example, are described by the nonconserved dynamics of a traceless, symmetric, tensor field. However, the presence of dominant string defects implies the same growth law as for the $O(2)$ model, namely $L(t) \sim t^{1/2}$, consistent with the simulations [36] (allowing for the predicted logarithmic corrections to scaling) and experiment [35].

The dominant role of topological defects also motivates approximate treatments of the pair correlation scaling function $f(x)$ [22-24,37], and the systematic treatment [41] discussed in section 5. All of these theories lead to the same scaling function (59), with the OJK scaling function (39) corresponding to the special case $n = 1$. The form (59) is a direct consequence of the non-linear mapping $\vec{\phi}(\vec{m})$, with $\vec{\phi} \to \hat{m}$ for $|\vec{m}| \to \infty$, and the gaussian distribution assumed for the field $\vec{m}$. The 'OJK-type' theories [22,23,37,41] and the 'Mazenko-type' theories [24,39,40] differ only in the equation for γ, the normalized pair correlation function for $\vec{m}$.

These approximate scaling functions all give good fits to experiment and simulation data (see, e.g., Figure 11). However, there is one important caveat. When fitting data to theoretical scaling functions, it is conventional to adjust the scale length $L(t)$ for the best fit. It has recently been noticed, however, that an *absolute* test can be obtained by calculating two *different* scaling functions and plotting one against the other [52]. For example, the normalized, connected correlator of the square of the field,

$$C^N_{\phi^2}(12) = \frac{\langle [1 - \phi^2(1)] [1 - \phi^2(2)] \rangle}{\langle [1 - \phi^2(1)] \rangle \langle [1 - \phi^2(2)] \rangle} - 1 , \qquad (60)$$

can also be calculated within 'gaussian' theories of the OJK or Mazenko type [20]. The result depends only on γ. Eliminating γ between $C(12)$ and $C^N_{\phi^2}(12)$ gives an absolute prediction for the function $C^N_{\phi^2}(C)$. When this function is compared to simulation results, however, the agreement is found to be rather poor [52]: C and $C^N_{\phi^2}$ can be fitted separately, as functions of $r/L(t)$, by choosing the scale length $L(t)$ independently for each fit, but not simultaneously. However, the agreement improves with increasing d, in agreement with the idea that these theories based on a gaussian auxiliary field become exact at large d [41]. Including the $1/N$ correction in the systematic approach of section 5 will presumably improve the fit at fixed d.

The calculation of scaling functions for *conserved fields* is a significantly greater challenge, especially for scalar fields, where even obtaining the correct $t^{1/3}$ growth law, within an approximate theory for the pair correlation function, is not straightforward. Mazenko has extended his approximate theory to conserved scalar fields [40], but the agreement with high quality simulation data is not as good as for nonconserved fields [53]. There is an additional complication that a naive application of Mazenko's method gives $t^{1/4}$ growth, which Mazenko argues corresponds to surface diffusion only. In order to recover the $t^{1/3}$ growth, he has to add an additional term to incorporate the effect of bulk diffusion. For conserved vector fields, the naive Mazenko approach gives the expected $t^{1/4}$ growth (see table 1), but without the logarithmic correction expected for $n = 2$ [46]. An approximate analytic treatment [46] of the equation for $C(12)$, valid for $n \gg 1$, gives good agreement with scaling functions extracted from simulations [32]. A systematic approach for conserved fields, generalizing the treatment of section 5, would be very welcome, although it is far from straightforward.

To summarize, we have focussed on the role of topological defects as a general way of deriving, through the Porod law (20) and the scaling hypothesis (12), the forms of the growth laws for phase ordering in various systems. The study of such defects also motivates, through the mapping to an auxiliary field that varies smoothly through the defect, approximate theories of scaling functions. For nonconserved fields, such methods are, in principle, systematically improvable (section 5). One of the challenges for the future is to try to develop comparable methods for conserved fields.

From a wider perspective, phase ordering dynamics is, perhaps, the simplest example of a scaling phenomenon controlled by a 'strong-coupling' RG fixed point (Figure 3). It may not be too much to hope that techniques developed here will find useful applications in other branches of physics.

Acknowledgements

My understanding of this subject owes much to discussions with R. E. Blundell, J. Filipe, K. Humayun, D. A. Huse, J. G. Kissner, A. J. McKane, M. A. Moore, T. J. Newman, A. D. Rutenberg, and N. Turok.

References

1. J. S. Langer, in *Solids Far From Equilibrium*, ed. C. Godrèche (Cambridge, Cambridge, 1992).

2. P. C. Hohenberg and B. I. Halperin, Rev. Mod. Phys. **49**, 435 (1977).

3. A. J. Bray, Phys. Rev. Lett. **62**, 2841 (1989); Phys. Rev. B **41**, 6724 (1990).

4. S. M. Allen and J. W. Cahn, Acta. Metall. **27**, 1085 (1979).

5. K. Binder and D. Stauffer, Phys. Rev. Lett. **33**, 1006 (1974).

6. J. Marro, J. L. Lebowitz and M. H. Kalos, Phys. Rev. Lett. **43**, 282 (1979).

7. H. Furukawa, Prog. Theor. Phys. **59**, 1072 (1978); Phys. Rev. Lett. **43**, 136 (1979).

8. The 1-d Glauber model can be solved exactly, and exhibits scaling: A. J. Bray, J. Phys. A **22**, L67 (1990); J. G. Amar and F. Family, Phys. Rev. A **41**, 3258 (1990).

9. The non-conserved $O(n)$ model can be solved for $n = \infty$ (see, e.g., A. Coniglio and M. Zannetti, Europhys. Lett. **10**, 575 (1989)) and exhibits scaling.

10. D. A. Huse, Phys. Rev. B **34**, 7845 (1986).

11. J. Amar, F. Sullivan and R. Mountain, Phys. Rev. B **37**, 196 (1988); T. M. Rogers, K. R. Elder and R. C. Desai, Phys. Rev. B **37**, 9638 (1988); R. Toral, A. Chakrabarti and J. D. Gunton, Phys. Rev. B **39**, 4386 (1989); C. Roland and M. Grant, Phys. Rev. B **39**, 11971 (1989).

12. I. M. Lifshitz and V. V. Slyozov, J. Phys. Chem. Solids **19**, 35 (1961); C. Wagner, Z. Elektrochem. **65**, 581 (1961).

13. J. D. Gunton, M. San Miguel and P. S. Sahni, in *Phase Transitions and Critical Phenomena*, Vol. 8, eds. C. Domb and J. L. Lebowitz (Academic, New York, 1983) p.267.

14. D. S. Fisher and D. A. Huse, Phys. Rev. B **38**, 373 (1988).

15. T. J. Newman and A. J. Bray, J. Phys. A **23**, 4491 (1990).

16. N. Mason, A. N. Pargellis, and B. Yurke, Phys. Rev. Lett. **70**, 190 (1993).

17. G. Porod, Kolloid Z. bf 124, 83 (1951); **125**, 51 (1952).

18. P. Debye, H. R. Anderson and H. Brumberger, J. Appl. Phys. **28**, 679 (1957); G. Porod, in *Small-Angle X-Ray Scattering*, edited by O. Glatter and O. Kratky (Academic, New York, 1982).

19. For a general discussion of topological defects, see e.g. M. Kléman, *Points, Lines and Walls, in Liquid Crystals, Magnetic Systems, and Various Ordered Media* (Wiley, New York, 1983).

20. A. J. Bray, Phys. Rev. E **47**, 228 (1993).

21. A. J. Bray and K. Humayun, Phys. Rev. E **47**, R9, (1993).

22. A. J. Bray and S. Puri, Phys. Rev. Lett. **67**, 2670 (1991).

23. H. Toyoki, Phys. Rev. B **45**, 1965 (1992).

24. Fong Liu and G. F. Mazenko, Phys. Rev. B **45**, 6989 (1992); A. J. Bray and K. Humayun, J. Phys. A **25**, 2191 (1992).

25. R. E. Blundell and A. J. Bray, unpublished.

26. A. J. Bray and A. D. Rutenberg, University of Manchester preprint.

27. A. J. Bray and A. D. Rutenberg, unpublished.

28. B. Yurke, A. N. Pargellis, T. Kovacs and D. A. Huse, preprint. See also A. N. Pargellis, P. Finn, J. W. Goodby, P. Panizza, B. Yurke, and P. E. Cladis, Phys. Rev. A **46**, 7765 (1992).

29. M. Mondello and N. Goldenfeld, preprint.

30. M. Mondello and N. Goldenfeld, Phys. Rev. A **45**, 657 (1992).

31. H. Toyoki, J. Phys. Soc. Jpn. **60**, 1433 (1991).

32. M. Siegert and M. Rao, Phys. Rev. Lett. in press.

33. T. J. Newman, A. J. Bray, and M. A. Moore, Phys. Rev. B **42** 4514, (1990) found $L \sim t^{1/4}$ for $d = 1$, $n = 2$, and $\mu = 0$.

34. Of course this means that our approach will not address systems with a potential-dependent growth law, e.g. $d = n$ for $n < 2$. We also do not address quenches in which thermal noise is essential, such as systems with static disorder (see D. A. Huse and C. L. Henley, Phys. Rev. Lett. **54**, 2708 (1985)), or quenches to a $T > 0$ critical point.

35. A. P. Y. Wong, P. Wiltzius and B. Yurke, Phys. Rev. Lett. **68**, 3583 (1992); A. P. Y. Wong, P. Wiltzius, R. G. Larson and B. Yurke, preprint.

36. R. E. Blundell and A. J. Bray, Phys. Rev. A **46**, R6154 (1992). Our results indicate that nematic liquid crystals, with line defects, should have the same growth law and leading correction as the $O(2)$ model.

37. T. Ohta, D. Jasnow and K. Kawasaki, Phys. Rev. Lett. **49**, 1223 (1982).

38. K. Kawasaki, M. C. Yalabik and J. D. Gunton, Phys. Rev. A **17**, 455 (1978).

39. G. F. Mazenko, Phys. Rev. B **42**, 4487 (1990).

40. G. F. Mazenko, Phys. Rev. B **43**, 5747 (1991).

41. A. J. Bray and K. Humayun, University of Manchester preprint.

42. Y. Oono and S. Puri, Mod. Phys. Lett. B **2**, 861 (1988).

43. M. Suzuki, Prog. Theor. Phys. **56**, 77 (1976); **56**, 477 (1976).

44. S. Puri and C. Roland, Phys. Lett. A **151**, 500 (1990).

45. Fong Liu and G. F. Mazenko, Phys. Rev. B **44**, 9185 (1991).

46. A. J. Bray and K. Humayun, Phys. Rev. Lett. **68**, 1559 (1992).

47. Eq. (47) only fixes $V(\phi)$ for $\phi^2 \leq 1$. Note that, for $T = 0$, $\phi^2(\mathbf{x}, 0) \leq 1$ everywhere implies $\phi^2(\mathbf{x}, t) \leq 1$ everywhere, so $\phi(\mathbf{x}, t)$ does not depend on the form of $V(\phi)$ for $\phi^2 > 1$. Of course, for stability against thermal fluctuations the points $\phi = \pm 1$ must be global minima of $V(\phi)$.

48. T. J. Newman and A. J. Bray, ref. 15; J. G. Kissner and A. J. Bray, J. Phys. A, in press.

49. E. D. Siggia, Phys. Rev. A **20**, 595 (1979).

50. H. Toyoki, J. Phys. Soc. Jpn. **60**, 1153 (1991).

51. A. J. Bray, S. Puri, R. E. Blundell and A. M. Somoza, submitted to Phys. Rev. E.

52. R. E. Blundell, S. Sattler and A. J. Bray, submitted to Phys. Rev. E.

53. A. Shinozaki and Y. Oono, Phys. Rev. Lett. **66**, 173 (1991).

Critical Wrinkling of Depinned Interfaces, Strings and Membranes.

Kim Sneppen and Mogens H. Jensen
Niels Bohr Institute
Blegdamsvej 17, DK-2100 Copenhagen Ø

Abstract: Interface dynamics governed by punktuated dynamics become self organized critical. In the critical state we observe temporal multiscaling of the interface profile and the activity pattern exhibits non trivial power law correlations in both space and time. This intermittent activity is characterized by avalanches which connect regions of recent local activity. These avalanches display scale invariance. An extension of the presented concepts to evolving strings leads to a new universality class.

The dynamical wrinkling of interfaces, strings and membranes is usually described in terms of the Kardar-Parisi-Zhang like equations although the theoretical exponents do not agree with any experiments. We recently proposed a dynamical depinning model [1] based on a rule similar to that of invasion percolation [2]. The key ingredient in the new model is punctuated dynamics simulated by a non-local rule. This is in contrast to the normally studied local KPZ dynamics where everything moves simultaneously and uncorrelated and exhibits no pinning effects. This type of punctuated depinning dynamics might also be applicable at the depinning transition of flux lines.

The model is defined on a lattice (x, h). In the 1-dimensional version a discrete interface $h(x)$ is defined on $x = 1, 2, 3...L$. Along this chain $(x, h(x))$ one initially distribute a sequence of uncorrelated random numbers η. We use periodic boundary conditions. The chain is updated by finding the site with the smallest random number $\eta(x, h(x))$ among all sites on the interface. On this site one unit is added to h. Then neighbouring sites are adjusted upwards $(h \rightarrow h + 1)$ precisely until all slopes $|h(x) - h(x-1)| \leq 1$. This induce a local burst of activity that is exponentially bounded. New random η's are assigned to all newly adjusted sites. Fig. 1 shows a snapshot of the interface where the shaded areas indicate the recent avalanches. A somewhat similar algorithm is mentioned in [3].

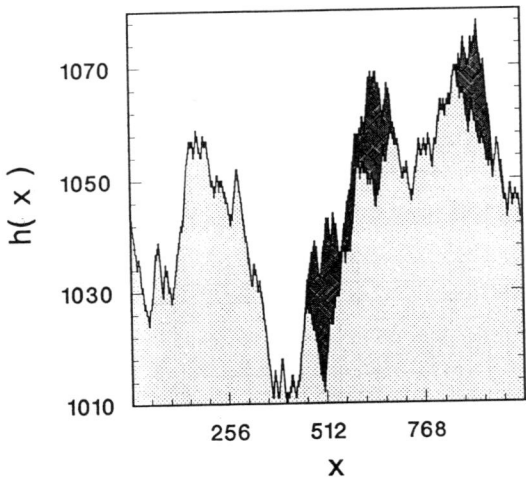

Figure 1: A snapshot of the wrinkled interface in a system of size $L=1000$. The shaded area indicates the sizes and locations of recent avalanches.

One prediction of the above model is the scaling of the saturated interface width $w_q = \langle (h - \langle h \rangle)^q \rangle^{1/q} \propto L^{0.63}$ with system size L, for all moments q [1] which is very close to the experimentally observed value for a fluid penetrating a porous medium[4, 5] and quite different from standard KPZ exponents [6]. This scaling is in [1, 7] connected to the transverse to longitudinal scaling for interfaces glued to directed percolating chains of effectively pinned sites. An interesting consequence of the analogy to directed percolation is that the dynamics induce a self organized threshold $\eta_{crit} = 1 - p_c = 0.4615$ below which the corresponding site spontaneously gets activated. Here p_c is the critical density of sites for directed percolation on a square lattice.

The dynamics of the model appear however much richer than expected from this simple analogy alone. Considering the activity along the chain, one observes initially that it is uncorrelated in space, but as time progresses subsequent activities gets correlated over larger and larger distances. Finally at saturation a critical state is build up, in the sense that the subsequent spatial activity seen in Fig. 2 shows a power law distribution:

$$P(X) \propto X^{-2.25 \pm 0.05} \qquad (1)$$

Therefore the model exhibits self organized critical behaviour. Notice that the same exponent is observed when one consider the spatial distribution function of distances between the two lowest η along a snapshot of a string [8]. Thus part of the observed dynamical correlation of activity is already an inherent property of static snapshots of the distribution of η in the self organized critical state. We stress however that other parts of the dynamical correlation do not appear connected to static correlations; E.g. is the decay of average activity in a given point after there has been activity ($\langle A(t) \rangle \propto t^{-0.6}$) slower than expected from the spatial correlations.

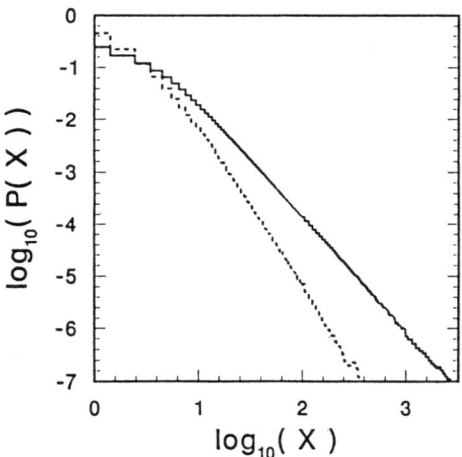

Figure 2: Spatial distribution of subsequent activity centers. The solid line is for the interface model, the dashed line for the string model. The activity appears more confined in the string model than in the interface model.

Another more dramatic consequence of the above is the observation of temporal multiscaling [9]. Denoting with τ a time at saturation we measure the height-height time correlations by

$$W_q(L,t) = \langle ((h(x, t+\tau) - h(x, \tau)) - \langle h(x, t+\tau) - h(x, \tau) \rangle)^q \rangle^{1/q} \quad (2)$$

where average $\langle \rangle$ over $x \in [1, L]$ and members of the ensemble. As seen in Fig.3a we, over a reasonable time interval, observe temporal multiscaling with $W_0 \propto t^{0.60}$, $W_2 \propto t^{0.69}$ and $W_\infty \propto t^{0.40}$ [9]. Thus, as is the case for most dynamical systems, the temporal (dynamical) scaling is much richer than the static scaling.

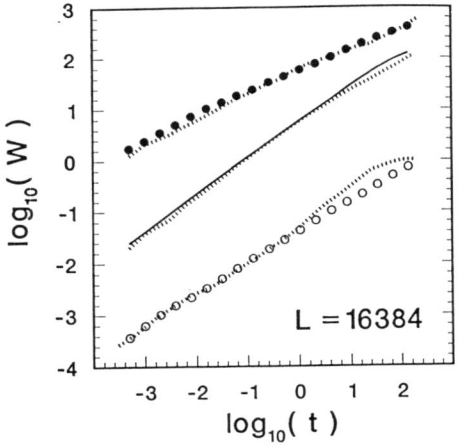 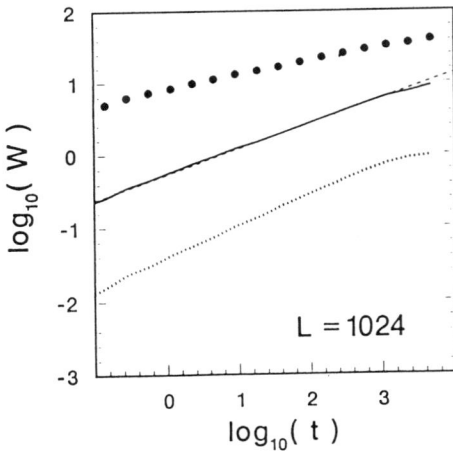

Figure 3: Scaling of $W_q(L,t)$ versus t for moments $q = 0, 2, \infty$. a): The interface model with $L=16384$. We find $W_0 \propto t^{0.60}$, $W_2 \propto t^{0.69}$ and $W_\infty \propto t^{0.40}$. The dotted lines are comparisons to Levy Flights models, see [9]. b): The string model with $L=1024$. We find $W_0 \propto t^{0.43 \pm 0.03}$, $W_2 \propto t^{0.35 \pm 0.02}$ and $W_\infty \propto t^{0.20 \pm 0.03}$.

In order to characterize the dynamic features of the present model further we now consider connected avalanches of activity. These are defined by regions of overlapped local bursts of activity, each initiated by localizing a minimal η as described above. When a new burst does not overlap with any of the previous, the connected avalanche is considered terminated and a new is started. In contrast to the locally confined activity at each timestep, these avalanches appear to display clear power law scalings. In fact defining the size S of the avalanche by the total number of invaded sites, we numerically observe a power law behaviour $P(S) = S^{-1.35 \pm 0.05}$ over tree decades, as seen in Fig. 4. As a consequence, the space-time plot of the activity exhibits intermittency with big regions of quiescence interrupted by scale invariant regions of high activity. So not only does the algorithm produce a self organized critical state, it in fact also predicts self organization of intermittent bursts of activity on all scales.

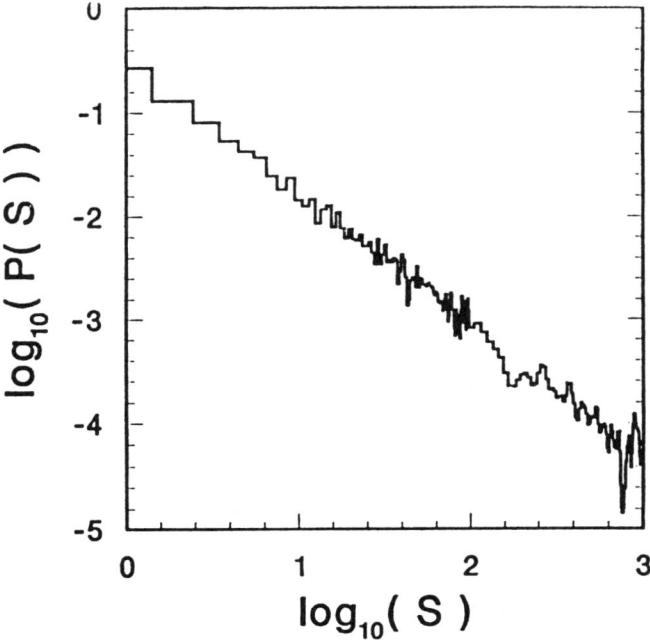

Figure 4: Size distribution of connected avalanches of activity, displayed for the interface model. A connected avalanche of activity is defined as a sequence of connected local bursts of activity, stopped by the first local burst that separates spatially from the previous. We observe a power law $P(S) \propto S^{-1.35 \pm 0.05}$, with small avalanches most probable but with nearly all the activity in the big avalanches.

It is interesting that one can define a closely related model similar to the above, with the exception that the up-down symmetry is restored. The symmetric model is defined on a lattice with sites (x, h). In the 1-dimensional version a discrete string $h(x)$ is defined on $x = 1, 2, 3...L$. Along the string a sequence of uncorrelated Gaussian random numbers η is distributed. We use periodic boundary conditions. The chain is updated, using a global comparison, by finding the site with the smallest random number η among all sites on the string. On this site one choses with equal probability to either add or subtract 1 from h. Then neighbouring sites are adjusted in the same direction precisely until all slopes $|h(x) - h(x-1)| \leq 1$. This create a local burst of activity that in simulations appears exponentially bounded. New random η's are assigned to all newly adjusted sites.

As is the case for the interface model, this string model also develops a self organized critical state with long range correlations of activity,

$$P(X) \propto X^{-3.1 \pm 0.2} \tag{3}$$

and temporal multiscaling both perpendicular and in fact also parallel to the string. The last is a new phenomena, which we in ref. [10] denote multidiffusion. The temporal multiscaling is shown in Fig. 3b for various moments.

Finally we would like to mention possible applications of the above developed models. The asymmetric model for interface propagation appears to mimic e.g. the interface of oil penetrating a porous medium. Here the penetration is often observed to occur in avalanches. Other possible and connected areas of interest are depinning of flux lines in type-2 superconductors.

The application of the string model, on the contrary, is not related to any experiment where one fluid displaces another. It's universality class may however turn out to have another and much broader range of applications, connected to an interpretation of the local η as barriers. Bak and Sneppen [11] has proposed to study evolution of a large number of systems, each with many metastable states, and organized such that neighbouring systems can modify each others barriers. For sufficiently low temperature this collection of weakly coupled systems is governed by fluctuations connected to passing the overall lowest barrier, like in the present model. The scaling exponents in the evolution model of ref. [11] furthermore appears identical to exponents of the present string model and the models therefore belong to the same new universality class.

K.Sneppen acknowledges financial support from the Carlsberg Foundation.

References

[1] K. Sneppen, Phys.Rev.Lett. **69**, 3539 (1992).

[2] D. Wilkinson and J.F. Willemsen, J.Phys.**A 16**, 3365 (1983).

[3] S. Havlin, A.-L. Barabási, S.V.C. Buldyrev, K. Peng, M. Schwartz, H.E. Stanley and T. Vicsek, in Proc. of Granada Conference on "Fractals", eds. P. Meakin and L. Sander (Plenum Press, New York, 1992).

[4] M.A. Rubio, C. Edwards, A. Dougherty and J.P. Gollup, Phys. Rev. Lett. **63**, 1685 (1989); **65**, 1389 (1990).

[5] V.K. Horváth, F. Family and T. Vicsek, Phys. Rev. Lett. **65**, 1388 (1990), J. Phys A **24**, L-25 (1991).

[6] M. Kardar, G. Parisi and Y.-C. Zhang, Phys.Rev.Lett. **56**, 889 (1986).

[7] Lei-Han Tang and H. Leschhorn, comment to appear in Phys. Rev. Lett.

[8] K. Sneppen and M.H. Jensen, reply to appear in Phys. Rev. Lett.

[9] K. Sneppen, M.H. Jensen, Phys.Rev.Lett., in press (1993).

[10] K. Sneppen, M.H. Jensen, NBI preprint 93/15, submitted to PRL.

[11] P. Bak and K. Sneppen, BNL preprint, Submitted to Nature.
Propose a model of co-evolution dynamics by a sequence of metastable complex systems each with a minimal random barrier for possible changes. The system evolves by locating the overall lowest barrier and updating this and its two nearest neighbours to new random barriers. This dynamics can be justified by exponential separation of timescales associated to an interpretation of the barriers as energies and the driving performed at very low temperature. Power laws will be observed for distances smaller than a correlation length which increases with decreasing temperature.

INDEX

α-process, 354
Abrikosov lattice, 27, 56-63, 95, 96,
    110, 112, 124
AC penetration depth, 45
    permeability, 119
    resistivity, 45
    susceptibility, 45
activation energy, 41, 43, 44, 352
aging effect, 344, 349, 352, 356
Allen-Cahn equations, 410, 424
Anderson localization, 107
anharmonicity, 239
anisotropic London theory, 28
anisotropy ratio, 28
Arnold tongue, 327
Arrhenius law, 260, 261, 269, 276-79,
    343, 350, 354
avalanche, 14, 438, 440

β-process, 354
back-flow, 219
Bean model, 39, 79, 87-90
benzyl chloride, 279
bifurcation, 208, 215, 226
Bitter pattern, 59-60
bond orientational order, 123
bose glass, 97
Brillouin peaks, 193

cage effect, 218, 226
Cahn-Hilliard equation, 408
cantorus, 336
charge density waves, 317-34, 339-58, 385
    broad band noise, 326
    glass, 354
    inhomogeneous velocity, 324-25
    metastability, 340
    phase deformation, 318
    phase slip, 318-26
    pinning energy, 322-23
    relaxation, 342
    self-organization, 331
Chern-Simon gauge field, 374
circle map, 337
coherence length, 95, 337
Cole-Cole distributions, 268
collective creep, 42, 79
collective transport, 15

colloidal suspension, 202, 222
columnar defect, 188
columnar pin, 97-117
commensurate-incommensurate transition, 335
complex conductivity, 388
conduction noise, 46
correlated disorder, 97
correlation function, 418
correlation length, 62, 123, 350, 381, 397
creep equation, 85
critical current, 20, 83
critical decay, 197, 210, 228
critical dimension, 338
critical dynamics, 177
critical field, 123
critical state, 39
critical wrinkling, 437
cross-over to equilibrium, 349

Davidson-Cole distribution, 268
Debye model, 266-69, 272, 277
Debye Waller factor, 308
defect
   energy, 126
   in flux lattice, 326-7
   in Wigner crystals, 326-7
   map, 61
   unbinding, 125, 126
density of states, 238, 241, 244
depinning, 42, 76, 437
   line, 45
   threshold, 397
   transition, 359, 360
   voltage, 381
depolarized light scattering, 198, 220, 230
diagonal resistivity, 375, 380, 383
dielectric susceptibility, 259, 266-80
differential resistance, 382
diffusion, 180, 263, 265
dirty boson, 188
discommensuration, 294
dislocation, 123
dislocation loop, 123
dislocation pairs, 378
domain
   coarsening, 405-36
   growth, 405-36
   walls, 4-10, 409-11, 427
donor, 372
dopant layer, 372
droplet excitations, 3
dynamic scaling, 1, 12, 21

elastic chain, 134
elastic matrix, 31
elastic moduli of flux lattice, 32
electron-electron interaction, 374
electron-phonon coupling, 339
electron solidification, 367-99
energy barriers, 11-13, 19
energy discontinuities, 133
entanglement, 99-117
entanglement length, 110
entropy, 259, 261-63, 264, 279

fast relaxation, 235-9, 243-5
filling factor, 376-97
fluids in random media, 15
fluctuations of flux lattice, 32, 37
flux
   creep, 40, 71-94
   crystal, 95-117, 125
   diffusion, 44
   flow, 38, 40
   lines, 95-117
   line lattice, 27
   liquid, 95-117, 125
   melting, 107, 109
   pinning, 119
   quantum, 380
fractal, 216, 221
fractional quantum Hall effect, 374, 376
fragile glass former, 260, 262, 265, 269, 270
Frenkel-Kontorova model, 290, 335
Fukuyama Lee Rice hamiltonian, 359

Ginzburg-Landau parameter, 27
glasses, 233, 239-43, 259-83
glass states, 207, 225
glass transition, 98, 189, 191-232, 233,
   259-83, 353, 354-6,
glassy anomalies, 241
glycerol, 265, 270, 271

Hall effect of flux lines, 47
Hall plateau, 374
Hall resistance, 372, 375, 385
hard-sphere system, 202
Havrilak and Negami, 268
heterostructure, 367-99
hexatic liquid, 123
hollandite structure, 288
hopping conduction, 378
Hubbard model, 104

incommensurability ratio, 288, 297
incommensurate structures, 286
inflection model, 337
instabilities, 136
insulating inclusions, 119
interface, 4-10, 438
intermittency, 440
interstitials, 123
interstitial proliferation, 123
invariant torus, 336
invasion percolation, 119
irreversibility, 120
irreversibility line, 45, 97
Ising ferromagnet, 406
Ising models, 2, 405

Josephson coupling, 36
Josephson vortices, 35
Juhario-Goldstein, 234, 259, 276-9

KAM theorem, 335
Kauzmann temperature, 259, 261-62,
   268, 271, 278
Kawasaki-Yalabik-Gunton method, 424
Kim-Anderson theory, 40
kinetics of ordering, 12
kink energy, 126
Kohlrausch (see stretched exponential)
Kosterlitz-Thouless transition, 36, 44,
   47, 368

Labusch parameter, 32, 45, 76
Landau level, 372, 381
Langevin equation, 169
Lawrence-Doniach model, 35
Lee-Rice energy, 323
Lee-Rice length, 322
Lindemann criterion, 32
Lindemann constant, 108, 126
localization lengths, 101
localized modes, 202
lock-in transition, 37
log-normal distribution, 277-79
London equation, 28
longitudinal resistance, 372, 377,
   379, 382
Lyapunov exponent, 337

magnetic localization, 376
magnetic penetration depth, 28
magnetophonon, 385, 387

magnetotransport coefficient, 384
mass anisotropy, 124
Mazenko's method, 425
Meissner phase, 124
melting, hierarchical, 313
melting of flux lattice, 32, 36, 76, 173
melting transition 63, 123, 368
membrane, 437
metastability, 15, 291, 340, 352
Mexican hat potential, 414, 429
mode coupling theory, 205, 218, 224, 233,
    247-50, 259, 275-77, 279
Monte Carlo calculations, 188, 405
Mountain resonance, 194, 203
multidiffusion, 442
multiscaling, 439, 442

nematic liquid crystal, 432
neutron scattering, 233
nonequilibrium dynamics, 1, 15
non-exponential decay, 348, 352
normal modes, 233
nucleation, 259

Ohta-Jasnow-Kawasaki theory, 424
order-parameter, 28
    field, 407
    conserved, 408
    non-conserved, 408
orientational glass, 353
Ostwald ripening, 406
oxygen vacancy, 95

pancake vortices, 32, 34
Peierls gap, 342, 343
Peierls-Nabarro barrier, 337
Peierls transition, 339, 355
permeability, 119
percolation, 438
perturbation theory, 157
phase
    diagram, 72, 162, 397
    excitations, 340
    ordering, 405, 413
    plot, 393, 394
    slip. 189, 360
    transition, 1, 4
phonons, 123
photon correlation, 202, 218
piezoelectric coupling, 388
pinned e-solid, 381

pinning, 38, 71-94, 98, 119, 123
    187, 385
  energy, 40, 149, 344
  force, 40, 130
  frequency, 387, 397
  mechanisms, 72, 73
  mode, 392, 394
plasmon-ripplon resonance, 370
point defects, 417
point disorder, 95
point vortices, 34
polymers, 98, 274
polymer glass, 113
Porod's law, 413, 416, 420-23
Porod tail, 430, 431
porous medium, 438
power-law tail, 416
primary relaxation, 259, 276, 277

quantum Hall effect, 372
quantum liquid, 388
quantum solid, 388
quasistatic response, 159
quenched randomness, 1

α-relaxation, 192
β-relaxation, 198
random ferromagnets, 2, 4
random friction model, 321
renormalization, 295, 297, 300
renormalization group, 3
rubber band model, 325

scaling, 42, 213, 216, 222, 227
    260, 272-5, 280
scaling functions, 412, 423-30
Schottky anomaly, 305, 306
secondary (Juhari-Goldstein)
    relaxation, 359, 276-79
selforganized criticality, 439
sheet conductivity, 388
sliding conductivity, 339
slow relaxation, 245-50
softening, 154
soft mode, 238-45
soft potential model, 238-41
sound attenuation, 393
specific heat, 287, 302, 346, 352
specific heat spectroscopy, 259, 261-65,
    267, 269-72, 275, 276, 279
spin density wave, 339-58

spin echo, 245-7
spin glass, 2, 4, 16, 275-280, 343, 353, 356
spinodal decomposition, 406
static scaling, 440
static structure factor, 287, 307
stiffness of ordered phases, 2
stretched exponential, 194, 214, 229, 264,
    268, 269, 275, 276, 341, 348, 363
stretching, 194, 203, 209, 221
string, 437
strong glass former, 260, 262
strong pins, 121
structure factor, 193, 202, 217, 416,
    418, 419
structural relaxation, 192, 218, 219, 221
superconductivity, 17-22, 27-190
supercooled liquids, 259-83
superfluid stiffness, 189
superposition principle, 213
supersolid, 125
surface acoustic wave, 387-97
susceptibility, 259-83

temporal scaling, 440
thermal conductivity, 263-65
thermoelectric effect, 47
thermomagnetic effect, 47
tilt modules, 189
time-dependent-Ginzburg-Landau
    equation, 408, 429
time-of-flight, 236-8, 249-50
Toda model, 336
topological defects, 407, 413-419
transducer, 388
translational order, 149
tunneling event, 125
twist map, 337

vacancy proliferation, 123-7
vibrating superconductors, 46
viscosity, 246-50, 252-3, 259-61,
    279, 355
Vogel-Fulcher law, 260, 269, 276, 354
Von Schweidler relaxation, 195, 211,
    221, 276
vortex
    creep, 43, 44
    dynamics, 63
    fluctuations, 95
    glass, 19-22, 42, 64-67, 77, 96, 113, 188
    lattice, 18

line, 95, 414
　　　loop, 189, 414
　　　melting, 121
　　　slush, 42
　　　tunneling, 44, 103

weak pins, 121
Wigner crystal, 392
Wigner glass, 381
Wigner solid 376
Wigner transition, 368
Williams-Watts (see stretched
　　　exponential)
winding number, 336